092060000

CW01336929

Empirical Asset Pricing

Empirical Asset Pricing

Models and Methods

Wayne Ferson

The MIT Press
Cambridge, Massachusetts
London, England

This book was set in Times New Roman by Westchester Publishing Services. Printed and bound in the United States of America.

Library of Congress Cataloging-in-Publication Data

Names: Ferson, Wayne E., author.
Title: Empirical asset pricing : models and methods / Wayne Ferson.
Description: Cambridge, MA : MIT Press, [2019] | Includes bibliographical
  references and index.
Identifiers: LCCN 2018020617 | ISBN 9780262039376 (hardcover : alk. paper)
Subjects: LCSH: Stocks—Prices. | Rate of return. | Econometric models. |
  Moments method (Statistics) | Estimation theory.
Classification: LCC HG4636 .F47 2019 | DDC 332.63/222—dc23 LC record available
at https://lccn.loc.gov/2018020617

10 9 8 7 6 5 4 3 2 1

Contents

Preface

This book is designed for PhD students in finance or advanced masters students interested in empirical asset pricing and as a reference for researchers. The main goal is to help readers build and understand a tool kit for empirical asset pricing research. As asset pricing has matured as a field, it has developed more breadth. This is a long book, but it leaves out many advanced topics. The aim is to present the material at an advanced introductory level. I try to provide enough references that a scholar can use the book as a jumping-off point for a research program. The structure is motivated from my experience in teaching this material in doctoral programs at more than a half-dozen universities where I have served as a regular or visiting faculty member.

The book has four main sections. The first section of the book, parts I–III (chapters 1–14), covers the theory of empirical asset pricing. This section is structured around three central paradigms: mean variance analysis, stochastic discount factors, and beta pricing models. I describe a relation among the three paradigms, which I call "triality." I develop the role of conditioning information in the implementation and interpretation of asset pricing tests. The second section of the book, part IV (chapters 15–24), examines empirical methods. I start with the generalized method of moments, emphasizing how to implement the technique in various types of problems. There is a long chapter on panel regression methods and a chapter on bootstrapping for inferences about cross sections, where multiple comparisons must be considered. The third section consists of part V (chapters 25–28) on fund performance evaluation and is perhaps the most comprehensive review of this important area that is currently available. Throughout the book, I combine classical foundations with more recent developments in the literature, and I relate some of the material to applications in investment management. The final section of the book, part VI (chapters 29–34), presents selected applied topics on empirical asset pricing. These include a long chapter on predictability in asset markets, including predicting the levels of returns, volatility and higher moments, and predicting cross-sectional differences in returns. There are shorter chapters on production-based asset pricing, long-run risk models, the Campbell-Shiller approximation, the debate on covariances versus characteristics, and the relation of volatility to the cross section of stock returns.

I owe a debt of gratitude to many people who helped me in the writing of this book. Several years of doctoral students at the University of Southern California suffered through being guinea pigs for early drafts. I am grateful to Wayne Chang, Suk Won Lee, and Ben Zhang for comments. Allison Kays and Min Kim found many typos in early drafts. Min Kim went beyond the call of duty, finding not only typos but several instances where my arguments were unclear or had holes in them. Davidson Heath contributed Matlab code for the appendix. Cyd Westmoreland did some of the most impressive copyediting that I have ever seen. And my wife, Nancy Ferson, deserves credit for putting up with my incessant typing far into many evenings, as I pursued this obsession.

Introduction

This book has four main sections. This introduction briefly describes the main sections and then goes through each part in more detail, including some background and suggestions for using the book in a PhD or masters level course. I then discuss how the material in this book compares with some of the competition.

The first section of the book, parts I–III, covers the theory of empirical asset pricing in 14 chapters. This is structured around three central paradigms: mean variance analysis, stochastic discount factors, and beta pricing models. The stochastic discount factor approach, which I call "m-talk," is a beautiful way to integrate and illustrate some of the main concepts and models in asset pricing. But it is only one of the three paradigms. I describe a relation among the three paradigms, which I call "triality." This idea has evolved from review chapters that I published in 1995 and 2003. This useful perspective is not emphasized in other books. I develop the role of conditioning information in the implementation and interpretation of asset pricing tests and in the triality relation in more depth than is done in other available texts.

The second section of the book, part IV, examines empirical methods in ten chapters. I start with the generalized method of moments (GMM), emphasizing how to implement the technique in various types of problems. I include material on some empirical methods that are becoming important for asset pricing research but don't get much coverage in other texts. This book has three chapters on regression methods, including a long chapter on panel regression methods and a shorter one on bootstrapping for inferences about cross sections, where multiple comparisons must be considered.

The third section of the book (part V) consists of four chapters on fund performance evaluation. The material on fund performance evaluation represents the most comprehensive review of this important area that is currently available. Throughout the book, I combine classical foundations with more recent developments in the literature, and I relate some of the material to applications in investment management.

The fourth section of the book (part VI) includes five chapters on selected applied topics in empirical asset pricing. There is a long chapter on predictability in asset markets that discusses predicting returns, predicting second moments, including GARCH-type models and realized volatility models, and predicting cross-sectional differences in stock returns. There are additional chapters on production-based asset pricing, long-run risk models, the Campbell-Shiller approximation, the debate on characteristics versus covariances in the cross section of portfolio returns, and the relation between volatility and the cross section of stock returns.

Parts I–III cover the theory of empirical asset pricing. Part I contains seven short chapters that review the basics, including stochastic discount factors and their relation to state pricing, utility maximization, and Euler equations. I include a description of a useful perturbation argument to derive Euler equations for estimating and testing models. I provide a taxonomy of the main asset pricing models in the empirical literature by describing the stochastic discount factors they imply. I then use m-talk to provide an overview of several important applications of asset pricing models. These applications include the term structure of interest rates, real interest rates and economic growth, real and nominal returns and inflation indexing, international finance (including the carry trade or forward premium puzzle), and overviews of conditional asset pricing and market efficiency and of investment performance evaluation. Part I concludes with a discussion of the three paradigms of empirical asset pricing and the triality relation. I teach part I at the PhD level to first- or second-year doctoral students over a three-week period, in about nine classroom hours. This material should also work at the advanced masters level, but it might take longer.

Part II discusses mean variance models in four chapters. The first covers the classical analysis using matrix algebra but goes beyond most textbook treatments to include discussions of modern portfolio management, the empirical performance of mean-variance optimized portfolios, short-sales constraints, and shrinkage. I also introduce the optimal orthogonal portfolio, which is used later to interpret classical regression tests of asset pricing models. I then revisit the mean variance analysis when there is explicit information to be conditioned on. This draws from my work on this problem with Andy Siegel (Ferson and Siegel 2001). It includes some recent extensions and a simplified discussion of the Hansen and Richard (1987) representation for minimum variance efficiency with conditioning information. The third chapter introduces the variance bounds of Hansen and Jagannathan (1991), and the fourth chapter discusses these types of bounds with conditioning information. I try to teach the material in part II over two weeks, three hours per week, finishing the first week with the classical material and starting the next week with the more advanced material with conditioning information. This gives the students time to assimilate the classical material before tackling the harder stuff.

Part III discusses multibeta models in three chapters. Chapter 12 discusses the arbitrage pricing theory of Ross (1976). It also discusses principal components analysis, dynamic factor models, and partial least squares, and relates the material briefly to limits to arbitrage. Chapter 13 presents a discrete time version of the intertemporal portfolio optimization problem and asset pricing model of Merton (1973). My treatment in discrete time provides basically the same results and insights as does the original model, cast in continuous time, but I use Stein's (1973) lemma and normality instead of diffusion processes. The focus is on empirical interpretations of the model, such as the determinants of risk premiums and hedging demands. The final chapter of this section draws from my work with Siegel and Xu (Ferson, Siegel, and Xu 2006), discussing multibeta models with explicit conditioning information.

Part IV presents the main econometric methods of empirical asset pricing. It starts with the generalized method of moments (GMM) and views the other methods as special cases of the GMM. I take the practical perspective of a user of the methods, sharing some hard-won tips and tricks. I offer a simple way to think about serial dependence in GMM moment conditions and survey the reasons serial dependence may arise and methods for dealing with it. The final chapter on the GMM reviews some asset pricing examples. This provides practice in formulating the moment conditions, which I find is the hardest part for many students, and some literature review. I include chapters on the classical, Fama-MacBeth (1973) style cross-sectional regression and on multivariate regression models, including well-known results from Gibbons, Ross, and Shanken (1989). I also provide a more up-to-date treatment of cross-sectional regressions, as recent work has made significant progress on characterizing the sampling distributions and dealing with issues like errors in the betas. This chapter includes a discussion of conditional asset pricing interpretations of cross-sectional regressions and a discussion of the relation between cross-sectional regressions and portfolio sorting. Chapter 23 is a long chapter on panel regressions. It provides a simple way to think about clustering, dummy variables, and fixed effects, a treatment of predictive panel regressions and causal inference, including differences in differences and regression discontinuity. Chapter 24 discusses bootstrapping for inferences about cross sections. These methods have become important recently for asset pricing research, but are not covered in the competing books.

Parts V and VI feature applied empirical topics. Part V is perhaps the most comprehensive overview of investment performance evaluation available. Chapter 25 reviews classical performance measures, their properties, and the empirical evidence. Chapter 26 reviews conditional performance evaluation, and chapter 27 discusses term structure models and bond fund performance. Chapter 28 integrates the field from a modern perspective.

Part VI presents a few chapters on additional applied empirical topics. A long chapter 32 on predictability presents a unique review, integrating the related topics along two dimensions. The first is what we are trying to predict, broken down into the levels of returns, conditional variance or volatility modeling, covariance and betas, higher moments, or the cross-section of stock returns through return differences or spread portfolios. The second dimension is what information we are trying to predict with. This is broken down into weak- or semistrong-form information, following the characterization from Fama (1970). Additional topics in section VI include production-based asset pricing, long-run risk models, the Campbell-Shiller approximation, the debate about characteristics versus covariances in the cross section of portfolio returns, and the relation of volatility to the cross section of stock returns.

The strengths of the book include its integrated treatment of the various theories and methods, which makes it easy for a reader to transition from one subject to the next. I use a conversational writing style, with a few brief stories from years in the profession to keep things, I hope, fun and entertaining. Throughout the book, I combine classical foundations with more recent developments in the literature and provide lots of recent references. The references alone comprise more than 700 references.

The choice of topics reflects the goal of providing a book that can be taught to second-year doctoral students in finance in a single semester. I typically devote about two weeks of a one-semester course to the topics section. In a one-quarter class, the topics section can be skipped. Parts I–III of this book can serve as a standalone text for a half-semester introductory theory course, or used in an advanced masters level class in finance or economics, combined with selected empirical topics from later sections of the book.

Given the goals of this book, there are several important topics that I do not cover. I do not cover stochastic calculus or derivatives pricing in any detail. There is a separate course for these topics at the institutions where I have taught this material. However, I do present a discrete-time analog to the continuous-time portfolio rules and intertemporal asset pricing model of Merton (1973), which provides basically the same results and insights, using Stein's (1973) lemma and normality instead of diffusion processes. I also discuss the issue of time aggregation and the link between continuous-time models and discrete-time data, and I discuss the quadratic variation of a process in the context of volatility modeling. I also do not try to cover market microstructure or the associated information economics, although I do discuss several microstructure-related biases that are of concern to empirical researchers in asset pricing.

The book fills several gaps in coverage left by the leading textbooks in this field. Much of the core material in parts I–III, covering the theory of empirical asset pricing, is of course treated in other texts. I use an integrated approach aimed at empirical application. My treatment of the three main paradigms of empirical asset pricing and how they are related to one another is unique. I provide deeper insights into the role of conditioning information in asset pricing models and tests than do previous treatments, drawing on my own research in this area. And the discussion of conditioning information in multibeta models is unique.

The main competitors to this book include Kerry Back's *Asset Pricing and Portfolio Choice Theory*, John Campbell's *Financial Decisions and Markets: A Course in Asset Pricing*, and the book by Turan Bali, Robert Engle, and Scott Murray, *Empirical Asset Pricing: The Cross-Section of Stock Returns*. Back's book covers much of the basic theory as in the first three parts of this book at a more rigorous level, including dynamic models, asymmetric information, and option pricing. Campbell's book covers intertemporal models and term-structure models in more detail, and includes some very nice and unique material on modern personal finance, but very little on empirical methods. I include 12 chapters focused on empirical methods.

In some ways, this book fits nicely between the Bali, Engle, and Murray book and the former two books in terms of the level of mathematical sophistication it requires of the reader. The level of mathematical sophistication is substantially less than in the textbooks by Darrell Duffie, *Dynamic Asset Pricing Theory*, or by Kenneth Singleton, *Empirical Dynamic Asset Pricing: Model Specification and Econometric Assessment*. I do assume the reader knows some differential calculus and algebra, including some matrix algebra. I also assume that the reader knows generally what stocks, bonds, and options are, about at the level of an undergraduate course in investments.

This book also fills gaps in the subject treatment. Bali, Engle, and Murray lack some of the statistical and econometric theory behind the various tests and methods that are described here. The Duffie and Singleton books include more limited discussions of empirical methods, compared with the applied approach taken here. My book provides a balanced emphasis on both the theory and methods of empirical asset pricing. The competing books do not provide treatments of panel regression methods for prediction or causality. My presentation of parametric bootstrap methods is unique. No other book provides the depth of coverage of fund performance evaluation or of predictability that is offered here.

There are several older books that may compete with this one in some ways, and I recommend them as supplementary readings in my course outline. These include John H. Cochrane, *Asset Pricing* (2005); John Campbell, Andrew Lo, and Craig MacKinlay, *The Econometrics of Financial Markets* (1997); Jon E. Ingersoll, *The Theory of Financial Decision-Making* (1987); and Chi-Fu Huang and Robert H. Litzenberger, *Foundations for Financial Economics* (1988). These are great reference books, especially for some of the classical theory in parts I–III of my book, and to a lesser extent for some of the empirical methods in part IV, but none of these books include the more modern subjects covered in my book, such as panel regressions, bootstrapping with multiple comparisons, long-run risk models, production-based asset pricing, solutions for mean variance analysis with conditioning information and in variance bounds, or modern investment performance evaluation.

I INTRODUCTION TO EMPIRICAL ASSET PRICING

1 Stochastic Discount Factors and *m*-Talk

This introductory chapter for the first section of the book introduces the theory of empirical asset pricing. That is not an oxymoron, even if you think that theory and empirical work are distinct efforts. There actually is a developed body of theory for empirical work in asset pricing. This chapter starts by introducing the all-encompassing equation, $E(m\mathrm{R}) = 1$, where m is the stochastic discount factor. It makes the point that the content of this equation is the assumptions that you make about the financial markets. The more you say, the more the equation means. I illustrate this first with concepts of arbitrage and then set things up for a discussion of state pricing in chapter 2.

1.1   The Main Equation (1.1)

Asset pricing models describe the prices or expected rates of return of *financial assets*, which are claims traded in financial markets, such as common stocks, bonds, options, and futures contracts. Virtually all asset prices are special cases of the fundamental equation:

$$\mathrm{P}_t = \mathrm{E}_t\{m_{t+1}(\mathrm{P}_{t+1} + \mathrm{D}_{t+1})\}, \tag{1.1}$$

where P_t is the price of the asset at time t, and D_{t+1} is the amount of dividends, interest, or other payments received at time $t + 1$. Here, $\mathrm{E}_t\{\cdot\}$ denotes the conditional expectation given the information in the market at time t. The marketwide random variable m_{t+1} is the *stochastic discount factor* (SDF). We say that an SDF "prices" the assets if this equation is satisfied. This equation is so important I will refer to it throughout the book.

   We can use recursive substitution in equation (1.1), substituting for the future price of the asset, to express the current price as a function of the future cash flows and future SDFs: $\mathrm{P}_t = \mathrm{E}_t\{\Sigma_{j>0}(\Pi_{k=1,\dots,j}\, m_{t+k})\mathrm{D}_{t+j}\}$. Thus, prices are obtained by discounting the payoffs, or multiplying by SDFs, so that the expected present value of the payoff is equal to the price.

   The term "stochastic discount factor" is usually attributed to Hansen and Richard (1987), although the idea goes back at least to Beja (1971). The SDF has other names in other contexts. It is sometimes called the "asset pricing kernel" or the "benchmark pricing variable." When the fundamental equation (1.1) is derived from an economic agent's optimization, it may be the intertemporal marginal rate of substitution. It is also proportional to the Radon-Nikodym derivative, equivalent martingale measure, or the "q-measure," if you are a fan of derivatives. More on this when we talk about state pricing in chapter 2.

For empirical work, we usually work with returns instead of prices, because prices are likely to be nonstationary, and this can be bad for the reliability of empirical methods. Define the gross return as $R_{t+1} = (D_{t+1} + P_{t+1})/P_t$. Then equation (1.1) is equivalent to

$$E_t(m_{t+1}R_{t+1}) = 1. \tag{1.2}$$

The first two equations of the book are equivalent given positive prices. All asset pricing models can be seen as a special case of equations (1.1) or (1.2). For short, let's call these equations "m-talk." There will be a lot of m-talk in this book.

The asset pricing models that equation (1.1) describes are based on two central economic concepts. The first is the *no-arbitrage principle*, which states that market forces tend to align the prices of financial assets so as to eliminate arbitrage opportunities. An arbitrage opportunity arises if assets can be combined in a portfolio with zero cost, no chance of a loss, and a positive probability of gain. Arbitrage opportunities tend to be eliminated in financial markets, because prices adjust as investors attempt to trade to exploit the arbitrage opportunity. For example, if there is an arbitrage opportunity because the price of security A is too low, then traders' efforts to purchase security A will tend to drive up its price, which will tend to eliminate the arbitrage opportunity. The Arbitrage Pricing Theory (APT; Ross 1976) is a well-known asset pricing model based on arbitrage principles. The arbitrage pricing approach is described in chapter 12.

The second central concept in asset pricing is *financial market equilibrium*. Investors' desired holdings of financial assets are derived from an optimization problem. In a market with no frictions, a necessary condition for financial market equilibrium is that the first-order conditions of the investors' optimization problem are satisfied. This requires that investors are indifferent at the margin to small changes in their asset holdings. Equilibrium asset pricing models follow from the first-order conditions for the investors' portfolio choice problem and a market-clearing condition. The market-clearing condition states that the aggregate of investors' desired asset holdings must equal the aggregate "market portfolio" of securities in supply. Differences among various asset pricing models arise from differences in their assumptions about investors' preferences, endowments, production and information sets, the process governing the arrival of news in the financial markets, and the types of frictions in the markets.

Without more structure, the m-talk equation (1.1) has no real content. In fact, you can always find an m that will "work," in the sense of pricing assets through equation (1.2). For example, let m_{t+1} be $[\underline{1}' \, (E_t\{\underline{R}_{t+1}\underline{R}'_{t+1}\})^{-1}]\underline{R}_{t+1}$, where $\underline{R}_{t+1}$ is the vector of all the assets' gross returns, and substitute into equation (1.1). As long as the raw second-moment matrix of the returns can be inverted, this m can be constructed. If there are redundant assets, so that the second-moment matrix is not invertible, we can just throw out the redundant ones and use what is left, until we get down to a nonredundant set of assets to construct the m. The result will also work on the redundant assets' gross returns that we threw out. To see this, let a redundant asset be $R_d = \underline{w}'R_{t+1}$, where w is the portfolio weight that sums to one: $\underline{w}'\underline{1} = 1$. Then, $E(mR_d) = E(m\underline{w}'R) = \underline{w}'\underline{1} = 1$.

We will see later that the ability to construct an SDF as a function of the returns that prices all included assets is essentially equivalent to the ability to construct a minimum variance efficient portfolio. In fact, the particular m constructed above has minimum second moment. We will see this once again, in the context of the Hansen-Jagannathan (1991) bounds in part II.

1.2   From the General to the Specific

While equation (1.1) by itself doesn't really say anything of substance, the more we say about the economy, the more we say about the SDF. Here are some examples where the amount of structure is gradually increased.

Most simply, dropping the time subscripts, the existence of some m that works implies that there is a "law of one price" in the economy. This law of one price says that a given payoff can only have one price. To see this almost trivially, imagine there are two payoffs, $x_1 = x_2$, with prices $p(x_1)$ and $p(x_2)$, then the existence of an m implies $mx_1 = mx_2$, so that $E(mx_1) = E(mx_2)$, and therefore, $p(x_1) = p(x_2)$. Suppose there are two ms that work in equation (1.1), m_1 and m_2, corresponding to the prices $p(x_1)$ and $p(x_2)$ for the two payoffs $x_1 = x_2$. Then equation (1.1) implies $E\{(m_1 - m_2)R\} = 0 = E\{(m_1 - m_2)x\}$ for all R and x. Letting $m_1 - m_2 = \varepsilon$, then $E\{\varepsilon x\} = 0$, and we have $p(x_1) = E(m_1 x_1) = E\{(m_2 + \varepsilon)x_2\} = E(m_2 x_2) = p(x_2)$. Thus, the law of one price says that a given payoff can only have one price.

Now, if the law of one price were to fail, it would generate an arbitrage opportunity in a frictionless market: Buy at the low price and sell the same thing at the higher price. Of course, that is what middlemen in the real world do for a living, but they live in a world with frictions.

1.3   Arbitrage and the SDF

With no frictions, it turns out that the existence of an m_{t+1} that works and is a strictly positive random variable makes equation (1.1) equivalent to the no-arbitrage principle, which says that all portfolios of assets with payoffs that can never be negative but are positive with positive probability, must have positive prices (Beja 1971; Rubinstein 1976; Ross 1977; Harrison and Kreps 1979; Hansen and Richard 1987). If this condition fails, Harrison and Kreps call the arbitrage opportunity a "free lunch," and the lack of arbitrage is the principle is that there are no free lunches (NFL). This has little to do with American football.

Here is a quick proof of the equivalence of a positive m and NFL. To show that a positive m implies NFL, consider $p = E(mx)$, $m > 0$; then NFL holds. That is, if the payoff x is positive, then p must be positive. The price of a positive payoff must be positive.

To show that NFL implies there is a positive m, most treatments involve theorems about separating hyperplanes. Here is a simpler argument. Consider the problem:

$$\text{Min}_w \ w'\underline{p} - \Sigma_s \lambda_s(w' \underline{X}_s - \text{Target}_s) \text{ with Target}_s > 0. \tag{1.3}$$

In this problem, Target_s is the desired payoff if state s occurs, $\underline{X}_s$ is the N-vector of assets' payoffs in state s, and $\underline{p}$ is the vector of current prices. If the target payoffs are positive, NFL implies that a solution to this problem must have $w'\underline{p}$ positive. This assumes that the price function $P(\cdot)$ is linear: $P(\underline{w}'X) = \underline{w}'P(X)$. But this is also an implication of NFL. To see this, suppose that the function is nonlinear, say, the price of the bundle $P(\underline{w}'X) > \underline{w}'p(X)$, the bundle of individual prices. You could then buy the pieces of X separately and sell the bundle $\underline{w}'X$ for more than it costs. Similarly, there is a free lunch if $P(\underline{w}'X) < \underline{w}'P(X)$, available from buying the bundle and selling off the pieces. Therefore, the price function must be linear under NFL. The first-order condition for problem (1.3) then implies $\underline{p} = \Sigma_s \lambda_s \underline{X}_s$, with Lagrange multipliers $\lambda_s > 0$. Write this equation as $\underline{p} = \Sigma_s \text{prob}_s(\lambda_s/\text{prob}_s)\underline{X}_s$, where prob_s is the probability that state s occurs, and let $m_s = \lambda_s/\text{prob}_s > 0$ for each state s. Now by construction, we have a stochastic discount factor, m_s, defined state-by-state, that satisfies $\underline{p} = E(m\underline{X})$, with $m > 0$.

This argument sneaks in the concept of state pricing, because we defined m_s as the ratio of the marginal value of a payoff in state s, λ_s, to the probability that the state is observed, prob_s. The next chapter takes up state pricing in more detail.

2 State Pricing

This chapter presents an introduction to state pricing, where the prices in securities markets are characterized in their most primitive form: as bundles of claims that pay off one unit of consumption if and only if a particular state of nature occurs at a future date. We explore how state pricing is related to consumer optimization, m-talk, and risk sharing when markets are complete, which means there are at least as many traded claims as there are future states of nature. That may seem fantastical, until you think about the fact that there are more mutual funds in the United States than there are individual common stocks. And markets may be becoming closer to complete over time. China's emerging insurtech industry has recently offered products like overtime-work insurance, mid-August moon insurance, traffic-jam insurance, parking tickets and package insurance (according to the *Wall Street Journal*, January 8, 2018). I don't think we have paper-rejected insurance yet, but I might want to buy it. In this chapter, we relate state prices to so-called risk-neutral pricing and the q-measure, which is at the core of derivatives pricing models. We review attempts to use state pricing to recover probabilities or investor preferences (or both) from derivatives prices, including the recent and somewhat controversial work in this area by the late Stephen Ross (2015), and wrap up the chapter with some comments on state pricing in multiperiod models.

2.1 Basic Setup

Imagine sitting down at the beginning of the period and listing all possible outcomes that might occur at the end of the period. These are the mutually exclusive and collectively exhaustive future states of the world, $s = 1, \ldots, S$. Imagine primitive securities that pay 1 unit if and only if state s happens at the end of the period. These securities implement state pricing. A primitive security is a stylized insurance contract, paying off only when the insured state occurs.

Why should you imagine such a strange and fantastic world? As an organizing principle, it is a powerful way to think about asset pricing models. Kenneth Arrow won a Nobel prize for the setup. It has the beautiful feature that the calculations associated with utility maximization under certainty, as in classical price theory, can be formally applied under uncertainty. Robert Litzenberger, one of the most prolific authors in the *Journal of Finance*, used to claim that finance wasn't really all that hard once you thought about it. He seemed to be able to think of everything in terms of state prices.

Real world securities make payoffs in many different states. Given any set of state-dependent payoffs, $\{X_s\}_s$, define the price of that set of payoffs as $P(X(s)) = \sum_s \text{pri}_s\, X_s$, where pri_s is the price of the primitive security that pays 1 unit when state s occurs. Note that NFL implies that state prices are positive, because 1 unit is a positive payoff. Note also that we have used linearity of the pricing function across states, as implied by NFL. To relate this to m-talk, we can write $P(X(s)) = \sum_s \text{prob}_s (\text{pri}_s / \text{prob}_s) X_s$, where prob_s is the probability that state s occurs. If $m_s = (\text{pri}_s / \text{prob}_s)$, we have $P(X(s)) = \sum_s \text{prob}_s (m_s) X_s = E(mX)$. With state pricing, it becomes clear why $m > 0$.

2.2   State Pricing with a Representative Maximizing Agent

Consider a representative agent who maximizes the expected utility of consumption:

$$\text{Max}_{\{C_0, C_s\}} \sum_s \text{prob}_s\, u(C_0, C_s),$$
$$\text{s.t. } W_0 = C_0 + \sum_s \text{pri}_s\, C_s. \tag{2.1}$$

The first-order conditions for this problem are:

$$\text{wrt } C_0: \quad u_0' = \lambda, \tag{2.2}$$
$$\text{wrt } C_s: \quad \text{prob}_s\, u_s' = \lambda\, \text{pri}_s, \text{ all } s,$$

where λ is the Lagrange multiplier on the wealth constraint (i.e., the marginal value of wealth). These conditions relate the previous definition of m in a state pricing context to the fundamentals of the investor's utility function:

$$\text{prob}_s\, u_s' = \text{pri}_s\, u_0', \text{ all } s, \tag{2.3}$$
$$m_s = \text{pri}_s / \text{prob}_s = u_s' / u_0', \text{ all } s. \tag{2.4}$$

These equations state that m_s is the marginal rate of substitution between current consumption at time 0 and future consumption in state s.

This representation provides a nice economic intuition for what m-talk does. Consider $p = E(mX) = \sum_s \text{prob}_s\, m_s X_s = \sum_s \text{prob}_s (u_s' / u_0') X_s$. Payoffs are highly valued in states where either the probability of occurrence is high or m_s is large. States where m_s is large command higher prices, because the marginal rate of substitution is large for these states. That is, the agent is willing to trade a lot of consumption at time $t = 0$ for a payoff in such a state tomorrow. Since the price is measured in terms of consumption at $t = 0$, the price of these payoffs is large.

The competitive equilibrium in a complete market state pricing setup inherits many of the properties of consumer equilibrium under certainty. In particular, the allocation is

Pareto optimal, meaning that it is not possible to change the allocation to make someone better off while hurting no one. The allocation of state-contingent consumption in a competitive equilibrium is equivalent to the allocation in a central-planning problem, where the planner maximizes a linear combination of the consumers' utility functions.

2.3   Heterogeneous Agents with Complete Markets

If markets are complete, there is a primitive security for each state. In this case, we can allow investors to be heterogeneous, but by trading in complete markets, they will equate their marginal rates of substitution. Consider heterogeneous agents i who maximize as follows:

$$\text{Max}_{\{C_{0i}, C_{si}\}} \sum_s \text{prob}_{is} \, u_i(C_{0i}, C_{si}),$$
$$\text{s.t. } W_{0i} = C_{0i} + \sum_s \text{pri}_s C_{si}. \tag{2.5}$$

The first-order conditions for this problem imply:

wrt C_{0i}: $u'_{i0} = \lambda_i$, all i. $\tag{2.6}$

wrt C_{si}: $\text{prob}_{is} \, u'_{is} = \lambda_i \, \text{pri}_s$, all s, i,

$\Rightarrow \text{prob}_{is} \, u'_{is}/u'_{i0} = \text{pri}_s$, all s, all agents i. $\tag{2.7}$

Agents can be completely heterogeneous in their preferences and beliefs. Still, when trading in complete markets, all agents end up with the same probability-weighted marginal rates of substitution in any given state s, because they face the same state price, pri_s, in the market. Now, suppose that we make the (heroic) assumption that agents share identical probability beliefs. It follows that

$$m_s = \text{pri}_s/\text{prob}_s = u'_{is}/u'_{i0}, \text{ all } s, i. \tag{2.8}$$

Note that the SDF is unique when markets are complete. That is, under the assumption of identical beliefs, since agents all face the same state prices, all agents will agree on m_s, state by state, and it will equal their marginal rates of substitution between current and future consumption in each state s.

The fact that agents equate their marginal rates of substitution has implications for how they share risk across the states. This is illustrated with a particular utility function that links the utilities to agents' consumption. Let $u_i(C_{0i}, C_{si}) = [C_{0i}^{(1-\alpha i)} + \beta C_{si}^{(1-\alpha i)}]/(1-\alpha_i)$, where α_i is the agent's relative risk aversion. Then the equation (2.8) states that $\beta[C_{si}/C_{0i}]^{(-\alpha i)}$ is equal for all i. Taking logs and then variances, this implies: $(\alpha_i)^2 \, \text{Var}\{\ln[C_{si}/C_{0i}]\} = (\alpha_j)^2 \, \text{Var}\{\ln[C_{sj}/C_{0j}]\}$ for any two agents, i and j. The more risk-averse agent has a smaller consumption growth variance. If the aggregate consumption growth variance is the aggregate risk to be shared in the economy, then more risk-averse agents take a smaller share of the aggregate risk, and the less risk-averse take a larger share.

2.4   Complete Markets, Risk Sharing, and a Constructed Representative Agent

That agents equate their marginal rates of substitution across states has two important implications. The first implication of complete markets is that we get optimal risk sharing. The second is the existence of a representative agent. (The latter idea comes from Constantinides 1982.)

Pick two states, say, s, k. The first-order conditions to the problems of agents i and j imply

$$m_s/m_k = u'_{is}/u'_{ik} = u'_{js}/u'_{jk}, \text{ all } i, j. \tag{2.9}$$

Suppose that the optimal consumption chosen by agent i in the two states is such that $C_{ki} < C_{si}$. Because agent i's marginal utility is decreasing in consumption, we have $u'_i(C_{ki}) > u'_i(C_{si})$. Because agents i and j equate their ratios of marginal utility in the two states, the same thing must be true for agent j: $u'_j(C_{kj}) > u'_j(C_{sj})$. By monotonicity of agent j's marginal utility, this implies $C_{kj} < C_{sj}$ for all j. This shows that in equilibrium, there is "optimal" risk sharing.

Why is the result that all agents order their consumption in the same way across the states optimal risk sharing? Consider a central planner who allocates consumption across agents so as to maximize a social welfare function; that is, a weighted average of the utility functions of the individual agents. Optimal risk sharing maximizes the weighted average utility. Suppose a subset of agents gets hit with a big shock in some state that lowers these agents' consumption by a great deal. Because the utility function is concave, the utility loss associated with a big negative hit is larger than the utility gain for a similar-sized increase in consumption. This means that the average utility across the agents will be lower when the allocation features greater heterogeneity across agents, other things equal, compared to an allocation that smooths out the shocks across the agents. The smoothest the shocks can be across agents is when each agent faces exactly the same shock. For example, all agents hold the market portfolio, whose return determines their final consumptions, in some version of the capital asset pricing model (CAPM; Sharpe 1964). Facing the same shocks, agents order their consumptions in the same way across states.

This idea of optimal risk sharing is related to international applications of m-talk, discussed in chapter 6. Here, we think of each country as an agent, and if risk sharing across countries is optimal, we would expect consumption to be highly correlated across countries. Empirically, this is far from true (see, e.g., Lewis 1999; Lewis and Liu 2015), which creates an international consumption correlation puzzle.

Summing over agents and letting A denote the aggregate consumptions, we have $C_{kA} = \sum_i C_{ki} < C_{sA}$ in the previous example, where state k is the low-consumption state relative to state s. Therefore, $u'(C_{kA}) > u'(C_{sA})$ for any state-separable, concave $u(\cdot)$. This allows us to construct a representative agent using the aggregate consumption. The SDF is then defined from the constructed agent's marginal rate of substitution: $m_s/m_k = u'(C_{sA})/u'(C_{kA})$.

Because the ratio must be the same for each agent in complete markets, the resulting m will correctly price all assets. This insight is due to Constantinides (1982).

2.5 Risk-Neutral Probabilities

Rearranging the basic state pricing relations, we can write:

$$
\begin{aligned}
P(X(s)) &= \sum_s \text{pri}_s X_s = (\sum_s \text{pri}_s) \sum_s (\text{pri}_s/(\sum_s \text{pri}_s)) X_s \\
&= R_f^{-1} \sum_s \text{prob}_s^* X_s \\
&= R_f^{-1} E^*(X),
\end{aligned}
\tag{2.10}
$$

where $R_f = (\sum_s \text{pri}_s)^{-1}$ is the gross rate of return on the combination of state claims that pays one unit in each future state: the risk-free interest rate.

In this equation, $\text{prob}_s^* \equiv (\text{pri}_s/(\sum_s \text{pri}_s))$ is called the risk-neutral probability. The name seems intuitive, as the last line shows the price as discounting the expected payoff at the risk-free rate, as if we were risk neutral, but where the expectation $E^*(\cdot)$ is taken using the risk-neutral probabilities. The prob_s^* does, in fact, define a valid probability measure. That is, prob_s^* is (1) positive and (2) sums to 1.0 across states. Moving from the original measure to the * measure is a change of measure, an operation described in continuous time with a Radon-Nikodym derivative. The prob_s^* measure is sometimes called the "q-measure," while the original is the "p-measure," or physical measure. The q-measure drives much of derivatives pricing, as it turns out that derivatives prices hinge on the q-measure. The q-measure is also known as the equivalent martingale measure. Each of these words means something. An equivalent measure has the same null sets as the original measure ($\text{prob}_s > 0 \Leftrightarrow \text{prob}_s^* > 0$). A martingale measure is one under which asset values follow a martingale. To see this, simplify by setting $R_f = 1$, and assume that dividends are zero. Then the value of an asset is $V_t = E_t^*(V_{t+1}) = E_t^*(E_{t+1}^*(V_{t+2})) = \cdots = E_t^*(V_{t+j})$ for all j. This is the definition of a martingale for the V_t process, under the $E^*(\cdot)$ measure.

The last argument uses the *law of iterated expectations* to infer that $E_t^*(E_{t+1}^*(V_{t+2})) = E_t^*(V_{t+2})$. The assumption is that more is known at time $t+1$ than at time t; that is, the time $t+1$ information refines the time t information. In this case, the law of iterated expectations states that the expectation of an expectation must be the expectation using the coarser of the two information sets. The famous economist Sanford Grossman once explained it to me this way: When you have an expectation of an expectation, the dumber one always wins out.

With complete markets, we can relate the risk-neutral probabilities to the SDF:

$$
\text{prob}_s^* = R_f \, \text{pri}_s = R_f (m_s \text{prob}_s).
\tag{2.11}
$$

This says that risk-neutral probabilities are high for a given state when either the objective probability of the state occurring is high or the value attached to payoffs in that state through the SDF is high. Thus, "risk-neutral" probabilities actually reflect both risk and risk aversion. Maybe the name is not so intuitive, after all!

2.6   Recovery

Since this is a book on empirical asset pricing, we could use an aside on related empirical work. There is a long empirical tradition of trying to extract forecasts from asset prices. If the market is pretty efficient, such forecasts could be pretty good, and in some cases there is evidence that they are. Much of this classical work assumes risk neutrality. For example, the expectations hypothesis of the term structure states that forward rates are forecasts of expected future "spot" interest rates. Futures prices, under risk neutrality, should equal the expected future spot prices of the underlying commodity, and forward currency prices should forecast future exchange rates. The difference between nominal and real bond yields is expected inflation under risk neutrality. The goal of recovery is to move beyond the assumption of risk neutrality that motivates these forecasts.

The relation between state prices and probabilities has been examined empirically, with the goal of recovering information about utility functions. Rewriting equation (2.11), we have

$$m_s = R_f^{-1} \, (\text{prob}_s^*/\text{prob}_s). \tag{2.12}$$

Equation (2.12) has been used to attempt to estimate the pricing kernel m from data. Typically, the heroic assumption is made that the return of a stock index summarizes the relevant states. (Think of the market as a proxy for the level of wealth in the economy, the relevant state variable.) Data on the underlying stock index, combined with a statistical model, are used to estimate the p-measure, or the probabilities, prob_s. Options prices on the market index can be used to infer the q-measure or prob_s^* (Banz and Miller 1978; Breeden and Litzenberger 1978). Combining these with a risk-free rate proxy delivers m. Examples include Jackwerth and Rubinstein (1996); Bakshi, Kapadia, and Madan (2003); Chernov (2003); and Barone-Adesi, Engle, and Mancini (2008).

One of the puzzles in this literature is that the fitted m, as a function of wealth, often is found to be U-shaped (e.g., Ait-Sahalia and Lo 2000). If marginal utility is declining in wealth, the upward sloped part is hard to understand. Bakshi, Madan, and Panayotov (2010) argue that a U-shape may be consistent with a model featuring heterogeneous beliefs and short selling. Shefrin (2008) argues that behavioral models featuring disagreement or sentiment can generate upward-sloping regions in the pricing kernel. Christoffersen, Heston, and Jacobs (2013) argue that if the pricing kernel depends on more than the level of the market, then even if the kernel is monotonic in the level of wealth, its projection onto the market level may not be monotonic. They use a negative dependence of the kernel on the market variance to illustrate their argument.

If we could observe state prices, m-talk states that $\text{pri}_s = m_s \text{prob}_s$, which suggests another decomposition. We observe in state prices the probability multiplied by the SDF. The literature just mentioned takes observations on pri_s, makes assumptions to model prob_s, and draws inferences about m_s. But if we could observe m_s, then we could extract the market forecasts, prob_s.

Ross (2015) takes on the problem of getting state prices to reveal the probabilities and the SDF. He assumes that the world is Markov, where only the state today matters in predicting the state tomorrow. The model has an $S \times S$ matrix of transition probabilities, $\pi = \{\text{prob}(i|j)\}_{i,j=1,\dots,S}$. There is a matching $S \times S$ matrix of state prices, P. Ross also assumes irreducibility, so that the S states of the world evolve over time in such a way that each state can and will be revisited again and again over time (sort of like in the movie *Ground-hog Day*). It is assumed that the SDF can be written as a ratio of (marginal utility) functions of the state variable, as in a time-separable model. Let M be an $S \times S$ matrix with $(1/\beta)$ $\{m_s, s = 1,\dots,S\}$ on the diagonal, where β is the rate of pure time preference. Then we have $P = \beta M \pi$, subject to $(1/\beta) M^{-1} P \underline{1} = \pi \underline{1} = \underline{1}$, as the probabilities must sum to 1, conditional on any starting state. Ross is able to transform this into an eigenvector-eigenvalue problem, where $M^{-1} \underline{1}$ is the eigenvector, and P is the matrix. This allows him to invoke a theorem from Perron-Frobenius theory to solve for the forecast probabilities as well as the SDF.

Barovicka, Hansen, and Scheinkman (2016) point out that, to recover both the probabilities and $\{m_s\}_s$ from state prices, Ross (2015) has ruled out permanent shocks to the SDF, which means that in the very long run, the economy becomes risk neutral. Thus, in the long run, the Ross recovery method goes back to the classical assumption of risk neutrality, as was used to motivate forward rates and futures prices as forecasts of the future. Barovicka, Hansen, and Scheinkman illustrate that this long-run risk neutrality assumption is not a feature in most of the standard asset pricing models.

2.7 Multiperiod *m*-Talk and State Pricing

I want to make some passing comments about extending the state preference approach to a multiperiod setting, where agents make consumption/investment choices and trade over several periods. In the simplest case, this can be handled formally by assuming that markets are complete over dated state claims to consumption. A representative agent would solve the problem:

$$\text{Max } \sum_{s,t} \text{prob}(st)\, U(Cs,t), \text{ s.t. } W_0 = \sum_{s,t} p(st)\, C_{s,t}, \tag{2.13}$$

where prob(st) is the probability that state s will be observed at time t, and p(st) is the price of a state claim that pays off only in state s at date t.

In this setup, most of the results of two-period state pricing carry over. In an exchange equilibrium, consumers equate their marginal rates of substitution over dates and states to the state price ratios. A central planning problem and a competitive equilibrium are both Pareto efficient and can implement the same allocation of dated state consumption claims. Given identical beliefs and a time- and state-separable utility function, one can construct a representative agent that consumes the aggregate endowment.

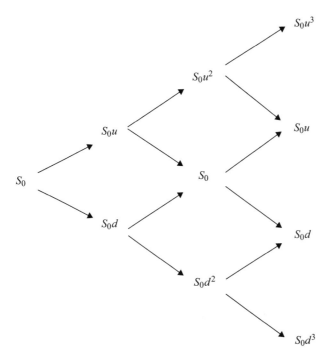

Figure 2.1
An event tree.

Now, let us define arbitrage in terms of trading strategies like x(s, t), which is a vector of holdings at time t if state s is observed at time t. An arbitrage is a trading strategy with zero initial cost, no interim investment ("drawdown"), but a positive payoff in some future state. The relations between arbitrage and pricing generalize from the two-period model. In particular, there is no arbitrage if and only if there exists an equivalent martingale measure.

This dated state setup makes things easy, but it does not reflect "time marching on." Usually, tomorrow comes after today. And things that were once possible are no longer possible. (This will make more intuitive sense when you get to be my age.) Conventionally, we can represent this situation with an event tree, as in figure 2.1.

This tree is the one used to represent the process assumed in the Black-Scholes option pricing model in discrete time. It imagines that only two states, up and down, can be observed at each date. More generally, we could have an event tree with more branches and different numbers of branches coming off of each node. The number of branches leaving a node corresponds to the amount of information at that node.

The tree in figure 2.1 has nine time-states, so nine linearly independent securities complete the market in terms of dated state claims. In this case, the market is "statically complete." However, nine securities are not needed if we allow for securities that live for multiple periods; say, paying off at all three dates in this example. The market is said to be "dynamically complete" when the number of linearly independent securities available at a node is equal to the number of branches leaving a node. In this tree, only two long-lived securities are needed to complete the market.

A dynamically complete market leads to what is known as a Radner (1972) equilibrium. A Radner equilibrium can attain the same equilibrium allocation of date-state consumption as can the equilibrium in the market that is complete over dates and states. It can be shown that the market is dynamically complete if and only if there is a unique equivalent martingale measure. This is the dynamic analog to the result that m is unique when markets are complete in the model with only two dates. In the Black-Scholes model, only a stock and a bond are used to span the state prices that determine options prices.

A dynamically complete market replicates the allocations of a statically complete market over dates and states, with the right prices and conditional probabilities at the intervening dates. The probabilities of the states evolve according to Bayes' rule. For example, consider the state probabilities at time $t=0$ and $t=1$ for state s_2 being observed at time $t=2$. These are related by $\text{prob}(s_2, 0) = \sum_{s_1 \text{ leading to } s_2} \text{prob}(s_1, 0)\,\text{prob}(s_2, s_1)$, where $\text{prob}(s_2, s_1)$ is the probability of state s_2 given that we are in state s_1. Similar relations may be derived for the state prices, too. But I don't want to get too far into this. Huang and Litzenberger (1988) provide a very good and more detailed treatment.

3 Maximization and the *m*-Talk Euler Equation

Many models in empirical asset pricing derive from the optimization problem of representative consumer-investors, and in particular from the first-order conditions for those problems. This chapter sets out the problem in general, illustrates the famous consumption-based CAPM as an example, and then provides a general approach to finding the stochastic discount factor in consumer-investor optimization problems. This trick, a version of a perturbation argument, was originally taught to me by George Constantinides (thanks, George), and I have used it in about a half dozen published papers. Maybe you will find it useful, too.

3.1 General Statement

Empirical work often explores the implications of equilibrium models for the SDF, based on investor optimization. Suppose quite generally that the representative investor maximizes a utility function for a lifetime consumption stream, $V(C_0, C_1, \ldots, C_t, \ldots)$, subject to a budget constraint: $W_{t+1} = (W_t - C_t) \underline{x}' \underline{R}_{t+1}$, where $\underline{x}$ is the portfolio weight vector, subject to $\underline{x}'\underline{1} = 1$. Solving this problem, we have

$$\text{Max } E_t\{V(\cdot)\} \Rightarrow P_{it} E_t\{\partial V/\partial C_t\} = E_t\{(P_{i,t+1} + D_{i,t+1})(\partial V/\partial C_{t+1})\}, \tag{3.1}$$

or

$$E_t(m_{t+1} R_{i,t+1}) = 1, \text{ with } m_{t+1} = (\partial V/\partial C_{t+1})/E_t\{(\partial V/\partial C_t)\}. \tag{3.2}$$

Condition (3.1) states that the utility cost of investing in an asset must equal the utility benefits of doing so. The current price (measured in consumption units) multiplied by the expected marginal utility of current consumption must equal the expected payoff of the asset multiplied by the marginal utility of the future consumption. In this formulation, it is clear that m_{t+1} is the representative agent's intertemporal marginal rate of substitution (IMRS) between consumption at time t and consumption at time $t+1$. This is similar to expressions we were looking at in the discussion of state pricing, if you think of the random future consumption C_{t+1} as driving the state of the world. The $E_t(\cdot)$ in the denominator of m_{t+1} in (3.2) drops out if the utility function $V(\cdot)$ is time separable, as $(\partial V / \partial C_t)$ is known at time t in this case. The $E_t(\cdot)$ in the denominator of m_{t+1} allows for cases where the future lifetime utility of consumption, not yet known at time t, may be influenced by the consumption choice at time t. For example, this occurs if there is habit persistence, as we will see below. In this case, a rational agent whose habit depends on past consumption would recognize that his future habit levels will be affected by his current consumption, and he would take this into account in his current consumption. This is an example of an "internal" habit, as opposed to an "external" one (or a habit that the agent takes as exogenous). For example, an agent may form his habit based on the aggregate consumption or a neighbor's consumption that he can't affect, in the case of external habit. These models are discussed more extensively in chapter 5.

3.2   The Consumption CAPM

Breeden (1979) derived a consumption-based asset pricing model in continuous time, assuming that the preferences are time-additive: $V = \sum_t \beta^t U(C_t)$, where β is a time preference parameter, and $U(\cdot)$ is increasing and concave in current consumption, C_t. Breeden's model is a linearization of equation (3.2) which follows from the assumption that asset values and consumption follow diffusion processes (shown by Bhattacharya (1981) and Grossman and Shiller (1982)). A discrete-time version follows Lucas (1978), assuming a power utility function:

$$U(C) = \frac{C^{1-\alpha} - 1}{1 - \alpha}, \tag{3.3}$$

where $\alpha > 0$ is the concavity parameter. This function displays constant relative risk aversion equal to α. *Relative risk aversion* in consumption is defined as $-Cu''(C)/u'(C)$, where u' and u'' denote the first and second derivatives. *Absolute risk aversion* is $-u''(C)/u'(C)$. Ferson (1983) studies a consumption-based asset pricing model with constant absolute risk aversion. These workhorse utility functions are members of the hyperbolic absolute risk aversion class, which displays linear risk tolerance. *Risk tolerance* is defined as the inverse of absolute risk aversion.

   Using the power utility function, equation (3.2) becomes

$$m_{t+1} = \beta (C_{t+1}/C_t)^{-\alpha}. \tag{3.4}$$

This asset pricing model featuring this m is often called the "consumption-based CAPM" (CCAPM). This workhorse model is featured in the equity premium puzzle of Mehra and Prescott (1985) and is used in many empirical papers.

3.3 Finding *m* Using a Perturbation

In a general consumer-investor optimization problem, we often have the Bellman equation:

$$J(W_t, s_t) \equiv \text{Max } E_t\{U(C_t, \cdot) + J(W_{t+1}, s_{t+1})\}, \tag{3.5}$$

where $U(C_t, \cdot)$ is the utility of consumption expenditures at time *t*. Equation (3.5) defines $J(\cdot)$ as the indirect utility of wealth. The notation allows the direct utility of current consumption expenditures to depend on other variables, such as past consumption expenditures or the current state variables. The state variables, s_{t+1}, are sufficient statistics, given wealth, for the utility of future wealth in an optimal consumption-investment plan. Thus, the state variables represent future consumption-investment opportunity risk. We will spend some time thinking about state variables in chapter 13, as the state variables are a crucial link to empirical work that uses different factors to describe state variables relevant to expected asset returns. Here, the problem is to find the *m* implied by some particular model for $U(\cdot)$ and/or $J(\cdot)$.

If the allocation of resources to consumption and investment assets is optimal, it is not possible to obtain higher utility by changing the allocation. This observation motivates the following perturbation argument. Suppose that an investor considers reducing consumption at time *t* to purchase more of (any) asset. The expected utility cost at time *t* of the reduced consumption is the expected product of the marginal utility of consumption expenditures, $U_c(C_t, \cdot) > 0$ (where the subscript c denotes a partial derivative), multiplied by the price of the asset, measured in the same units as the consumption expenditures. The expected marginal utility gain of selling the investment asset and consuming the proceeds at time $t+1$ is $E_t\{(P_{i,t+1} + D_{i,t+1}) J_w(W_{t+1}, s_{t+1})\}$, where $J_w(\cdot)$ is the partial derivative of $J(\cdot)$ with respect to wealth. If the allocation maximizes expected utility, the following must hold: $P_{i,t} E_t\{U_c(C_t, \cdot)\} = E_t\{(P_{i,t+1} + D_{i,t+1}) J_w(W_{t+1}, s_{t+1})\}$, which is equivalent to equation (1.1), with

$$m_{t+1} = J_w(W_{t+1}, s_{t+1})/E_t\{U_c(C_t, \cdot)\}. \tag{3.6}$$

This is about the most general equation for *m* in a model with a representative consumer-investor. The consumption-based CCAPM of equation (3.4) can be derived as a special case from the intertemporal problem and the envelope condition, $U_c(\cdot) = J_w(\cdot)$, which states that the marginal utility of current consumption must be equal to the marginal utility of current wealth if the consumer has optimized the trade-off between the amount consumed and the amount invested. We will work through the intertemporal consumption-investment problem more carefully in chapter 13.

Asset pricing models typically focus on the relation of security returns to aggregate quantities. To obtain equilibrium expressions, it is therefore necessary to either adopt the fiction of a representative agent or aggregate the first-order conditions of the individuals. Then equation (3.6) may be considered to hold for a representative investor who holds all the securities and consumes the aggregate quantities. Theoretical conditions that justify the use of aggregate quantities are discussed by Gorman (1953), Wilson (1968), Rubinstein (1974), and Constantinides (1982), among others; and we saw in chapter 2 that the assumption of complete markets allows the use of a representative agent who holds the aggregate quantities in the model. When these conditions fail, investors' heterogeneity will affect the form of the asset pricing relation. In some cases, heterogeneous agents' first-order conditions can be added up (i.e., aggregated) to obtain pricing relations in terms of aggregate quantities. Early examples include Lintner (1965), who worked with mean-variance preferences, and Brennan and Kraus (1978), who worked with the hyperbolic absolute risk aversion (HARA) utility class. The effects of heterogeneity are examined by Lee, Wu, and Wei (1990), Constantinides and Duffie (1996), Sarkissian (2003), and Shefrin (2010), among many others, and the effects of heterogeneity remains an active area of research.

4 Expected Risk Premiums and Alphas

This chapter introduces expected excess returns or risk premiums, the most common objects examined in empirical asset pricing. The most general versions of these follow directly from m-talk. We discuss the expected abnormal return or *alpha*, and again the most general version follows from m-talk. By now you are seeing a pattern. Everything in asset pricing is m-talk. Alphas are related to betas, and beta pricing models, through maximum correlation or "mimicking" portfolios. I briefly introduce some applications of alphas, and we finish up with an introduction to the volatility risk premium.

4.1 General Statement

Commonly, empirical work in asset pricing focuses on expressions for expected returns and excess rates of return. The expected excess returns are modeled in relation to the risk factors that create variation in m_{t+1}. Consider any asset return $R_{i,t+1}$ and a reference asset return, $R_{0,t+1}$. Define the excess return of asset i, relative to the reference asset as $r_{i,t+1} = R_{i,t+1} - R_{0,t+1}$. If equation (1.1) holds for both assets, it implies:

$$E_t\{m_{t+1}r_{i,t+1}\} = 0 \text{ for all } i. \tag{4.1}$$

Expand equation (4.1) into the product of expectations plus the covariance, obtaining:

$$E_t\{r_{i,t+1}\} = \text{Cov}_t(r_{i,t+1}; -m_{t+1})/E_t\{m_{t+1}\}, \text{ for all } i, \tag{4.2}$$

where $\text{Cov}_t(\cdot\,;\,\cdot)$ is the conditional covariance at time t. Equation (4.2) is a general expression for the expected excess return or "risk premium," from which most of the expressions in the literature can be derived. Just plug in your favorite model for m and there you have it! Note that when we work with excess returns, because equation (4.2) can be multiplied by any constant, we do not determine the scale of the returns. To determine the scale, we need to use equation (1.1). Then $E_t(m_{t+1}) = R_{0t}^{-1}$ determines the scale of returns, if R_{0t} is chosen to be uncorrelated with m, and in that sense is risk free.

Equation (4.2) implies that the covariance of return with m_{t+1} is a general measure of *systematic risk*. This risk is systematic in the sense that it is related to the SDF and thus to the marginal utility of aggregate consumption or wealth in an equilibrium model where agents optimize. In any model, systematic risk is the risk that is related to the factors that drive m_{t+1}. Any fluctuations in the asset return that are uncorrelated with fluctuations in the SDF are not "priced," meaning that they do not command a risk premium. For example, consider the conditional regression $r_{it+1} = a_{it} + \beta_{it}m_{t+1} + u_{it+1}$, with $Cov_t(u_{it+1}, m_{t+1}) = 0$; then β_{it} is the "beta" for asset i on m_{t+1}. Only the beta part of the risky asset return is correlated with the SD, and the risk premium for the asset is proportional to the beta.

Equation (4.2) states that a security will earn a positive risk premium if its return is negatively correlated with the SDF. Negative correlation means that the asset is likely to pay off more when the marginal utility is low, so the value of a given amount to be paid out at such times is relatively low. This is not a desirable pattern of payoffs, and agents will require a premium in expected return for holding such an asset. For a given expected payoff, if the expected return is higher, then the price in the market is lower.

4.2 The SDF Alphas and Abnormal Returns

Finance researchers and practitioners alike have an easy familiarity with alpha, which is supposed to measure the expected abnormal return of an investment. Alpha is so ubiquitous that it has become a generic. Just like we "Google" something to search the internet, we add investment value through alpha. Studies refer to CAPM alpha, three-factor alpha, four-factor alpha, or even five-factor alpha, assuming that the reader hardly requires a definition. Investment practitioners routinely discuss their strategies in terms of their quest for alpha, and the number of investment firms with alpha in their names is staggering. Alpha can be active, conditional, or even portable. All of this follows easily from m-talk. Define the SDF alpha as:

$$\alpha_p = E_t\{m_{t+1}R_{pt+1}\} - 1, \tag{4.3}$$

where R_p is a tested asset, strategy, or managed portfolio gross return. Let $r_p = R_p - R_F$ be the fund's excess return, where R_F is a risk-free rate with $E_t\{m_{t+1}R_{pt+1}\} = 1$. Of course, m is the stochastic discount factor, which depends on the model. Suppressing the time subscripts, we can see that the SDF alpha is proportional to the difference between the expected excess return of the fund and the expected excess return predicted by the model:

$$\alpha_p = E(mr_p) = E(m)E(r_p) + Cov(m, r_p) = E(m)[E(r_p) - Cov(-m/E(m), r_p)], \tag{4.4}$$

because $Cov(-m/E(m), r_p)$ is the expected excess return predicted by the model (any model). Thus, alpha captures the expected abnormal return, relative to the model's predictions.

4.3 Relation to Beta-Pricing: The Maximum Correlation Portfolio

The SDF alpha is related to the more traditional alpha in a "beta-pricing" formulation of a model. A beta-pricing formulation involves a benchmark portfolio with excess return, r_B. The excess returns of the evaluated portfolio are regressed over time on the benchmark, and alpha is the intercept of the regression:

$$r_p = \text{alpha}_p + \beta_p r_B + u_p. \tag{4.5}$$

To see the relation between the beta pricing alpha and the SDF α_p, we introduce the *maximum (squared) correlation portfolio* to an SDF, and let it be the benchmark r_B. Consider a regression of the SDF on the vector of excess returns:

$$m = A + B'r + \varepsilon, \tag{4.6}$$

where the definition of a regression implies that $E(r\varepsilon) = 0$. From this regression, the benchmark portfolio with excess return $r_B = B'r$ is the maximum (squared) correlation portfolio to m, because the fitted values of the regression maximize the R-squared of the regression. Letting the fitted regression be $m^* = A + B'r$, we can write $m = m^* + \varepsilon$, and we see that $\text{Cov}(r, m) = \text{Cov}(r, m^*) = \text{Cov}(r, r_B)$, because the constant A does not change the covariance. This implies that r_B and m have the same implications for expected risk premiums. This result is quite general. Breeden (1979) was the first to note that a (particular) SDF and its maximum correlation portfolio have identical asset pricing implications, so we can always substitute the portfolio for m in asset pricing models. This can be useful, for example if the original m depends on data that are noisy or incomplete and where the asset return data are better, a fact exploited in a consumption model by Breeden, Gibbons, and Litzenberger (1989).

Consider the beta-pricing alpha, using the maximum correlation portfolio as the benchmark:

$$\begin{aligned}
\text{alpha}_p &= E(r_p) - \beta_p E(r_B) \\
&= E(r_p) - [\text{Cov}(r_p, r_B)/\text{Var}(r_B)]\ E(r_B) \\
&= E(r_p) - [\text{Cov}(r_p, m)/\text{Var}(r_B)]\ E(r_B) \\
&= \alpha_p/E(m),
\end{aligned} \tag{4.7}$$

where the last line plugs in $E(r_B) = \text{Cov}(-m, r_B)/E(m) = \text{Cov}(-m^*, r_B)/E(m) = -\text{Var}(r_B)/E(m)$. Thus, the two versions of alpha determine the same expected excess returns.

4.4 Applications of Alpha

We will encounter many applications of alpha. When we describe empirical tests of asset pricing models, we will examine the null hypothesis that the vector of alphas for a sample of test assets is zero; H_0: $\underline{\alpha} = 0$. When we discuss investment performance evaluation, we will use the SDF alphas or beta-pricing alphas to judge the expected abnormal returns of the funds' managed portfolios.

The regression-based approach illustrates an alternative measure, the (ex post) abnormal return, defined at each date t to equal to the intercept plus the residual, $\text{alpha}_p + \text{u}_{pt}$, of the regression (4.5), for the excess returns on the benchmark excess returns. Abnormal returns like this have long been used in "event studies" (Fama et al. 1969, discussed in more detail in chapter 23). The abnormal return may have more power to detect departures from the pricing model than does the alpha alone, because there may be patterns in the residuals that are ignored by the intercept. The residuals average to zero by construction over the full sample, so the alpha is the expected abnormal return. However, there may be times in the sample when the residuals are positive or negative, depending on the state of the economy, firm-specific events, or other factors that are not considered in the regression.

4.5   Volatility Risk Premiums

A relatively recent topic in asset pricing is the volatility risk premium (e.g., Stein and Stein 1991; Bansal, Shaliastovich, and Yaron 2011). If the future volatility or variance of the market portfolio is a factor correlated with the SDF, it may have a risk premium. Buying options is a good way to capture that risk, as option prices are sensitive to changes in market volatility. To see the intuition, drop the time subscripts, let $\hat{m} = -m/\text{E}(m)$ be the scaled SDF, where we set $R_f = 1$, and consider the payoff $X_{t+1} = \text{Var}_{t+1}(\text{r}_m)$. Its expected return is

$$
\begin{aligned}
\text{E}(\text{r}_x) &= \text{Cov}(\hat{m}, \text{Var}(\text{r}_m)) \\
&= \text{E}\{\hat{m}\,\text{Var}(\text{r}_m)\} - \text{E}\{\hat{m}\}\text{E}\{\text{Var}(\text{r}_m)\} \\
&= \text{E}\{\hat{m}\,\text{Var}(\text{r}_m)\} + \text{E}\{\text{Var}(\text{r}_m)\} \\
&= -[\text{E}^*\{\text{Var}(\text{r}_m)\} + \text{E}\{\text{Var}(\text{r}_m)\}] \\
&\equiv -\text{VRP}.
\end{aligned}
\tag{4.8}
$$

The last line of equation (4.8) defines the *volatility risk premium* (VRP), the difference between an expected option-implied variance, $\text{E}^*\{\text{Var}(\text{r}_m)\}$, calculated under the risk-neutral distribution (or q-measure), and $\text{E}\{\text{Var}(\text{r}_m)\}$, which uses the physical or p-measure. For example, the sample variance of daily returns over the month (a proxy for the realized variance) is used to get an empirical proxy for the monthly p-measure variance. The VRP can be evaluated empirically by comparing the two variances. Typically, when computed using stock market indexes, the risk-neutral or option-implied variance is larger than the sample variance of returns, so VRP tends to be positive, and the premium for an asset that pays off the variance is negative.

Zhou (2010), Bollerslev, Gibson, and Zhou (2011), and Bollerslev et al. (2014) find evidence that the VRP for a stock market index computed using information at time t can predict future market rates of return. Bali and Zhou (2016) find that the VRP is correlated with other measures of economic uncertainty. Bakshi and Madan (2006) and Chabi-Yo (2012) show that the skewness and kurtosis of the market index are determinants of the VRP. Kilic and Shaliatovich (2018) break the VRP down into downside and upside second moments, or "good" and "bad" volatility-related premiums; these authors find that the moments' joint predictive ability for market returns exceeds that of a single measure. Pyun (2018) finds that the regression coefficient for the market return on contemporaneous shocks to the realized variance can replace the regression coefficient of market returns on the VRP, leading to improved predictive ability using the VRP. Gonzáles-Urteaga and Rubio (2016) find that portfolios of stocks sorted by their betas on VRP can predict differences in average returns in the cross section.

5 So Many Models, So Little Time

By now you are beginning to appreciate how powerful the m-talk equation is for organizing asset pricing models. Each model corresponds to an expression for the SDF. This chapter provides a taxonomy of asset pricing models, based on their specifications for m. Sorry if I left your favorite model out.

5.1 Polynomials in Consumption or Wealth

We start with the simplest cases: single-period models where m is a polynomial in end-of-period wealth or consumption. These are implied by models where the representative investor maximizes a utility function that is also a polynomial, but of one higher order. The first example is almost trivial.

5.1.1 The SDF Is Constant

The SDF is a constant when the utility function is linear: $u(x) = a_0 + b_0 x$. Then we have $u'(x) = b_0$, and $m_{t+1} = u'(x_{t+1})/u'(x_t) = m$ is a constant. A linear utility function exhibits risk neutrality. If markets don't care about risk, then the expected returns of all assets are the same, independent of their risk: $E(R_{t+1}) = 1/m$ for all assets.

5.1.2 The SDF Is Linear

When the utility function is quadratic, the SDF is linear. A quadratic utility function is risk averse, at least over a range of its argument, and it is intimately connected to the CAPM. For example, Berk (1997) shows that if you want the CAPM to hold without placing restrictions on the distributions of asset returns, then you have to assume that agents have quadratic utility functions. It is easy to show that a representative agent with quadratic utility implies the CAPM. Consider the following representative agent's problem:

$$\text{Max } U(C_0) + E\{U(W_1)\},$$
$$W_1 = (W_0 - C_0) R_m, \tag{5.1}$$
$$U(x) = a + bx - (d/2)x^2.$$

The first-order condition to this problem is equation (1.1), with

$$
\begin{aligned}
m &= U'(W_1)/U'(C_0) \\
&= [b - dW_1]/[b - dC_0] \\
&= [b - d\{(W_0 - C_0)R_m\}]/[b - dC_0] \\
&\equiv \theta_0 - \theta_1 R_m.
\end{aligned}
$$

(5.2)

In the last line, we use the fact that the variables at time 0 are constants from the point of view of the agent, who knows their values when making the decision at time 0. The coefficients θ_0 and θ_1 in the linear relation between the market return and m are functions of C_0. When prices are set and expectations are formed at time 0, the consumption C_0 is known and is treated as a constant. In this sense, the coefficients are conditional on the current consumption.

Later in the chapter we discuss other examples of conditional CAPMs, where the coefficients are conditioned on other information that is known when prices are set at the beginning of the period. We will see in chapter 14 that the condition that m is conditionally linear in the market portfolio return is equivalent to the condition that the CAPM describes assets' conditional expected returns. This was first observed by Dybvig and Ingersoll (1982). The intuition is that assets' expected returns are always determined by their covariance with m; in this case, the covariance with m is proportional to the covariance with R_m, which is proportional to the famous beta coefficient of the CAPM.

Beta pricing models with multiple betas and factor risk premiums are popular in the literature. We will see in chapter 7 that a multiple-beta model for expected returns is equivalent to the statement that m is linear in the multiple factors. In this case, the covariance with m becomes a linear combination of the covariances with the various factors.

5.1.3   The SDF Is Quadratic

This is the case when the utility function is cubic: $U(x) = a + bx - (d/2)x^2 + (f/3)x^3$. Typically, $f > 0$, so that the agent has a preference for positive skewness. Kimball (1990) calls a positive third derivative of the utility function "prudence." Skewness preference models are examined by Kraus and Litzenberger (1976), Harvey and Siddique (2000), and others, where calculations similar to the last case show that $m = \theta_0 - \theta_1 R_m + \theta_2 R_m^2$. Because risk premiums are always proportional to their covariance with m, risk premiums are determined in this model by their covariances with the market and the squared market return. The covariance with the squared market return is the systematic *co-skewness* of the asset.

Kraus and Litzenberger (1976) show theoretically that if the market portfolio's skewness is positive, agents like the co-skewness and apply a negative premium for it. If the market skewness is negative, agents don't like co-skewness with the market, so it requires a positive premium. Empirically, they find that co-skewness and beta are highly correlated in a cross section of stocks, so that systematic co-skewness does not provide a strong empirical explanation for returns, once the market beta is in the model. Harvey and Siddique (2000) consider the skewness premium conditional on market states. Conrad, Dittmar, and Ghysels (2013) find that high idiosyncratic skewness corresponds to low average returns in the cross section, and this is not completely captured by the systematic co-skewness.

Recent studies examine premiums for volatility risk exposure. Co-skewness and volatility risk exposure are closely related. The covariance of an asset return with the market variance looks a lot like the co-skewness of the asset return with the market. Chabi-Yo (2011) uses approximation to derive a stochastic discount factor that displays both effects separately. In his model, $m_{t+1} = E_t(m) + D_0[R_{mt+1} - E_t(R_{mt+1})] + D_1[R_{mt+1}^2 - E_t(R_{mt+1}^2)] + D_2[\sigma_{mt+1}^2 - E_t(\sigma_{mt+1}^2)]$. The D coefficients depend on agents' risk tolerances and skewness preferences and are constants. Since $D_0 < 0$, the market premium is positive, as in the CAPM. Since $D_1 > 0$, the premium for co-skewness is negative according to this model. The sign of D_2 determines the premium for variance risk exposure and its sign depends on the risk-tolerance weighted average of the skewness preference parameters of investors.

5.1.4 The SDF Is Cubic

The SDF is cubic when the utility function is quartic: $U(x) = a + bx - (d/2)x^2 + (f/3)x^3 - (g/4)x^4$, and it expresses an aversion to kurtosis if $g > 0$. Gollier and Pratt (1996) call a negative fourth derivative of the utility function "temperance." It seems somewhat intuitive that in practice, investors care about the market variance, and the trading in market index options seems to support that view. Investors might care about kurtosis, too, which intuitively is like the variance of variance, which makes fat tails in the distribution of returns. Since 2012, the Chicago Board Option Exchange has produced the VVIX, an index of the volatility of volatility. This is the implied volatility of the 30-day VIX. It is calculated in the same way that the implied volatility or VIX is calculated from options on the market index.

Dittmar (2002) studies models with kurtosis preference, $m = \theta_0 - \theta_1 R_m + \theta_2 R_m^2 - \theta_3 R_m^3$. Agents are assumed to dislike kurtosis, which implies a positive premium for co-kurtosis, that is, the covariance between the return and the cube of the market return. Dittmar finds some evidence in support of such a model. Bakshi and Madan (2006) study a quartic pricing kernel, and Chabi-Yo (2012) derives a model featuring preferences over stochastic kurtosis.

5.1.5 The SDF Is a Higher-Order Polynomial

We could go on forever like this, adding more terms in a polynomial expansion, and eventually we would be able to approximate any reasonable function using the polynomials, at least at the local point of the approximation. Some researchers have taken polynomials for the utility function out to fifth order. For example, Eeckhoudt and Schlesinger (2006) call a positive fifth derivative "edginess," a preference over the fifth moment or "pentosis" risk. Chabi-Yo (2012) provides both theory and empirical work involving polynomials to the fifth order, and Chabi-Yo, Leisen, and Renault (2014) study the aggregation of heterogeneous investors that can imply a representative agent with preferences over the fifth moment. Scott and Horvath (1980) provide conditions under which an agent will be averse to the even moments (e.g., variance and kurtosis) and have a positive preference for the odd moments (e.g., skewness). Of course, if we add a term like $(h/5)x^5$ to the utility function, h is proportional to $(1/5!)$ in a Taylor series, so the higher-order terms matter less.

5.2   Habits and Durability

We discussed the consumption-oriented CCAPM in chapter 3. Many papers in asset pricing now generalize the consumption-based model to allow for *nonseparabilities* in the lifetime utility function $V(\cdot)$, as may be implied by the durability of consumer goods, habit persistence in the preferences for consumption, and other refinements. Time nonseparabilities mean that today's marginal utility of consumption is affected by previous decisions in time. In the case of durable goods, this is because you get a flow of services from your previous purchases. In the case of habit, it is because past consumption may determine a reference level of consumption, which becomes habitual. This idea has been used in the study of addiction. For example, I might evaluate the utility of a shot of fine scotch (or heroin) relative to the one I had last time. If the utility function is defined over the difference between what I get today and the habit level, and if the function has infinite marginal utility at zero, then I would do just about anything to get as many shots as before. Habit may also be considered as getting used to a standard of living. Agent's past decisions may influence future utility through habit. Constantinides (1990) studies a model in which the representative agent's utility function is of the form: $V = (1-\alpha)^{-1} \sum_t \beta^t (C_t - hX_t)^{1-\alpha}$, $0 \le h < 1$. Here, hX_t is the habit or subsistence level, and X_t is assumed to be an exponentially weighted average of past consumptions C_{t-k}, $k > 0$.

I might evaluate the utility of my car relative to my neighbor's car. If I can't influence my neighbors' choices, I have to take this kind of habit level as exogenous or "external." Abel (1990) studies a form of habit in which the consumer evaluates current consumption in part relative to the aggregate consumption in the previous period, which he or she takes as exogenous. The idea is that people care about "keeping up with the Joneses." The utility function in Abel's model is $V = E\{(1-\alpha)^{-1} \sum_t \beta^t (C_t / [C_{t-1}^d C_{At-1}^{1-d}]^\gamma)^{1-\alpha}\}$, where the subscript A denotes the aggregate consumption that the agent cannot affect, and the C_t is the agent's own consumption. Thus, one component of the habit is external and one is internal, and only the internal part can be influenced by the consumer in the optimization. In equilibrium, the aggregate consumption is that of the representative agent, so the two consumptions become one.

Campbell and Cochrane (1999) develop another model in which the habit stock is taken as exogenous or external by the consumer. The habit stock in this case is modeled as a highly persistent weighted average of past aggregate consumptions, similar to Constantinides (1990). However, in this model, the agent assumes that her choices have no effect on her future habit. Taking the habit as external results in a simpler and more tractable model. By modeling the habit stock as an exogenous time series process, the Campbell and Cochrane model provides more degrees of freedom to match asset market data, compared with the simpler consumption-based model. In this model, we have $V = E\{(1-\alpha)^{-1} \sum_t \beta^t (C_t - X_{t-1})^{1-\alpha}\}$, where X_{t-1} is the habit level, so that $\partial V/\partial C_t = \beta_t (C_t - X_{t-1})^{-\alpha}$, and $m = (\partial V/\partial C_{t+1} / \partial V/\partial C_t) = \beta\{(C_{t+1} - X_t) / (C_t - X_{t-1})\}^{-\alpha}$.

Dunn and Singleton (1986), Eichenbaum, Hansen, and Singleton (1988), and Yogo (2006) develop consumption models with durable goods. Durability introduces nonseparability over time, since the actual consumption at a given date depends on the consumer's previous expenditures. The consumer optimizes over the current expenditures C_t, accounting for the fact that durable goods purchased today increase consumption at future dates.

Such studies as Dunn and Singleton (1986) and Eichenbaum, Hansen, and Singleton (1988) find mixed evidence for the importance of durability in consumption-based asset pricing models. Allowing nonseparability seems to improve goodness-of-fit tests of the model, but in some cases, the estimated coefficients on the lagged expenditures are negative, which is the wrong sign for durability. Durability in expenditures reduces the volatility of the flow of services relative to the lumpy measured expenditures, and therefore reduces the volatility of the implied marginal utility of the services, compared with a model that uses the same data but ignores the durability. In the equity premium puzzle, Mehra and Prescott (1985) found that the volatility of the implied marginal utility is too low in a model without durability. Therefore, durability has the wrong implication for the volatility of marginal utility.

With habit, the fitted IMRS is more volatile than it is in a time-separable model or a model with durability, based on the same expenditure data. With habit, a consumer will optimally smooth consumption more than in a time-separable model. Constantinides (1990) found that habit persistence allows the model to successfully match certain sample moments of consumption and returns that Mehra and Prescott (1985) were unable to match in a time-separable model.

Ferson and Constantinides (1991) model both durability and habit persistence in consumption expenditures in a simple model. They show that the two combine as opposing effects. In an example where "memory" is truncated at a single lag, the derived utility of expenditures is

$$V = (1-\alpha)^{-1} \sum_t \beta^t (C_t + bC_{t-1})^{1-\alpha}, \tag{5.3}$$

where the coefficient b is positive and measures the rate of depreciation if the good is durable and there is no habit persistence. Habit persistence implies that the lagged expenditures enter with a negative effect ($b < 0$). In this model, $E_t\{\partial V/\partial C_t\} = \beta^t (C_t + bC_{t-1})^{-\alpha} + \beta^{t+1} bE_t \{(C_{t+1} + bC_t)^{-\alpha}\}$. This is an example, as alluded to earlier, where $\partial V/\partial C_t$ needs the $E_t(\cdot)$ term, capturing the expectation at time t of a function of future consumption. Ferson and Constantinides find that estimates of this model imply that the b parameter is negative, so that habit dominates durability in the data. Further empirical evidence on similar habit models is provided by Heaton (1993) and Braun, Constantinides, and Ferson (1993), who find evidence for habit in international consumption and returns data.

Consumption expenditure data are highly seasonal, and Ferson and Harvey (1992) argue that the Commerce Department's seasonal adjustment program may induce spurious time series behavior in the seasonally adjusted consumption data that most empirical studies have used. Using data that are not adjusted, they find strong evidence for a seasonal habit model. In a seasonal habit model, I might for example evaluate the utility of my winter ski vacation, relative to what I experienced last year on my ski vacation. Kroencke (2017) takes this idea further, reverse engineering the smoothing function used to create the consumption data in the national income and product accounts. He finds that the unsmoothed data work better in several empirical asset pricing models.

Some studies compare the empirical performance of internal habit, as in Constantinides (1990) and Ferson and Constantinides (1991), with external habit, as in Campbell and Cochrane (1999). In particular, Grishchenko (2010) and Chen and Ludvigson (2009) find that internal habit models provide a better fit to the cross section of stock returns. Ferson and Constantinides (1991) find that, although they did not reject the habit model when their tests focus on the dynamic properties of returns, when the tests emphasize the cross-sectional structure of average returns, the models can be rejected. Allen (1991) finds that failure to reject the model in some cases is attributable in part to the higher volatility of the IMRS under habit persistence, which results in less precise estimates of the moments that are used to form the test statistics. Telmer (1993) and Ni (1997) come to similar conclusions using different approaches.

Habit models remain an active area of research. Van Binsbergen (2017) studies models in which habit is formed separately over multiple goods. He emphasizes how demand elasticities for the goods depend on consumers' habits, and how that feeds into a model with production. Gomez, Priestly, and Zapatero (2016) examine a model with habit in which the desire to hedge the systematic part of local relative wealth concerns, as in "keeping up with the Joneses," induces a new risk factor in the model.

5.3   State Nonseparability

Epstein and Zin (1989, 1991) consider a class of recursive preferences that can be written as $V_t = F(C_t, \text{CEQ}_t(V_{t+1}))$. In this equation, $\text{CEQ}_t(\cdot)$ is a time t "certainty equivalent" for the future lifetime utility V_{t+1}. The function $F(\cdot, \text{CEQ}_t(\cdot))$ generalizes the usual expected utility function. Epstein and Zin study a special case of the recursive preference model in which the preferences are

$$V_t = [(1-\beta)C_t^\rho + \beta E_t(V_{t+1}^{1-\alpha})^{\rho/(1-\alpha)}]^{1/\rho}. \tag{5.4}$$

Why state nonseparability, as opposed to time nonseparability? Consider the second term of equation (5.4), which would be a linear function of future discounted utility of consumption in the simple, separable, consumption model: $E_t E_t(V_{t+1}) = E_t(\sum_{j \geq 1} \beta^j C_{t+j}^{1-\alpha}) = \sum_{j \geq 1} \sum_s \text{prob}_s \beta^j C_{st+j}^{1-\alpha}$, if s indexes the states that may occur at the future dates, conditional on the information at time t, and prob_s is the probability that state s will occur. The simple model is separable across states, in the sense that its derivative with respect to C_{st+1} is not a function of the other states for $t+1$. In contrast, in (5.4), because of the various terms raised to a power, the derivative of the utility with respect to a given state at $t+1$ does depend on the other states that may occur at time $t+1$. This is state nonseparability: Your marginal utility in a given state depends on the other states that might occur. For example, I might be happier on a blue-sky day in Florida if I knew that a hurricane might, but did not, occur.

In the Epstein-Zin model, we have two preference parameters, α and p, that describe the utility function. The coefficient of relative risk aversion for timeless consumption gambles is α, and the elasticity of substitution for deterministic consumption is $(1-p)^{-1}$. To see this, consider a case where there is no time (only $t+1$) but there is risk. Then we have only the second term in equation (5.4), and the parameter p cancels out. The parameter α captures the concavity of the utility function, or the risk aversion. Now consider a case where there is no risk, but there is time. Iterating equation (5.4) forward, we get that $V_t = (1-\beta)\sum_{j\geq 0}\beta^j C_j^p$. Thus, the parameter p controls the intertemporal elasticity of substitution for consumption.

More generally, if $\alpha = 1-p$, the model reduces to the time-separable power utility model, in which intertemporal substitution and risk aversion must work together in a fixed relation. If $p=0$, then $\alpha=1$, and the log utility model of Rubinstein (1976) is obtained. Epstein and Zin show that the m with a representative agent in this model becomes

$$m_{t+1} = \beta[(C_{t+1}/C_t)^{p-1}]^{(1-\alpha)/p}[V_{t+1}^{1-\alpha}/E_t(V_{t+1}^{1-\alpha})]^{(1-\alpha-p)/(1-\alpha)} \tag{5.5}$$

(when $p\neq 0$ and $1-\alpha\neq 0$). Evidently, the CCAPM is the leading term of the SDF, and there is an additional term related to the future utility. This term encapsulates nonseparable preferences, and in particular, an aversion to the risk pertaining to the future expected utility, for certain parameter values.

Intuitively, at time t, the $E_t(\cdot)$ term in equation (5.5) is treated like a constant. If $(1-\alpha-p)<0$, then the second term is convex in V_{t+1}. The agent is averse to variance in the future expectation, in the sense that m is larger when the variance of the future expectation is larger. In this sense, the agent prefers an early resolution of the uncertainty. The condition $(1-\alpha-p)<0$ states that the elasticity of intertemporal substitution is small relative to the effect of risk aversion, so the agent is relatively unwilling to substitute future for present consumption. Aversion to variance in the future expectation occurs in this model when the agent is relatively unwilling to substitute consumption across time. This feature of the model looms large in the long-run risk literature, which we discuss in chapter 31. In that literature, we have expectations about objects in the future, the agent cares a lot about changes in the expectations, and changes in the expectations become factors with risk premiums.

Epstein and Zin (1991) study a special case of the model in which the future utility is captured by the market portfolio return:

$$m_{t+1} = [\beta(C_{t+1}/C_t)^{p-1}]^{(1-\alpha)/p} \{R_{m,t+1}\}^{(1-\alpha-p)/p}. \tag{5.6}$$

Risk premiums, which depend on covariances with m, are functions of covariances with consumption growth rates and also of covariances with the market return. The model reduces to either the market-based CAPM or the simple CCAPM as special cases. Roussanov (2014) finds empirically that the covariances of some stocks with consumption depend on the level of wealth relative to consumption, which he interprets as supporting a model with both a consumption factor and a wealth factor. Campbell (1993) shows that the Epstein-Zin model can be transformed into an empirically tractable model without consumption data. He uses a linearization of the budget constraint that makes it possible to substitute for consumption in terms of the factors that drive the optimal consumption function. Expected asset returns are then determined by their covariances with the market and the underlying factors.

5.4 Heterogeneity

When markets are not complete and agents are heterogeneous or face heterogeneous constraints on their optimizations, markets will not produce optimal risk sharing in general. There are two main approaches to obtaining SDFs in this case. The first is to find an unconstrained agent who is at an interior optimum and to use that agent's marginal rate of substitution to price assets. This is the approach of Brown (1988), Zeldes (1989), and Moskowitz, Malloy, and Vissing-Jorgensen (2009). These studies use data on heterogeneous households from the Panel Study of Income Dynamics in an attempt to measure the m of an unconstrained agent.

The second approach is to try to aggregate the heterogeneous agents' preferences in a model with heterogeneity, deriving an aggregate utility function. Early work on this goes back to Lintner (1965) in a mean-variance context, and Brennan and Kraus (1978) with hyperbolic absolute risk-aversion utility functions. More recently, Chabi-Yo, Leisen, and Renault (2014) studied heterogeneous agents with polynomial utility functions. In incomplete markets, utility functions generally aggregate to a representative agent with stochastic weights (Magill and Quinzii 1996), which complicates the analysis.

One way to simplify the problem is to reverse engineer an aggregate m that will price assets correctly in the model with heterogeneity. The reverse engineering here means to find processes to assume for the shocks in the model such that a function of the aggregate consumption can still serve as the SDF. A classical example is Constantinides and Duffie (1996), who derive: $m_{t+1} = \beta(C_{t+1}/C_t)^{-\alpha} e^{(1/2)\gamma(\gamma+1)(\text{cross-sectional consumption variance})}$. In this model, m is higher when there is more heterogeneity (or dispersion across the agents) in the economy.

This makes sense if you think about a central planner, maximizing a weighted average of investors' utility functions. Consider starting in an economy with identical agents and identical allocations, and then taking consumption from some agents and giving it to other agents, thereby inducing heterogeneity. Assuming declining marginal utility (risk aversion), the loss in utility from the reductions is higher than the gain from the increases, and the overall social utility (any weighted average of the individuals' utility functions) is lower when there is more heterogeneity. Therefore, the aggregate marginal utility and risk premiums are higher with more heterogeneity. The cross-sectional variance of consumption can serve as a priced risk factor. However, in the United States, it seems that the cross-sectional dispersion in consumption, as it has been measured so far, may not be large enough to fit the data on asset market returns. Sarkissian (2003) takes this model to international data, finding some support for its predictions in that setting.

5.5 General Affine Models

In affine models of the SDF, the log of m is linear in the main variables of interest. These models can be written as

$$m_{t+1} = \exp(R_{ft} - (1/2)\Lambda_t'\Lambda_t + \Lambda_t'\epsilon_{t+1}). \tag{5.7}$$

The ϵ_{t+1} are the priced shocks in the model, often assumed to be normal $(0, 1)$. Affine models are popular in studies of the term structure of interest rates, because in these models, the natural logarithms of bond prices (the expected values of m) are linear in the relevant variables, resulting in very tractable models (see, e.g., Dai, Singleton, and Yang 2004). We examine models of this type in chapter 27 when we discuss the term structure and bond fund performance.

5.6 Ambiguity Aversion and Knightian Uncertainty

So far we have discussed models in which there are probabilities attached to future outcomes, and in which agents are presumably able to figure out those probabilities. This is distinct from a situation where we don't even understand the model or the probabilities that describe future events. Uncertainty where we can't even describe the probabilities is termed "Knightian uncertainty" (Knight 1921). We can think of this as uncertainty about the model, in contrast to the uncertainty in a given model, as we have so far discussed. Some authors have argued that agents should behave conservatively in this situation, as if the worst reasonable model is the true model (see the review by Epstein and Schneider 2010). Such agents are ambiguity averse. This motivates behavior similar to "maximin" problems in engineering, where we make choices to do the best we can, on the assumption that the worst possible outcome will prevail. Think of a game against Murphy, as in "Murphy's law," where you know that Murphy will pick the model that causes you the most trouble, and you have to make choices recognizing that Murphy is afoot. Hansen and Sargent (2001a, 2001b) argue for maximin, given the worst reasonable model in a set of models, and formalize the reasonableness of an alternative model, given a reference model, in terms of the differences in the state probabilities of the two models. This is captured with a measure of relative entropy. Hansen and Sargent suggest, inter alia, that we might pick the worst reasonable model from the set of models that can be rejected at some significance level—say, 5%—given a sample of a given size generated from the reference model. Agents who maximize expected utility in this setup are more risk averse than otherwise similar agents who do not face the uncertainty over models. They also show how Epstein-Zin style preferences can be motivated by such considerations in special cases.

Ambiguity aversion straddles the divide between purely rational models and behavioral models. The conservative response prescription given ambiguity can be thought of as a behavioral characteristic of investors, or as a normative prescription for obtaining rational decision rules that are robust to misspecified models.

5.7   Behavioral SDFs

As behavioral finance models have become more sophisticated, various utility functions and SDFs have become associated with behavioral finance. Behavioral finance focuses on human imperfections, and in that sense is similar to the study of taxes, transactions costs, and other market imperfections. Since the main goal of this book is to study the friction-less models that make up the foundations of asset pricing, models with frictions are not emphasized. Still, a nice example of a behavioral SDF is derived by Shefrin (2010), where

$$m_{t+1} = \beta (C_{t+1}/C_t)^{-\alpha} \, e^{(\text{weighted-average belief errors})}. \tag{5.8}$$

In Shefrin's model, investor sentiment is defined as a deviation between investors' beliefs and the beliefs implied by an objective probability distribution. Baker and Wurgler (2006) seem to have a similar idea in mind in their sentiment index. Note that by analogy to equation (5.6), it is not hard to imagine that sentiment is a priced risk factor in some behavioral models, and indeed, it can be (e.g., Shleifer and Vishny 1997).

Behavioral models for m often feature a kink in the utility function, motivated by the idea that investors dislike losses more than they like gains (e.g., Kahneman and Tversky 1979). The reference point, where the kink occurs, is an open question at this point. In various studies, it is taken to be a risk-free return, zero, a fixed point less than zero, or other values. Often the kinky utility function is combined with other behavioral biases, such as (incorrect) probability weighting, based on observed behavior in laboratory experiments with human subjects. Delikouras (2017) studies a consumption model that modifies the Epstein-Zin preferences described above to incorporate a kink at the endogenous certainty equivalent for future consumption and finds empirically that this helps explain cross-sectional variation in stock portfolios' average returns.

An interesting empirical anomaly, which has been given a behavioral spin in recent studies, is evidence of SDFs that are not decreasing in wealth or at least in the return of the stock market, as discussed earlier. A first-order intuition about m is that it should be smaller in good times and larger in bad times. Ait-Sahalia and Lo (2000) and Jackwerth (2000) estimate the SDF as a function of the market return using nonparametric methods. They find evidence that it is not a monotonically declining function. Jackwerth calls this the "pricing kernel puzzle." Shefrin (2008) provides examples where the behavioral m can have regions that are not declining in wealth, which he attributes to overconfidence. Bakshi and Madan (2008) and Bakshi, Madan, and Panayotov (2010) study the pricing kernel puzzle and attribute it to heterogeneous beliefs interacting with short sales. Christoffersen, Heston, and Jacobs (2013) present a model with a variance risk premium, where the SDF is monotonic in the market return but also depends on the variance, and its projection on only the stock market is not monotonic. This suggests that the pricing kernel puzzle can be explained by an omitted variables bias, but as of this writing, the jury is not yet in on this question.

5.8 Multigood Models

Early studies of the consumption based CCAPM consider multiple goods. Here is an example of a two-good utility function that frequently appears in the literature:

$$U(C_1, C_2) = [(1-\alpha)C_1^p + \alpha C_2^p]^{1/p}. \tag{5.9}$$

In this Cobb-Douglas model, one can derive an Euler equation and identify m using either of the two goods. The two expressions are linked by the atemporal substitution between the two goods, or the cross-partial derivative of the utility function, which determines the relative prices of the two goods. Examples include Dunn and Singleton (1986), Eichenbaum, Hansen, and Singleton (1988), and Gomez, Kogan, and Yogo (2009).

Piazzesi, Schneider, and Tuzel (2007) consider a two-good model in which one good is a standard commodity and the other is a durable good, the services delivered from housing. In that model, a power utility function is assumed, defined over the composite commodity $[C^{(\varepsilon-1)\varepsilon} + wS^{(\varepsilon-1)/\varepsilon}]^{\varepsilon/(\varepsilon-1)}$, where C is the standard good, and S is the housing services. The parameter w is the weight given in utility to housing services, and ε is the elasticity of housing relative to consumption. In this model, the SDF is given by

$$m = \beta(C_{t+1}/C_t)^{-\alpha} \times \text{CompositionRisk}^{\{(1/\alpha)\varepsilon/((1/\alpha)\varepsilon-1)\}}, \tag{5.10}$$

where CompositionRisk is given by $[1 + w(S_{t+1}/C_{t+1})^{(\varepsilon-1)/\varepsilon}]/[1 + w(S_t/C_t)^{(\varepsilon-1)/\varepsilon}]$, and it depends on the expenditure share, S/C. These authors argue that the expenditure share is much easier to measure with accuracy than is the flow of housing services as appears, for example, in the Consumer Price Index (CPI), because it is hard to account for quality changes over time (e.g., Boskin 1996). They find that proxies for composition risk suggest that the ratio of housing to other types of consumption has strong business cycle-related patterns, and the additional risk factor can help explain the equity premium puzzle.

5.9   Representation Using Future Consumption and Returns

Parker and Julliard (2005) present an interesting version of the standard power utility model with consumption, where m is stated using future consumption and returns. Start with $E_t\{m_{t+1}R_{t+1}\} = 1$, $m_{t+1} = \beta(C_{t+1}/C_t)^{-\alpha}$, and substitute $C_t^{-\alpha} = E_t\{\beta C_{t+1}^{-\alpha} R_{t+1}\}$ and $C_{t+1}^{-\alpha} = E_{t+1}\{\beta C_{t+2}^{-\alpha} R_{t+2}\}$ into the Euler equation to obtain

$$1 = E_t\{\beta[E_{t+1}\beta C_{t+2}^{-\alpha} RF_{t+2}]/(C_t^{-\alpha} R_{t+1})\} = E_t\{\beta^2(C_{t+2}/C_t)^{-\alpha} RF_{t+2}R_{t+1}\}. \tag{5.11}$$

Thus we can use $m \equiv \beta^2(C_{t+2}/C_t)^{-\alpha} RF_{t+2}$ to price the returns R_{t+1}. This can be motivated from errors in the consumption data associated with time aggregation, as discussed in chapter 17. Parker and Julliard find that this approach works better than the standard power utility consumption model for pricing some portfolios of stocks.

6 Applications of *m*-Talk

Since every asset pricing model on planet earth (and probably other planets, too) is a description of the stochastic discount factor, the number of applications of *m*-talk is out of this world. In fact, most of the literature that studies financial markets illustrates applying *m*-talk. This chapter starts with classical models for real interest rates, which apply the normal moment generating function to *m*-talk with a power utility function for consumption. Using *m*-talk to price bonds with different times to maturity, we have the general underpinnings of the term structure of interest rates and models of spot and forward prices. Distinguishing between real and nominal payoffs leads us to the role of price indexes and inflation in asset pricing models. Allowing for payoffs denominated in different currencies leads to international asset pricing in *m*-talk and such issues as international diversification, risk sharing, interest rate parity, the forward premium anomaly, and the carry trade. After setting up "conditional" asset pricing, I offer a version of the concept of market informational efficiency and the famous "joint hypothesis" problem of Fama (1970) using *m*-talk. I finish the chapter with suggestions on how to think about market efficiency in delegated portfolio management, where there are securities markets and managerial labor markets.

6.1 Real Interest Rates

One of the first applications of the CCAPM in finance was to the study of real interest rates. Generally, equation (1.1) implies $E_t(m_{t+1}R_{Ft}) = 1 = R_{Ft}E_t(m_{t+1})$ when the interest rate is known at time t, so that $R_{Ft} = 1/E_t(m_{t+1})$. Assuming power utility, so that $m_{t+1} = \beta(C_{t+1}/C_t)^{-\alpha}$, and assuming that $\ln(C_{t+1}/C_t) \sim N(\mu_c, \sigma_c^2)$, then

$$\ln(R_F) = -\ln\beta + \alpha\mu_c - (1/2)\alpha^2\sigma_c^2. \tag{6.1}$$

Equation (6.1) is derived using the normal moment generating function, which implies that if x is normal, then $E(e^x) = \exp(E(x) + (1/2)\text{Var}(x))$. This model for real interest rates was first studied empirically by Ferson (1983), Hansen and Singleton (1983), and Hall (1987), and it is one of the first examples of an affine model for interest rates. More current work has extended this approach using entropy, defined as the difference between the log of the expected value and the expected value of the log, to generalize the normal moment generator (e.g., Backus, Chernov, and Martin 2011).

Equation (6.1) displays the determinants of the real interest rate in three terms. The first is a pure time discount effect. The second is related to intertemporal substitution, and the third to "precautionary savings." To understand the intertemporal substitution term, consider a world with no risk, in which $\sigma_c^2 = 0$. Then equation (6.1) corresponds to the Fisherian diagram that equates the bond price to the marginal rate of substitution between consumption at time t and consumption at time $t+1$. That is, $R_F^{-1} = \beta (C_{t+1}/C_t)^{-\alpha}$, and the coefficient $\alpha > 0$ regulates the intertemporal substitution of the consumer. Other things being equal, the higher the growth of consumption is, the higher the risk-free interest rate will be (the lower the bond price will be). The intuition is that when there will be a lot of consumption in the future, the price to buy consumption in the future is low, so the interest rate is high. Holding fixed the consumption growth rate, the larger the parameter α is, the higher the interest rate will be. A higher α implies that the consumer has a smaller marginal rate of substitution for consumption through time: She is more willing to give up current for future consumption. Thus, the price in current consumption of a bond that pays one unit of future consumption is lower and the interest rate is higher.

To understand the risk effect, recall that the price of a bond, given the current consumption, is proportional to the expected future marginal utility. Other things being equal, when the future consumption is riskier, the expected marginal utility of the future consumption is higher. Future consumption is more valued at such time, and the greater demand for future consumption pushes up the bond price, which lowers the interest rate. This corresponds to a precautionary savings motive to buy bonds. More formally, think of a convex marginal utility function (assuming nonincreasing absolute risk aversion, $u''' > 0$); the expected value of the marginal utility function is above the function at the expected value. Because the expected marginal utility is higher when there is more risk, the price of the bond is higher, thus the risk-free rate is lower. The larger the risk aversion is, measured through the parameter α, the larger will be the consumption variance effect on the risk-free rate. This term may also be thought of as a "flight to quality" effect, where when things get scary, investors flock to safe assets, driving up the price of the safe bond and thus driving down its yield. Hartzmark (2016) finds empirically that the variance term is more strongly related to the levels of interest rates than is the expected consumption growth term in equation (6.1).

The fact that the parameter α in this simple consumption model has to capture both agents' intertemporal substitution, even in the absence of risk, and risk aversion in the face of risk, illustrates one of the weaknesses of the model that motivated studies to introduce time and state nonseparabilities, such as habit persistence and Epstein-Zin preferences, as discussed in chapter 5.

6.2   Term Structure

The term structure of interest rates refers to the set of bond prices or yields in a cross section of bonds with different times to maturity. One of the core questions in this literature is: Why do some bonds return more than other bonds do? An intriguingly equivalent question is: How is today's yield curve related to expected future interest rates? *m*-talk provides a simple way to characterize the answers to these questions. Let P_t^j be the price at time t of a bond with j periods to maturity, and paying off one unit at maturity only. In particular, *m*-talk says that $P_t^1 = E_t\{m_{t+1}\}$, and $P_t^2 = E_t\{m_{t+1} \, E_{t+1}[m_{t+2}]\}$. Now, if we assume a time-separable model, so that m_{t+1} is known at time $t+1$, we can write

$$P_t^2 = E_t\{E_{t+1}(m_{t+1}m_{t+2})\} = E_t\{m_{t+1}m_{t+2}\}, \tag{6.2}$$

where we use the law of iterated expectations to remove the expectation at time $t+1$. Intuitively, equation (6.2) says that a long-term bond is priced as if it paid off m_{t+2} at time $t+1$. Recall that the law of iterated expectations says, crudely, that when you have an expectation of an expectation, the "dumber" one wins out. This assumes that we learn more as time marches on, a standard (if somewhat optimistic) characterization. Equation (6.2) illustrates a more general result, which is that in time-separable models, the multiperiod *m* is the product of the single-period *m*.

Studies of the term structure focus on the returns of bonds of different maturities. The gross return of a two-period bond over the first period t to $t+1$ is P_{t+1}^1/P_t^2, because the two-period bond at time t becomes a one-period bond at time $t+1$. Define the expected *term premium* as the expected gross return of the two-period bond divided by the gross return of the one-period bond:

$$E_t\{(P_{t+1}^1/P_t^2)/(1/P_t^1)\} = E_t\{m_{t+2}\}E_t\{m_{t+1}\}/[E_t\{m_{t+2}\}E_t\{m_{t+1}\} + \text{Cov}_t(m_{t+1}, m_{t+2})]. \tag{6.3}$$

This is but one of many ways to define a term premium, but it is the simplest approach for present purposes and is approximately equivalent to other definitions in the literature discussed later in this chapter. We see that term premiums are zero, meaning that the expected returns of long- and short-term bonds are equal in a model where the SDF has zero auto-covariance. A model produces a positive term premium (the ratio is greater than one) if and only if the autocorrelation of its SDF is negative. When the SDF has negative autocorrelation, the long-term bond has a risky payoff at time $t+1$, negatively correlated with the SDF, and thus the long bond requires a positive risk premium.

The term structure is often characterized in terms of forward prices or forward rates, and their relation to spot prices or spot rates. We can see this by using the one-period forward price for a one-period bond. This price, F_t, is set at time t and is the payment to be made at time $t+1$ in exchange for a one-period bond at time $t+1$ that matures at time $t+2$. If you agree to this exchange, you are long the forward contract. Forward contract prices relate bonds of different maturities, in the sense that holding the two-period bond is essentially equivalent to holding a one-period bond and just the right-sized long position in the forward contract. In fact, this is how we compute the implied forward price from the prices of one- and two-period bonds: $P_t^2 = P_t^1 F_t$. Since there is no cash flow at time t for a pure forward contract, the value of entering into the contract at time t must be zero:

$$0 = E_t\{-F_t m_{t+1} + (m_{t+1} m_{t+2})\ 1\}.   \tag{6.4}$$

Solving for the forward price, we find

$$F_t = E_t(m_{t+2}) + \text{Cov}_t\{m_{t+1},\ m_{t+2}\}/E_t(m_{t+1}).   \tag{6.5}$$

The forward price is equal to the expected future spot price, $E_t\{P_{t+1}^1\} = E_t\{E_{t+1}(m_{t+2})\}$, plus a risk premium. The forward price is lower than the current expectation of the future spot price if the SDF in the model has negative autocorrelation.

The same relationships can be described in terms of forward rates and spot rates instead of forward prices and bond prices, since prices are the inverse of one plus rates. Consider the one-period spot rate r_1, based on the price of a one-period bond: $(1+r_1)^{-1} = P_t^1$, and the two-period spot rate, stated (per period) based on the price of a two-period bond: $(1+r_2)^{-2} = P_t^2$. The forward rate is found from $(1+r_2)^2 = (1+r_1)(1+f)$, which states that the return to buying the one-period bond and rolling over the proceeds at the forward rate for the second period is the same as the return over two periods from holding the two-period bond.

Since the expected gross return on the two-period bond over the first period is $E_t\{(P_{t+1}^1/P_t^2)\}$, it follows from the preceding equations that the forward rate is the expected value of the future one-period spot rate to be observed at $t+1$, plus a risk premium, where the risk premium is the expected return of the two-period bond over the first period, in excess of that of the one-period bond.

If one assumes that the risk premium is zero, then the forward rate is equal to the expected future spot interest rate. This is the classical *expectations hypothesis* of the term structure (Kessel 1965). This hypothesis is relevant if the stochastic discount factor has zero autocorrelation. This claim is simple and potentially very powerful, because it means we could compute forward rates and we would have the "market's expectation" of the future spot interest rates in hand. Of course, the world refuses to be that simple, and we find nonzero expected risk premiums. The literature goes back a long way. For example, Hicks (1946) proposes a liquidity premium hypothesis, where longer-term bonds offer a premium for their relative illiquidity. Culbertson (1957) argues that segmentation in the bond markets can lead to different premiums for different maturities that may vary with economic conditions.

The fact that the forward rate is the expected future spot rate plus a risk premium allows a puzzling possibility. Suppose that there is some factor whose fluctuations influence expected future spot rates and also influence expected risk premiums, but it increases one and decreases the other by the same amount. Such a latent factor would not be seen to be related to current forward rates at all, and it could not be extracted from them using models of the yield curve. But it might nevertheless have a strong relation to risk premiums. As unlikely as it may seem, the literature finds evidence that such "unspanned" latent factors exist and are related to the macroeconomy (e.g., Joslin, Priebsch, and Singleton 2014; Duffee 2011).

The basic idea of a term structure extends beyond simple bonds and has been applied to indexed bonds, exchange rates, volatility measured over different horizons, and more. Van Binsbergen et al. (2013) study the term structure of common stock dividend payout yields. They find a downward-sloping term structure, unlike the commonly upward-sloping term structure of bond returns, kicking off a currently active literature that attempts to explain it. Backus, Boyarchenko, and Chernov (2016) provide an analysis of several kinds of term structures in *m*-talk.

6.3 Real and Nominal *m*-Talk

So far, we have implicitly been talking about real payoffs and SDFs, related to real consumption or wealth. However, most assets pay off in nominal units like dollars. Indeed, a short-term Treasury security, even if risk free in dollar terms, is not quite risk free in real terms if inflation is uncertain. Therefore, it is often important to distinguish between real and nominal in *m*-talk. We start with the premise that real things drive the model. Denote the real price of a security as $P_t^r = E_t\{m_{t+1}^r X_{t+1}^r\}$, where X_{t+1}^r is the real payoff (e.g., ears of corn), and the superscript r denotes real units. The nominal (e.g., dollar) payoff is $X_{t+1}^n = X_{t+1}^r \pi_{t+1}$, where π_{t+1} denotes the price level at $t+1$ (e.g., dollars per unit of corn). Similarly, the nominal price of the security is $P_t^n = P_t^r \pi_t = E_t\{m_{t+1}^r X_{t+1}^r \pi_t\} = E_t\{m_{t+1}^r (\pi_t/\pi_{t+1})X_{t+1}^n\}$. There are two ways to think of this. We can use our old friend the real SDF to get the real price P_t^r of the real payoffs X_{t+1}^n/π_{t+1}. Equivalently, we can use a nominal SDF, $m_{t+1}^r \pi_t/\pi_{t+1}$, to get the nominal price of the nominal payoffs. Note that the ratio π_t/π_{t+1} is the inverse of the rate of price inflation between the two dates, so the nominal SDF is the real SDF deflated by one plus the rate of price inflation.

These expressions also describe the relation between real and nominal bonds. The price of the one-period nominal bond is

$$P1_t^n = E_t\{m_{t+1}^n\} = E_t\{m_{t+1}^r \ \pi_t/\pi_{t+1}\}$$
$$= E_t\{m_{t+1}^r\}E_t\{\pi_t/\pi_{t+1}\} + \text{Cov}_t\{m_{t+1}^r, \ \pi_t/\pi_{t+1}\}. \tag{6.6}$$

When the final covariance is zero, we see that the nominal bond price is the real bond price $E_t\{m_{t+1}^r\}$ adjusted by the expected value of the inverse of the rate of inflation. If we take natural logarithms in this case, then the interest rate on the nominal bond is approximately the interest rate on the real bond plus the expected rate of inflation. This is known as the "Fisher effect," in honor of the famous economist Stanley Fisher, who studied interest rates almost a century ago.

The Fisher equation (when the final covariance term in equation (6.6) is zero) approximately states that nominal interest rates equal real rates plus expected inflation. If we could observe both real and nominal rates, we could extract inflationary expectations from them. Studies have used inflation-indexed bonds and Treasury bonds for this purpose. To extract inflationary expectations, Fama (1972) uses nominal bonds and assumes that the expected real interest rates are constant. Perhaps studies of the term structure of inflationary expectations will become more prominent again when inflation levels and volatility return to higher levels, as occurred in the United States in the late 1970s and early 1980s.

The Fisher effect fails if the final covariance term in equation (6.6) is not zero. In this case, inflation has real effects, in the sense that it is correlated with the real SDF, and there will be a risk premium for inflation. The nominal bond is exposed to inflation risk, so the nominal bond price will be higher or lower, depending on the sign of the covariance. If inflation is a bad thing (in the sense that it is correlated with a higher real m), then the covariance term is negative and nominal bond prices are lower (rates are higher) than they would be under the Fisher equation.

6.4   International m-Talk

In international finance, the study of exchange rates is central, and we have m-talk in each country, linked by their exchange rates. Let the values for the foreign country have the superscript y (e.g., yen), where $E(m^y R^y) = 1$. Let the United States (i.e., dollar) be the domestic country, where $E(m^\$ R^\$) = 1$. Imagine integrated capital markets, meaning that agents from each country can trade either the yen-denominated or the dollar-denominated securities freely. There is an exchange rate $e_t = \$/y$, the dollar price of one yen.

A US investor can trade dollars for yen at time t, purchase a yen-denominated asset, sell it at time $t+1$, and convert the gross yen return R^y into dollars at the time $t+1$ exchange rate. This implies

$$1 = E(m^\$ R^\$) = E(m^\$(e_{t+1}/e_t)R^y).   \tag{6.7}$$

International asset pricing studies often assume complete markets in this setting, which allows the derivation of a relation between the two countries SDFs. Let $R^y(s)$ be the yen return if state s is realized at time $t+1$. Equation (6.7), $E(m^y R^y) = 1$, and complete markets then imply the following:

$$0 = E\{[m^y - m^\$(e_{t+1}/e_t)]\, R^y(s)\}.   \tag{6.8}$$

Given complete markets, we can choose $R^y(s)$ to be the state claim, paying one yen for a particular state s. Then the above equation implies $0 = \text{prob}_s\,[m^y - m^\$(e_{t+1}/e_t)]R^y(s)$. Since prob_s and $R^y(s)$ are positive numbers, the term in square brackets must be zero for each state s, which implies $m^y = m^\$(e_{t+1}/e_t)$ on a state-by-state basis.

We probably should be talking about real *m*s and real exchange rates here, not dollars and yen. To do that, we would have to introduce inflation rates in the two countries, as done earlier in the chapter, to talk about nominal yen and dollar securities. The notation starts to get complicated. Of course, it's OK to work with the nominal *m*s and payoffs here, but that only makes it look easier. We still need the two price indexes to compute the nominal *m*s, given that we start from a model with real *m*s.

Brandt, Cochrane, and Santa Clara (2006) decompose the variance of log exchange rates under complete markets: $\ln(e_{t+1}/e_t) = \ell n(m^y) - \ell n(m^\$)$, which implies that

$$\text{Var}\{\ln(e_{t+1}/e_t)\} = \text{Var}\{\ln(m^\$)\} + \text{Var}\{\ln(m^y)\} - 2\text{Cov}\{\ln(m^y), \ell n(m^\$)\}. \qquad (6.9)$$

These authors fit the right-hand side of this decomposition using annual data for 1975–1998 and simple power utility consumption models for the *m*s, and find that in the data, the right-hand side is larger than the left-hand side variance of actual real exchange rates for most developed countries. This reflects a low correlation of the measured consumption growth rates across countries. They show that if you set the correlation to be high, say 1.0, then the left- and right-hand sides line up much better. This observation turns on its head previous work (e.g., Brennan and Solnik 1989; Lewis 1999), which found that exchange rates are too volatile to be explained by most international macroeconomics models, and that the correlation across countries' consumption growths seems too low (say, in the range of 0.2–0.4) to be consistent with optimal risk sharing. This conclusion implies perfect dependence of the consumptions, as we saw in chapter 2. Brandt, Cochrane, and Santa-Clara (2006) conclude that either international risk sharing (consumption correlation) is higher than previously thought, or exchange rates are too smooth.

Stathopoulos (2016) rationalizes these results by using a model with habit that is common across countries. Then the consumption correlations can be low, while at the same time the correlations of the SDFs can be high because of the common habit. Lewis and Liu (2015) make a related argument using a model with a common persistent long-run risk component shocking consumption growth across countries. (Long-run risk models are discussed in chapter 31 of this book.)

Without too much more work, we can use these equations to describe the *forward premium*, or *carry trade anomaly*. Starting with $m^y = m^\$(e_{t+1}/e_t)$, we take natural logs and then expectations. The risk-free rate, say in yen, satisfies $-\ln(R_f^y) = \ln(E(m^y))$, and similarly for the risk-free rate in dollars. Assuming normality or using a Taylor series, we can write $E(m) = E\{e^{\ln(m)}\} = e^{E\{\ln(m)\} + (1/2)\text{Var}\{\ln(m)\}}$. Taking logs and substituting, we arrive at an expression for expected exchange rate depreciation:

$$E\{\ell n(e_{t+1}/e_t)\} = \ell n(R_f^\$) - \ell n(R_f^y) + (1/2)\,[\text{Var}\{\ell n(m^\$)\} - \text{Var}\{\ell n(m^y)\}]. \qquad (6.10)$$

Imagine that the interest rate in the United States ($) is higher than in Japan (y). If the difference in the two variances in square brackets is small and/or constant over time, equation (6.10) states that the high interest rate currency, the dollar, is expected to depreciate in value. Ignoring the variance term, this is called "uncovered interest rate parity." What you earn on the higher interest rate, you expect to lose on the currency. The anomaly is that on average, the high interest rate currency is seen to appreciate in value, just the opposite of what the formula says should happen. Thus we have the profitability of the carry trade: borrow at the low (yen) interest rate to invest at the high (dollar) interest rate, then pay back the loan with dollars that have appreciated relative to yen, not depreciated like equation (6.10) says they should. Such a trade is said to be "uncovered," because it is exposed to exchange risk.

How does this relate to forward rates? Through the (covered) *interest rate parity theorem*, which notes that if you have forward contracts on currencies, there are two equivalent ways to invest a dollar today and get dollar payoffs tomorrow. Let's say that f is the forward price in dollars per yen, for delivery in one period. You can invest in the US Treasury and earn $R_f^\$$, or you can trade your dollars for yen today, invest the yen at R_f^y, and sell the future yen payoff back into dollars at the forward rate f. This trade is covered, because the use of the forward contract removes the exposure to exchange rate risk. To avoid arbitrage, the two risk-free returns must be the same:

$$\ln(f/e_t) = \ln(R_f^\$) - \ln(R_f^y). \tag{6.11}$$

This covered interest rate parity theorem states that the high interest rate country must have a high forward exchange rate. It is based on ruling out arbitrage in trades that can be and are conducted at high frequency, at high volumes, and at low transaction costs in the interbank exchange market. The equation holds so well in the actual data that researchers sometimes use it to create forward rate data from existing spot rate series and use the results for other research purposes.

Substituting equation (6.11) into (6.10) relates the forward exchange rate to the expected future (spot) exchange rate. For example, if we assume risk neutrality and ignore the variance terms in square brackets in equation (6.10), the resultant equation states that the forward rate must equal the expected future spot exchange rate. The high interest rate country has a high forward rate, which predicts a high future exchange rate, or as we have set it up here, a depreciation of the currency. The *forward premium anomaly* (Fama 1984a) is the finding that the high forward rate countries tend to experience currency appreciation, not depreciation. Hence, the carry trade anomaly and the forward premium anomaly are the same thing, given that covered interest rate parity holds.

Many attempts have been made to explain the carry trade returns, and this remains an active area of research. Colacito and Croce (2013) find that the puzzle becomes worse in more recent data (after 1970). One obvious approach is to find the source of the risk premium in the returns to the carry trade. Most of the standard "factors" in the literature have failed at this (see the review by Burnside, Eichenbaum, and Rebelo 2011). Another tack is to consider that the carry trade carries episodic event risk, such as "crash" risk (Brunnermeier, Nagel, and Pedersen 2008; Jurek 2008). Recently, Chen, Lustig, and Naknoi (2016) question the market completeness assumption that relates the two exchange rates to the two countries' stochastic discount factors. There is more work to be done.

6.5   Conditional Asset Pricing

Conditional asset pricing studies the predictability in the returns of financial assets and the ability of asset pricing models to accommodate this predictability. To do this, we need to distinguish the information used by agents in the model from the information used by researchers testing the model. The notation $E_t\{\cdot\}$ in equation (1.1) denotes the conditional expectation, given a marketwide information set (let's call it Ω_t). Empiricists don't get to see Ω_t, so it is convenient to consider expectations conditional on instruments, Z_t, an observable subset of Ω_t. These expectations are denoted by $E(\cdot|Z_t)$. When Z_t is the null information set, we have the unconditional expectation, denoted by $E(\cdot)$.

Empirical work on conditional asset pricing models typically relies on *rational expectations*, which is the assumption that the expectation terms in the model are mathematical conditional expectations. This carries two important implications. First, it implies that the *law of iterated expectations* can be invoked. This law states that the expectation, given coarser information, of the conditional expectation, given finer information, is the conditional expectation given the coarser information. For example, taking the expected value of equation (1.1), rational expectations implies that versions of (1.1) must hold for the expectations $E(\cdot|Z_t)$ and $E(\cdot)$. Second, the rational expectations assumption implies that the differences between realizations of the random variables and the expectations in the model should be unrelated to the information that the expectations in the model are conditioned on. This leads to implications for the predictability of asset returns.

The return of the asset i may be predictable. For example, a linear regression over time of R_{it+1} on Z_t may have a nonzero slope coefficient. Equation (1.1) implies that the conditional expectation of the product of m_{t+1} and R_{it+1} is the constant 1.0. Therefore, $1 - m_{t+1}R_{it+1}$ should not be predictably different from zero using any information in the public information at time t, Ω_t. If there is predictability in a return R_{it+1} using any lagged instruments Z_t, the model implies that the predictability is removed when R_{it+1} is multiplied by the correct m_{t+1}. This is the sense in which conditional asset pricing models may accommodate predictable variation in asset returns.

Studies of predictability in stock and bond returns often report regressions that attempt to predict future returns using lagged variables. Chapter 32 studies in detail the evidence for and issues with predictability. Predictive regressions for shorter horizon (monthly or annual holding periods) returns for stocks and long-term bonds typically have small R-squares, as the fraction of the variance in long-term asset returns that can be predicted with lagged variables over short horizons is small. But because stock returns are very volatile, small R-squares can mask economically important variations in expected returns. To illustrate, consider a version of the Gordon (1962) constant-growth model for a stock price: $P = kE/(r-g)$, where P is the stock price, E is the earnings per share, k is the dividend payout ratio, g is the future growth rate of earnings, and r is the discount rate. The discount rate is the required or expected rate of return of the stock, and the model assumes that the discount rate and expected growth rate are fixed. Stocks are long-duration assets, so a small shift in the expected return can lead to a large fluctuation in the asset value. Consider an example where the price/earnings ratio is $P/E = 15$, the payout ratio is $k = 0.6$, and the expected growth rate is $g = 3\%$. The expected return r is 7%. Suppose there is a shock to the expected return, ceteris paribus. In this example, a change of 1% in r leads to approximately a 20% change in the asset value. This example suggests that small changes in expected returns can produce large and economically significant changes in asset values. Campbell (1991) generalizes the Gordon model to allow for stochastic changes in growth rates using the Campbell-Shiller approximation (Campbell and Shiller 1988a) reviewed in chapter 30. Campbell estimates that changes in expected returns through time may account for about half of the variance of equity index values.

6.6   Market Efficiency

The informational efficiency of markets is famously described by Fama (1970, 1991). I offer an updated interpretation using the SDF approach. As emphasized by Fama, any analysis of market efficiency involves a "joint hypothesis." There must be a hypothesis about the equilibrium model and also one about the informational efficiency of the markets. These can be described using the m-talk equation (1.1), with expectations conditioned on particular information, Z_t, that is used in the tests.

If X_{t+1} is the payoff of an asset and P_t is its market price, then $R_{t+1} = X_{t+1}/P_t$, and the equilibrium model states that $P_t = E\{m_{t+1}X_{t+1}|\Omega_t\}$. The equilibrium price is the mathematical conditional expectation of the payoff given Ω_t, discounted using m_{t+1}. In the language of Fama (1970), this says that the price "fully reflects" Ω_t. By the law of iterated expectations, prices must also fully reflect information that is courser than Ω_t.

The joint hypotheses in tests of asset pricing and market efficiency include a hypothesis about the model and one about the information. The hypothesis about the model of market equilibrium amounts to a specification of the stochastic discount factor m_{t+1}. We have already studied many examples. The second part of the joint hypothesis is informational efficiency. This may be described simply as the hypothesis that the Z_t used in the tests is contained in the market's information set Ω_t.

Fama (1970) describes increasingly fine information sets in connection with market efficiency. Weak-form efficiency uses only the information in past stock prices and returns. Semistrong-form efficiency uses variables that are obviously publicly available, and strong-form efficiency uses anything else. The different information sets described by Fama amount to different assumptions about what information is contained in Ω_t and what is therefore legitimately used as empirical instruments in Z_t. For example, weak-form efficiency states that past stock prices may be used in Z_t, semistrong includes public information, and strong-form efficiency includes all information.

Informational efficiency is related to whether trading strategies can earn so-called abnormal returns, whose expected values we know as alpha. If the SDF prices the primitive assets $\underline{R}_{t+1}$, then α_{pt} will be zero for a portfolio of the primitive assets formed using only information Z_t contained in Ω_t at time t. The portfolio return is $R_{p,t+1} = \underline{x}(Z_t)'\underline{R}_{t+1}$, and

$$\alpha_{pt} = \mathrm{E}\{[\mathrm{E}(m_{t+1}\underline{x}(Z_t)'\underline{R}_{t+1}|\Omega_t)] - 1|Z_t\} = \mathrm{E}\{\underline{x}(Z_t)'[\mathrm{E}(m_{t+1}\underline{R}_{t+1}|\Omega_t)] - 1|Z_t\}$$
$$= \mathrm{E}\{\underline{x}(Z_t)'\underline{1} - 1 |Z_t\} = 0,$$

since the portfolio weights sum to 1.0. Thus, informational efficiency states that you can't get an alpha different from zero using any information Z_t that is fully reflected in market prices. Since alpha depends on the model through m_{t+1}, there is always a joint hypothesis at play. Indeed, any evidence in the literature on market efficiency can be described in terms of the joint hypothesis; that is, the choice of m_{t+1} and the choice of the information Z_t.

If we find any investment strategy that has a positive alpha in a model that controls for public information, this rejects a version of the joint hypothesis with semistrong-form efficiency. If we don't question the model for m_{t+1}, then we can interpret such evidence as a rejection of informational efficiency. Alternatively, the model for m_{t+1} could be wrong. This view of market efficiency applies to conditional asset pricing. If a conditional model fails to explain predictability as described above, there are two possibilities. Either the specification of m_{t+1} in the model is wrong, or the market prices don't rationally reflect the particular information Z_t. The first instance motivates research on better conditional asset pricing models. The second possibility motivates research on human departures from rationality, market frictions, and how they show up in asset market prices.

6.7 Investment Performance Evaluation

The application of *m*-talk to the problem of investment performance evaluation goes back to at least Glosten and Jagannathan (1994), but recent work shows that *m*-talk still has insights to offer on this problem, reviewed in chapter 28. Much of the evaluation of a portfolio manager involves inferences about how much information his or her investment process is using. Now we must distinguish the public information Ω_t from the information used by the manager to form the portfolio, say, Z_{pt}, which likely differs from the public information Z_t used in the evaluation.

Imagine that an investment manager forms a portfolio of the assets with gross return $R_{pt+1} = \underline{x}(Z_{pt})'\underline{R}_{t+1}$, where $\underline{x}(Z_{pt})$ is the vector of portfolio weights at the beginning of the period (time t). If Z_{pt} is more informative than the market's information set Ω_t, the portfolio

R_{pt+1} may not be priced through the m-talk equation (1.1). That is, the manager may record "abnormal performance," or nonzero alpha. (In what follows, I will drop the time subscripts.) We previously defined the SDF alpha for portfolio p as $\alpha_p = E(mR_p|\Omega) - 1$, where Ω is the public information set that is fully reflected in market prices. If R_{Bt+1} is any zero-alpha benchmark, the SDF alpha may also be written as $\alpha_p \equiv E(m[R_p - R_B]|\Omega)$.

Consider how a manager with superior information Z_p, which is finer than Ω, generates alpha. Substitute $R_p = \underline{x}(Z_p)'\underline{R}$ into equation (1.1), and use the definition of covariance to see that

$$
\begin{aligned}
\alpha_p &= E(m\underline{R}'\underline{x}(Z_p)|\Omega) - 1 \\
&= E(m\underline{R}|\Omega)'E(\underline{x}(Z_p)|\Omega) - 1 + \text{Cov}(m\underline{R}'\underline{x}(Z_p)|\Omega) \qquad\qquad (6.12)\\
&= \text{Cov}(m\underline{R}'\underline{x}(Z_p)|\Omega),
\end{aligned}
$$

where the notation $\text{Cov}(\underline{x}'\underline{y})$ denotes the sum of the covariances across the elements of the conformable vectors. Moving from the second to the third equality, we use the facts that $E(m\underline{R}|\Omega)$ is a vector of ones and the weights $\underline{x}(Z_p)$ sum to one.

Equation (6.12) shows that the SDF alpha is the sum of the covariances of the manager's weights with the future abnormal returns of the assets, $m\underline{R}$, conditional on the public information. If the manager does not have access to better than public information, the alpha is zero. (Of course, it is possible to produce a negative alpha through high expenses, trading costs, or fraud.) This result is pretty intuitive from an m-talk perspective, but it is not what most of the literature has typically done. Typically, when the portfolio weights are measured, the covariance has been computed without multiplying by m, which ignores the risk adjustment.

What if the agent measuring the performance does not have access to the Z_p that the fund manager uses? In many performance measures, only the managed portfolio returns are observed. The preceding analysis goes through, in the sense that, to produce alpha, the manager must have more information than the Z used by the evaluator. One of the issues in conditional performance evaluation is to decide on the Z that is used in the tests, and thus decide on how to set the bar for the manager to record performance. These issues and more are discussed in part V of the book.

7 Three Paradigms of Empirical Asset Pricing

This chapter defines the three main paradigms of empirical asset pricing and the relations among them. The three paradigms are (1) m-talk, or equation (1.1); (2) minimum variance efficiency; and (3) beta pricing. Historically, each paradigm has been associated with different empirical methods. However, the three paradigms are essentially equivalent, so any such association is unnecessary. You know what I mean by m-talk, so let's define the other two paradigms carefully and then describe their interrelations. In this chapter, we ignore conditioning information for simplicity, so everything is stated in terms of unconditional moments. In chapter 14 we revisit this topic with conditioning information.

7.1 Minimum Variance Efficiency

A portfolio $\underline{x}'\underline{R}$, formed from the asset vector $\underline{R}$, using portfolio weights $\underline{x}$, is *minimum variance efficient* if it solves the following problem:

$$\text{Min}_{\underline{x}} \ \text{Var}(\underline{x}'\underline{R}),$$
$$\text{subject to } E(\underline{x}'\underline{R}) = \text{target}, \tag{7.1}$$
$$\underline{x}'1 = 1.$$

The properties of this problem are reviewed in chapter 8. For now, we use an interesting fact from Gonzales-Gaverra (1973), Fama (1973), and Roll (1977), who show that a portfolio return $R_p = \underline{x}'\underline{R}$ is minimum variance efficient if and only if there exist (cross-sectional) constants γ_0 and γ_1 such that

$$E(R_i) = \gamma_0 + \gamma_1 \text{Cov}(R_i, R_p), \tag{7.2}$$

for all i (chapter 8 provides a proof). This is the representation of minimum variance efficiency that we use for now.

7.2 Beta Pricing

Beta pricing models are a class of asset pricing models that imply that the expected returns of securities are related to their sensitivities to changes in the underlying factors measuring the state of the economy. Sensitivity is measured by the securities' beta coefficients, which are just the regression slope coefficients from regressing the returns on the factors. The factors arise because they drive variation in m, so covariance with the factors determines expected risk premiums. For each of the relevant state variables, there is a marketwide price of beta, measured in the form of an increment to the expected return (a "risk premium") per unit of beta.

The CAPM is the premier example of a single-beta pricing model. Multiple-beta models are developed in continuous time by Merton (1973), Breeden (1979), and Cox, Ingersoll, and Ross (1985). Long (1974), Sharpe (1977), Cragg and Malkiel (2009), Dybvig (1983), Grinblatt and Titman (1983), Connor (1984), Shanken (1987), and chapter 13 of this book provide multibeta interpretations of equilibrium models in discrete time. The canonical beta pricing model states that the expected return is a linear function of possibly several betas,

$$E(R_i) = \lambda_0 + \sum_{j=1,\dots,K} \beta_{ij} \lambda_j, \tag{7.3}$$

where the β_{ij}, $j = 1, \dots, K$, are the multiple regression coefficients of the return of asset i on K economywide risk factors F_j, $j = 1, \dots, K$, where the first factor is often taken to be the market portfolio, with return R_m. The coefficient λ_0 is the expected return on an asset that has $\beta_{0j} = 0$, for $j = 1, \dots, K$; that is, it is the expected return on a zero (multiple) beta asset. If there is a risk-free asset, then λ_0 is the return of this asset. The coefficient λ_k, corresponding to the kth factor, has the following interpretation: it is the expected return differential, or premium, for a portfolio that has $\beta_{ik} = 1$ and $\beta_{ij} = 0$ for all $j \neq k$, measured in excess of the zero-beta asset's expected return. In other words, it is the expected return premium per unit of beta risk for the risk factor k.

The beta-pricing relation (7.3) can equivalently be written in a *price of risk* form:

$$E(R_i) = \lambda_0 + A\,\text{Cov}(R_i, R_m) + \sum_{j=1,\dots,K} B_j \text{Cov}(R_i, F_j), \tag{7.4}$$

where the F_j are the factors; A is the price of market variance risk, $A = \lambda_m / \text{Var}(R_m)$; and the B_j are the prices of factor covariance risk. Equations (7.3) and (7.4) are equivalent. The price of risk formulation is often useful, because A and the B_j may be related to parameters of agents' utility functions in a model. For example, the parameter A is a coefficient of risk aversion in the model of Merton (1980), discussed in chapter 13.

7.3 Relation among the Three Paradigms

Basically, the three paradigms are equivalent, a beautiful result that one could call "triality" (even better than duality), depicted in figure 7.1.

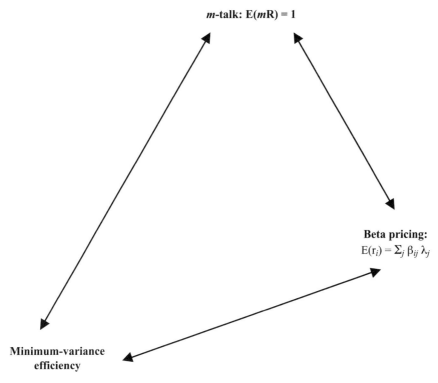

Figure 7.1
Triality in empirical asset pricing.

We now discuss this equivalence. First, the multibeta expected return model of equation (7.3) is equivalent to the m-talk equation (1.1) when the SDF is linear in the factors: $m_{t+1} = a_t + \sum_j b_{jt} F_{jt+1}$. Actually, we can allow for an error term in this relation, if it is uncorrelated with asset returns. This equivalence was first discussed, for the case of a single-factor model, by Dybvig and Ingersoll (1982). The general, multifactor case follows from Ferson and Jagannathan (1996). Dropping the time subscripts, let λ_0 be the expected return on the zero beta portfolio. The m-talk equation (1.1) is equivalent to

$$E(R_i) = \lambda_0 + \text{Cov}(R_i; -m)/E(m), \tag{7.5}$$

where $\lambda_0 = 1/E(m)$. Substituting $m = a + \sum_j b_j F_j$ we arrive at

$$E(R_i) = \lambda_0 + \sum_{j=1,\dots,K} -b_j [\text{Cov}\{F_j, R_i\}/E(m)]. \tag{7.6}$$

which is a version of the beta-pricing equation (7.3). Substituting for the betas of asset i, $\beta_{ij} = \text{Cov}(R_i, F_j)/\text{Var}(F_j)$, the marketwide risk premium per unit of type-j beta is $\lambda_j = [-b_j \text{Var}(F_j)/E(m)]$. This is consistent with our most general equation for risk premiums discussed in chapter 4. In the special case where the factor F_j is a traded asset return, applying equation (7.3) to that return implies that $\lambda_j = E(F_j) - \lambda_0$; the expected risk premium equals the factor portfolio's expected excess return.

If a risk factor F_j is negatively correlated with m (i.e., $b_j < 0$), the model implies that a positive risk premium is associated with that factor beta. This is the case for the market portfolio return in the CAPM. A factor that is positively related to marginal utility should carry a negative premium, because the big payoffs come when the value of payoffs is high.

The steps that take us from (1.1) to (7.3) can be reversed, so the SDF and multibeta representations are, in fact, equivalent. The formal statement is as follows.

Lemma 7.1 (Ferson and Jagannathan 1996). *The stochastic discount factor representation (1.1) and the multibeta model (7.6) are equivalent, where*

$$m_{t+1} = c_0 + c_1 F_{1t+1} + \cdots + c_K F_{Kt+1},$$

with

$$c_0 = [1 + \Sigma_k \{\lambda_k E(F_{k,t+1}) / \mathrm{Var}(F_{k,t+1})\}] / \lambda_0, \tag{7.7}$$

and

$$c_j = -\{\lambda_j / \lambda_0 \, \mathrm{Var}(F_{j,t+1})\}, \, j = 1, \ldots, K.$$

If the factors are not traded asset returns, then it is typically necessary to estimate the expected risk premiums for the factors λ_k. These may be identified as the conditional expected excess returns on *factor-mimicking portfolios*. These portfolios were introduced briefly in chapter 4. A factor-mimicking portfolio is defined as a portfolio whose return can be used in place of a factor in the model. There are several ways to obtain factor-mimicking portfolios. Perhaps the simplest way to see this is to consider a regression of the factor on the vector of asset returns and let F^* be the fitted value. The fitted value is a linear function of the test asset returns in R. Since the regression coefficients are chosen to maximize the R-squared, the portfolio with weights proportional to the regression coefficients has the maximum squared correlation with the factor, among all portfolios of the assets. Since $F = F^* + u$, with $E(uR) = 0$ for all assets in R, we can substitute F^* for the factor in equation (7.3) and obtain the same pricing results. Breeden (1979, footnote 7) was the first to derive these maximum correlation mimicking portfolios. Grinblatt and Titman (1987), Shanken (1987), Lehmann and Modest (1987), and Huberman, Kandel, and Stambaugh (1987) provide further characterizations of mimicking portfolios when there is no conditioning information. Ferson and Siegel (2015) and Ferson, Siegel, and Xu (2006) consider cases with conditioning information available, as discussed in chapter 14.

The next leg of the triality is the relation between beta pricing and minimum variance efficiency. Beta pricing models imply that combinations of the factor-mimicking portfolios are minimum variance efficient. Equation (7.3) is actually equivalent to the statement that a combination of K factor-mimicking portfolios is minimum variance efficient. This result is proved by Grinblatt and Titman (1987), Huberman, Kandel, and Stambaugh (1987), and Shanken (1987). The correspondence between multibeta pricing and minimum variance efficiency is exploited by Jobson and Korkie (1982), Gibbons, Ross, and Shanken (1989), Kandel and Stambaugh (1989), and Ferson and Siegel (2009), among others, to develop tests of multibeta models that we discuss in chapter 14.

Lemma 7.2. *There is beta pricing with respect to a given set of K factor portfolios if and only if a combination of those factor portfolios is minimum variance efficient.*

Proof. Consider the portfolio with return $R_p = \sum_{j=1,\ldots K} w_j F_{j,t+1}$, where the weight in the jth factor portfolio is $w_j = [\lambda_j / \text{Var}(F_j)]/[\sum_j \lambda_j / \text{Var}(F_j)]$. By linearity of the covariance, the beta pricing relation (7.3) is equivalent to $E(R_i) = \lambda_0 + \lambda_1 \text{Cov}(R_i, R_p)$, with $\lambda_1 = [\sum_j \lambda_j / \text{Var}(F_j)]$. Thus, by mean covariance linearity or equation (7.2), R_p is a minimum variance efficient portfolio. QED.

We have now established that the SDF representation in equation (1.1) is equivalent to a multibeta expression for expected returns, and a multibeta model is equivalent to a statement about minimum variance efficiency. It follows that the SDF representation is equivalent to a statement about minimum variance efficiency. But let's complete the loop explicitly.

Lemma 7.3. *A portfolio that maximizes the squared correlation with m_{t+1} in equation (1.1) is a minimum variance efficient portfolio.*

Proof. Define the portfolio return $R_p = (B/B'\underline{1})'R_{t+1}$, where B is the regression slope coefficient vector obtained by regressing m on the asset return vector R. The portfolio R_p maximizes the squared correlation with m if and only if it uses the regression coefficients B. The regression of m on the vector of returns can be written as $m = a + (B'\underline{1})R_p + \varepsilon$. The error term ε is uncorrelated with R_i for all i, and therefore with R_p. Substituting for m in equation (7.2) shows that the N-vector E(R) is a linear function of Cov(R, R_p), and thus R_p is a minimum variance efficient portfolio by equation (7.2). QED.

Note that the fitted values of the regression above will have the same pricing implications as m_{t+1}. That is, $m_{t+1}^* = (B'\underline{1})R_{p,t+1}$ can replace m_{t+1} in equation (1.1). Note that when the covariance matrix of asset returns is nonsingular, m_{t+1}^* is the unique SDF (i.e., satisfies (1.1)), which is also linear in an asset return. An SDF that satisfies (1.1) is not in general an asset return, nor is it unique, unless markets are complete.

We have seen that multibeta pricing is equivalent to the statement that $E(mR_i) = 1$ for all i, under the assumption that m is a linear function of the factors, and also equivalent to the statement that a portfolio of the factors is a minimum variance efficient portfolio. Thus, we have equivalence among the three paradigms: multibeta pricing, stochastic discount factors, and mean variance efficiency. What a beautiful result!

II MEAN VARIANCE MODELS

8 Mean Variance Analysis

Mean variance efficiency is the second of the three paradigms of empirical asset pricing (*m*-talk, of course, being the first), and historically it is one of the foundations of financial economics. Portfolio optimization problems, along with the famous Modigliani and Miller (1958) capital structure theorems, kicked off finance as an academic discipline. We start in this chapter with the classical mean variance analysis, done with simple matrix algebra. We relate this algebra to a representation from Hansen and Richard (1987) that was developed in more complex, infinite dimensional spaces. The last part of the chapter reviews some practical extensions to the classical problem and discusses some of the related empirical evidence.

8.1 Deriving the Minimum Variance Frontier: The Classics

Markowitz (1952) explored portfolio problems that maximize the tradeoff between the expected return and the portfolio risk (variance), and Sharpe (1964) put the mean variance optimum portfolio in a market equilibrium, arriving at the CAPM. It has been a long time since 1964, but mean variance problems are still very important for asset pricing. As we saw in chapter 7, something is always minimum variance efficient in an asset pricing model; it's just that different models identify different portfolios that should be efficient.

A classical version of the minimum variance problem is to find a portfolio weight vector x, such that the portfolio $R_p = x'R$ satisfies

$$\text{Min}_x \ \text{Var}(R_p) \ \text{s.t.} \ E(R_p) = \mu_p, \ 1'x = 1, \tag{8.1}$$

where μ_p is some target expected return level, and the weights sum to 1.0. Let Σ be the variance matrix of the vector of returns R, let μ be the vector of the expected returns E(R). To solve this problem, form a Lagrangian L by appending the constraints with their multipliers, and take the first-order conditions:

$$\text{Min } L = x'\Sigma x + 2\lambda_1(1 - x'\underline{1}) + 2\lambda_2(\mu_p - x'\mu)$$
$$\Rightarrow 2\Sigma x = 2\lambda_1\underline{1} + 2\lambda_2\mu, \tag{8.2}$$

and we obtain

$$x = \lambda_1 \Sigma^{-1} \underline{1} + \lambda_2 \Sigma^{-1} \mu. \tag{8.3}$$

Note that we could have solved the dual problem—maximize the expected portfolio return subject to a target portfolio variance—and derived the same solution with different multipliers. We could also have replaced the criterion Min $\mathrm{Var}(R_p)$ with Min $E(R_p^2)$, because $E(R_p^2) = \mathrm{Var}(R_p) + E(R_p)^2$, and the constraints fix the value of $E(R_p)$.

If we multiply the preceding equation by the vector of ones to obtain one new equation and multiply it by the vector of means to obtain a second, setting the results equal to 1.0 and the portfolio target mean μ_p, respectively, we have two equations in two unknowns:

$$\begin{aligned}
\underline{1}'x = 1 &= \lambda_1(\underline{1}'\Sigma^{-1}\underline{1}) + \lambda_2(\underline{1}'\Sigma^{-1}\mu), \\
\mu'x = \mu_p &= \lambda_1(\mu'\Sigma^{-1}\underline{1}) + \lambda_2(\mu'\Sigma^{-1}\mu),
\end{aligned} \tag{8.4}$$

where the unknowns are the Lagrange multipliers. The solution for the multipliers in this 2×2 system is naturally expressed in terms of the *efficient set constants:*

$$a = (\underline{1}'\Sigma^{-1}\underline{1}), \, b = (1'\Sigma^{-1}\mu), \text{ and } c = (\mu'\Sigma^{-1}\mu). \tag{8.5}$$

This expression is natural, because the efficient set constants depend only on the mean vector and covariance matrix of the asset returns in the problem, not on the target expected return. The solution for the multipliers is

$$\begin{aligned}
\lambda_1 &= (c - b\mu_p)/(ac - b^2), \\
\lambda_2 &= (a\mu_p - b)/(ac - b^2).
\end{aligned} \tag{8.6}$$

Thus, for different target mean returns, we have different multipliers, and as we vary the target return, we vary the multipliers and trace out a set of solutions for the optimal portfolio weights x, given by (8.3).

8.2   The Minimum Variance Boundary

From equations (8.3) and (8.5), we have

$$\begin{aligned}
\mathrm{Var}(R_p) = x'\Sigma x = x'\Sigma\left[\lambda_1\Sigma^{-1}\underline{1} + \lambda_2\Sigma^{-1}\mu]\right] &= \lambda_1 + \lambda_2\mu_p \\
&= \{(c - b\mu_p) + (a\mu_p - b)\mu_p\}/(ac - b^2).
\end{aligned} \tag{8.7}$$

This equation shows that for minimum variance portfolios, the variance is a quadratic function of μ_p. That is, the relation between the mean and the variance of the portfolio returns is a parabola. (The relation between the mean and standard deviation describes a hyperbola.) An example is shown in figure 8.1.

Figure 8.1 shows several curves, indicating how various combinations of the securities might determine the means and standard deviations of portfolios. The outer envelope of

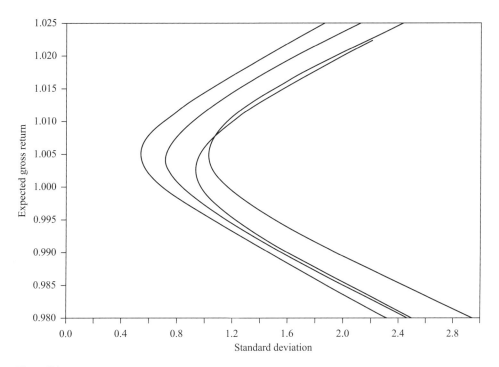

Figure 8.1
Minimum variance boundaries. Draw the tangent on this figure.

those curves is the minimum variance boundary. This is the set of points with the lowest possible standard deviation of return for a given mean return. The positively sloped portion of this curve is populated with portfolios that are *mean variance efficient*. That, is, they maximize mean return given a target variance or standard deviation. The entire region enveloped by the envelope was famously called the mean variance "bullet," I think first by Bob Haugen. I like that term. The bullet is dense with portfolios, but only those on its outer skin are efficient.

The minimum variance boundary has several interesting properties. One is *two-fund separation*, which states that any minimum variance efficient portfolio can be expressed as a combination of two portfolios on the bullet. This result comes from staring at equation (8.3), which states that x is a combination of two portfolios on the minimum variance boundary. One of the portfolios corresponds to the target mean $\mu_p = b/a$, where $\lambda_2 = 0$ and $\lambda_1 = (1/a)$ from equation (8.7), and it has the portfolio weight vector $x_1 \equiv (1/a)\Sigma^{-1}\underline{1} = \Sigma^{-1}\underline{1}/(\underline{1}'\Sigma^{-1}\underline{1})$, so its weights sum to 1.0. The second portfolio corresponds to the target mean $\mu_p = c/b$, where $\lambda_1 = 0$ and $\lambda_2 = (1/b)$, and it has the portfolio weight vector $x_2 \equiv (1/b)\Sigma^{-1}\mu = \Sigma^{-1}\mu/(\underline{1}'\Sigma^{-1}\mu)$, so its weights also sum to 1.0. The two portfolios are combined so that the overall weight sums to 1.0, since from (8.3) and

(8.6), $\underline{1}'x = \lambda_1(\underline{1}'\Sigma^{-1}\underline{1})\underline{1}'x_1 + \lambda_2(1'\Sigma^{-1}\mu)1'x_2 = \lambda_1(\underline{1}'\Sigma^{-1}\underline{1}) + \lambda_2(\underline{1}'\Sigma^{-1}\mu) = \lambda_1 a + \lambda_2 b = (c - b\mu_p)/(ac - b^2) + b(a\mu_p - b)/(ac - b^2) = 1$. This is just an example, because any two portfolios on the boundary can be combined to produce any third portfolio. Algebraically, equation (8.3) states that minimum variance portfolio weights are in a two-dimensional subspace of $\mathbb{R}^n$.

Two-fund separation is part of the intellectual foundation for index mutual funds. Taken literally it states—on the heroic assumption that agents have the same probability beliefs and thus the same μ and Σ, although perhaps differences in risk aversion and therefore different target means—that all agents can find their optimal portfolios as combinations of only two mutual funds on the minimum variance boundary. When we discuss multibeta models in chapters 12–14, we will see there is a k-fund separation result in those models, where k may be larger than two but should not be a huge number. This makes the huge number of actual mutual funds and exchange-traded funds (ETFs) that exist (more than 9,000 in the United States as this is written; more than the number of listed common stocks) a bit of a puzzle.

There is a simpler version of the minimum variance problem, and a simpler version of two-fund separation, when we assume the existence of a risk-free asset. The mean variance separation result, called "two-fund monetary separation" by Cass and Stiglitz (1970), uses the risk-free asset as one of the two portfolios (the risk-free asset is the "money"). The second portfolio is uniquely determined as the one on the tangent line from the risk-free rate to the positively sloped part of the bullet. Formally, if we redefine terms so that μ and Σ refer to returns in excess of the risk-free rate, we can drop the restriction that the weights sum to 1.0 (putting the discrepancy in the risk-free asset). Then the solution to the problem becomes $x = \lambda\Sigma^{-1}\mu$, where λ is the multiplier on the target (now excess) mean return. This much simpler solution is often analytically useful.

8.2.1   The Global Minimum Variance Portfolio

Minimize equation (8.7) over the choice of μ_p to discover that the global minimum variance (GMV) portfolio has $\mu_p = b/a$. It is the portfolio defined above as portfolio x_1. Note that the GMV portfolio weight vector is $\Sigma^{-1}\underline{1}/(\underline{1}'\Sigma^{-1}\underline{1})$, and it does not depend on the vector of mean returns. This portfolio has an interesting property. Take any portfolio with weight w. Its covariance with the GMV portfolio is $w'\Sigma x_1 = w'\underline{1}/(\underline{1}'\Sigma^{-1}\underline{1}) = 1/(\underline{1}'\Sigma^{-1}\underline{1}) = (1/a)$. (Aren't these efficient set constants amazing?) The covariance of any asset or portfolio with the GMV portfolio is the same number, 1/a. This actually makes some sense, if you think of the variance of a portfolio as the covariance with itself, $Cov(R_p, R_p)$, and thus equal to the portfolio-weighted covariance of each asset with the portfolio, $\Sigma_i w_i Cov(R_i, R_p)$. Suppose that there were two assets or portfolios with different covariances with the GMV portfolio. This means we could move some money out of the high-covariance asset and into the low-covariance asset, lowering the variance of the GMV portfolio, which would be a contradiction.

8.2.2 Quadratic Utility Maximization

A quadratic utility function of the wealth at the end of a single period is $u(w) = a + bW - cW^2$. The problem with this equation is that as W gets large, the squared term dominates the function. The marginal utility, $b - 2cW$, can turn negative. We have to be careful to apply quadratic utility only in the relevant range, or the agent will start throwing money away. Ignoring any initial consumption, the end-of-period wealth is $W = W_0 R_p$, where R_p is the gross return of the optimal portfolio, and W_0 is the initial investment. The quadratic utility agent likes the expected return and dislikes the second moment of the return. For a given expected return, this is equivalent to disliking the variance, and the quadratic utility maximization solves $\text{Max}_{\{x:x'1=1\}} Eu(R_p) = E(R_p) - (A/2)\text{Var}(R_p)$ for the right constant, A. Imposing that the portfolio weights add up to 1.0, the maximization is

$$\text{Max}_x \{x'\mu - (A/2)(x'\Sigma x) - \lambda(x'\underline{1} - 1)\}. \tag{8.8}$$

The quadratic utility agent thus chooses a mean variance efficient portfolio.

Mean variance analysis is classical stuff, but clever researchers keep coming up with new twists on the theme. A recent behavioral twist is proposed by Van Nieuwerburgh and Veldkamp (2010). Here, investor attention is a limited resource, and it takes attention to invest in stocks. Investors maximize the utility of their portfolios, allocating their wealth and limited attention across the stocks they invest in.

8.2.3 The Minimum Second-Moment Portfolio

Consider the following problem:

$$\text{Min}_{\{x:x'1=1\}} E(R_p^2) = \mu_p^2 + \text{Var}(R_p). \tag{8.9}$$

The solution must minimize the portfolio variance given its mean, so the optimal portfolio in this problem must be on the minimum variance bullet for some target mean return. We can find the target mean return by searching over portfolios on the minimum variance boundary, where $\text{Var}(R_p)$ is expressed as a function of the target mean using equation (8.6):

$$\text{Min}_{\{\mu_p\}} E(R_p^2) = \{\mu_p^2 + (c - b\mu_p + a\mu_p^2 - b\mu_p)/(ac - b^2)\}. \tag{8.10}$$

The solution to this problem is $\mu_p = b/(1 + a) < b/a$, where b/a is the mean of the GMV portfolio. This shows that the minimum second-moment portfolio appears on the minimum variance bullet, at a target mean below that of the GMV portfolio, and thus on the negatively sloped portion of the boundary. This is not something a utility-maximizing agent would be directly interested in. But we will see this portfolio again when we discuss the Hansen-Jagannathan variance bounds on SDFs in chapters 10 and 11.

8.2.4   Zero-Beta Portfolios

Take any minimum variance portfolio with weights x (except the global minimum variance) satisfying (8.3) with mean μ_p, and consider a portfolio with weights x_0 that have zero correlation to it:

$$x_0' \Sigma x = x_0' \Sigma [\lambda_1 \Sigma^{-1} \underline{1} + \lambda_2 \Sigma^{-1} \mu] = 0. \tag{8.11}$$

It follows that the weights of this zero-beta portfolio to x must satisfy $0 = \lambda_1 + \lambda_2 x_0' \mu$. Its expected return is $E(R_{0p}) = x_0' \mu$ and can be found from

$$E(R_{0p}) = -\lambda_1 / \lambda_2 = (b\mu_p - c)/(a\mu_p - b) \equiv \gamma_0(\mu_p). \tag{8.12}$$

This equation defines the zero-beta rate relative to a minimum variance portfolio with target mean equal to μ_p. For any value on the y-axis of the minimum variance bullet, call it γ_0, the portfolio on the minimum variance boundary with a target mean μ_p that satisfies $\gamma_0 = (b\mu_p - c)/(a\mu_p - b)$ solves $\mathrm{Max}_x [x'(\mu - \gamma_0 \underline{1})]^2 / (x' \Sigma x)$ and has the solution $x^* = \Sigma^{-1}(\mu - \gamma_0 \underline{1})/\underline{1}' \Sigma^{-1}(\mu - \gamma_0 \underline{1})$. Therefore the maximizing portfolio has mean return $\mu' x^* = (c - b\gamma_0)/(b - a\gamma_0)$.

   All this has a famous graphical interpretation. Pick a value on the y-axis of the mean-standard deviation graph, and call it γ_0. Draw a tangent line to the minimum variance boundary. The portfolio on the boundary has the weights x^* and a target mean $\mu_p = (c - b\gamma_0)/(b - a\gamma_0)$. For any given γ_0, the portfolio with this x^* and μ_p maximizes the expected excess return (squared) per unit of variance. We can also do this starting with any portfolio on the minimum variance bullet. This nails down a target mean μ_p. Draw a tangent line from the point on the bullet to the y-axis. This line intercepts the y-axis of the graph at γ_0. Draw this line on figure 8.1 for fun. In the special case of a risk-free asset, its return would equal γ_0, because a risk-free asset has zero variance and thus no covariance with anything.

8.2.5   The Sharpe Ratio

The Sharpe ratio for a portfolio with weight x relative to the risk-free or zero-beta rate γ_0 is defined as

$$SR = (x'\mu - \gamma_0)/[x' \Sigma x]^{1/2}. \tag{8.13}$$

For a given γ_0, a line drawn from that point on the y-axis, tangent to the mean-standard deviation bullet, has a slope equal to the maximum SR for that zero-beta rate. That is, $\mathrm{Max}_{\{x:x'\underline{1}=1\}} SR \Rightarrow \mathrm{Max}_x \{x'\mu - \gamma_0 - \lambda_1(x' \Sigma x) - \lambda_2(x' \underline{1} - 1)\}$. Note that the maximized value of the squared Sharpe ratio, found by plugging the solution to this maximization into the definition of the Sharpe ratio squared, is $(\mu - \gamma_0 \underline{1})' \Sigma^{-1}(\mu - \gamma_0 \underline{1})$. We will see this expression for the maximum squared Sharpe ratio again many times, in asset pricing tests and in portfolio management problems.

Note that the expected utility obtained by the maximum quadratic utility in (8.8) is proportional to $(\mu - \lambda_2 \underline{1})\Sigma^{-1}(\mu - \lambda_2 \underline{1})$, a maximum squared Sharpe ratio. This result gives an economic interpretation to the maximum squared Sharpe ratio as the maximal utility of a quadratic utility investor, as discussed by Kan and Zhou (2007).

8.2.6 Return Covariance (or Beta) Linearity and the CAPM

Take x to be minimum variance efficient, $R_p = x'R$, and take any other portfolio weight w, where $E(R_w) = w'\mu$. Then from equation (8.3), we have:

$$\text{Cov}(R_w, R_p) = w'\Sigma x = w'\Sigma[\lambda_1 \Sigma^{-1} \underline{1} + \lambda_2 \Sigma^{-1} \mu] = \lambda_1 + \lambda_2 E(R_w). \tag{8.14}$$

This result is the expected return-covariance linearity, as proved by Fama (1973), Gonzalez-Gaverra (1973), and Roll (1977). (The intellectual property rights to this important result are a matter of debate, so I had better cite all three.)

We can go further. The previous expression must also hold when $w = x$, implying that

$$\text{Cov}(R_p, R_p) = \lambda_1 + \lambda_2 E(R_p), \tag{8.15}$$

and it must hold when w corresponds to a zero-beta portfolio to R_p (call it R_{0p}):

$$0 = \lambda_1 + \lambda_2 E(R_{0p}). \tag{8.16}$$

Combining the last three equations, we arrive at

$$E\{R_w - R_{0p}\} = [\text{Cov}(R_w, R_p)/\text{Var}(R_p)] E\{R_p - R_{0p}\}. \tag{8.17}$$

Now if we argue, as Sharpe (1964) did, that in a model with mean variance optimizers, the market portfolio must be mean variance efficient, then we can say that the market portfolio R_m is an efficient portfolio R_p, and we have the CAPM. Thus, the CAPM simply says that the market portfolio is mean variance efficient. We recognize the famous beta coefficient as equal to $[\text{Cov}(R_w, R_m)/\text{Var}(R_m)]$. Of course, any likely reader of this book already knows this.

The mean-covariance linearity result holds for any minimum variance efficient portfolio R_p, which implies some more insights about the graph that you modified in figure 8.1. Take any portfolio on the bullet and its zero-beta value. Suppose the zero-beta rate is below μ_p. A horizontal line through the bullet at the zero-beta value divides the bullet into two pieces. The part above this line has expected return larger than the zero-beta rate, and the part below the line contains portfolios with expected returns below the zero-beta rate. Thus, equation (8.17) states that assets and portfolios above the line are positively correlated with R_p, and those below the line have negative R_p betas. This also works if the tangency from the y-axis to R_p is in the bottom portion of the bullet, but then $E\{R_p - R_{0p}\} < 0$ and portfolios below the line are positively correlated with R_p.

Let's get one more result from this analysis. Consider any arbitrary portfolio, depicted as a point inside the bullet. Draw one line from the zero-beta rate on the y-axis to this point and compare it to the line from the zero-beta rate that is tangent to the bullet at the efficient portfolio. Writing the covariance as the product of the correlation and the two standard deviations, equation (8.17) implies that for any point inside the bullet, the ratio of the slopes of the two lines is equal to the correlation of the arbitrarily chosen portfolio with the efficient portfolio. The ratio of the slopes is the ratio of the two Sharpe ratios. Thus, the ratio of the two Sharpe ratios is the correlation between the point inside the bullet and the return of the boundary portfolio. This relation is used in tests of asset pricing models, where we test for the efficiency of a portfolio, and the two slopes tell us how far away from efficient the test subject is.

8.3   The Hansen and Richard Representation

Hansen and Richard (1987) prove a version of the following decomposition in a classic *Econometrica* paper. For any minimum variance portfolio R_{mv} and for any return R_i, there exist scalar constants w, w_i, such that

$$R_{mv} = R^* + w\ R_{e*},$$
$$R_i = R^* + w_i\ R_{e*} + n_i,\ E(n_i) = 0 = E(n_i R^*) = E(n_i R_{e*}),$$

(8.18)

where R_{e*} is an "excess" return (weights sum to zero), and R^* is a portfolio return (weights sum to 1.0) on the minimum variance boundary. The R^* and R_{e*} may be chosen to be orthogonal. The first expression in (8.18) states that any minimum variance portfolio return is some combination of the two frontier returns, R^* and R_{e*}, with unit weight on R^*. This is a version of two-fund separation. The second expression states that any return R_i may be expressed as a return on the bullet with the same mean, plus orthogonal noise. Hansen and Richard prove versions of this decomposition using mathematics for infinite-dimensional spaces.

Versions of the Hansen-Richard representation also flow from "the classics" of mean variance analysis. First, from two-fund separation, we can find w such that

$$R_{mv} = wR_p + (1-w)R_{0p} = R_{0p} + w(R_p - R_{0p}) \equiv R^* + w\ R_{e*}.$$

(8.19)

Thus, a version of the first statement in (8.18) works using any R_p on the boundary and its zero-beta portfolio. Now consider a regression:

$$R_i = a_i + \beta_i R_p + c_i R_{0p} + n_i,$$

(8.20)

where

$$E(n_i) = 0 = E(n_i R_p) = E(n_i R_{0p}).$$

Return-beta linearity using R_p restricts the regression coefficients to be $a_i = 0$ and $c_i = (1 - \beta_i)$. The restricted regression is

$$R_i = \beta_i R_p + (1 - \beta_i) R_{0p} + n_i,$$
$$\quad = R_{0p} + \beta_i (R_p - R_{0p}) + n_i, \tag{8.21}$$
$$\quad = R^* + w_i R_{e*} + n_i,$$

where $E(n_i) = 0 = E(n_i R^*) = E(n_i R_{e*})$.

8.4   Generalizations, Extensions, and Evidence

8.4.1   Optimal Orthogonal Portfolios

An optimal orthogonal portfolio combines with a benchmark portfolio that lies inside the frontier, to form a minimum variance efficient portfolio. Such portfolios are featured in tests of asset pricing models and are important in active portfolio management problems. In asset pricing problems, the optimal orthogonal portfolio represents the "most mispriced" portfolio by a benchmark, if we use the benchmark as if it were the market portfolio in the CAPM. Its squared Sharpe ratio is the difference between the squared Sharpe ratio of the benchmark portfolio and the maximum squared Sharpe ratio in the sample of assets (Jobson and Korkie 1982). In modern portfolio management the optimal orthogonal portfolio shows how to actively tilt away from a given benchmark portfolio to achieve portfolio efficiency (Gibbons, Ross, and Shanken 1989; or Grinold and Kahn 2000). Optimal orthogonal portfolios are studied by Roll (1980), MacKinlay (1995), Campbell, Lo, and MacKinlay (1997), and others. Here is a brief introduction.

Consider a regression of excess returns r on a benchmark: $r = \alpha + \beta r_B + u$. The optimal orthogonal portfolio solves

$$\text{Max}_x \{ (x'\alpha)^2 / \text{Var}(x'r) \}, \tag{8.22}$$

and the solution is proportional to $\Sigma^{-1} \alpha$.

It should be clear from the definition in (8.22) that the optimal orthogonal portfolio can be interpreted as the most mispriced portfolio by the benchmark r_B. The vector α captures the mispricing of the tested asset returns when evaluated using the benchmark, and the optimal orthogonal portfolio has the largest squared alpha relative to its variance. This interpretation also reveals why the portfolio is of central interest in active portfolio management. Given a benchmark portfolio r_B, an active portfolio manager places bets by deviating from the portfolio weights that define the benchmark. The manager is rewarded for bets that deliver higher returns and is penalized for increasing the volatility. The optimal orthogonal portfolio describes the active bets that achieve the largest amount of extra return for the variance. Thus, the solution is also referred to as the *active* portfolio by Gibbons, Ross, and Shanken (1989). See Grinold and Kahn (2000) for an in-depth treatment of modern portfolio management.

When the weights of the optimal orthogonal portfolio sum to 1.0, the portfolio weight vector is $x_c = \Sigma^{-1}\alpha/(1'\Sigma^{-1}\alpha)$, and the return is $R_c = x_c'R$. Using this solution, several well-known properties of the optimal orthogonal portfolio follow. For example, a combination of the optimal orthogonal portfolio and the benchmark portfolio is *optimal*; that is, it is minimum variance efficient (Jobson and Korkie 1982). The portfolio is *orthogonal* to the benchmark portfolio in the sense that $Cov(r_c, r_B) = 0$. Orthogonality means that we can replace $\Sigma = Var(r)$ with $Var(u)$ in the maximization problem (8.22), because $Var(x'u)$ and $Var(x'r)$ are equal for this portfolio.

The optimal orthogonal portfolio is central for the interpretation of tests of portfolio efficiency. Classical test statistics for the hypothesis that $\alpha = 0$ for all test assets in the factor model regression of excess returns on the benchmark can be written in terms of squared Sharpe ratios involving the optimal orthogonal portfolio. In this context, a property of the optimal orthogonal portfolio, which Ferson (2003) calls the "law of conservation of squared Sharpe ratios," is invoked. Simply stated, the law says that

$$\text{Max}_x\, S^2(x'r) = S^2(r_B) + S^2(x_c'r), \tag{8.23}$$

where $S^2(\cdot)$ denotes the squared Sharpe ratio. Like energy, a given set of assets has a law of conservation: There is only so much squared Sharpe ratio available in the system. That amount is $\text{Max}_x\, S^2(x'r)$. If the tested portfolio has some amount of squared Sharpe ratio $S^2(r_B)$, then its optimal orthogonal portfolio must have the rest. We develop tests of mean variance efficiency for a given benchmark using these properties in chapter 21.

The optimal orthogonal portfolio problem is related to quadratic utility maximization as follows. Maximizing a quadratic utility defined over the active bets $(r_p - \beta_p r_B)$ implies $\text{Max}_{x:x'1=1}\{(x'\alpha)^2/Var(x'u)\}$. This provides a framework for applications of the optimal orthogonal portfolio in active portfolio management.

8.4.2 Modern Portfolio Management

If an active portfolio manager is evaluated relative to a benchmark, this induces a utility function over the return relative to benchmark, $r_p - \beta_p r_B$. If the utility function can be approximated by a quadratic, then the optimal orthogonal portfolio is the optimal solution to the active managers' problem. This is the foundation for modern portfolio management, as reviewed by Grinold and Kahn (2000). In active portfolio management a special case is often considered, where the manager's preferences are defined over the *tracking errors*: $r_p - r_B$. This is obviously a special case where the regression coefficient β_p is assumed to equal 1.0. Then the mean variance solution corresponds to a quadratic utility function defined over the tracking error. This special case, among others, is studied by Roll (1992), Jorion (2003), and others.

A couple of performance measures arise in the modern portfolio management context that should be introduced at this point. The first is the *information ratio* (IR), defined for an active portfolio r_p as

$$\text{IR: } \alpha_p/\sigma(u_p), \tag{8.24}$$

where $\sigma(\cdot)$ denotes the standard deviation and the subscripts refer to the values for portfolio p in the factor model regression of its excess returns over that of the benchmark. The optimal orthogonal portfolio maximizes the information ratio, and its maximized value is $[\alpha'\Sigma_u^{-1}\alpha]^{1/2}$, where Σ_u is the residual covariance matrix. The maximum information ratio (squared) is proportional to the maximum of the expected utility for the quadratic utility function defined over return net of benchmark, which provides an economic interpretation for the information ratio. The information ratio is also called the "appraisal ratio" (Treynor and Black 1973).

A related measure in modern portfolio management is the *information coefficient*, which is the squared correlation between an active manager's signal and the future abnormal returns. Grinold (1989) presents a model of an active portfolio manager maximizing the appraisal ratio, conditioning the analysis on a signal, z, that is correlated with the vector of stocks' idiosyncratic returns. On the assumption that the idiosyncratic returns are uncorrelated, then the maximum squared IR for the portfolio is $[\alpha'\Sigma_u^{-1}\alpha] = \Sigma_i[\alpha_i^2 / \mathrm{Var}(u_i)] = \Sigma_i[\mathrm{IR}_i^2]$. Assuming homoskedasticity, $\mathrm{Var}(u_i|z) = \mathrm{Var}(u_i)(1 - \mathrm{IC}_i^2)$, where IC_i is the unconditional correlation between the signal z and the abnormal return u_i (i.e., the information coefficient). Assuming that the IC_i for each stock is the same, Grinold shows that the first-order condition implies that for the portfolio, the IC is approximately the IR multiplied by the square root of the number of stocks or investment picks: $\mathrm{IC} = \mathrm{IR}\sqrt{n}$. He calls this the "fundamental law of active management."

8.4.3   Empirical Evidence

The empirical evidence for the practical utility of classical mean variance optimal portfolio solutions is not very comforting. In an "out-of-sample" exercise, securities means and variances are estimated using past data, and the estimates are used to form portfolio weights. The portfolio weights are applied to future returns, and the "out-of-sample" returns of the portfolios are examined. Mean variance solutions rarely deliver the performance predicted by the in-sample parameter values. In fact, Jobson and Korkie (1980) and DeMiguel et al. (2009) find that mean variance optimal solutions rarely beat an equally weighted portfolio out of sample. (This result is criticized on technical grounds by Kirby and Ostdiek [2012] and Kan, Wang, and Zhou [2016].)

Michaud (1989) attributes the poor performance to the tendency of mean variance solutions to be very sensitive to errors in estimating the means. To see this, consider $\mathrm{Max}_x\{x'\mu - (A/2)(x'\Sigma x)\}$ for excess returns, implying a solution: $x^* = (1/A)\Sigma^{-1}\mu$. Taking the derivative with respect to the mean vector, we see that $\partial x^*/\partial\mu' = (1/A)\Sigma^{-1}$. This sensitivity can be very high. For example, consider a single risky asset with a monthly volatility of 7%, then $(1/A)\Sigma^{-1} \approx 204/A$. Even for fairly risk-averse utility functions, where A might be on the order of 4–10, this is a large number.

When the out-of-sample tests are rolled through time, mean variance efficient portfolios typically experience high rates of turnover. When more than a few assets are used, the portfolios often call for extreme long and short positions. Green and Hollifield (1992) attribute extreme portfolio weights to the factor structure in stock returns. Intuitively, if it is possible to form two portfolios with similar betas but different estimated means, an extreme long and short portfolio has a low variance relative to its mean and would be attractive to the sample mean variance problem.

Several solutions are considered in the literature to treat the extreme sensitivity of mean variance portfolio solutions, in an attempt to improve their out-of-sample performance. Explicit constraints on the portfolio weights are examined by Frost and Savarino (1986), DeMiguel et al. (2009), and Jagannathan and Ma (2003), discussed below, among others. Bayesian shrinkage estimators—where the estimated sample means are replaced with a convex combination of the sample estimate and some prior value, such as the grand mean—are examined by Karolyi (1992), Jorion (1986), Ledoit and Wolf (2003), and Kan and Zhou (2007), among others. More general Bayesian solutions are examined by Brown (1976), Kandel and Stambaugh (1996), Pástor and Stambaugh (2000), and Johannes, Korteweg, and Polson (2014), among others. Kan, Wang, and Zhou (2016) derive mean variance optimal portfolio rules designed to maximize the out-of-sample performance, accounting for estimation error. The GMV portfolio pays no attention to the means and is thus robust to errors in their estimation. Haugen and Baker (1991) find that the GMV solution has superior out-of-sample performance, compared to mean variance portfolios that use sample means.

8.4.4   Short Sales Constraints and Shrinkage

It turns out that constraints on the vector of means that are used as inputs for the mean variance solution are in some cases isomorphic to shrinkage estimation. In shrinkage estimation, the sample estimates are shrunk toward some conservative value, such as the grand mean. Jagannathan and Ma (2003) present an example. Consider the global minimum variance problem with short-sales constraints: $\text{Min}_x\{(1/2)x'\Sigma x + \gamma(1-x'1) + x'\lambda\}$, with solution $x^* = \Sigma^{-1}(\gamma\underline{1} + \underline{\lambda})$. This solution is the same as ignoring the constraint and replacing μ with $(\gamma\underline{1} + \underline{\lambda})$ in the classical problem for excess returns. It is a shrinkage toward the grand mean $\gamma\underline{1}$. Jagannathan and Ma show that the solution is also the same as ignoring the short-sale constraint and replacing Σ with $\Sigma - (1'\underline{\lambda} + \underline{\lambda}'1)$. This is a shrinkage of the covariance matrix.

Frazzini and Pedersen (2014) consider margin constraints. The Lagrange multiplier is added to the risk-free rate, appearing as a higher zero-beta rate or a higher borrowing cost, as in Black's (1972) constrained-borrowing CAPM. In these models, the linear relation between expected stock returns and their market betas has a flatter slope than in the Sharpe (1964) CAPM with a risk-free rate. Thus, high-beta stocks earn lower expected returns than predicted by the classical model (negative alphas), while low-beta stocks earn higher expected returns (positive alphas). Frazzini and Pedersen examine a "betting against beta" portfolio, long the low-beta stocks and short the high-beta stocks, and find that its payoff varies over time with a measure of aggregate funding constraints, the U.S. Treasury-Eurodollar (TED) spread.

8.4.5 Multiperiod Mean Variance Analysis

Consider an optimization that recognizes that you will be able to trade for several periods, $t = 0, 1, \ldots, T-1$, and imagine that you consume the proceeds at time T. Such a problem is useful in retirement planning problems, for example:

$$\text{Max } Eu(W_T), \quad W_T = W_0(\Pi_{t=1,\ldots,T}R_t). \tag{8.25}$$

Several utility functions $u(\cdot)$ have been studied (e.g., power, log, exponential, and quadratic), but let's talk quadratic (mean variance problems). If you can only trade once, picking x_0, just replace R in the mean variance problem with $(\Pi_{t=1,\ldots,T}R_t)$, and off we go. This works if returns are independent and identically distributed (iid) and utility is constant relative risk averse, which is called the "myopic solution" by Mossin (1968). In general, the optimal portfolio weight at $t = 1$ will reflect the option to trade again at $t = 2, 3$, and so on, and this will matter if returns have serial dependence. Intuitively, if a return has negative serial correlation, the return for $t = 1$ is a hedge for the return at $t = 2$, which reduces the risk over the two periods. This reduced risk could justify a larger overall holding of a negatively serially correlated asset. See Basak and Chabakauri (2010) and their references for examples. For continuous-time solutions, see Cvitanić and Zapatero (2004), who assume complete markets.

8.4.6 Multiperiod Consumption and Portfolio Choice

When investors can optimize their consumption and portfolio choices over a multiperiod horizon, we typically have to move outside the mean variance setting. The ability to trade and consume at the periods between now and the terminal date induces a state dependency in the value function for the problem. This is often represented by a Bellman equation:

$$J(W_t, s_t) \equiv \text{Max } E_t\{U(C_t) + J(W_{t+1}, s_{t+1})\}, \tag{8.26}$$

where $U(C_t)$ is the direct utility of consumption expenditures at time t, and $J(\cdot)$ is the indirect utility of wealth. The indirect utility is dependent on the state variables, s_t. The state variables are sufficient statistics, given wealth, for the utility of future wealth in the optimal consumption-investment plan. Thus, changes in the state variables represent future consumption-investment opportunity risk. This problem is a root of equilibrium multibeta models for expected returns, which are discussed in chapter 13 of this book.

9 Mean Variance Efficiency with Conditioning Information

This chapter explores mean variance efficiency when there is information about the future expected returns or covariances of the securities that may be explicitly conditioned on. There are a couple of different ways to do this, but the most subtle and interesting is when the conditioning information is used to optimize the unconditional moments of the portfolio whose weights may vary with the conditioning information. This is interesting when you think that agents in the market may form portfolios using more information than we, the poor empiricists, can actually measure. It is also interesting from a portfolio management perspective, if you think the manager knows more about security returns than the client does. Closed-form solutions to the problem from Ferson and Siegel (2001) are presented and interpreted. I introduce a perturbation argument, not unlike the one we met in chapter 3, that delivers solutions to problems like this. The chapter finishes up by describing some extensions of the mean-variance analysis with conditioning information, and some of the empirical evidence on portfolios that are mean variance optimized with respect to conditioning information.

9.1 Conditional and Unconditional Efficiency with Information

We explicitly bring conditioning information to the mean variance optimization problem. Consider a given set of returns R_{t+1} and lagged information Ω_t, (think of the public information reflected in asset prices), and consider all portfolios that can be formed with weights that depend on the information: $R_{pt+1} = x(\Omega_t)'R_{t+1}$. There are different ways to define minimum variance efficiency in this case. These two will be our main focus:

Conditionally efficient (CE) portfolios:

$$CE(\Omega): \operatorname{Min}_{x(\Omega)} \operatorname{Var}(R_p | \Omega) \text{ subject to } E(R_p | \Omega) = \mu(\Omega). \tag{9.1}$$

Unconditionally efficient (UE) portfolios with respect to Ω:

$$UE(\Omega): \operatorname{Min}_{x(\Omega)} \operatorname{Var}(R_p) \text{ subject to } E(R_p) = \mu. \tag{9.2}$$

The first definition of minimum variance efficiency is straightforward in principle. We can just work with conditional moments given Ω, and all the results for classical minimum variance efficiency, discussed in chapter 8, hold. For example, graphically we would have a mean-standard deviation bullet diagram with the conditional means and standard deviations on the y- and x-axes, respectively, and the graph would change with each realization of the information.

In the second definition, UE portfolios use the information Ω to generate trading strategies that are efficient in terms of the unconditional moments. There are at least two motivations for this notion of efficiency. First, imagine a portfolio manager who has information Ω that is better than the information of the client investor (suppose the client has no information). Suppose that the manager does not or cannot communicate her information to the client. Assume that the client desires a mean variance efficient portfolio. The UE problem is then the one that the client would wish the manager to solve on his behalf.

A second motivation for UE is tests of asset pricing models. Recall that any asset pricing model—that is, a specification of m in equation (1.1)—always implies that something should be minimum variance efficient. Suppose that we wish to test whether the particular portfolio specified by the model is efficient. Since the tester of the models cannot measure all the public information that is reflected in market prices, he or she cannot test whether a portfolio is CE(Ω), but can in principle test whether it is UE, or whether it is efficient, given some observed subset of Ω. This is because the tester can estimate the unconditional means and variances but cannot estimate the moments conditional on the unobserved information Ω. This brings us to a big result from Hansen and Richard (1987). It says that a UE portfolio must be CE, but a CE portfolio does not have to be UE.

Proposition (Hansen and Richard 1987). UE$(\Omega) \Rightarrow$ CE(Ω), but CE$(\Omega) \Rightarrow$ UE(Ω).

Proof. To show that UE implies CE, we proceed by contradiction. We show that not CE $\Rightarrow$ not UE. Suppose R_p is not CE. Then there exists R^* such that $\text{Var}(R^*|\Omega) < \text{Var}(R_p|\Omega)$ with $E(R^*|\Omega) = E(R_p|\Omega)$, which implies $E(R^{*2}|\Omega) < E(R_p^2|\Omega)$ and $E(R^*|\Omega) = E(R_p|\Omega)$. Taking unconditional expectations, we have $E(R^{*2}) < E(R_p^2)$ and $E(R^*) = E(R_p)$. Thus, $\text{Var}(R^{*2}) < \text{Var}(R_p^2)$, while $E(R^*) = E(R_p)$, so R_p is not UE.

To show that CE does not imply UE, we present a general counterexample. R_p is CE if and only if there are scalars $a(\Omega)$, $b(\Omega)$ such that $E(\underline{R}|\Omega) = a(\Omega) + b(\Omega)\,\text{Cov}(\underline{R}, R_p|\Omega)$. This is a version of mean-covariance linearity, as discussed in chapter 8 and proved by Hansen and Richard (1987), but it should seem pretty obvious to you. By iterated expectations, $E(\underline{R}) = E[a(\Omega)] + E[b(\Omega)]E[\text{Cov}(\underline{R}, R_p|\Omega)] + \text{Cov}\{b(\Omega), \text{Cov}(\underline{R}, R_p|\Omega)\}$. Now, letting $E[a(\Omega)] = a$, $E[b(\Omega)] = b$, and using the fact that $E[\text{Cov}(\underline{R}, R_p|\Omega)] = \text{Cov}(\underline{R}, R_p) - \text{Cov}\{E(\underline{R}|\Omega), E(R_p|\Omega)\}$, we have $E(\underline{R}) = a + b\text{Cov}(\underline{R}, R_p) + \underline{\alpha}$, where the value of $\underline{\alpha} \neq 0$ follows by substitution. Since $\underline{\alpha}$ is not zero in general, R_p is not in general UE. QED.

The last argument shows that when R_p is CE(Ω), even though it will price the assets in the sense of "explaining" their conditional expected returns as functions of their conditional covariance with R_p, it will not in general price the unconditional mean returns using the unconditional covariances. There will be a nonzero alpha in the unconditional relation. Note that these arguments go through if we replace the unconditional moments with moments conditioned on Z, a subset of Ω that can be measured by an empirical researcher. It is important that we are working in the set of all portfolios that can be formed using the information Ω. Thus, a portfolio that is efficient conditional on Z in this set of portfolios must be efficient conditional on Ω, but a portfolio that is efficient conditional on Ω need not be efficient with respect to Ω, conditional on Z. Got it?

The previous result has profound implications for asset pricing research in general, and for tests of asset pricing models in particular. Recall that the three paradigms of empirical asset pricing are fundamentally equivalent statements. In particular, any model that we write down using *m*-talk is equivalent to the statement that some portfolio is minimum variance efficient. It is natural to interpret this to mean efficient given Ω, the public information that is fully reflected in market prices. Thus, with this interpretation, asset pricing models state that some portfolio is CE(Ω). However, we can't measure all of Ω, just some subset of it, Z. We can test if the portfolio is efficient given Z. Suppose that we reject this hypothesis. The problem is, the portfolio could be efficient given Ω and not be efficient given Z. Even if our tests reject efficiency, the asset pricing model could be correct. Specifically, if we interpret the CAPM to say that the market portfolio is efficient given Ω, and if we admit that we can't measure Ω, then in principle the CAPM could be the correct model for expected returns, despite half a century of research studies that have rejected the efficiency of the market portfolio conditioning on something less than Ω.

This problem of unobservable Ω has been called the "Hansen-Richard critique" (Cochrane 2005). This is distinct from the famous Roll (1977) critique, which states that we can't test the CAPM, because we can't hope to accurately measure the market portfolio of all assets. Even if we could measure the market portfolio, we are stumped by the problem of the unobservable Ω.

You might say: Why not just condition down equation (1.2) to $E(mR|Z)=1$ and test the model in this form? Sometimes this works. For example, it works for the power utility consumption model *m*, assuming that consumption growth can be observed, because that model's *m* depends only on observable data and parameters that can be estimated. However, what if $m=m(\Omega)$ depends on something we can't observe? Now we are stuck, because we can't test $E(m(\Omega)R|Z)=1$ without observing Ω.

There remains the question of the empirical bite of the Hansen-Richard critique. If there were no information in Ω that could predict returns or their second moments, then the unconditional means and covariance matrix would be the same as the conditional ones given Ω, and we could use sample averages and covariances to conduct valid tests of the models. Until about the 1980s, this view of asset pricing models was taken seriously, but now we have lots of evidence that the moments of asset returns are time varying with information. Chapter 32 provides a review.

Suppose that the conditional CAPM is the correct asset pricing model, which we state in terms of expected returns and betas as

$$E(r \mid \Omega) = \beta(\Omega) E(r_m \mid \Omega). \tag{9.3}$$

What are the implications of the conditional CAPM for the unconditional expected returns?

$$E(r) = E(\beta(\Omega)) E(r_m) + Cov(\beta(\Omega),\ E(r_m \mid \Omega)). \tag{9.4}$$

The last term is approximately an induced alpha. This point was first made by Chan and Chen (1988) in the context of the firm size effect. They suggested that the cross section of betas for firms of different sizes (market capitalization) would expand in recessions (when expected market returns are high) and shrink at the top of the business cycle (when expected returns are low). Thus the covariance between beta and the market premium is positive for small stocks (delivering a positive alpha in the unconditional CAPM) and negative for large firms (delivering negative alphas in the unconditional model). Santos and Veronesi (2005) make a similar observation, replacing small firms with value firms (those with low prices relative to the book value of equity) and replacing large firms with growth firms (those with high prices relative to the book value of equity).

Lewellen and Nagel (2006) address the question of whether there is enough variation in conditional betas to generate covariances with the market premium that are large enough to explain the value-growth and momentum effects. They use daily data for a month to estimate market betas that vary month by month, on the assumption this provides good estimates of $\beta(\Omega)$ even when Ω cannot be observed. Nelson and Foster (1990) argue that, under certain conditions, a "continuous record asymptotic" shows that high frequency betas are consistent estimates for conditional betas. They suggest that daily data might be a high enough frequency for a good approximation. Lewellen and Nagel find that the variation in betas is not large enough to generate covariances with the market premium that are large enough to explain the momentum and value-growth anomalies. However, the continuous record asymptotic consistently captures the quadratic variation of the process, ignoring the variation in the conditional means during the month, so Lewellen and Nagel's analysis seems to ignore that variation. Gilbert et al. (2014) argue that daily betas are biased for so-called opaque firms. These issues muddy the water a little, but Lewellen and Nagel also examine betas estimated using weekly data over a quarter, and they provide further calibrations and simulations to support their conclusions about the value-growth and momentum anomalies. Bali and Engle (2010) examine a conditional CAPM with dynamic conditional correlations and claim that it can explain the value-growth anomaly and the firm size effect, but not the momentum anomaly.

9.2 Relation to Utility Maximization

The UE optimal portfolio solution turns out to be an optimal solution for an agent maximizing a quadratic utility function with particular utility function parameters, conditional on the information Ω, as shown by Ferson and Siegel (2001). A quadratic utility optimally chooses a CE(Ω) solution, as can be inferred by reinterpreting the results in chapter 8 for the conditional moments given Ω. But other utility functions also will choose a CE(Ω) solution, such as an exponential utility function under the assumption that the conditional distributions are normal. Those solutions, however, will not be the same as for the quadratic utility and will not be UE(Ω). This once again illustrates the fact that UE(Ω) solutions are a subset of the CE(Ω) solutions.

Quadratic utility provides one way to resolve the Hansen-Richard critique for the CAPM. We can assume quadratic utility maximization in the conditional CAPM, so that model implies that the market portfolio is both CE(Ω) and UE(Ω). Then we can test whether the measured market proxy is UE, and if it is not, we reject the theory even though it is cast in terms of Ω.

The assumption of quadratic utility is motivated by a conditional CAPM. Berk (1995) shows that if you want the CAPM to hold without restricting the probability distributions, you must assume that a representative agent has quadratic utility. Quadratic utility also resolves the modern portfolio management problem discussed in chapter 8 with the Hansen-Richard critique. In modern portfolio management, the portfolios formed are conditionally efficient given a manager's signal about the stock returns. Yet unconditional performance statistics are employed that examine unconditional efficiency. In general, as emphasized by Dybvig and Ross (1985a), a manager's portfolio can be conditionally efficient but appear unconditionally inefficient to the client. Not if the manager and the client have quadratic utility functions. In this case the optimal portfolios for both agents are UE.

9.3 Explicit UE Solutions

Portfolio weights for efficient portfolios in the presence of conditioning information are derived by Ferson and Siegel (2001). They consider the case with no risk-free asset and the case with a fixed risk-free asset whose return is constant over time. Theorem 9.1 presents a generalization from Ferson and Siegel (2015) of the case with a risk-free asset whose return is known at the beginning of the period (and thus is included in the information Z) but may vary over time. Then theorem 9.2 reproduces the case with no risk-free asset from Ferson and Siegel (2001). In this section, we replace the information Ω with the observable information Z, and the optimization is over the unconditional moments of the dynamic strategy returns that depend on Z.

Consider N risky assets with returns R. In $N \times 1$ column-vector notation, we have $R = \mu(Z) + \varepsilon$. The noise term ε is assumed to have conditional mean zero given Z and a nonsingular conditional covariance matrix $\Sigma_\varepsilon(Z)$. The conditional expected return vector is $\mu(Z) = E(R|Z)$. Let the $N \times 1$ column vector $x'(Z) = (x_1(Z), \ldots, x_N(Z))$ denote the portfolio share invested in each of the N risky assets, investing (or borrowing) at the risk-free rate the amount $1 - x'(Z)\underline{1}$, where $\underline{1} \equiv (1, \ldots, 1)'$ denotes the column vector of ones. We allow for a conditional risk-free asset returning $R_f = R_f(Z)$. The rate over the next period is known and varies over time as part of the information set Z. The return on the portfolio is $R_s = R_f + x'(Z)(R - R_f\underline{1})$, with unconditional expectation and variance as follows:

$$\mu_s = E(R_f) + E\{x'(Z)[\mu(Z) - R_f\underline{1}]\},$$
$$\sigma_s^2 = E(R_s^2) - \mu_s^2 = E[E(R_s^2|Z)] - \mu_s^2, \tag{9.5}$$

$$\sigma_s^2 = E(R_f^2) + E[x'(Z)Q^{-1}x(Z)] + 2E\{R_f x'(Z)[\mu(Z) - R_f\underline{1}]\} - \mu_s^2, \tag{9.6}$$

where we have defined the $N \times N$ matrix

$$Q = Q(Z) \equiv \{E[(R - R_f\underline{1})(R - R_f\underline{1})'|Z]\}^{-1} = \{[\mu(Z) - R_f\underline{1}][\mu(Z) - R_f\underline{1}]' + \Sigma_\varepsilon(Z)\}^{-1}. \tag{9.7}$$

Also, define the following constants:

$$\zeta \equiv E\{[\mu(Z) - R_f\underline{1}]'Q[\mu(Z) - R_f\underline{1}]\},$$
$$\varphi \equiv E\{R_f[\mu(Z) - R_f\underline{1}]'Q[\mu(Z) - R_f\underline{1}]\}, \tag{9.8}$$
$$\psi \equiv E\{R_f^2[\mu(Z) - R_f\underline{1}]'Q[\mu(Z) - R_f\underline{1}]\}.$$

Theorem 9.1. Given a target unconditional expected return μ_s, N risky assets, instruments Z, and a conditional risk-free asset with rate $R_f = R_f(Z)$ that may vary over time, the unique portfolio having minimum unconditional variance is determined by the weights

$$x_s(Z) = \left(\frac{\mu_s - E(R_f) + \varphi}{\zeta} - R_f\right)Q[\mu(Z) - R_f\underline{1}]$$
$$= [(c+1)\mu_s + b - R_f]Q[\mu(Z) - R_f\underline{1}], \tag{9.9}$$

and the optimal portfolio variance is

$$\sigma_s^2 = a + 2b\mu_s + c\mu_s^2, \tag{9.10}$$

where

$$a = E(R_f^2) + \frac{[E(R_f) - \varphi]^2}{\zeta} - \psi, \quad b = \frac{\varphi - E(R_f)}{\zeta}, \quad \text{and } c = \frac{1}{\zeta} - 1.$$

When the risk-free asset return is constant, then these formulas simplify to theorem 2 of Ferson and Siegel (2001) with

$$x_s(Z) = \frac{\mu_s - R_f}{\zeta}Q[\mu(Z) - R_f\underline{1}], \tag{9.11}$$

and with optimal portfolio variance $\sigma_s^2 = \dfrac{1 - \zeta}{\zeta}(\mu_s - R_f)^2$.

Proof. Our objective is to minimize, over the choice of $x_s(Z)$, the portfolio variance $\mathrm{Var}(R_s)$ subject to $E(R_s) = \mu_s$, where $R_s = R_f + x'(Z)(R - R_f\underline{1})$ and the variance is given by equation (9.6). We form the Lagrangian,

$$L[x(Z)] = E[x'(Z)Q^{-1}x(Z)] + 2E\{R_f x'(Z)[\mu(Z) - R_f\underline{1}]\}$$
$$\qquad + 2\lambda E\{\mu_s - R_f - x'(Z)[\mu(Z) - R_f\underline{1}]\}, \tag{9.12}$$

and proceed using a perturbation argument. Let $q(Z) = x(Z) + dy(Z)$, where $x(Z)$ is the conjectured optimal solution, $y(Z)$ is any regular function of Z, and d is a scalar. Optimality of $x(Z)$ follows when the partial derivative of $L[q(Z)]$ with respect to d is identically zero when evaluated at $d = 0$. Thus,

$$0 = E(y'(Z)\{Q^{-1}x(Z) + (R_f - \lambda)[\mu(Z) - R_f\underline{1}]\}) \tag{9.13}$$

for all functions $y(Z)$, which implies that $Q^{-1}x(Z) + (R_f - \lambda)[\mu(Z) - R_f\underline{1}] = 0$ almost surely in Z. Solve this expression for $x(Z)$ to obtain equation (9.9), where the Lagrange multiplier λ is evaluated by solving for the target mean, μ_s. The expression for the optimal portfolio variance follows by substituting the optimal weight function into equation (9.9). Formulas for fixed R_f then follow directly. QED.

When the risk-free asset's return is time varying and contained in the information set Z at the beginning of the portfolio formation period, the *conditional* mean variance efficient boundary varies over time with the value of $R_f(Z)$. In general, all conditional means and variances will vary as the risk-free rate and the other conditioning information varies over time. A lagged Treasury bill rate, the usual risk-free rate proxy in empirical work, is an especially powerful empirical predictor of time varying means and covariances (e.g., see Breen, Glosten, and Jagannathan 1989; Ferson 1989), so there will be variation over time in the boundary.

A zero-beta parameter, γ_0, may be chosen to fix a point on the boundary that is *unconditionally* efficient with respect to Z. The choice of the zero-beta parameter corresponds to the choice of a target unconditional expected return μ_s. For a given value of γ_0, the target mean maximizes the squared Sharpe ratio $(\mu_s - \gamma_0)^2/\sigma_s^2$ along the mean variance boundary, which implies that $\mu_s = -(a + b\gamma_0)/(b + c\gamma_0)$. The same result appears in the classical mean variance analysis of chapter 8.

When there is a risk-free asset that is constant over time, the unconditionally efficient-with-respect-to-Z boundary is linear (a degenerate hyperbola) and reaches the risk-free asset at zero risk. In this case, we use $\gamma_0 = R_f$ and can obtain any μ_s larger or smaller than R_f, levering the efficient portfolio up or down with positions in the risk-free asset.

When there is no risk-free asset, we let $x' = x'(Z) = [x_1(Z), \ldots, x_N(Z)]$ denote the shares invested in each of the N risky assets, with the constraint that the weights sum to 1.0 almost surely in Z. The return on this portfolio, $R_s = x'(Z)R$, has expectation and variance as follows:

$$\mu_s = E[x'(Z)\mu(Z)], \tag{9.14}$$

$$\sigma_s^2 = E\{x'(Z)\Lambda^{-1}x(Z)\} - \mu_s^2, \tag{9.15}$$

where we have defined the $N \times N$ matrix

$$\Lambda = \Lambda(Z) \equiv \{E[RR'|Z]\}^{-1} = [\mu(Z)\mu'(Z) + \Sigma_\varepsilon(Z)]^{-1}. \tag{9.16}$$

Also, define the following constants:

$$\delta_1 = E\left(\frac{1}{\underline{1}'\Lambda\underline{1}}\right),$$

$$\delta_2 = E\left(\frac{\underline{1}'\Lambda\mu(Z)}{\underline{1}'\Lambda\underline{1}}\right), \quad \text{and} \tag{9.17}$$

$$\delta_3 = E\left[\mu'(Z)\left(\Lambda - \frac{\Lambda\underline{1}\underline{1}'\Lambda}{\underline{1}'\Lambda\underline{1}}\right)\mu(Z)\right].$$

Theorem 9.2 (Ferson and Siegel 2001, theorem 3).   Given N risky assets and no risk-free asset, the unique portfolio having minimum unconditional variance and unconditional expected return μ_s is determined by the weights

$$x_s'(Z) = \frac{\underline{1}'\Lambda}{\underline{1}'\Lambda\underline{1}} + \frac{\mu_s - \delta_2}{\delta_3}\mu'(Z)\left(\Lambda - \frac{\Lambda\underline{1}\underline{1}'\Lambda}{\underline{1}'\Lambda\underline{1}}\right) = \frac{\underline{1}'\Lambda}{\underline{1}'\Lambda\underline{1}} + [(c+1)\mu_s + b]\mu'(Z)\left(\Lambda - \frac{\Lambda\underline{1}\underline{1}'\Lambda}{\underline{1}'\Lambda\underline{1}}\right),$$

$$\tag{9.18}$$

and the optimal portfolio variance is $\sigma_s^2 = a + 2b\mu_s + c\mu_s^2$, where $a = \delta_1 + \delta_2^2/\delta_3$, $b = -\delta_2/\delta_3$, and $c = (1 - \delta_3)/\delta_3$.

The UE boundary is formed by varying the value of the target mean return μ_s in (9.18). Note that the second term on the right-hand side of equation (9.18) is proportional to the vector of weights for an excess return, or zero net investment portfolio (postmultiplying that term by a vector of ones implies that the weights sum to zero). The first term in (9.18) is the weight of the global minimum conditional second-moment portfolio. Thus, equation (9.18) illustrates two-fund separation: Any UE(Z) portfolio can be found as a combination of the global minimum conditional second-moment portfolio and some weight on the unconditionally efficient excess return described by the second term.

9.3.1   Properties of the UE Solution

The UE portfolio weights have some interesting properties as a function of the realization of the conditioning information (or signal) Z. Consider an example with two assets, one risk free and one risky. The solution for the weight in the risky asset is

$$x(Z) = [(\mu_P - R_F)/\zeta] [\mu(Z) - R_F]/[\sigma_\varepsilon^2 + [\mu(Z) - R_F]^2], \tag{9.19}$$

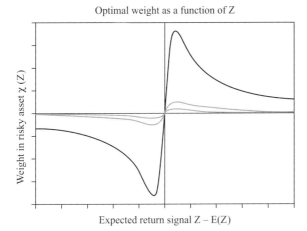

Figure 9.1
UE(Z) portfolio weights.

where μ is the risky asset's unconditional mean, $\mu(Z)$ is its conditional mean given Z, and

$$\zeta = E\{[\mu(Z) - R_F]^2/[\sigma_\varepsilon^2 + [\mu(Z) - R_F]^2]\}. \tag{9.20}$$

Figure 9.1 illustrates the behavior of x(Z) as a function of Z.

The figure shows three curves for different levels of the R-square of the regression of the risky stock return on the lagged variable Z. The R-squares are 0.45%, 10.5%, and 35.3%, to capture the range in which reality may reside. As the figure illustrates, the risky asset weight is approximately linear in the conditional mean for values close to the center of the distribution. However, for more extreme signals, the quadratic term in the denominator dominates, and the investment in the risky asset shrinks toward zero.

The conservative behavior of the curve for large values of the signal is similar to how one might weigh the data in robust statistics. By putting less weight on extreme values, the statistic is robust to extreme values. It is interesting that the strategy that maximizes the unconditional Sharpe ratio would have this property.

9.3.2   Representation

The UE portfolio satisfies a version of two-fund separation, according to equation (9.18). Any UE portfolio R_p may be written as

$$R_p = R^* + w_p R^{e^*}, \tag{9.21}$$

where R^* has the portfolio weights $\Lambda\underline{1}(\underline{1}'\Lambda\underline{1})^{-1}$ and R^{e^*} is an excess return, weight summing to zero: $[\Lambda - \Lambda\underline{1}\underline{1}'\Lambda(\underline{1}'\Lambda\underline{1})^{-1}]\,\mu(Z)$ and multiplied by the constant $w_p = (\mu_p - \alpha_2)/\alpha_3$. The first portfolio is the minimum conditional second-moment portfolio. The two portfolios are orthogonal, in that $E\{[\Lambda\underline{1}(\underline{1}'\Lambda\underline{1})^{-1}]'RR'[\Lambda - \Lambda\underline{1}\underline{1}'\Lambda(\underline{1}'\Lambda\underline{1})^{-1}]\mu(Z)\} = 0$. This is sometimes called the Hansen and Richard (1987) representation.

Hansen and Richard show that if the weight can depend on Z, as in $w_p(Z)$, then all conditionally efficient portfolios given Z can be represented by equation (9.21), substituting in the time-varying weight. This shows, once again, that UE portfolios are a special case of CE portfolios, where the weight $w_p(Z)$ is a constant for the UE portfolio, and the same CE portfolios can be used to represent both the CE and the UE portfolios.

9.4   Extensions

This section reviews a couple of extensions of UE portfolios. The first is (so far) a curiosity in the literature, while the second is related to our subsequent discussion of tests of asset pricing models based on minimum variance efficiency.

9.4.1   The Residual Frontier

Penaranda and Sentana (2016) describe portfolios $R_p = x(Z)'R$ to be on the residual frontier if they solve the following problem:

$$\text{Min } E[\text{Var}(R_{pt}|Z_{t-1})], \text{ subject to } E(R_p) = \mu_p. \tag{9.22}$$

To understand the name "residual frontier," consider a regression: $R_t = E(R_t|Z_{t-1}) + u_t$, where the fitted value is the conditional mean. The regression error satisfies $E(u_t^2) = E[\text{Var}(R|Z_{t-1})]$, so minimizing the unconditional variance of the regression error is equivalent to minimizing the expected value of the conditional variance. Penaranda and Sentana show that the "residual" minimum variance efficient portfolios are a subset of CE portfolios. We know UE portfolios are another subset. They show that the two subsets are disjoint, unless they touch at a constant expected return portfolio.

9.4.2   Optimal Orthogonal Portfolios with Conditioning Information

We discussed the optimal orthogonal portfolio in chapter 8 for the classical case with no conditioning information. To briefly review, consider a regression of excess returns on a benchmark: $r = \alpha + \beta r_B + u$; then the optimal orthogonal portfolio solves $\text{Max}_x\{(x'\alpha)^2/\text{Var}(x'r)\}$, and the solution x_c is proportional to $\Sigma^{-1}\alpha$. The optimal orthogonal portfolio r_c can be interpreted as the most mispriced portfolio by the benchmark r_B. It is optimal in the sense that a combination of $r_c = x_c'r$ and r_B is minimum variance efficient, and it is orthogonal in the sense that $\text{Cov}(r_c, r_B) = 0$. Orthogonality means that we can replace $\text{Var}(x'r)$ with $\text{Var}(x'u)$ in the maximization problem.

Ferson and Siegel (2015) develop optimal orthogonal portfolios with conditioning information and show there are analogies to each of these results. The optimal orthogonal portfolio in this case has weights that depend on Z, and it maximizes its squared unconditional alpha divided by the unconditional variance of return. A combination of the optimal orthogonal portfolio using Z and the benchmark is a UE(Z) portfolio. The portfolio is orthogonal to the benchmark portfolio return.

Zhou (2008) and Chiang (2015) study UE versions of the modern portfolio management problem, which we reviewed in chapter 8 for the case with no explicit conditioning information. In this case, the manager has information Z, which the client does not see, and observes conditional alphas and perhaps betas as a function of Z. The manager uses the information to find dynamic trading strategies that maximize the unconditional alpha relative to the unconditional residual variance. Chiang considers, inter alia, a tracking error minimization problem that sets $\beta = 1$ in the factor model regression and solves the UE problem subject to various constraints, motivated by the practitioner literature. The solutions to the constrained problems turn out to be isomorphic to multifactor minimum variance problems with conditioning information, which are examined in chapter 14.

9.5 Empirical Evidence

9.5.1 Out-of-Sample Performance of UE Portfolios

The empirical evidence on the out-of-sample performance of traditional mean variance efficient portfolios is summarized in chapter 8 as not very comforting. Michaud (1989) attributes the poor out-of-sample performance to the sensitivity of mean variance solutions to errors in estimating the means. As noted above, the UE portfolio weight does not react so aggressively to extreme values of the estimated mean returns, suggesting that it may have better out-of-sample performance. The evidence suggests that it does.

Chiang (2015) examines the out-of-sample performance of UE versions of the modern portfolio management problem with conditioning information and compares it to cases of no conditioning information and to a CE(Z) portfolio. He finds that the out-of-sample performance, as measured by the unconditional information ratio, is best for the UE version of the tracking error problem. Abhay, Basu, and Stremme (2002) study UE portfolio strategies and find that using lagged term spreads and credit spreads leads the strategies to outperform buy-and-hold in out-of-sample exercises; these strategies also produce lower turnover and trading costs than CE strategies. Fletcher and Basu (2016) examine the out-of-sample performance of UE portfolios in samples of UK closed-end funds. They also find high Sharpe ratios and lower turnover for the UE strategies compared with the other strategies. Penaranda and Wu (2017) examine five different UE-type portfolio strategies to learn how various combinations of mean and variance predictability determine their relative performance as measured by the unconditional Sharpe ratios.

In unpublished work that was removed from Ferson and Siegel (2001) prior to publication before the days of on-line journal appendixes, we examine monthly returns data for 25 industries and the S&P 500 index for 1963–1994. The predictor variables Z include lagged dividend yields, short-term Treasury yields, a default-related corporate yield spread (Baa–Aaa), a 10-year minus 1-year Treasury yield spread, and a 3-month less 1-month lagged Treasury bill return. In a weak signal case, only the dividend yield is used as an instrument, and the R-squared for the CRSP market portfolio monthly return is 0.45%. In a medium signal case, where all five instruments are used, the R-squared is 10.5%. These are the cases depicted in figure 9.1. The strong signal example in figure 9.1 is hypothetical, as we don't know of data that produces such a high R-squared. If you find some, please let me know!

Ferson and Siegel compare the UE(Z) portfolio strategy with a CE(Z) strategy maximizing an exponential utility function assuming normality. Using the subsequent realized returns, Ferson and Siegel compute the average utility that the exponential function experiences "out-of-sample" (really, one step ahead in the sample) using the CE(Z) strategy and using the UE(Z) strategy. We also compute a cost C, which, if subtracted from the returns achieved by the CE(Z) strategy, results in the same after cost utility as achieved out of sample by the UE(Z) strategy with no cost. Thus, the cost is what the exponential agent would pay to avoid using the UE(Z) solution, since his optimal solution is CE(Z). The out-of-sample utility cost is negative, indicating that the UE(Z) strategy beats even the optimal CE(Z) strategy out of sample. Ferson and Siegel observe higher out-of-sample Sharpe ratios for the UE(Z) solution. Rolling the out-of-sample procedure through time, we examine the time series of the portfolio weights for the 25 industries and observe that the UE(Z) solution presents less turnover than the CE(Z) solution.

9.5.2   Empirical Performance of the Optimal Orthogonal Portfolio

Ferson and Siegel (2015) is the only paper I know of that empirically examines optimal orthogonal portfolios with conditioning information. The paper uses the returns of common stocks sorted according to market capitalization and book-to-market ratios. Details about the data are given in the note to table 9.1. Three different versions of the solutions for the optimal orthogonal portfolios treat the risk-free rate in different ways. In the first case, the risk-free rate is assumed to be a fixed constant. Here we do not include the Treasury return as a lagged instrument, and we set $\gamma_0 = 3.8\%$ to be the average Treasury return during the sample. The target mean μ_s of the efficient-with-respect-to-Z portfolio is set equal to the sample mean return of the market index, which determines the amount of leverage the portfolio uses at the fixed risk-free rate. In the second case, no risk-free asset exists. Here we again set $\gamma_0 = 3.8\%$ to pick a point on the mean variance boundary, and we do not allow the portfolio to take a position in a risk-free asset. We do allow the lagged Treasury return as conditioning information in Z, which highlights the information in lagged Treasury returns about the future risky asset returns. In the third case, there is a conditionally risk-free return that is contained in Z. Here we use the lagged Treasury return in the conditioning information, and we allow the portfolio to trade the subsequent Treasury return in addition to the other risky assets. This highlights the effects of market timing, or varying the amount of "cash" in the portfolio, in addition to varying the allocation among the risky assets.

Table 9.1
Optimal orthogonal portfolios.

Panel A: Summary statistics

Asset	CAPM α	σ_u	Information Ratio	Fixed R_c	Average Active Portfolio Weights [X_c(Z)]		
					Fixed R_f	No R_f	Varying R_f
Market index	0	0	0.000	−1.260	0.179	−0.210	0.308
Small stocks	3.95	25.5	0.155	0.353	−0.262	0.133	0.097
Value stocks	3.14	17.8	0.177	0.345	0.137	0.079	0.142
Growth stocks	−1.00	8.3	−0.121	0.155	0.028	0.098	0.020
Bonds	1.66	9.1	0.183	1.410	0.918	0.900	0.432
Fixed R_c	4.66	16.5	0.282	1.000			
X_c(Z) Portfolios							
Fixed R_f	8.66	25.5	0.340				
No R_f	3.12	10.5	0.297				
Varying R_f	8.77	9.2	0.955				

Panel B: Squared Sharpe ratios

Case	$S^2(R_p)$	$S^2(R_c)$	Sum
Fixed R_c	0.189	0.079	0.267
X_c(Z) Portfolios			
Fixed R_f	0.189	0.115	0.304
No R_f	0.189	0.088	0.277
Varying R_f	0.182	0.909	1.191

Source: Ferson and Siegel (2015).
Notes: Annual returns on portfolios of common stocks and long-term US Government bonds cover the period 1931–2007, focusing on value-weighted decile portfolios of small-capitalization stocks, value stocks (high book/market), and growth stocks (low book/market), as provided on Ken French's website (mba.tuck.dartmouth.edu /pages/faculty/ken.french/data_library.html). They also include a long-term US Government bond return, splicing the Ibbotson Associates 20-year US Government bond return series for 1931–1971 with the CRSP greater than 120-month US Government bond return after 1971. The market portfolio, measured as the CRSP value-weighted stock return index, is the benchmark or tested portfolio R_p. The risk-free return is the return from rolling over one-month Treasury bills from CRSP. Its sample average, 3.8% per year, is taken as the fixed zero-beta rate in all examples. As conditioning information in Z, its return is lagged by one year. All returns are discretely compounded annual returns, and the sample period is 1931–2007.

The lagged instruments Z are the lagged Treasury return and the log of the market price/dividend ratio at the end of the previous year. In calculating the price/dividend ratio, the stock price is the real price of the S&P 500, and the dividend is the real dividends accruing to the index over the past year. These data are from Robert Shiller's website (www.econ.yale.edu/~shiller/data.htm).

The CAPM α refers to the intercept in the regression of the portfolio return in excess of the zero-beta rate, on that of the broad value-weighted stock market index (market index). The symbol σ_u refers to the standard deviation of the regression residuals. The fixed portfolio R_c ignores the conditioning information. The alphas and residual standard deviations are annual percentages. The information ratio is the ratio α/σ_u.

Table 9.1 summarizes the results. The rows show results for the benchmark index ("Market" in the table), three equity portfolios, and the government bond return. The CAPM α refers to the intercept in the regression of the portfolio returns in excess of the Treasury bill returns, on the excess return of the market index. The symbol σ_u refers to the standard deviation of the regression residuals. The small stock portfolio has the largest σ_u or non-market risk, at more than 25% per year.

The bottom four rows of panel A summarize the optimal orthogonal portfolios when the market index is the benchmark. The fixed-weight portfolio R_c uses no conditioning information. Its alpha is larger than any of the separate assets, at 4.66% per year, and its residual standard deviation is also relatively large, at 16.5% per year. Since the portfolio is orthogonal to the market index, its residual standard deviation is the same as its total standard deviation (or volatility of return). The ratio of the alpha to the residual volatility is the appraisal ratio or information ratio. Optimal orthogonal portfolios try to maximize the square of this ratio.

The fixed-weight orthogonal portfolio R_c delivers an information ratio of 0.282, substantially larger than those of the small stock or value portfolios, at 0.155 and 0.177, respectively, and also larger than the bond portfolio, which has an information ratio of 0.183 by virtue of its relatively small volatility.

Table 9.1 also summarizes performance statistics for the optimal orthogonal portfolios with conditioning information. The information ratios in all three cases are larger than those of the individual assets or the fixed-weight orthogonal portfolio. The averages over time of the optimal orthogonal portfolios' weights on the risky assets are shown in the right-hand columns of panel A. The weights are normalized to sum to 1.0. The weights $X_c(Z)$ of the optimal orthogonal portfolios combine with the benchmark (whose weights x_p are 100% in the market index) to determine an efficient portfolio. The overall efficient portfolio weights x_s therefore vary over time. In the fixed risk-free rate case, $x_s = 0.60 X_c(Z) + 0.40 x_p$. In the no risk-free rate case, $x_s = 0.23 X_c(Z) + 0.77 x_p$. In the time-varying risk-free rate case, $x_s = 0.44 X_c(Z) + 0.56 x_p$. Thus the efficient portfolio is formed as a convex combination of the benchmark and the active portfolio, with reasonable weights in each case.

The four rightmost columns of panel A show that two of the active portfolios take short positions in the market benchmark, indicating an optimal tilt away from the market index. The fixed-weight portfolio R_c takes an extreme short position of -126%, consistent with previous evidence of wild portfolio weights in these kinds of problems, while on average the portfolio using the conditioning information but no risk-free asset takes a position of -21% in the market index. These short positions finance large long positions in the US government bond, and also long positions (in most cases) in small stocks, value stocks, and growth stocks. It is interesting that all portfolios tilt positively, although by small amounts, into growth stocks, even though growth stocks have negative CAPM alphas. This occurs because of the correlations among the asset classes. All versions of the optimal orthogonal portfolio suggest strong tilts into government bonds. The bond tilt is the most extreme for the fixed-weight solution, at 141%, and is relatively modest (43.2%) for the portfolio assuming a time-varying conditional risk-free rate.

The squared Sharpe ratios in panel B of table 9.1 indicate how far the stock market index is from efficiency. The squared Sharpe ratio for the market is 0.189, measured relative to the fixed zero beta rate of 3.8%. The market portfolio's squared Sharpe ratio is slightly smaller, at 0.182, for returns measured in excess of a time-varying risk-free rate. This is because of the negative covariance between the risk-free rate and stock returns. For the fixed-weight orthogonal portfolio R_c, the squared Sharpe ratio is 0.079, and for the portfolios using conditioning information, it varies between 0.088 and 0.909.

According to the *law of conservation of squared Sharpe ratios*, the sum of the index and optimal orthogonal portfolios' squared Sharpe ratios is the squared slope of the tangency from the given zero-beta rate to the relevant mean variance frontier. The sum is 0.267 when no conditioning information is used, and it is 0.304–1.091 when the information is used. With a fixed risk-free rate, we condition only on the lagged dividend yield, which is a relatively weak predictor for stock returns. In the case with no risk-free rate, the portfolio is not allowed to hold short-term Treasuries, which substantially weakens the performance. In the time-varying risk-free rate case, the portfolio strategy is allowed to "market time" by holding short-term Treasuries. In the other two cases, we use the information in the lagged Treasury rate, which is a relatively strong predictor, and the efficient-with-respect-to-Z boundary is far above the mean variance boundary that ignores the conditioning information. (Ferson and Siegel (2009) present an analysis of the statistical significance of differences like these.)

10 Variance Bounds

Bounds on the variance of SDFs are first derived by Hansen and Jagannathan (HJ; 1991). These bounds are dual to the classical mean variance analysis and thus serve a useful pedagogical role. Studies also develop tools to conduct inferences about asset pricing models based on the bounds. This chapter develops the original HJ bounds and illustrates the duality with the classical mean variance analysis, using the maximum correlation portfolio from chapter 4. Before we envisioned the mean-standard deviation "bullet"; now we rotate it 90 degrees and have the mean-standard deviation "cup." We briefly relate the ideas to the earlier bounds on the volatility of stock prices from LeRoy and Porter (1981) and Shiller (1981), and we review further extensions that have been developed.

10.1   A Simple Development

While visually the bullet is rotated 90 degrees to produce the cup, in another sense the HJ bounds turn the canonical asset pricing problem on its head. In the usual asset pricing problem, we have a model that specifies some m that should work in equation (1.1), and the goal is to use a sample of asset returns to test the model. In the HJ bounds, we take as given the set of asset returns, and we ask: What can we say about the set of ms that might be able to price these assets? The answer in the HJ bounds comes in the form of restrictions between the mean and variance of the SDF. The restrictions are depicted in figure 10.1.

The figure depicts the HJ bounds, a cup formed by the graph of the standard deviation of m versus the mean of m. The basic idea is that $E(mR) = 1$ implies that the point $\{E(m), \sigma(m)\}$ must plot inside the cup, if the m is to price assets. This suggests a diagnostic for candidate asset pricing models. If we have a lot of ms and a given set of returns R that we want to price, we could imagine plotting all the ms in figure 10.1 and ruling out those that do not fall inside the cup from further consideration.

To see how the cup arises, first simplify to unconditional moments and posit the existence of a hypothetical constant risk-free asset with price: $E(m) = R_F^{-1}$. Letting $r = R - R_F$, we know that expected excess returns are given by $E(r_i) = Cov(-m, r_i)/E(m)$, so that

$$E(r_i)/\sigma(r_i) = [Cov(-m, r_i)/\sigma(m)\sigma(r_i)]/[E(m)/\sigma(m)]. \qquad (10.1)$$

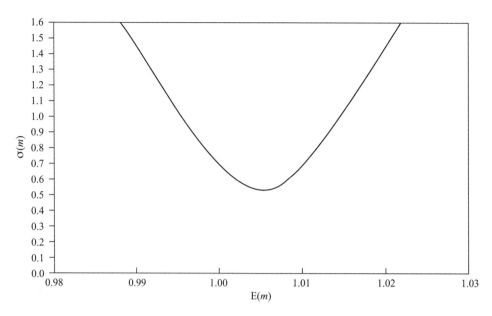

Figure 10.1
A Hansen-Jagannathan bound.

Substituting the correlation $\rho(-m, r_i)$ times the product of the standard deviations for the covariance in the above equation, and because the correlation must be less than 1.0, we have that $\sigma(m)/E(m) \geq E(r_i)/\sigma(r_i)$ for all of the assets i. Therefore, we have the HJ bound:

$$\sigma(m) \geq E(m) \, \text{Max}_i \, |E(r_i)/\sigma(r_i)|. \tag{10.2}$$

Let $S(R_F) \equiv \text{Max}_i \, |E(r_i)/\sigma(r_i)|$ be the maximum absolute Sharpe ratio, which depends on R_F or $E(m)$. We can write the bound as

$$\sigma(m) \geq E(m)S(R_F). \tag{10.3}$$

10.2   Duality with the Usual Mean-Variance Diagram

The Sharpe ratio of any boundary portfolio R^* on the usual mean-standard deviation diagram (the bullet) is the slope of a line from the notional risk-free rate R_F, tangent to the minimum variance boundary at R^*, and the slope of that line is the maximum Sharpe ratio for that R_F, denoted by $S(R_F)$. Both the tangent portfolio R^* and its Sharpe ratio $S(R_F)$ depend on the value of R_F or $E(m)^{-1}$. As we vary R_F, we move the tangent point around the minimum variance boundary in figure 10.2. If R_F is above the expected return for the global minimum variance portfolio, then R^* is on the lower half of the boundary and is not mean variance efficient, although it is minimum variance efficient.

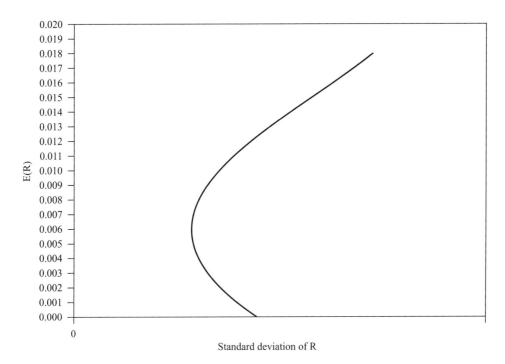

Figure 10.2
A minimum variance bullet.

The boundary of the HJ region for $\{E(m), \sigma(m)\}$ depicts the minimum value of $\sigma(m)$ for each value of $E(m)$, as specified by $\sigma(m) \geq R_F^{-1} S(R_F)$, where $R_F = E(m)^{-1}$. As we vary $E(m)$ in figure 10.2, we move around the $\{E(m), \sigma(m)\}$ boundary, tracing out the HJ cup. Figures 10.1 and 10.2 are duals to each other, in the sense that any point $\{\mu_p, \sigma_p\}$ on the bullet corresponds to a point on the cup, which can be determined by setting $\{E(m), \sigma(m)\} = \{R_F^{-1}, R_F^{-1} | S(R_F)|\}$, where R_F is taken to be the zero-beta rate for the point on the bullet with mean μ_p. We found an expression for this zero-beta rate in terms of the efficient set constants in chapter 8: $\gamma_0 = (b\mu_p - c)/(a\mu_p - b)$. Therefore, starting with any point $\{E(m), \sigma(m)\}$ on the cup, its corresponding point on the bullet is

$$\{\mu_p = (c - b/E(m))/(b - a/E(m)), \ \sigma_p = [\mu_p - 1/E(m)]/[\sigma(m)/E(m)]\}. \tag{10.4}$$

Just as well, any point $\{\mu_p, \sigma_p\}$ on the bullet corresponds to a point on the cup, which can be determined as functions of $\{E(m), \sigma(m)\}$ and the efficient set constants. In summary, the asset pricing problem is turned on its head by the HJ bounds, but graphically, the mean variance bullet is turned on its side, and the bullet becomes the cup.

Now, the bullet changes when you add more assets to the mix. For a given target mean return, the standard deviation must get (weakly) smaller when more assets are added, because the optimization to minimize variance can always choose to ignore the new assets. Thus, the maximum Sharpe ratio gets higher as more assets are included. This translates via the duality to the shape of the cup. When you add more assets, the Sharpe ratio gets higher, and the minimum $\sigma(m)$ for a given $E(m)$ becomes larger. Graphically, the cup gets skinnier and moves higher up in the picture. Intuitively, when you are asked to price more assets, fewer ms will be able to do the job. The taller and skinnier cup holds fewer ms. If we think of the cup as a diagnostic tool for ruling out ms, this is a good thing. The standard of beauty for HJ cups is similar to that for humans in southern California—taller and thinner is better.

Recall the "equity premium puzzle" of Mehra and Prescott (1985), who observe that the average returns of stocks seem too high to be priced by simple consumption model ms. There is a similar puzzle in bonds returns, illustrated here using HJ bounds. In figure 10.3, the HJ bounds are drawn for quarterly data and a sample period similar to Hansen and Jagannathan (1991), consisting of 3-, 6-, 9-, and 12-month Treasury bill returns for 1964–1986. The X symbols in the figure denote the sample means and standard deviations of the intertemporal marginal rate of substitution from the consumption model with power utility: $m_{t+1} = \beta(C_{t+1}/C_t)^{-\alpha}$, where C_t is per capita US real consumption at time t, β is the rate of time discount (which is set equal to 1.0 in this example), and α is the concavity parameter of the power utility function. The X symbols are shown for values of α set equal to 1, 2, 3, ..., 100. (They actually go off the plot on the left-hand side and then curve back.) The implied m values do not enter the HJ cup until $\alpha = 71$. Thus, without resorting to implausibly large values of risk aversion, the consumption model's SDF can't price these returns.

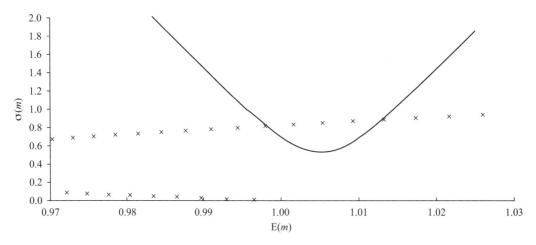

Figure 10.3
The equity premium puzzle for bonds.

10.3 Computing the HJ Bounds

There is a more direct approach for calculating the HJ bounds. A regression approach is shown and related to the maximum correlation portfolio.

10.3.1 A Regression Approach

Consider a regression of m_{t+1} on $\underline{R}_{t+1}$:

$$m_{t+1} = \text{const.} + \beta' \, \underline{R}_{t+1} + \varepsilon_{t+1}, \tag{10.5}$$

or $m_{t+1} - E(m_{t+1}) = \beta'[\underline{R}_{t+1} - E(\underline{R}_{t+1})] + \varepsilon_{t+1}$, $E(\varepsilon_{t+1}\underline{R}_{t+1}) = E(\varepsilon_{t+1}) = 0$, and $\beta = V(R)^{-1}\text{Cov}(\underline{R}, m)$, where $V(R)$ is the covariance matrix of the given returns R. From the regression, we can decompose the variance of the SDF as

$$
\begin{aligned}
\text{Var}(m) &= \beta'V(R)\,\beta + \text{Var}(\varepsilon) \\
&= \text{Cov}(m, \underline{R}')V(R)^{-1}\text{Cov}(\underline{R}, m) + \text{Var}(\varepsilon) \\
&\geq \text{Cov}(m, \underline{R}')V(R)^{-1}\text{Cov}(\underline{R}, m).
\end{aligned}
\tag{10.6}
$$

Now using $E(m\underline{R}) = \underline{1}$, $\text{Cov}(\underline{R}, m) = E(\underline{R}m) - E(\underline{R})E(m) = \underline{1} - E(\underline{R})E(m)$. Substituting this in for the covariance, we get

$$\text{Var}(m) \geq [\underline{1}' - E(m)E(\underline{R}')]V(R)^{-1}[\underline{1} - E(m)E(\underline{R})]. \tag{10.7}$$

The HJ cup can be computed directly from equation (10.7). You only need estimates for the vector $E(\underline{R}')$ and the matrix $V(R)$. As you vary $E(m)$, you trace out the cup.

The equivalence between this expression and the previous one based on maximum Sharpe ratios in equation (10.3) becomes clear when you recall the expression for a maximum squared Sharpe ratio. A general result for any vector $E(\underline{R}')$ and matrix $V(R)$ is

$$\text{Max}_x \, [x'E(\underline{R})]^2/[x'V(R)x] = E(\underline{R}')'V(R)^{-1}E(\underline{R}). \tag{10.8}$$

Applying this result shows that the right-hand side of equation (10.7) is equal to $[1/E(m)]^2\text{SR}(E(m)^{-1})^2$, where $\text{SR}(\cdot)^2$ is the squared Sharpe ratio. Thus this representation is equivalent to (10.3).

10.3.2 Relation to the Maximum Correlation Portfolio

In the previous regression approach, $m_{t+1} - E(m_{t+1}) = \beta'[\underline{R}_{t+1} - E(\underline{R}_{t+1})] + \varepsilon_{t+1}$. Define the fitted value: $m^*_{t+1} = E(m_{t+1}) + \beta'[R_{t+1} - E(R_{t+1})]$. Then $m_{t+1} = m^*_{t+1} + \varepsilon_{t+1}$, where $E(\varepsilon_{t+1}\underline{R}_{t+1}) = E(\varepsilon_{t+1}m^*_{t+1})$ $= E(\varepsilon_{t+1}) = 0$. Thus, m^*_{t+1} is a linear function of a portfolio of the returns R_{t+1}, with weights proportional to β. This portfolio is the *maximum correlation portfolio* for m. It is really the maximum squared correlation portfolio, because the regression chooses the coefficients to minimize the sum of squared residuals, or maximize the R-squared, which is the squared correlation. For a given $E(m)$, the variance of m can be no smaller than $\text{var}(m^*)$:

$$\mathrm{var}(m) = \mathrm{var}(m^* + \varepsilon) \geq \mathrm{var}(m^*) = [\underline{1}' - \mathrm{E}(m)\mathrm{E}(\mathrm{R}')]\mathrm{V}(\mathrm{R})^{-1}[\underline{1}' - \mathrm{E}(m)\mathrm{E}(\mathrm{R}')]'.$$

This is the same as equation (10.7).

Another fun fact about the maximum correlation portfolio to m follows by writing $m = m^* + \varepsilon$, which implies $\sigma(m^*)/\sigma(m) = \rho(m, m^*) < 1$. This has a nice graphical interpretation. A given m that plots in the HJ cup plots at a distance above the boundary of the cup, and this distance depends on the correlation of the m with its projection onto asset returns, m^*. SDFs that are close to the boundary have correlations close to 1.0. The higher in the cup the SDF plots, the lower is its correlation with asset returns. Cochrane and Hansen (1992) discuss this relation.

10.4   Drawing Inferences with HJ Bounds

To use the bounds as diagnostic tools, we need to estimate them with data and characterize their sampling distributions. One attractive way to estimate the bounds is by maximum likelihood (ML). For a given $\mathrm{E}(m)$, the bound is given by

$$\mathrm{Var}(m^*) = [\underline{1} - \mathrm{E}(m)\mu]' \Sigma^{-1} [\underline{1} - \mathrm{E}(m)\mu]. \qquad (10.9)$$

An ML estimator for the left-hand side is obtained when the ML estimates for (μ, Σ) are plugged into the above equation. Given ML estimators for the mean vector and the covariance matrix Σ, a standard error for the estimate of $\mathrm{Var}(m^*)$ can be derived using the delta method. This will all be clear when we discuss these methods in part IV of the book.

A related and more general approach is to use the generalized method of moments (GMM; Hansen 1982), which is developed in detail in part IV. The GMM estimator for $\mathrm{Var}(m^*)$ comes with expressions for the asymptotic standard errors. Asymptotic standard errors for the HJ bounds are derived by Hansen, Heaton, and Luttmer (1995), Jagannathan and Wang (1996), Kan and Zhou (2004), and Kan and Robotti (2008). Several studies use simulations to evaluate the finite sample properties of the bounds, including Tierens (1993); Burnside (1994); Ferson and Siegel (2003), discussed in chapter 11; and Bekaert and Liu (2004). Gordon et al. (1996) present a Bayesian approach.

10.5   Related Measures and Bounds

10.5.1   Volatility Bounds on Stock Prices

Shiller (1981) and LeRoy and Porter (1981) develop bounds on the prices of stocks, which have an interesting relation to the preceding analysis. We saw that $m = m^* + \varepsilon$, with $\mathrm{E}(m^*\varepsilon) = 0$, which implies $\mathrm{Var}(m) \geq \mathrm{Var}(m^*)$.

These papers use a similar idea. Assume that the expected return of a stock is a constant. The present value model (when m is replaced by the constant discount rate R) states:

$$P_t = E_t \{ \textstyle\sum_{j>0} (R)^{-j} D_{t+j} \}. \tag{10.10}$$

Define the "ex post rational" stock price as

$$P_t^* = \{ \textstyle\sum_{j>0} (R)^{-j} D_{t+j} \} \tag{10.11}$$

Then rational expectations implies: $P_t = E_t(P_t^*)$, or $P_t^* = P_t + \varepsilon$, with $E(P_t \varepsilon) = 0$. Thus,

$$\text{Var}(P_t^*) \geq \text{Var}(P_t). \tag{10.12}$$

This is the simplest Shiller bound. Campbell, Lo, and MacKinlay (1997, chapter 7) provide a nice review and their own versions of this type of bound.

Shiller (1981) finds that stock prices are, in fact, more volatile than the P_t^* values that he constructed, and he concludes that prices are too volatile to be the product of rational stock markets. He considered models where the discount rate in the above equations (or the expected return of the stock) can vary over time. For example, he used the variation in the short-term interest rate as one proxy for this variation. However, Cochrane (1991) argues that he might not have gone far enough in this regard. Cochrane argues that, in fact, the statement that stock prices are too volatile in Shiller's bounds is equivalent to the statement that m is "too smooth." Shiller, being a clever fellow, did not pick the latter characterization. To see the intuition for Cochrane's claim, consider a single-period model for the stock market as a whole. Using $R_{t+1} = D_{t+1}/P_t$, suppose that I set $m_{t+1} = (R_{t+1})^{-1}$, in which case $P_t^* = P_t$ and the bound is satisfied, but the SDF is not very smooth.

10.5.2 The HJ Distance

Hansen and Jagannathan (1997) extend the analysis of the HJ bound to consider the distance between ms that work in equation (1.1) from candidate ms that may not work. Let $m(\varphi)$ be a candidate stochastic discount factor proposed by an asset pricing model, with φ the model parameters. The model states that $E(m(\varphi)R - 1|Z) = 0$, but the model is wrong. How close is $m(\varphi)$ to a stochastic discount factor that works?

We know that $m^* = [\underline{1}' E(RR')^{-1}]R$ works for pricing R. How close is $m(\varphi)$ to m^*? Project $m(\varphi)$ on the returns R to get $m^\wedge = [E(m(\varphi)R') E(RR')^{-1}]R$. Measure the mean square distance between $\hat{m}$ and m^*. This is the HJ distance (HJD) measure:

$$\text{HJD} = E\{(\hat{m} - m^*)^2\}. \tag{10.13}$$

Note that $\hat{m} - m^* = [E(m(\varphi)R - \underline{1}]' E(RR')^{-1}R$, so we can write the HJD as

$$\text{HJD} = [E(m(\phi)R) - \underline{1}]' E(RR')^{-1} [E(m(\phi)R) - \underline{1}]. \tag{10.14}$$

This has some nice interpretations. If $\alpha = E(m(\varphi)R) - \underline{1}$ is the vector of expected pricing errors from the candidate model, then letting $V = E(RR')$, we have $\text{HJD} = \alpha' V^{-1} \alpha$. The HJD measures a version of the "most mispriced" return available, because $\text{Max}_x\{(x'\alpha)^2\}/\alpha' V \alpha = \alpha' V^{-1} \alpha$.

10.5.3   HJ Bounds with Conditioning Information: A First Cut

Hansen and Jagannathan (1991) considered lagged conditioning information in their bounds, bringing in the lagged information Z in a clever way. The model states that $E(m(\varphi)R - 1|Z) = 0$, which implies

$$E[m(\varphi)(R \otimes Z) - (1 \otimes Z)] = E(0 \otimes Z) = 0, \qquad (10.15)$$

where $\otimes$ is the Kronecker product. If we let $\acute{z} \equiv Z./E(Z)$, where $./$ denotes element-by-element division, the previous equation is equivalent to

$$E[m(\varphi)(R \otimes \acute{z})] = \underline{1}. \qquad (10.16)$$

So if we define $R^{\#} \equiv R \otimes \acute{z}$ as a vector of multiplicative gross returns, we have $E[m(\varphi)R^{\#}] = \underline{1}$, and we can proceed as before, using $R^{\#}$ as the returns. In chapter 11, we refer to Hansen and Jagannathan's approach as a *multiplicative approach* to the conditioning information, and we discuss alternative approaches that have been subsequently developed.

10.5.4   Positivity

The original HJ bound solves Min $\sigma(m)$ s.t. $E(mR) = 1$. Hansen and Jagannathan (1991) also consider positivity: Min $\sigma(m)$ s.t. $E(mR) = 1$, $m > 0$. This is loosely motivated by no-arbitrage considerations, as the existence of some $m > 0$ is equivalent to a lack of arbitrage. (However, in incomplete markets, there can be ms that are not strictly positive but still correctly price the traded assets without implying arbitrage.) This constrained bound requires the solution of an N-dimensional numerical search problem, which has limited its popularity.

The first HJ distance is $\|m - m^{*}\|$ s.t. $E(m^{*}R) = 1$. Hansen and Jagannathan also consider a second HJD: $\|m - m^{*}\|$ s.t. $E(m^{*}R) = 1$, $m > 0$. The motivation is similar, and numerical solutions are once again needed.

Kan and Robotti (2008) show that there is a duality between the constrained SDF bounds and the mean-variance portfolio return analysis, analogous to the simpler, unconstrained bounds. They provide an easier way to compute the constrained bounds. Under normality, their approach replaces the N-dimensional numerical search with a well-behaved, one-dimensional search problem. Under normality, they derive the exact, finite sample distributions for constrained HJ bounds. They also extend the bias adjustment of Ferson and Siegel (2003), discussed in chapter 11, for the constrained bound, and they evaluate its performance with simulations. Similar to Ferson and Siegel, they find that the bias adjustment works well and that the adjusted estimator of the bounds is less volatile.

10.5.5   Bounds on the Autocorrelations of SDFs

We saw in chapter 6 that the expected excess returns or term premiums in pure, default-free bonds depend on the autocovariance of the SDF. Chrétien (2012) develops a bound on the autocorrelation of the SDF based on term premiums. Let P_t^2 be the price of a two-period bond and P_t^1 be the price of a one-period bond at time t. It follows from m-talk with time-separable ms, that

$$E(P_t^2) - E(P_t^1)^2 = E(m_{t+1}m_{t+2}) - E(m_{t+1})^2 = \text{Cov}(m_{t+1}m_{t+2}) = \rho_m \text{Var}(m) \geq \rho_m \text{Var}(m^*), \quad (10.17)$$

where m^* is the projection of the SDF onto the space of asset returns that we used to derive the HJ bounds at the beginning of this chapter, and ρ_m is the autocorrelation of the SDF. This assumes that m is stationary, so that $E(m_{t+1}) = E(m_{t+2})$.

Using one- and two-quarter Treasury bonds for 1959–2000, Chrétien finds that the left-hand side of equation (10.7) is close to −0.001, which bounds the autocorrelation of the SDF to be between −0.005 and zero. Consumption models with habits produce autocorrelations closer to −0.50, while many popular linear factor models for SDFs, aimed at pricing equities in the original studies, produce positive autocorrelations, so both types of model violate the autocorrelation bounds. This illustrates the difficulty of finding ms that can price both bond and stock returns.

10.5.6   It Takes a Model to Beat a Model Bounds

The original HJ bounds are for ms that price assets exactly; that is, all alphas are zero. But no model is perfect. Liu (2008) extends the HJ bounds to consider imperfect models, developing the "takes a model to beat a model" (TMBM) bounds. The idea is that we can use a model, even if it is imperfect, and that we would like to compare imperfect models. The TMBM bound is a characterization of models that can "beat" a given model in the sense of delivering smaller pricing errors. Let $\alpha = E(mR) - 1$. If the N-vector $\alpha = 0$, then the model works perfectly, and m plots inside the original HJ cup. Regress an m with pricing error α on R. Its fitted value $m^\#$ has the same α vector and can be represented as

$$m^\# = E(m) + \{[\alpha + 1 - E(m)E(R)]' V^{-1}[R - E(R)]\}, \quad (10.18)$$

where V is the covariance matrix of R. A benchmark model m_B has quadratic form in its alphas or pricing errors given by $\alpha_B' V^{-1} \alpha_B$. Models that beat m_B have $\alpha' V^{-1} \alpha \leq \alpha_B' V^{-1} \alpha_B$. Liu (2008) derives the TMBM bounds for model m_B:

$$\text{Min}_\alpha \, \text{Var}(m^\#) \text{ s.t. } \alpha' V^{-1} \alpha \leq \alpha_B' V^{-1} \alpha_B. \quad (10.19)$$

To beat the model m_B, alternative models must plot above the TMBM bound for m_B. Figure 10.4 demonstrates this idea. The (a) figure uses a treasury bill and twelve industry portfolios as the test assets, while the other two figures replace the industry portfolios with 25 size and book-to-market portfolios (b) or the 30 Dow Jones industrial stocks (c). Liu (2008) finds that a class of popular models, including the CAPM, linearized consumption CAPM, Campbell and Cochrane (1999), and the Fama and French (1996) and Carhart (1997) factor models all have TMBM bounds that appear very close together, compared

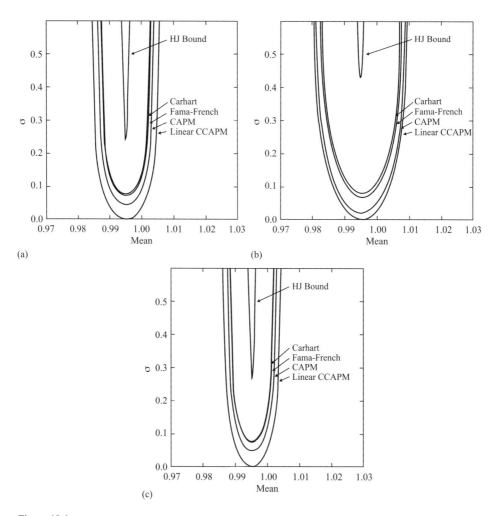

Figure 10.4
TMBM bounds.
Source: Liu (2008).

with their distance from the HJ bound for perfect models. This illustrates that asset pricing research has a long way to go.

10.5.7 More Moment Bounds

Several studies generalize the HJ bounds to consider moments beyond the mean and variance of the SDF. Bansal and Lehmann (1997) consider a generalization of the HJ bounds for the log of the SDF, which implicitly involves all of the moments. Snow (1991) develops bounds based on the Holder inequality:

$$E\{|mR|\} \leq E\{|m|^p\}^{1/p} \, E\{|R|^q\}^{1/q}, \tag{10.20}$$

where p, q > 1 and $1/p + 1/q = 1$. He applies this to the small-firm effect and finds that before 1975, small firms' stock returns had information in them about higher moments of the SDF that are not found in large-firm stock returns. In more recent data, however, this result weakens.

Balduzzi and Kallal (1997) refine the HJ bounds by placing restrictions on the risk premiums that an SDF implies for a nontraded factor. Starting with $E(r_i) = \mathrm{Cov}(r_i, -m/E(m))$, let $q = -m/E(m)$; then $\mathrm{Var}(q) = \mathrm{Var}(m)/E(m)^2$. Project q onto returns R, obtaining $q = q^* + \varepsilon$, with $E(\varepsilon R) = 0 = E(\varepsilon) = E(\varepsilon q^*)$. Expected excess returns are the same under q and q^*. This implies a bound:

$$\mathrm{Var}(q) = \mathrm{Var}(q^*) + \mathrm{Var}(\varepsilon) \leq \mathrm{Var}(q^*). \tag{10.21}$$

However, $E(\varepsilon f)$ may be nonzero for some risk factor f not in the traded returns, so a bound can be developed by exploiting the explanatory power of the nontraded factor. Let the premium on such a factor be $\lambda = \mathrm{Cov}(f, q)$, and $\lambda^* = \mathrm{Cov}(f, q^*)$, and express:

$$\lambda - \lambda^* = \mathrm{Cov}(f, \varepsilon) = \rho \sigma(f) \, \sigma(\varepsilon). \tag{10.22}$$

Here I have related the regression coefficient of ε on the nontraded factor to risk premiums and used the explanatory power of that regression to explain some of the $\mathrm{Var}(\varepsilon)$. Plug in for $\mathrm{Var}(\varepsilon)$ from this equation into equation (10.21) to obtain

$$\mathrm{Var}(q) \leq \mathrm{Var}(q^*) + (\lambda - \lambda^*)^2 / \sigma^2(f). \tag{10.23}$$

Kan and Zhou (2006) derive a multifactor bound that tightens the HJ bound, in terms of the relation between *m* and some empirical factors. Suppose that $m = m(f)$ is a function of some factors. Kan and Zhou show that

$$\mathrm{Var}(m(f)) \geq [1/\rho(f, m^*)^2] \mathrm{Var}(m^*), \tag{10.24}$$

where $\mathrm{Var}(m^*)$ is the usual HJ bound for a given $E(m)$. The term $\rho(f, m^*)^2$ is the R-squared from regressing m^* on f. Kan and Zhou find that when f is US consumption growth, $\rho(f, m^*)^2 < 0.008$, and the new bound is 128 times higher than the HJ bound. Using the Campbell and Cochrane (1999) surplus consumption and consumption growth as factors, $\rho(f, m^*)^2 < 0.15$, the new bound is 6 times higher than the classical HJ bound.

Cochrane and Saa-Requejo (2000) develop "good deal" bounds that admit models are imperfect. In particular, the problem they consider is to price some hard-to-value payoff x_c relative to the prices of other assets. Consider a set of basis asset returns R and the following problem:

$$\mathrm{Min}_{\{m\}} E(m x_c) \text{ subject to } E(mR) = 1, \; m > 0, \; \sigma(m) \leq h/R_F. \tag{10.25}$$

The problem is to find the *m* that gives the payoff x_c the smallest price while correctly pricing the underlying basis assets in R, satisfying positivity, and subject to a restriction

that looks like the HJ bound, with h as the maximum Sharpe ratio. The parameter h is specified exogenously. If the Sharpe ratio h is "too large," the model provides "good deals" that
are ruled out. They apply their bounds in an option pricing context.

The literature continues to supply new refinements and variations on the HJ bounds.
Bakshi and Chabi-Yo (2012) develop bounds that decompose the SDF into "permanent"
and temporary components, as proposed by Alvarez and Jermann (2005). Backus, Chernov,
and Martin (2011) develop bounds in terms of the entropy of the SDF. Almeida and Garcia
(2016) derive nonparametric bounds that generalize variance, entropy, and higher moments.
This is starting to get pretty abstract! In chapter 11, we describe the role of lagged conditioning information in HJ-style bounds.

11 Variance Bounds with Conditioning Information

This chapter further studies the HJ type bounds on stochastic discount factors when we measure and use lagged conditioning information Z. Now we want models to price with $E(mR|Z) = 1$, not just $E(mR) = 1$. This is a tougher requirement that expands the set of returns that we ask the model to price, now including dynamic trading strategies that may depend on Z. Examples in the literature differ according to the set of trading strategies priced. This is analogous to how chapter 9 refined the classical mean variance analysis of chapter 8. The classical mean variance bullet has a beautiful dual relation with the classical HJ cup, and most of the cups with conditioning information have dual mean variance bullets. Closed-form solutions can be derived using perturbation arguments. The latter parts of this chapter spend some time on issues that arise in empirical applications, including asymptotic standard errors for the bounds, finite-sample bias adjustment, and the finite-sample properties of the bounds and their standard errors.

11.1 Refining HJ Bounds with Conditioning Information

The question being asked in this chapter is a refinement of the question asked by the original HJ bounds. That question was: Given a set of asset returns R_{t+1}, what can we say about the moments of an m_{t+1} that prices those assets through equation (1.1)? Now we consider a set of assets and a given set of observable lagged conditioning variables Z_t, and we ask what these imply about the moments of an m_{t+1} that "conditionally prices" the assets through the equation

$$E\{m_{t+1}R_{t+1}|Z_t\} = 1. \tag{11.1}$$

11.1.1 The General Idea

The way that conditioning information refines the HJ bounds is most easily seen if we work with excess returns:

$$E\{m_{t+1}r_{i,t+1}|Z_t\} = 0, \; r_i = R_i - R_f, \tag{11.2}$$

where this should hold for all assets i and for all realizations of Z_t. The latter statement is true if and only if

$$\mathrm{E}\{[m_{t+1}\mathrm{r}_{i,t+1}]\mathrm{f}(Z_t)\} = 0, \tag{11.3}$$

for all i, and for all functions $\mathrm{f}(Z_t)$, such that the unconditional expectation exists. Ingersoll (1987) provides a proof of this equivalence. It is easy to see how the first expression leads to the second, as it implies that $\mathrm{E}\{m_{t+1}\mathrm{r}_{i,t+1}\mathrm{f}(Z_t)|Z_t\} = 0$ almost surely in Z_t. Now take the unconditional expectation. Conversely, if there were some $\mathrm{f}(Z)$ for which equation (11.3) did not hold, then the expected values of the product $m_{t+1}\mathrm{r}_{i,t+1}$ would be nonzero for some values of Z_t.

Equation (11.3) has a nice interpretation. Think of $[\mathrm{r}_{i,t+1}\mathrm{f}(Z_t)]$ as the excess return of a *dynamic trading strategy*. This strategy is a very simple one. It scales the long and the short sides of the excess return r_{it+1} according to the realized value of $\mathrm{f}(Z_t)$. Perhaps it should be called a dynamic "leverage" strategy, if returns are measured in excess of a short-term interest rate. This was one of the great insights of Hansen and Jagannathan (1991). Note that, for a given $\mathrm{f}(\cdot)$, we end up with a version of the pricing equation that is based on unconditional expectations. Thus, we can use the same methods discussed in chapter 10 on HJ bounds, where we ignored the conditioning information. But now we have an expanded set of excess returns to price.

We bring conditioning information to the HJ bounds by asking the model to price not just the basic asset returns but also the dynamic trading strategies that trade based on Z_t. Just as for the original HJ bounds, when we ask the model to price more assets, fewer ms can do the job. The cup with more assets becomes taller and skinnier. The main point is that the larger the set of returns or dynamic trading strategies is, the tighter the HJ bounds will be. Hansen and Jagannathan (1991) used $\mathrm{f}(Z) = I \otimes Z$, resulting in a *multiplicative approach* to the conditioning information. However, other functions can obviously be used, and this is how the literature has progressed. If we could figure out how to use all functions $\mathrm{f}(\cdot)$, we would challenge the model to the maximal extent, and it would be equivalent to testing the expression in equation (11.2).

11.1.2   Flavors of the Bounds with Conditioning Information

Studies have advanced the cause, building HJ cups with different functions $\mathrm{f}(\cdot)$. This is a relatively simple way to provide a taxonomy of the main approaches to date. The more general the set of functions $\mathrm{f}(\cdot)$ is, the larger the set of dynamic trading strategies and the tighter the bounds will be. There has been a bit of an arms race. The original Hansen and Jagannathan (1991) paper started with $\mathrm{f}(Z) = I \otimes Z$, resulting in what Andy Siegel and I call "multiplicative bounds." Ferson and Siegel (2003) imposed the pricing restriction for $[\mathrm{r}'_{t+1}\mathrm{f}(Z_t)]$, all $\mathrm{f}(Z)'\underline{1} = 1$. This says that the functions $\mathrm{f}(\cdot)$ describe portfolio weights that sum to one (almost surely in Z) but can otherwise depend almost arbitrarily on the information Z.

This results in the *efficient-portfolio (UE) bounds.* Bekaert and Liu (2004) did not require $f(Z)'\underline{1} = 1$ and called the resulting cup the "optimally scaled" bounds. The bounds are optimally scaled because $f(\cdot)$ is a vector-valued function with the same number or elements as r, so each return can be scaled, but the weights don't have to add up to 1.0. Finally, the winner of the arms race is Gallant, Hansen, and Tauchen (1990), who derive a cup that asks a model to price the payoffs $[r_{t+1} xf(Z_t)]$ for all possible $f(Z)$. Let's call the winner the "optimal bounds," because you just can't make a cup that is any taller or skinnier for a given set of returns and conditioning information.

11.1.3   HJ Cups and Their Dual Mean Variance Bullets

Just as the classical mean variance analysis has a beautiful dual relation with the classical HJ cup, most of the fancier cups with conditioning information have a dual mean variance boundary. The first example is pretty obvious. If we interpret all the means and covariances as conditional on the lagged information Z, then the conditional minimum variance boundary and a conditional HJ cup are duals in virtually the same sense as described in chapter 10. All we have to do is relabel the axis to read $\mu(R|Z)$ and $\sigma(R|Z)$ for the bullet, and to read $E(m|Z)$ and $\sigma(m|Z)$ for the cup. For the other cases, things get only a little more involved.

In chapter 9, we discussed unconditionally efficient portfolios with respect to conditioning information, as derived by Ferson and Siegel (2001). Given how research programs evolve, it is perhaps not surprising to find that these are dual to the efficient portfolio bounds studied by Ferson and Siegel (2003). The relation between the two is the same as the relation between the classical HJ bounds and the classical mean variance frontier with no conditioning information. All moments on the graphs are unconditional moments; $\mu(R)$ and $\sigma(R)$ for the bullet, and $E(m)$ and $\sigma(m)$ for the cup. The only difference is that the set of returns is expanded beyond the basic returns in R to include the dynamic portfolio strategies: all portfolios $x(Z)'R$ with $x(Z)'\underline{1} = 1$.

This theme continues with the other flavors of the bounds. In each case, we have the unconditional means and variances on the axes of the cup, but we put different payoffs into the bullet and raise the cup. In the Bekaert and Liu (2004) bounds, the set of payoffs that we ask the model to price includes all portfolios $x(Z)'R$, where the weights $x(Z)$ don't have to add up to 1.0. This relaxes the constraint on the UE bounds that the weights always add up to 1.0. We can normalize the weights in the Bekaert and Liu case so that they add up to 1.0 on average. Divide equation (11.3) by $E(x(Z)'\underline{1})$, and you can see that the weights sum to 1.0 in expectation. In the Gallant, Hansen, and Tauchen (1990) optimal bounds, the payoffs to be priced are all $[r_{t+1} xf(Z_t)]$ for all functions $f(\cdot)$. Abhay, Basu, and Stremme (2002) and Penaranda and Sentana (2016) provide analyses of the duality between mean-variance bounds and stochastic discount factor bounds with information. Chabi-Yo (2007) develops HJ cups for pricing kernels with higher-order moments.

11.2   Empirical Application

This section outlines some of the issues with empirical application of HJ bounds with conditioning information. Some of the basics were described in chapter 10, but there are some new subtleties. First, it is hard to compute the HJ bounds with a regression of m_{t+1} on $\underline{R}_{t+1}$ when the returns vector is infinite dimensional. Estimating an expression like $[\underline{1}' - E(m)E(R')]V(R)^{-1}[\underline{1}' - E(m)E(R')]'$ looks challenging when R is an infinite-dimensional vector of dynamic-strategy returns. This section describes how these problems are finessed, discusses standard errors for the bounds with conditioning information, and adjustments for finite-sample biases. Some empirical evidence on the sampling properties of the bounds is then reviewed.

11.2.1   Closed-Form Solutions

When there is a possibly infinite-dimensional space of strategy returns, the key to estimating HJ bounds is to find closed-form solutions for the projection of m onto asset returns that defines the boundaries of the relevant HJ cup. It is helpful to recall the case with no conditioning information. In this case, Hansen and Jagannathan (1991) show that the HJ bounds are obtained by solving $\text{Min}_m \text{Var}(m)$ s.t. $E(mR) = \underline{1}$, $E(m)$ given. The solution is

$$m^* = E(m) + [\underline{1} - E(m)\mu]'\Sigma^{-1}[R - \mu]. \tag{11.4}$$

The solution m^* is a linear function of a portfolio return; in particular, the maximum correlation portfolio to m. The sample variance of m^* defines a point on the sample cup for each E(m) on the x-axis of the graph. Now suppose that we did all this for conditional moments given Z. We would then have the *conditional HJ bounds*:

$\text{Min}_m \text{Var}(m|Z)$ s.t. $E(mR|Z) = \underline{1}$, $E(m|Z)$ given, with solution
$$m^\# = E(m|Z) + [\underline{1} - E(m|Z)\mu(Z)]'\Sigma(Z)^{-1}[R - \mu(Z)], \tag{11.5}$$

where $\mu(Z) = E(R | Z)$, $\Sigma(Z) = \text{Var}(R | Z)$. $\text{Var}(m^\#|Z)$ establishes the minimum conditional variance of the SDF for a given E(m|Z). This version of the conditional HJ bound was the main subject of the Gallant, Hansen, and Tauchen (1990) paper, which spent a lot of effort estimating conditional moments. The optimal bound mentioned above only appears briefly in the paper, sketched out in a paragraph. It just goes to show that you can never really predict what part of your paper will turn out to be the most important, once hindsight has been applied!

Why belabor the solution to the conditional cup problem? Well, it goes back to the dualities described above, combined with the Hansen and Richard (1987) result on conditional and unconditional efficiency with respect to Z. There we saw that a UE(Z) portfolio is a special case of a CE(Z) portfolio. The duality suggests that a cup like the efficient portfolio bound, since it is based on UE(Z) portfolios, should have a solution that is a special case of the conditional cup above, based on CE(Z) portfolios. It turns out that is precisely the case. This result is very helpful, because the conditional cup problem above requires only the N-vector of conditional means and a model for the $N \times N$ conditional covariance matrix. That seems a lot easier than dealing with infinite-dimensional vectors of dynamic strategy returns. Let's flesh this idea out and derive expressions for HJ cups with conditioning information that can be applied to data. We start with the most general version of the problem, which corresponds to the optimal bounds of Gallant, Hansen, and Tauchen (1990).

The problem that we want to solve is

$$\text{Min}_m \text{ Var}(m) \text{ s.t. } E(mR|Z) = \underline{1}, \text{ given } E(m). \tag{11.6}$$

This roughly says that we want to describe a cup with unconditional means and variance of m on the axes, but we want to impose that m prices the returns R conditionally, given Z. Writing the unconditional variance as the sum of the expected conditional variance and the variance of the conditional mean, we can rewrite the problem as

$$\text{Min}_m E\{\text{Var}(m|Z)\} + \text{Var}\{E(m|Z)\} \text{ s.t. } E(mR|Z) = \underline{1}. \tag{11.7}$$

Note that if $E(m|Z)$ is set to some function $g(Z)$, then the conditional cup $m^{\#}$ in equation (11.5) solves this problem for that $E(m|Z)$. This is because if the solution minimizes $\text{Var}(m|Z)$, as does the $m^{\#}$ in (11.5), then it must also minimize $E\{\text{Var}(m|Z)\}$. (Think by contradiction: If the expectation were not minimized, there must be some values of Z for which the variance could be made smaller.)

The upshot is that we can recast the problem as that of finding an "optimal" conditional mean function $g(Z) = E(m|Z)$ to use in

$$m^{\#} = g(|Z) + [\underline{1} - g(|Z)\mu(Z)]' \Sigma(Z)^{-1}[R - \mu(Z)]. \tag{11.8}$$

Note that the $m^{\#}$ in equation (11.8) satisfies the constraint that $E(m^{\#}R|Z) = 1$, all Z, for any function $g(Z) = E(m|Z)$, so we don't need to impose that constraint explicitly again. A perturbation argument from the calculus of variations can be used to derive the optimal conditional mean function $g(Z)$. The approach is similar to the one we used to solve for efficient (UE) portfolios with respect to Z before. The problem is

$$\text{Min}_{g(Z)} E\{\text{Var}(m^{\#}|Z)\} + \text{Var}\{g(Z)\}, \text{ subject to } E(g(Z)) = E(m), \text{ given } E(m). \tag{11.9}$$

Define: $\beta(Z) = \underline{1}' \Sigma(Z)^{-1}\mu(Z)$, and $\gamma(Z) = \mu(Z)' \Sigma(Z)^{-1}\mu(Z)$. The optimal solution is found by Ferson and Siegel (2003) and is given by the expression

$$g(Z) = \beta(Z)/\{1 + \gamma(Z)\} + \{1 + \gamma(Z)\}^{-1} [E(m) - E(\beta(Z)/\{1 + \gamma(Z)\})]/E(1/\{1 + \gamma(Z)\}). \tag{11.10}$$

Note that we have not yet restricted in any way the set of payoffs that are priced, requiring the conditional pricing expression $E(mR|Z) = 1$ to hold. As we saw above, this corresponds to requiring all possible dynamic strategies that may depend on Z to be priced. Thus, plugging $g(Z)$ back into $m^{\#}$ gives the "optimal" bounds of Gallant, Hansen, and Tauchen (1990).

Further restrictions reduce the solution to special cases, including the Bekaert and Liu (2004) "optimally scaled" bounds and the Ferson and Siegel (2003) efficient portfolio bounds. (The algebra, however, gets messy.) To obtain the efficient portfolio bounds, we can use the duality of the cup to the bullet, together with the closed-form solutions for the UE portfolio weights from chapter 9. Consider the solution presented in theorem 9.2 for the case with no risk-free asset. The portfolio weight of the UE portfolio is stated in terms of a notional zero beta rate γ_0. This portfolio maximizes the squared Sharpe ratio at a point on the bullet drawn from the zero-beta rate to the boundary. By varying the zero-beta rate, we trace out the UE bullet. Call the squared Sharpe ratio on the UE boundary $SR_{ue}^2(\gamma_0)$. The efficient portfolio bounds can be drawn by mapping a point on the bullet to a point on the cup: $\{E(m) = 1/\gamma_0,\ \sigma(m) = [E(m)^2 SR_{ue}^2(\gamma_0)]^{.5}\}$.

11.2.2   Finite-Sample Bias Adjustment

It turns out that applying the expressions from section 11.2.1 to data results in estimates of the HJ bounds that are upwardly biased in finite samples. The bias is similar to that of a sample maximum Sharpe ratio, which is also upwardly biased in finite samples (Jobson and Korkie 1980). Since we use the Sharpe ratio for so many things in asset pricing, it is useful to have an unbiased estimate for it. By the duality between bullets and cups, a sample Sharpe ratio that is too high means that a sample cup will look taller and skinnier than the true cup based on the population moments.

Figure 11.1 presents an example of HJ bounds based on the quarterly dataset from Ferson and Siegel (2003), which is very similar to the example in Hansen and Jagannathan (1991).

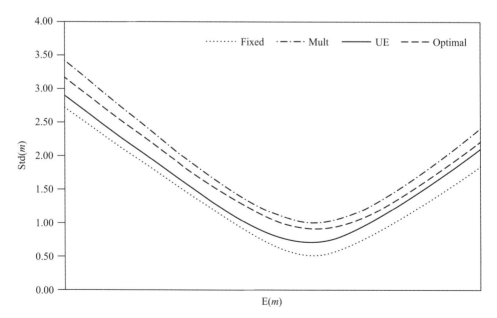

Figure 11.1
HJ bounds with conditioning information.

It covers 1963–1987 (T = 93 quarters) and uses $N = 5$ test assets, which are US Treasury bills that start the quarter with 3, 6, 9, or 12 months to maturity. The conditioning information Z is the first lag of the four returns and the growth rate of aggregate per capita consumption.

The figure shows that something is wrong with the finite-sample estimates. The optimal bound should be the highest, but it is not. The figure also foreshadows the finding that the finite-sample bias in the bounds is worst for the multiplicative bounds.

To understand the finite-sample bias, it helps to start with the classical HJ cup with no conditioning information. Let $\bar{R}$ be the sample mean of R, and let s denote the sample estimate of the covariance matrix (dividing by T, as in maximum likelihood). The sample estimate of the HJ boundary is

$$s_m^2 = [\underline{1} - \mathrm{E}(m)\bar{R}]'\mathrm{S}^{-1}[\underline{1} - \mathrm{E}(m)\bar{R}]. \tag{11.11}$$

The estimator is biased in finite samples, and for a given E(m), it is related to the bias in a sample maximum squared Sharpe ratio, $\bar{R}'\mathrm{S}^{-1}\bar{R}$. The squared Sharpe ratio is studied assuming normally distributed returns by Jobson and Korkie (1980), who derive the bias (see also Shanken (1982, 1987) for generalizations). Under normality, the bias is easy to represent as shown by Ferson and Siegel (2003) in their proposition 4:

$$\mathrm{E}(s_m^2) = \sigma_m^2[T/(T-n-2)] + \mathrm{E}(m)^2[n/(T-n-2)]. \tag{11.12}$$

This leads to a simple adjustment. Just solve expression (11.12) for something whose expected value is the true variance:

$$\sigma_{m,\text{adjusted}}^2 = [(T-N-2)/(T)]\, s_m^2 - \mathrm{E}(m)^2\,(N/T). \tag{11.13}$$

The adjustment only requires E(m), the number of time series T and the number of assets N. Adjusting for finite-sample bias, the adjusted SDF variance estimate reduces to s_m^2 when T gets large. The adjustment displays the importance of the number of assets N relative to the number of time series. Note how simple the adjustment for finite-sample bias is to use. Imagine that your code produced a value for s_m^2, the sample estimate of the HJ bound for a given E(m). All you would have to do is add one line in your code, given by expression (11.13), to make the finite-sample adjustment.

The finite-sample adjustment may be applied directly to the multiplicative bounds with conditioning information, using the products of the returns and the lagged instruments as the test assets. The other bounds with conditioning information require a bit more work.

Recall that all bounds with conditioning information are defined by a version of $\mathrm{Var}(m^{\#})$, where

$$m^{\#} = \mathrm{E}(m\,|Z) + [\underline{1} - \mathrm{E}(m\,|\,Z)\mu(Z)]'\,\Sigma(Z)^{-1}[R - \mu(Z)], \tag{11.14}$$

and where E(m|Z) depends on which bound we are talking about. From this expression, the variance is

$$\text{Var}(m^{\#}) = \text{Var}\{E(m|Z)\} + E\{[\underline{1} - E(m|Z)\mu(Z)]' \Sigma(Z)^{-1}[\underline{1} - E(m|Z)\mu(Z)]\}, \tag{11.15}$$

because the two additive terms in equation (11.14) are uncorrelated. Substituting the ML estimates of $\mu(Z)$ and $\Sigma(Z)$, call them $\hat{u}(Z)$ and $S(Z)$, gives the estimated bound. Ferson and Siegel's (2003) proposition 5 provides a bias adjustment for this case:

$$\text{Adj Var}(m^{\#}) = (T-N-2)/Ts^2(m^{\#}) - (N/T)E(m^{\#}) + (2/T)\text{Var}\{E(m|Z)\}. \tag{11.16}$$

The term $(2/T)\text{Var}\{E(m|Z)\}$ depends on the bound (e.g., optimal versus efficient portfolio bound) but is relatively small and can often be ignored in practice. (Think of a Treasury bill rate, whose variance is small relative to stock returns.) Where this term is small, the same adjustment can be used for all bounds with conditioning information.

11.2.3   Asymptotic Standard Errors

If you want to take inference using HJ bounds seriously, you need some standard errors. Actually, you need a standard error for $s(m)$, a variance for the estimator of a variance! Sounds daunting, but it's not too bad. The first pass at this problem comes from Hansen, Heaton, and Luttmer (1995) for the case with no conditioning information. They very cleverly define

$$\Phi_t = -[(\underline{1} - E(m)\mu)' \Sigma^{-1}(R_t - \mu)]^2 - 2(\underline{1} - E(m)\mu)' \Sigma^{-1}[E(m)R_t - \underline{1}], \tag{11.17}$$

where consistent (e.g., ML) estimates are plugged in for Σ and μ. The estimate of the classical HJ variance bound at the point $E(m)$ is $s(m) = \Sigma_t \Phi_t / T$.

   We can estimate the variance of the mean consistently, plugging in consistent estimators for the mean and covariance matrix in equation (11.17), such as the ML estimates. This gives us standard errors for the fixed-weight bounds, and if applied to the expanded set of payoffs, for the multiplicative bounds as well. We can also get standard errors using the GMM, as described in chapter 18. Kan and Robotti (2008) provide distribution theory when the SDF is constrained to be positive, in the case of no conditioning information.

   Since $s(m) = \Sigma_t \Phi_t / T$, the variance of the estimate $s(m)$ is the variance of $\Sigma_t \Phi_t / T$. The variance of $\Sigma_t \Phi_t / T = (1/T)^2 \text{Var}[\Sigma_t \Phi_t]$ depends on the autocovariances of Φ_t. For example, if the autocovariances are zero, then the variance of the HJ bound simplifies to $(1/T)$ times the sample variance of the Φ_t. We discuss how to deal with these autocovariances in chapter 17.

11.3   Sampling Performance

This section summarizes some simulation evidence on HJ bounds with conditioning information. Many of the results are from Ferson and Siegel (2003). The data in that study consist of a monthly, a quarterly, and an annual dataset. The quarterly data on Treasury bill returns are similar to that in Hansen and Jagannathan (1991) and have already been described. The monthly data consist of 25 industry portfolio returns with a lagged dividend yield for the

S&P 500 and a lagged Treasury bill yield as Z, 1963–1994. The annual data consist of the stock market return and a Treasury bill return, with the lagged returns as the instruments, 1891–1985. This provides a range of values for the number of time series observations T (93–393), the number of assets N (2–25), and the number of lagged instruments L (3–6).

11.3.1 Parametric Bootstrap Resampling

The sampling properties of the various bounds are evaluated by simulation in Ferson and Siegel (2003). They use a version of the parametric bootstrap that is described in more detail in chapter 24. The first step is to model the conditional mean vector and covariance matrix of the returns, R. The conditional mean is modeled by regressing the returns on the lagged information variables in the actual data:

$$R_t = a + \delta Z_{t-1} + U_{rt}. \tag{11.18}$$

Since the lagged information variables can be highly persistent, it is also useful to preserve their persistence in the simulations. This is done by modeling the lagged information as a vector first-order autoregression:

$$Z_{t-1} = A Z_{t-2} + U_{zt}. \tag{11.19}$$

The regression model parameters $\{a, \delta, A\}$ are retained and used for the data-generating process of the simulations. This is what makes the approach parametric. Let $X_t = (U_{rt}, U_{zt})$ be the fitted values of the residuals from the return regression and the lagged instrument regression. Here is where the bootstrapping comes in. Artificial samples $\{R_t, Z_{t-1}\}$ are generated by randomly resampling from the rows of the residual matrix X and building up the samples recursively by using the fitted regression equations.

Note that by keeping together the N and L vectors, U_r and U_z, and drawing the dates randomly, the bootstrap method draws from an "urn" that preserves the dependence between the elements of the vectors but destroys any time-series dependence. Thus, the sample covariance matrix of the N returns is the population matrix for the data generation. Similarly, the covariance matrix of the Z values is preserved, as is dependence between the returns and the lagged instruments, such as conditional heteroscedasticity. However, any serial dependence in the residuals is lost.

Many artificial samples are produced, each one with the same T, N, and L values as in the original sample, and the various HJ-style bounds are estimated using each sample. The result is a sampling distribution of the bounds for a given finite sample size.

11.3.2 Finite-Sample Performance

Figure 11.1 shows sample estimates of the various bounds using the quarterly dataset. No adjustment for finite-sample bias has been applied. This figure illustrates that the HJ bounds have a serious finite-sample bias. Figure 11.2 shows how the finite-sample bias adjustment changes things. At least now the optimal bounds curve is the highest, as it should be.

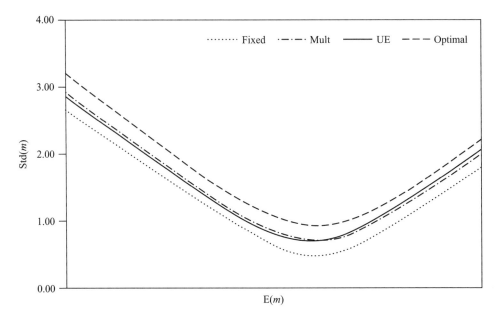

Figure 11.2
HJ bounds with bias adjustment.

The sampling distributions of the bounds with and without finite-sample adjustments is examined in Ferson and Siegel (2003). They find that the simple fixed-weight bounds (no conditioning information) work pretty well in the quarterly dataset, even without any bias adjustment, in the sense that the expected location of the cup is just about right. The multiplicative bounds have a large upward bias in the quarterly dataset. About 95% of the time, the sample bounds will be above the true bound in the multiplicative case. Adjustment for bias produces a substantial improvement in the locations of the bounds.

Table 11.1 summarizes results from Ferson and Siegel (2003), comparing the simulation results for fixed weight (no conditioning information), multiplicative, efficient portfolio (UE), and optimal bounds. The true and mean values of the bounds are given in the first two data columns of the table. The true bounds are found by letting T be 1 million. The mean values are the average finite sample values in the simulation. All values in the table are summarized at the point $E(m) = 1$. The true value of the optimal bounds is the largest. The multiplicative and UE bounds tussle to be the second tallest cups, depending on the dataset.

Comparing the mean value from the simulation (column 2) with the true bounds (column 1) reveals the expected finite-sample bias in the bounds. All bounds are upwardly biased, and so they rule out more models than they should. The third data column of the table lists the standard deviation across the 5,000 simulation trials. It shows that the optimal and multiplicative bounds tend to have the most sampling variability.

The fourth and fifth columns show the mean and standard deviation of the bounds after the finite-sample bias adjustment is applied. In most cases, the adjusted means are much closer to the true values, indicating that the adjustment does a good job placing the cups in their proper locations.

Finally, the right-hand column of the table summarizes the bite of the various bounds, showing how low a two-standard error confidence band on the variance bounds extends. If it extends to zero, there is no restriction on asset pricing models, and the bound would be useless in practical terms as a diagnostic tool. This is the case for the fixed and multiplicative bounds in the annual dataset. The UE and optimal bounds restrict the volatility of the SDF to be about 1% per year or larger—not a very strong restriction. The quarterly and monthly datasets are much more restrictive on SDF variances, and the table shows the substantial impact of using conditioning information in the bounds.

11.3.3 Performance of Asymptotic Standard Errors for HJ Bounds

The asymptotic standard errors for HJ bounds, derived by Hansen, Heaton, and Luttmer (1995) are evaluated by simulation in Ferson and Siegel (2003). The results are summarized here

Table 11.1
Precision of the bounds.

Type of bound		Simulated:					
		Simulated true	Simulated mean	Standard deviation	Adjusted mean	Adjusted standard deviation	$\mu - 2\sigma$
Panel A: Annual data							
fixed	bound	0.197	0.223	0.103	0.194	0.099	<0
mult	bound	0.211	0.273	0.121	0.190	0.112	<0
UE	bound	0.203	0.248	0.108	0.216	0.104	0.008
optimal	bound	0.212	0.265	0.112	0.232	0.110	0.010
Panel B: Quarterly data							
fixed	bound	0.488	0.561	0.193	0.488	0.182	0.124
mult	bound	0.914	1.564	0.414	0.992	0.319	0.354
UE	bound	0.915	1.167	0.306	1.051	0.287	0.477
optimal	bound	1.144	1.509	0.336	1.369	0.315	0.739
Panel C: Monthly data							
fixed	bound	0.104	0.200	0.058	0.121	0.054	0.013
mult	bound	0.313	0.626	0.114	0.305	0.092	0.121
UE	bound	0.329	0.523	0.096	0.421	0.089	0.243
optimal	bound	0.386	0.615	0.115	0.506	0.107	0.292

Source: Expanded version of table 3 in Ferson and Siegel (2003). Copyright Society for Financial Studies.
Note: Mult, multiplicative; UE, unconditionally efficient.

in table 11.2. The empirical standard error (data column 1 in the table) is the standard deviation of the estimated standard deviations for the HJ bound, taken across the 5,000 simulation trials. The column labeled "Adjusted empirical" applies the finite-sample adjustment to the HJ bound estimators. Think of the first two columns as what you want the asymptotic standard deviations to predict. The two right-hand columns are the average results across the 5,000 trials of estimating the asymptotic standard error, with and without bias adjustments in the estimator for $\sigma(m)$. This shows how accurate the standard error estimators are expected to be. The simulations show that the asymptotic standard error is a pretty good proxy for the actual sampling variability, slightly underestimating the variability in quarterly and monthly data. The adjusted asymptotic is expected to be an even better proxy for the sampling variation in the adjusted estimator, with a slightly smaller downward bias.

11.4 Practical Summary

Let's summarize the horse race of the different version of the HJ bounds with conditioning information, in terms of their overall empirical attractiveness. We have discussed four versions.

The multiplicative approach in the original Hansen and Jagannathan (1991) paper is certainly easiest to use. Sometimes the simulation evidence shows that it is not far from

Table 11.2
Evaluation of asymptotic standard errors for HJ bounds.

Bound	Empirical standard error	Adjusted empirical standard error	Average asymptotic standard error	Adjusted asymptotic standard error
Panel A: Annual data				
Normal	0.103	0.099	0.100	0.097
Non-normal, homoskedastic	0.108	0.105	0.102	0.099
Non-normal, heteroskedastic	0.112	0.108	0.105	0.101
Panel B: Quarterly data				
Normal	0.193	0.182	0.174	0.165
Non-normal, homoskedastic	0.166	0.157	0.154	0.145
Non-normal, heteroskedastic	0.172	0.163	0.157	0.149
Panel C: Monthly data				
Normal	0.058	0.054	0.048	0.044
Non-normal, homoskedastic	0.061	0.057	0.050	0.047
Non-normal, heteroskedastic	0.057	0.053	0.032	0.030

Source: Table 5 in Ferson and Siegel (2003). Copyright Society for Financial Studies.
Note: The values in the final column were calculated using proposition 5 from Ferson and Siegel (2003).

optimal. However, it can be terribly biased in finite samples and is often the most biased of the three bounds we have discussed. Finite-sample bias adjustment helps. However, the bounds have relatively poor precision, and they are the least restrictive of SDFs, accounting for sampling error.

The optimal bounds of Gallant, Hansen, and Tauchen (1990) have the advantage that they are optimal! However, they are harder to use, requiring the conditional mean vector and variance matrix of returns, $\mu(Z)$ and $\Sigma(Z)$. The bounds are also badly biased in finite samples. However, accounting for their precision, they are the most restrictive of SDFs. One disadvantage of these bounds is that they are not valid if you get $\mu(Z)$ or $\Sigma(Z)$ wrong.

The efficient portfolio bounds of Ferson and Siegel (2003) are about as difficult as the optimal bounds to implement, requiring conditional means and variances of returns to be modeled. However, unlike the optimal bounds, they are robust to the misspecification of those moments. This is because they are formed from estimated dynamic strategy weights. If you get $\mu(Z)$ or $\Sigma(Z)$ wrong, you don't have the optimal weights, but you still have a valid dynamic strategy. Their precision is better than that of the multiplicative bounds, and they are more restrictive of SDFs than the multiplicative bounds. Of all the bounds considered in Ferson and Siegel (2003), they produce the smallest sampling bias before adjustment.

Finally, the Bekaert and Liu (2004) bounds are similar in difficulty to implement as the optimal or efficient portfolio bounds. Simulations in that paper show that they are often near optimal. Like the efficient portfolio bounds, they are robust to getting $\mu(Z)$ and $\Sigma(Z)$ wrong.

III MULTIBETA PRICING

12 Arbitrage Pricing and Factor Analysis

Some doctoral programs no longer educate their students very well about the arbitrage pricing theory (APT), associated mainly with Ross (1976). If you are in such a program, perhaps you can skip this chapter, but I would not advise it. While APT papers do pop up from time to time, it is true that the literature has moved beyond this pure frictionless model to consider the costs that prevent arbitrage relations from holding exactly in real world data. (This work is reviewed in section 12.2 of this chapter. Maybe you should at least read that section.) However, I believe that an understanding of the basic ideas of the APT is a good part of one's liberal education in asset pricing. From a methodology standpoint, factor models are used in several areas of empirical research, where it is useful to reduce datasets of high dimensionality. As bigger datasets become increasingly the norm, these models are likely to remain relevant. Finally, some of the concepts introduced here, like rotational indeterminacy, are important for empirical work on factor models in general. It might be a good idea to review appendix section A.1 on elements of linear algebra before reading this chapter.

12.1 Overview

12.1.1 MBA-Level Intuition for APT

This intuition is found in various MBA-level textbooks. It may not be entirely rigorous, but it illustrates the basic ideas. Since MBAs can be pretty good at finding arbitrage opportunities when they exist, this approach can be effective. Consider one de-meaned factor, $f \equiv F - E(F)$, and regressions of asset returns on the factor:

$$R_i = E(R_i) + b_i f + u_i. \tag{12.1}$$

In an MBA curriculum, you would have to spend some time discussing the difference between a common factor f and the idiosyncratic residual u_i and give lots of examples. The important thing is that the residual variance can be diversified away in a large, well-diversified portfolio. For example, consider an equally-weighted portfolio with weights $x_i = 1/n$ in each of the n securities (this is an example of a well-diversified portfolio as defined more generally below), and assume that the residuals in equation (12.1) are independent across the securities. Then the portfolio residual variance is $\mathrm{Var}(\sum_i x_i u_i) = (1/n)^2 \mathrm{Var}(\sum_i u_i) \leq (1/n)\mathrm{Max}_i \mathrm{Var}(u_i) \to 0$ as n gets large, assuming that the residual variance is bounded as n gets large. (This is a special case of the bounded eigenvalue restriction in the large-markets APT, described below.) While this is a central intuition, you would not dare present these equations in an MBA class. Instead, try illustrating the idea with a spreadsheet.

The assumption in the intuitive development of the APT is usually that, since the residual variance can be eliminated in a large diversified portfolio, it doesn't matter, so we can just ignore it for the determination of expected returns. That is typically acceptable to MBA students, although it turns out to be the trickiest part of the argument in the academic literature on APT.

The next step is to graph $E(R_i)$ versus b_i and argue, graphically, that since portfolios allow betas and expected returns to combine linearly, then there would be arbitrage forces unleashed if $E(R_i)$ were not linear in b_i.

As figure 12.1 illustrates, the assets present an arbitrage opportunity, because you can combine the risk-free rate and asset 2 in a portfolio, achieve the same factor beta as asset 1, but get a higher return. Since we have assumed that the beta is the only relevant risk, traders will spring into action. As traders sell asset 1 and buy asset 2 and the bond, prices would have to adjust until asset 1 is on the same line as asset 2. The equilibrium condition that all assets have to plot on the same line can be represented as a "pricing" equation (OK, expected return equation), where the slope of the line is the risk premium λ for the risk factor f: $E(r_i) = b_i \lambda$. We have a most primitive form of the APT.

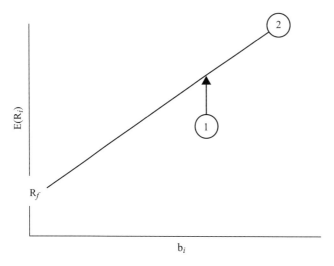

Figure 12.1
MBA-level argument for APT.

12.1.2 The Approximate Linear Algebra

Ross (1977), in a famous book chapter, developed the APT with an argument similar to the one above but using more rigorous linear algebra. Here is a summary of the argument. Let β be the $n \times K$ matrix of n assets' regression betas on K factors, let $\underline{1}$ be an n-vector of ones and μ an n-vector of expected returns. Consider a portfolio with (1) no net investment and (2) no (factor) risk. Assert that this implies (3) zero expected return. The conclusion is that there must exist constants γ_0, $\gamma_{(K \times 1)}$ such that $\mu = \underline{1}\gamma_0 + \beta\gamma$, which is a K-factor version of the previous APT expression.

The linear algebra goes like this. For any portfolio weight $x \in R^n$, (1) states that $x'\underline{1} = 0$, or zero net investment. Ross (1977) calls such portfolios "arbitrage portfolios." Harrison and Kreps (1979) call them "zero-net-investment-portfolios," or ZENIPs. Hansen and Richard (1987) call them "zero-price assets." Cochrane (2005) calls them "excess returns." In modern portfolio management, they are called "tilts." Whatever you call these portfolios, condition (2) states that $x'\beta = 0$, the portfolio has no factor risk. (If you can't find a portfolio weight x that satisfies the conditions (1) and (2), Ross suggests that you can simplify the factor model by combining the factors linearly and work with fewer factors.) The proof then insists that this implies (3) $x'\mu = 0$; the expected return must be zero. Let $((1, \beta))$ denote the span (the set of all linear combinations) of these columns. Write $R^n = ((1, \beta)) + ((1, \beta))^{\perp}$, where $^{\perp}$ denotes the orthogonal complement. Note that conditions (1) and (2) say x is orthogonal to $((1, \beta))$. Thus, $x \in ((1, \beta))^{\perp}$. Condition (3) states that μ is orthogonal to x, so μ must be in $((1, \beta))$; that is, there exist constants γ_0, γ such that $\mu = \underline{1}\gamma_0 + \beta\gamma$.

While the preceding argument has a veneer of rigor, it is not a rigorous proof of the APT, because it assumes that we can ignore the residual. Ross thought that it should be possible to establish an APT by assuming only that agents prefer more to less, without relying on their risk aversion. Ross (1976) was unable to prove that the expected utility of a portfolio was independent of the portfolio's residual as n gets large, without restricting agents' risk aversions to be bounded. The problem is that, even while the variance of the residual shrinks to zero as n gets large, if agents' risk aversion also gets large as the number of assets grows, the residual might still matter in the limit.

12.1.3 Huberman's Simple Proof of the APT

Huberman (1982) presented a brilliantly simple argument for the APT as an approximate expected return model in large markets. What is so brilliant about this, to my mind, is the way it uses cross-sectional regressions. We will see more of the connection in chapter 22.

The following factor model is assumed to hold when there are n assets (for all n):

$$\underline{R}^n = \underline{\mu}^n + \underline{\beta}^n f + \varepsilon^n, \ E(\varepsilon^n \varepsilon^{n'}) = D^n, \tag{12.2}$$

where the assumption of a diagonal covariance matrix D^n defines this as an *exact factor model*. Consider a cross-sectional regression of the expected returns μ^n on a constant and the K-vector of betas for each asset:

$$\mu^n = \underline{1}\gamma_0 + \beta^n\gamma + \alpha^n. \tag{12.3}$$

Note the clever notation, referring to the regression residual as an alpha. That's exactly what it is: the vector of alphas in the linear beta expected return model using the K factors. By the definition of an ordinary least squares (OLS) regression, the residuals must sum up to 0.0 across assets and the residuals must be orthogonal to the right-hand side betas. Thus: $\underline{1}'\alpha^n = 0$, $\alpha^{n\prime}\beta^n = 0'_{(1\times k)}$. Dropping the superscript n notation, denoting the number of assets, consider a portfolio with weight $x = \alpha/(\alpha'\alpha)$. The portfolio weight vector x has: (a) $\underline{1}'x = 0$ (zero net investment), and (b) $x'\beta = 0$ (no factor risk). These are the same as conditions (1) and (2) in the Ross (1977) algebra above. (You can see why Huberman's use of cross-sectional regression is so clever.) The expected return of this portfolio is $x'\mu = (1/\alpha'\alpha)$ $\alpha'[\underline{1}\gamma_0 + \beta^n\gamma + \alpha^n] = (\alpha'\alpha)/(\alpha'\alpha) = 1 > 0$. If the variance of this portfolio did go to zero with n, it would be an asymptotic arbitrage: in the limit it would have no risk but a positive expected excess return. Huberman's proof asserts that such asymptotic arbitrage opportunities cannot happen. Thus, APT states that the variance of the portfolio $x'R$ must *not* go to zero in n. Let $\lambda_0^1 \backslash \lambda_n^0$ be an $n \times n$ diagonal matrix with the eigenvalues λ_1 to λ_2 on the diagonal. Then

$$\begin{aligned}
\text{Var}(x'R) &= (1/\alpha'\alpha)^2\,\text{Var}(\alpha'R) = (1/\alpha'\alpha)^2\alpha'\,D\alpha\,(\text{since } \alpha'\beta = 0)\\
&= (1/\alpha'\alpha)^2\alpha'[H'(\lambda_0^1\backslash\lambda_n^0)H]\alpha \le (1/\alpha'\alpha)^2\lambda_{\max}(D)\alpha'[H'I_nH]\alpha\\
&= (1/\alpha'\alpha)\lambda_{\max}(D).
\end{aligned} \tag{12.4}$$

Asserting that the variance can't go to zero, then, on the assumption that the maximum eiegenvalue of D is bounded above zero, implies that $(1/\alpha'\alpha)$ cannot go to zero, which implies $(\alpha^{n\prime}\alpha^n) < \infty$ for all n. This is the primitive implication of the large-markets APT: The alphas representing the expected pricing errors of the factor model must be bounded as the number of assets grows without bound.

Ingersoll (1984) generalizes Huberman's proof to use generalized least squares (GLS), instead of an OLS regression. This allows him to replace the diagonal matrix D with a nondiagonal matrix V. Transforming the GLS problem to an equivalent OLS problem, the pricing error bound becomes $(\alpha^{n\prime}V^{-1}\alpha^n) < \infty$.

The problem with the APT as a theory is that Huberman, similar to Ross, had to assert the no-arbitrage condition but could not prove it without referring to agents' risk aversion. In Huberman's case, this is the condition (3) that a portfolio with no factor risk and a positive return must have a limiting variance that is nonzero. Huberman could not prove that if there were a limiting variance of zero, then all investors with strictly increasing utility functions would want to trade it. Again, an assumption of bounded risk aversion is required.

Connor (1984) developed an equilibrium version of the APT, making stronger assumptions to show that the pricing errors should go to zero for each asset as n gets large. In his model, agents optimally hold well-diversified portfolios, so their SDFs, m, depend only on the common factor part of securities returns as n gets large. Assuming that the residuals are independent of the explanatory factors in the factor model, the residuals are uncorrelated with m, so they get zero-risk premiums in equilibrium as the number of assets grows without bound.

12.1.4   Interpretation

Saying that the pricing errors of the factor model must be bounded in n might seem to be a weak implication. But it does imply that the average pricing error, taken across assets, must go to zero. By the Cauchy-Schwartz inequality, $[\underline{1}'\alpha]^2 \le (\underline{1}'\underline{1})(\alpha'\alpha) = n(\alpha'\alpha)$. Therefore, the average squared pricing error $[(1/n)\underline{1}'\alpha]^2 = (1/n)^2[\underline{1}'\alpha]^2 \le (1/n)(\alpha'\alpha) \to 0$ as n gets large. The claim that the average pricing error must be zero sounds better, but the weakness of the restriction on pricing errors has been one of the disadvantages of the APT.

The main result of the APT is related to the information ratio in modern portfolio management, described in chapter 8. We can think of modern portfolio managers as solving $\text{Max}_x[(x'\alpha)^2/x'Dx] = \alpha'D^{-1}\alpha$. Then, the APT says that they cannot have infinite information ratios.

We can, of course, relate the APT to equation (1.1), or m-talk. Let $\alpha = \mu - \underline{1}\gamma_0 - \beta\gamma$, and the APT states that $\alpha'\alpha < \infty$, or $\alpha'V^{-1}\alpha < \infty$ per Ingersoll (1984). Let $m = a + b'f$ be linear in the factors, and recall that $E(mR - \underline{1}) = E(m)[\alpha]$. The APT says that the weighted sum of squared pricing errors of a model where m is linear in the APT factors is bounded as n gets large: $\alpha'V^{-1}\alpha = (1/E(m))^2\{E(mR-1)'V^{-1}E(mR-1)\} < \infty$. This further implies that the Hansen and Jagannathan (1997) distance, $\{E(mR-1)'E(RR')^{-1}E(mR-1)\}$, is bounded.

12.1.5   APT with Approximate Factor Structures

Ross (1976) and Huberman (1982) use an exact (or strict) factor structure, assuming D^n is diagonal. Chamberlain and Rothschild (1983) develop an APT with an approximate factor structure:

$$\Sigma^n \equiv \text{Cov}(R^n) = B^nB^{n\prime} + V^n,$$
$$\lambda_{\max}(V^n) < \infty, \quad \text{for all } n, \lambda_{\min}(B^{n\prime}B^n) \to \infty, \quad \text{as } n \to \infty. \tag{12.5}$$

Equation (12.5) normalizes the covariance matrix of the factors to be an identity matrix, which is without loss of generality because of rotational indeterminacy, as discussed below. In this setup, diversifiable risks correspond to bounded eigenvalues, while pervasive or "systematic" risks correspond to unbounded eigenvalues. The first task is to see why it works this way. Consider a portfolio with weight vector x: $R_p = x'R^n$. A portfolio is defined by Chamberlain and Rothschilds to be *well diversified* iff $x'x \to 0$ as $n \to \infty$. An example of a well-diversified portfolio is the equal-weighted portfolio used earlier in the chapter for the MBA-level intuition ($x = (1/n)\underline{1}$). For any portfolio, the residual variance is given as

$$\text{Var}(u_p) = x'V^nx = x'[H'\Lambda H]x \le \lambda_{\max}(V^n)x'[H'IH]x = \lambda_{\max}(V^n)x'x. \tag{12.6}$$

So if the portfolio is well-diversified and $\lambda_{\max}(V^n)$ is bounded, the residual variance goes to zero as n grows. Thus, bounded eigenvalues correspond to diversifiable risks. Note that the condition that $\lambda_{\max}(V^n)$ is bounded generalizes the assumption used at the beginning of this chapter that the maximum residual variance is bounded.

Why do we need K unbounded eigenvalues in an approximate factor structure? The answer is that we don't want the total variance of well-diversified portfolios to go to zero, only the idiosyncratic part. If the total variance went to zero, the limiting economy would be risk free and pretty boring. The variance of the portfolio with weights x is

$$\text{Var}(R_p) = x'\sum{}^nx = x'[B^nB^{n\prime} + V^n]x = x'(B^nB^{n\prime})x + x'V^nx. \tag{12.7}$$

The second term in the last line approaches zero for large n, if x is well diversified. Assume there is some well-diversified portfolio whose total variance does not go to zero. That is, $x'\sum{}^n x \to \sigma^2 > 0$ as $n \to \infty$. Then, $\sigma^2 = x'\sum{}^n x \to x'(B^n B^{n\prime})x \ge \lambda_{\min}(B^{n\prime}B^n)x'x$. Since $x'x \to 0$, we need $\lambda_{\min}(B^{n\prime}B^n) \to \infty$ to avoid the boring, risk-free limiting economy.

12.1.6   Conditional APT

Stambaugh (1983b) extends Chamberlain and Rothschild's version of the APT to consider conditional moments. He assumes that the asset returns, R_{t+1} and a vector of conditioning variables, Z_t, are jointly normal or Student's-t distributed. He derives an APT pricing restriction, similar to the bound in Ross and Huberman, but where the bound on a version of the conditional alphas, $\alpha(Z)'\alpha(Z)$, is some function of Z_t. The idea is that there is no realization of the information Z such that you would know there is an arbitrage opportunity as n gets large. This bound is stated in terms of the conditional expected returns given Z, and the unconditional loadings or betas. The more informative is the conditioning information, the tighter the bound will be. He also derives a weaker bound based on the unconditional means and alphas, concluding that either conditional or unconditional expected returns are appropriate for empirical work.

Ferson and Korajczyk (1995) examine empirically conditional versions of beta pricing models using APT factors. They find that the models do a pretty good job of conditional asset pricing in the sense of "explaining" predictability in returns. Ferson and Korajczyk find that a five-factor APT model captures most of the predictable variation in some standard portfolio returns, using lagged predictors that were standard at the time.

12.2   Limits to Arbitrage and Costly Arbitrage

This is a book on the "Newtonian physics" of empirical asset pricing, but a brief digression into the world of frictions is worthwhile. Physical frictions in the process of international trade are well known in the literature, playing a large role in understanding the deviations from pure purchasing power parity across countries, for example. Purchasing power parity states that the cost of a good should be the same in any country, after you adjust for the exchange rates. But if haircuts in Hong Kong get cheap relative to the United States, it does cost something to import the barber, or to move the head between countries, and this allows the haircut to cost different amounts in the two countries. The difference can't get arbitrarily large, as eventually the profits for Hong Kong barbers moving to the United States will justify the move, and the supply of haircuts in the United States will increase, driving down the price. The pure arbitrage relation should serve as a measure of central tendency for prices and asset returns.

There are, of course, various frictions in the process of trading in financial markets as well. The presence of these costs allows arbitrage-based relations between different asset prices to fail to hold. Examples of obvious frictions include trading costs like bid-ask spreads, the costs of short sales, the costs of taxes on investment gains, and various costs associated with the adverse effects of differential information in trading. There may be institutional portfolio constraints that impede arbitrage trades. The literature on costly arbitrage asks whether the departures from the central tendency implied by a frictionless model can be justified by the frictions involved. This can get subtle. Gromb and Vayanos (2010) provide a review.

Professional arbitrageurs face financing frictions, in that arbitrage trades require (often, other people's) money. It may be hard, as a money manager, to convince your investor that you have a profitable arbitrage position if it may lose money in the short run (e.g., Shleifer and Vishny 1997). If the investor jerks your funding before the trade has paid off, you may suffer real costs. The availability of arbitrage capital—and leverage financing for arbitrageurs in particular—likely varies with economic conditions. There is some evidence that times of greater arbitrage capital availability are times when stock return anomalies are smaller (e.g., Akbas et al. 2016).

Pontiff (2006) argues that residual risks can measure holding costs that impede arbitrage. Consider a regression on some benchmark, $r_i = \alpha_i + \beta_i r_B + u_i$, and assume that the residuals are independent across the securities. Suppose further that an asset manager is aware of the mispricing, denoted by the alpha, and behaves according to modern portfolio theory. As we described in chapter 8, this is equivalent to maximizing a quadratic utility function defined over the portfolio abnormal return, $x'[\alpha + u]$. The optimal amount to place at risk in the arbitrage trade for asset i is given as $x_i^* = (1/\gamma)[\alpha_i/\sigma(u_i)]$, where $(1/\gamma)$ is a measure of risk tolerance. The higher is the residual risk, the smaller the optimal arbitrage trade will be. Pontiff further argues that residual risk might be the most important indicator of arbitrage costs.

Duan, Hu, and MacLean (2005) find that shorted stocks with high idiosyncratic risk experience more negative abnormal returns than shorted stocks with low idiosyncratic risk. Brav, Heaton, and Li (2010) compare the strength of anomalous returns in low versus high residual risk portfolios and find that overvaluation anomalies (growth stocks, momentum losers, and negative earnings surprises) are stronger when there is higher residual risk, but undervaluation anomalies (small stocks, value stocks, and momentum winners) are not.

Stambaugh, Yu, and Yuan (2015) relate idiosyncratic volatility and short sales costs to a large set of stock return anomalies. They argue that the cost of short selling stocks compared to buying them is an important asymmetry in arbitrage costs that interacts with idiosyncratic volatility in predicting the returns to the long and short sides of many anomalies. Limits to arbitrage remains an active area of both theoretical and empirical research.

12.3   Factor Analysis

Starting with Schipper and Thompson (1981), a large empirical literature evolved, motivated by the APT, that uses factor analysis and principal components analysis. There are various techniques and controversies associated with these studies, and this section reviews some of them. Current work still uses versions of factor analysis and principal component analysis, and we review some current examples at the end of the chapter.

12.3.1   Finding and Numbering the Factors

The first job of any empirical test of a factor pricing model is to identify the factors. Basically, four approaches have been used in the context of asset pricing. The first approach is the cleanest: A theory to be tested sometimes states what the factors should be. Examples include the CAPM and early factor models using consumption growth as the factor (Ferson 1983; Hansen and Singleton 1983). I would place the recent investment-related factors of Hou, Xue, and Zhang (2015) and Fama and French (2015) loosely in this category, as production-based asset pricing models loosely motivate these factors. The second approach is purely statistical, based on factor analysis or principal component analysis; this approach is the most closely aligned with the APT. The third approach is to pick the factors using economic intuition, an approach taken by Chen, Roll, and Ross (1986), although these authors tied their intuition to the APT by looking for pervasive factors that should matter for many or most firms. A fourth approach is to find factors based on patterns in a cross section of securities returns, illustrated by Fama and French (1996).

The second task, closely related to the first, is to determine how many factors should be studied. When the theory tells you what the factors are, the answer is clear. When the factors are based on intuition or cross-sectional patterns, it is less clear. The number of factors seems to have been growing over time, as new factors are proposed at the margin.

When statistical methods like factor analysis and principal components are used to extract the factors from asset returns or from large sets of economic variables, several statistical approaches have been popular. The most primitive is the scree test (Cattell 1966), where the eigenvalues λ are graphed against the number of factors K, and we look for an obvious drop-off in the eigenvalues to signal the number. Since the eigenvalues in a factor analysis represent the average R-squared of all the assets, regressed on the associated factor, a drop in the eigenvalue indicates that the marginal explanatory power of the factor model by adding another factor has dropped off.

Specific to the APT is the asymptotic principal components approach of Connor and Korajczyk (1986), discussed in the next section. In that model, the first K eigenvalues should explode to infinity, and the remainder should stay bounded as the number of assets is increased. Bai and Ng (2006) use this idea to identify the number of pervasive factors. Ahn and Horenstein (2013) develop tests for the asymptotic APT based on the ratio of successive eigenvalues, and find that it has better sampling properties.

12.3.2 Factor Analysis and Asymptotic Principal Components

Early empirical work on the APT (Schipper and Thompson 1981; Roll and Ross 1980) used factor analysis of the $n \times n$ covariance matrix of stock returns to find the common factors in returns. They start with an exact factor structure assumption, $\Sigma^n \equiv \mathrm{Cov}(R^n) = B^n B^{n\prime} + D^n$, with a diagonal residual covariance matrix, D^n. For a given number of factors K, the covariance matrix of returns is approximated by $B^n B^{n\prime} \approx \sum_{i=1,...,K} \lambda_i(x_i x_i')$, using the first K eigenvectors x_i and eigenvalues λ_i of the sample covariance matrix. The loading matrix B^n is formed using the first K eigenvectors. Since this is still the theory part of this book, I won't go into too much detail. Good reviews of factor analysis are available in many statistics books.

Connor and Korajczyk (1986) develop the asymptotic principal components method of factor extraction and show that it has the elegant feature of closely following the large markets assumptions of the APT. The approximate factor structure is assumed:

$$R_{(T \times n)} = F_{(T \times K)} \beta_{(K \times n)} + e_{(T \times n)}, \tag{12.8}$$

where the covariance matrix of e has bounded eigenvalues as n grows, but the covariance matrix of R has K unbounded eigenvalues. Connor and Korajczyk suggest using the eigenvectors of RR'/n as the factors. Their argument is summarized as follows. Consider the expression

$$RR'/n = (F\beta\beta'F')/n + F\beta e'/n + e\beta'F'/n + ee'/n. \tag{12.9}$$

It is assumed that the last term converges in probability to a diagonal matrix with σ^2 on the diagonal as $n \to \infty$. The two middle terms converge in probability to zero, because the regression residual should be uncorrelated with the right-hand side variables. The conclusion is that the first term on the right-hand side should approximate the left-hand side, apart from a diagonal matrix with identical entries, as n grows. This diagonal matrix difference will affect the eigenvalues but not the eigenvectors. Thus, the eigenvectors RR'/n deliver the factors F to within a rotation. Since, as we will see in the next section, you can never do any better than identifying factors to within a rotation, you can use the eigenvectors of RR'/n to determine the APT factors. Connor and Korajczyk (1986) prove the consistency of the principal components estimator for the factors for fixed T, as n goes to infinity.

Note that the asymptotic principal components approach is based on the $T \times T$ matrix RR'/n and essentially averages across the stocks. This trait is appealing in other contexts as well, when the number of variables n is large relative to the number of time series observations that are available. If the residuals are only weakly correlated across the variables or stocks, they average out to zero as n gets large, according to a weak law of large numbers, and we are left with just a few common factors. Indeed, such studies as Stock and Watson (2002) find that it only takes a few common factors to summarize a large number of economic time series.

12.3.3 APT Tests

As mentioned above, early studies started with a factor analysis of $\Sigma_{(n \times n)} = \mathrm{Cov}(r)$ to get the factor-loading matrix $B_{(n \times K)}$. Then, cross-sectional regressions each month deliver factors f_t, from the regression $r_t = Bf_t + u_t$. Connor and Korajczyk (1986) reverse that procedure. They start with K principal components of the matrix $(rr'/n)_{(T \times T)}$, using the de-meaned returns for r. Let the first K principal components be the matrix of factors $F_{(T \times K)}$. In a second step, they run time series regressions of r on F to get loadings β, as $r = F\beta + v$.

It turns out that these two approaches are asymptotically equivalent as both n and T go to infinity. Here is a loose but hopefully intuitive argument. Start with the fact that for two matrices rr' and $r'r$, the eigenvector problem may be written, first for $(rr')_{(T \times T)}$ as $[(rr') - \lambda I]x = 0$, where each of the K vectors in x is $(T \times 1)$. Multiply through by r' and then by $(r'r)^{-1}$, and letting $x = rx^*$, where x^* is $N \times 1$, we have that $[(r'r) - \lambda I]x^* = 0$. We see that the two matrices rr' and $r'r$ have the same eigenvalues, and their eigenvectors are related as $x = rx^*$. Roll and Ross (1980) start with the first K principal components of x^* as the loadings B. The factors are then given by the cross-sectional regression of r on B. Thus, the fitted factors are $F = rB(B'B)^{-1} = rx^*(B'B)^{-1} = x(B'B)^{-1}$. Connor and Korajczyk start directly with the first K columns of x as the factors F. Thus, the two approaches use the same factors to within a $K \times K$ rotation. Since the factors are the same, the loadings are the same as well. The two sets of factors and loadings should be the same as n and T go to infinity. (The ambiguity about "to within a rotation" is clarified in the next section.)

The classical factor analysis approach and the asymptotic principal component approach deliver the same results asymptotically, but the two approaches are likely to differ for finite samples. The classical approach starts with an eigenvalue-eigenvector decomposition of the $n \times n$ covariance matrix of returns. For an accurate result, you need a good estimate of the covariance matrix. This means that T should be large relative to n, and it limits the number of securities n in practice. Indeed, most of the work along these lines has used portfolios instead of individual stocks, to keep n relatively small. Portfolio formation introduces another decision in the empirical design that can alter the results. If the loadings are poorly estimated, then the cross-sectional regression of returns on the loadings to get the factors suffers from an errors-in-variables problem, which biases the factor estimates, usually toward zero. Thus, the risk premiums are underestimated.

The asymptotic principal components approach starts with an eigenvalue-eigenvector decomposition of the $T \times T$ matrix that is estimated by averaging across the n stocks. The larger is the n, the better this estimate will be, so the approach may be used on individual stocks, avoiding the need to form portfolios. In fact, if n is large enough, the estimates of the time series of the factors can be used in applications as if they were data, as the sampling errors can be negligible. This is attractive in other applications, such as forecasting and instrumental variables (e.g., Stock and Watson 2002). The time series regressions of returns on the factors to obtain the factor loadings should work well if the factors are estimated accurately in the first step. All of this assumes that the loadings are constant parameters over time, but the relevant sample period T does not have to be large.

Let me finish this section on APT tests with an interesting angle from Zhang (2009b). Writing $E(rr') = \mu\mu' + \Sigma = \mu\mu' + BB' + V$, let K be the number of eigenvalues of the matrix $(BB' + V)$ that explode as n gets large. Zhang notes that if $\mu\varepsilon((B))$ as $n \to \infty$ (as predicted by the large markets APT), then K^*, defined as the number of eigenvalues of the matrix $(\mu\mu' + BB' + V)$ that explode as n gets large, should be equal to K. If the APT fails and the vector of means is not spanned by the loadings, then $K^* = K + 1$. He constructs tests of the APT based on this insight.

12.3.4   Rotational Indeterminacy

In a multifactor expected returns model, the factors can never be unique, and this is not unique to the APT. It applies to any beta pricing model. Start with K "true" excess return factors, f: $r = \beta f + u$, $E(u) = 0$, $E(uf') = 0$, with $E(r) = \beta\lambda$, and $\lambda = E(f)$. We can transform the model to work equally well with a different set of K factors. Let $f^* = Af$, where A is any nonsingular $K \times K$ rotation matrix. The f^* are the new factors. We have $E(r) = \beta AA^{-1}\lambda$. The loadings of the asset returns on the new factors will be

$$\beta(r \text{ on } f^*) = \text{Cov}(r, f^{*\prime})V(f^*)^{-1} = \text{Cov}(r, f')A'V(Af)^{-1}$$
$$= \text{Cov}(r, f')A'A'^{-1}V(f)^{-1}A = \beta A. \tag{12.10}$$

The risk premiums on the new factors will be $A^{-1}\lambda$. Thus, we get the same expected returns, and even the same regression residual u, using either f or f^*. With more than one factor, we can never determine the factors uniquely.

Lewellen, Nagel, and Shanken (2010) present a version of rotational indeterminacy aimed at the Fama and French (1996) three-factor model. Suppose there exist three factors f that "work" for the portfolio design of Fama and French, so the portfolio returns satisfy the restricted regression $r = \beta f + u$, with no intercept or alpha. Suppose there exist three variables x with: $\text{Cov}(x, f') \neq 0$ and $\text{Cov}(x, u') = 0$. The first condition is recognized as the "inclusion restriction" in instrumental variables, where x is an instrument for f, which is easy to satisfy. The second is the "exclusion" restriction, which is hard to satisfy. Lewellen, Nagel, and Shanken argue that while this only has to hold approximately, we are likely to be able to find such variables, because the Fama-French portfolio design has a strong factor structure. This means that the factors capture most of the correlation in the portfolio returns, leaving the residuals roughly independent.

Regress x on f: $x = \mathrm{Cov}(x, f')V(f)^{-1}f + v$. Because v is a regression residual, it is uncorrelated with f, and it is assumed to be uncorrelated with u by the exclusion restriction, which implies that v is uncorrelated with r. Thus, $\mathrm{Cov}(r, x') = \mathrm{Cov}(r, f')V(f)^{-1} \mathrm{Cov}(f, x') = \beta \mathrm{Cov}(f, x')$. Thus, we can use x as the "factors." That is, solving the last expression for β and plugging it in, we have

$$E(r) = \beta\lambda = \mathrm{Cov}(r, x')\mathrm{Cov}(f, x')^{-1}\lambda = \mathrm{Cov}(r, x')V(x)^{-1}V(x)\mathrm{Cov}(f, x')\,\lambda. \qquad (12.11)$$

We can identify $\mathrm{Cov}(r, x')V(x)^{-1}$ as the matrix of betas of the asset returns on the new x factors, and define $V(x)\mathrm{Cov}(f, x')\lambda$ as the new vector of risk premiums for the new x factors. The upshot of all this is that it might be "too easy" to find three-factor models that work as well as the Fama and French (1996) factors, given the strong factor structure in the portfolio design for which those factors were constructed. Lewellen, Nagel, and Shanken conclude that we should look beyond the size and book-to-market portfolios when we test models in competition with the Fama and French factors, because the additional portfolios will likely have stronger correlation with the residuals u, breaking the exclusion restriction.

12.3.5   A Testability Debate about the APT

Shanken (1982) characterizes empirical studies as embracing *empirical APT*. That is, if $R = a + \beta F + u$, with $E(uu') = D$ (a diagonal matrix), then studies test the hypothesis that $\mu = \gamma_0 1 + \beta\gamma$. The approximation of the APT pricing restriction (that the alphas are only bounded and not actually zero) is ignored. He argues that this makes no sense. A simplified version of the argument follows. Work in excess returns r. Transform them to $r^* = \mathrm{Cov}(r)^{-1/2}r$, then $\mathrm{Cov}(r^*) = I_n$. The transformed excess returns follow a "zero factor structure," and the empirical APT logic then implies $E(r^*) = 0$, or $E(R^*) = \gamma_0 1$, that is, all the transformed expected returns are the same! But this implies that all the original expected returns must be the same, which is a contradiction!

Dybvig and Ross (1985c) point out that Shanken's transformation, $r^* = \mathrm{Cov}(r)^{-1/2} r$, is not well diversified. There is some debate about whether the APT can be applied, ignoring the approximation, to individual stocks or to large, well-diversified portfolios. Shanken (1985b) responds that equilibrium models are better, as they indicate which portfolios (e.g., the market) should be well diversified.

12.3.6   Dynamic Factor Models

So far we have been discussing static factor models, where any serial dependence in the data is ignored. Dynamic factor models are often applied when this assumption fails. Such models are used by Ludvigson and Ng (2009) to model the relation between the term structure of interest rates and a bunch of macroeconomic variables. In a dynamic factor model, the factors are commonly assumed to follow some autoregressive model, and their lagged values can influence the dependent variables r_t (here, de-meaned or excess returns). An example posits a VAR(1) model:

$$f_t = A f_{t-1} + \eta_t,$$
$$r_t = \beta_0 f_t + \beta_1 f_{t-1} + e_t, \qquad\qquad (12.12)$$

where (β_0, β_1) are the dynamic factor loadings, and $E(e_t \eta_{t-j}) = 0$, all j. In an "exact" dynamic factor model, the residuals are assumed to be uncorrelated across the stocks at all leads and lags, so the autoregressive structure of the factors captures all the dynamic structure in the system. One appeal of a dynamic factor model is in forecasting, because if the model is well specified, you only need the lagged factors and perhaps the own-lagged returns to efficiently forecast, no matter how many returns are in the system.

We can transform a dynamic factor model to look like the previously discussed static model by defining the static factors $F_t = (f_t, f_{t-1})$, perhaps with more lags if a higher order autoregressive model is used. Estimation for dynamic factor models is reviewed in Stock and Watson (2010). The new twist is that the idiosyncratic error terms e_t now depend on lagged values of the factors. Here is an example that illustrates the idea, following Stock and Watson (2002), who provide conditions under which the estimators are n-consistent. Each month, each stock is first regressed on its lagged daily returns to estimate an autoregressive parameter δ. Then the residuals $e_t = r_t - \delta'(r_{t-1}, \ldots, r_{t-j})$ are formed. These residuals should be approximately purged of their serial dependence. Let the $T \times n$ matrix of their sample values be denoted by V. Asymptotic principal components analysis of the $T \times T$ matrix VV'/n can be used to get the initial set of factors. The returns r_t are then regressed on the lagged returns and the fitted factors to get a new estimate of δ. This procedure is iterated to convergence.

12.3.7 *Partial Least Squares*

Kelly and Pruit (2012, 2013) develop an alternative to principal components analysis in the context of predicting asset returns. Principal components analysis summarizes a bunch of variables by finding factors that are linear combinations of them, such that the average regression R-square of the variables on the combinations is maximized. The method of partial least squares summarizes a bunch of variables by finding linear combinations of them, where each variable is weighted according to its covariance with the future value of the object to be predicted. Both approaches assume that the variables follow a factor structure. Kelly and Pruit (2013) use the book/market ratios from a cross section of common stocks, and they find a combination of them to predict the future stock market return. Let Z_{it} be the predictor variable for stock i observed at time t. The approach proceeds in two steps. First, a time series regression is run for each predictor variable on the future market return:

$$Z_{it} = \varphi_{oi} + \varphi_i \, r_{mt+1} + u_{it}, \quad t = 1, \ldots, T. \qquad\qquad (12.13)$$

The idea of this regression is to extract the covariance between Z_{it} and the expected value at time t of the market return at time $t+1$. Because the actual future return is on the right-hand side, the estimated slope coefficient suffers from an errors-in-variables bias, where the measurement error is the difference between the actual market return and its conditional expected value. For example, under standard assumptions, the OLS estimator converges in probability to $\varphi_i\{\mathrm{Var}(E_t(r_{mt+1}))/\mathrm{Var}(r_{mt+1})\}$, but the multiplicative bias is the same for all stocks i.

In a second step, the slope coefficients for each stock are used as right-hand side variables in cross-sectional regressions:

$$Z_{it} = a_t + Z_t^* \varphi_i + \epsilon_{it}, \ i = 1, \ldots, N, \tag{12.14}$$

where Z_t^* is the regression coefficient in the regression for month t. The second-stage regression is run for each month, similar to the Fama and MacBeth (1973) style cross-sectional regressions that we study in chapter 22. The estimated slope coefficient from the second step is used to predict the market return. The estimated slope is a linear combination of the Z_{it}, weighted according to their covariances with the future market return, as captured in φ_i. Because the estimated φ_i is "too small" given the errors in variables bias, the fitted Z_t^* is "too large," but by a constant that should wash out of the predictive regression, because in the predictive time series regression, the slope coefficient on Z_t^* will be scaled to maximize the R-square in predicting the market return.

To make sure this method is legitimate for prediction, Kelly and Pruit (2013) estimate the coefficients recursively, using only data up to time t to get the Z_t^* to predict market returns for $t+1$ or later. Using a cross section of firms' book/market ratios, they find significant predictive power for the stock market return, better than using the aggregate book/market or other predictor variables.

13 Multibeta Equilibrium Models

Multibeta empirical asset pricing models arise in various contexts. We have seen that they are equivalent to a linear model for the SDF. It is useful to consider in more detail the situation where the model flows from an intertemporal consumption-investment problem with a representative agent. This is the main subject of this chapter. I present the model in discrete time using normality and Stein's (1973) lemma, which delivers essentially the same results as in the original continuous-time model of Merton (1973). Optimal hedging demands are discussed. An important focus for empirical work is the intuition the model gives us for the types of factors likely to arise in a multibeta equilibrium model and the determination of the signs of the risk premiums attached to the factors.

13.1 Multiperiod Consumption-Investment Problems

The general structure of the classical Merton (1973) style problem, the solution for the optimal portfolio weights, and the asset pricing implications are similar in continuous time and in discrete time under normality. This chapter takes the latter tack, which is much simpler, especially when we will be using discrete horizon returns data in empirical work. The reason is that formally, a continuous-time model must be time aggregated to handle returns measured over more than an instant, and this requirement can change the model materially (e.g., Longstaff 1989), as is discussed in chapter 27. The goal here is to elucidate the basic structure of the solutions in a way that can be useful to empiricists. So this is not a deep theoretical treatment. Mathematical formulations of optimal consumption and investment rules in multiperiod models appear in a long and frequently complex literature. This simplified presentation focuses on the main empirical implications of the model.

Consider a representative agent maximizing a time-additive utility function for real consumption:

$$\text{Max}_{\{c,x\}} \, E \sum_t \beta^t u(C_t), \, u' > 0, \, u'' < 0. \tag{13.1}$$

The gross portfolio return is $R_p = \sum x_i R_i + (1 - \sum x_i) R_F = \sum x_i (R_i - R_F) + R_F \equiv x'r + R_F$, where x is the n-vector of portfolio weights in the risky assets, with $(1 - \sum x_i)$ invested in the risk-free asset. The budget constraint determines the wealth level entering into any period t as

$$W_t = (W_{t-1} - C_{t-1})(x'r_t + R_{F,t-1}).$$ (13.2)

13.1.1 Backward Induction and the State Variables

Assuming a terminal date T, we study the dynamic program at date $T-1$, defining

$$J(W_{T-1}, S_{T-1}) \equiv \text{Max}_{\{C,x\}} \, u(C) + E_{T-1}\{\beta u(W_T)\},$$ (13.3)

subject to the budget constraint $W_T = (W_{T-1} - C)(x'r_T + R_{F,T-1})$. The state vector S_{T-1} is defined implicitly here as the information needed, in addition to wealth, to summarize the expected maximum at time $T-1$. The first-order conditions to this problem with respect the consumption and portfolio weight choices are

$$C^*: \quad 0 = u_C(C) - E_{T-1}\{\beta \, u_W(W_T)(x'r_T + R_{F,T-1})\},$$ (13.4)

$$x^*: \quad \underline{0} = E_{T-1}\{\beta u_W(W_T)r_T\}(W_{T-1} - C),$$ (13.5)

where the W and C subscripts on $u(\cdot)$ denote the partial derivatives with respect to wealth or consumption. The first equation determines the optimal consumption choice as

$$C^* = u_C^{-1}(E_{T-1}\{\beta u_W(W_T)(x'r_T + R_{F,T-1})\}).$$ (13.6)

Campbell (2003) shows how to solve this expression for the optimal consumption approximately, using log linearization and particular preferences. See also Ingersoll (1987) for several solutions.

The second first-order condition (13.5) determines the optimal portfolio weight x^* implicitly. Brandt (1999) uses this equation, substituting in from the budget constraint for W_T, and finds the optimal portfolio weights numerically for particular utility functions.

Substituting the optimal solutions back into equation (13.3), we can write it as

$$J(W_{T-1}, S_{T-1}) = u(C_{T-1}^*) + E_{T-1}\{\beta u([W_{T-1} - C_{T-1}^*][x^{*\prime}r_T + R_{F,T-1}])\}.$$ (13.7)

Writing the problem in this way yields insights into the identity of the state variables S_{T-1}. Evidently, the risk-free rate is one of the necessary ingredients, in addition to the current wealth level. We should also include a set of sufficient statistics for the distribution of the next period's excess returns on the vector of risky assets.

13.1.2 Multiple-Good Models and Inflation Risk

Models with more than a single consumption good are developed by Long (1974), Breeden (1979), and others. Think of C_{T-1}, the real consumption expenditures, as induced by the choice over multiple goods at each date: At each date t, the vector of goods consumed $\underline{c}_t$ is optimally chosen via

Max v($\underline{c}$) s.t. $\underline{c}'p = C^n$, C^n = nominal ($) expenditures, (13.8)

where v($\cdot$) is the period utility function for the vector of goods. The solution is a function $\underline{c}^*(C^n, p)$. Now define a price index I($\cdot$) so that real consumption is $C \equiv C^n/I \equiv v(\underline{c}^*(C^n, p))$. This way, the real consumption obtained by deflating the nominal expenditures by the price index corresponds to the maximum utility obtained through the multiple goods for a given nominal consumption expenditure and price vector p. Now we can use the real consumption in the original problem.

There are two main implications of multiple goods for the asset pricing problem. First, relative prices become part of the state vector S_t. (Recall from basic price theory that $\underline{c}^*(\alpha C^n, \alpha p) = \underline{c}^*(C^n, p)$ for any scalar α, because demand is homogeneous of degree zero, so only relative prices matter.) Long (1974) is explicit about using the vector of relative prices as the state variables in his intertemporal model.

The second implication of multiple goods refines the discussion of real versus nominal m-talk in chapter 6. Note that to price a nominal payoff, x^n in "dollar" terms, we must have by the chain rule:

$$P_t^n[(\partial U(C_t)/\partial C_t)(\partial C_t/\partial C_t^n)] = E_t\{x_{t+1}^n[(\partial u(C_{t+1})/\partial C_{t+1})(\partial C_{t+1}/\partial C_{t+1}^n)]\}. \qquad (13.9)$$

Rewriting this equation in the form of the m-talk equation (1.1), we see that the real SDF for a time-separable utility is $m = [(\partial u(C_{t+1})/\partial C_{t+1})/(\partial U(C_t)/\partial C_t)]$, and the SDF to price nominal payoffs is $m^n = m[(\partial C_{t+1}/\partial C_{t+1}^n)/(\partial C_t/\partial C_t^n)]$. From the definition of the price index, $\partial C/\partial C^n = (1/I)[1 - (C^n/I)\partial I/\partial C^n]$. This generalizes the inflation measure used in the previous discussion (which was basically I) to allow for an elasticity of the price index with respect to the nominal consumption expenditure budget that may differ from zero. Generally, people will change their expenditure shares for the various goods as their nominal expenditure budget changes. The CPI, however, is constructed by holding the basket of goods largely fixed over time. This approach is justified if the preference for the vector of multiple goods is homothetic (Samuelson and Swamy 1974). (A function F(c) is *homothetic* if it can be written as h(g(c)), where g($\cdot$) is homogeneous, and h($\cdot$) is continuous, nondecreasing, and positive.) In this case, the ratio of price indexes becomes I_t/I_{t+1}, as we used before. Inflation in the United States has been low and stable for much of the twenty-first century, so the literature typically ignores these concerns. Remember this discussion when the situation changes.

13.2 The Classical Multibeta Model

Merton (1973) developed the intertemporal multibeta model in continuous time, assuming (as we do here) a representative agent with a time-additive utility function for consumption. Under the diffusion processes as assumed by Merton, the random variables are locally normally distributed. The development here also uses normality, but in discrete time.

13.2.1 Recursion

Consider the dynamic program at any date t, where the indirect value function is written

$$J(W_t, S_t) \equiv \mathrm{Max}_{\{C,x\}} \ u(C) + E_t\{\beta J(W_{t+1}, S_{t+1})\}. \tag{13.10}$$

The first-order conditions with respect to the consumption and portfolio choices are

$$C^*: \quad u_C(C) = E_t\{\beta J_W(t+1)(x'r_{t+1} + R_{F,t})\}, \tag{13.11}$$

$$x^*: \quad \underline{0} = E_t\{J_W(t+1)r_{t+1}\} \ (W_t - C_t). \tag{13.12}$$

13.2.2 The Envelope Condition

Taking the derivative of equation (13.10) with respect to wealth and combining with condition (13.11), we can see that

$$J_W(t) = E_t\{\beta J_W(t+1)(x'r_{t+1} + R_{F,t})\} = u_C(C_t), \tag{13.13}$$

that is, $u_C = J_W$: The marginal utility of consumption must equal the marginal utility of wealth. This shows that, on the assumption that the utility function $u(C)$ is increasing in C, the indirect value function $J(W)$ must be increasing in W. Taking another derivative with respect to wealth, we get $u_{CC}C_W = J_{WW}$. This equality shows that if $u(C)$ is concave in C, and if consumption is a normal good ($C_W > 0$), then $J(\cdot)$ must be concave in W. Combining these two expressions implies that

$$\{(W/C)C_W\}\{-Cu_{CC}/u_C\} = \{-WJ_{WW}/J_W\}. \tag{13.14}$$

The first term is the elasticity of consumption with respect to wealth, which should be positive, and the second term is the relative risk aversion for consumption. The right-hand side is a relative risk aversion stated in terms of the indirect utility of wealth. If the elasticity of consumption is less than 1, it implies that the relative risk aversion in terms of wealth is lower than the risk aversion for consumption. This makes sense if the fraction of wealth optimally consumed gets smaller as the wealth rises. Cox, Ingersoll, and Ross (1985) show that if the utility function $u(\cdot)$ features constant relative risk aversion, then the indirect utility function $J(W, S, t)$ exhibits a separation between wealth and the state variables: $J(W, S, t) = f(S, t)U(W, t) + g(S, t)$. In this case, the relative risk aversion stated in terms of the indirect utility function is also constant and independent of both wealth and the state variables.

13.2.3 Multibeta Pricing via Stein's Lemma

Instead of diffusions implying local normality, this discrete time development assumes normality and uses Stein's lemma.

Stein's Lemma (Stein 1973).

$$(Y, x) \sim N \Rightarrow \mathrm{Cov}(f(Y), x) = E(\nabla f)'\mathrm{Cov}(Y, x), \tag{13.15}$$

where ∇ denotes the gradient, or the vector of partial derivatives of the function with respect to the elements of the vector Y.

We use Stein's lemma to express covariances with marginal utility functions in terms of covariances with the underlying variables on which the marginal utility functions are defined. This is a useful trick in many models. Dropping the t notation, the first-order condition to the intertemporal optimization implies that

$$0 = E(J_w r) = E(J_w)E(r) + Cov(J_w, r), \tag{13.16}$$

where $J_w = J_w(W_{t+1}, S_{t+1})$. Stein's lemma allows us to write the covariance as

$$Cov(J_w, r) = E(J_{ww})Cov(W, r) + \sum_s E(J_{ws})Cov(S, r), \tag{13.17}$$

where the sum is taken over the elements of the state vector S_{t+1}. Plugging this back into equation (13.16), we can solve for the vector $E(r)$:

$$E(r) = \{-E(J_{ww})/E(J_w)\}Cov(r, W) + \sum_s \{-E(J_{ws})/E(J_w)\}Cov(r, S). \tag{13.18}$$

This is the famous Merton (1973) multibeta model in discrete time. In this formulation, we have covariances with wealth and the state variables and *market prices of risk* associated with each covariance. Implicitly, the expectations are conditioned on the information at time t, and the random variables are realized at time $t+1$. Many applications and empirical papers are based, at least loosely, on this model. Fama and French (1996) boldly suggest that their three-factor model using the market size and book/market spread portfolios is a version of this model. Practitioners sometimes talk about "alternative risk premiums," meaning exposures to factors other than the market portfolio.

We can rescale the expressions in (13.18) in various ways. Multiplying and dividing by the lagged wealth in the first term, and by the lagged levels of the state variables in each of the summed terms, we can express the risks as the covariances with the relative changes in wealth and the state variables. Then the price of covariance risk for wealth becomes a relative risk aversion measure due to Rubinstein (1976). We can multiply and divide by the variances of wealth and each state variable in the summation, in which case we have betas and the risk premiums associated with the betas on wealth and the state variables. This version is stated with simple regression betas, as if the state variables were uncorrelated, but by rotational indeterminacy, as discussed in chapter 12, this is without loss of generality.

13.2.4 Breeden's Consumption-Based Model

Breeden (1979) manipulated Merton's model to obtain a consumption-based intertemporal asset pricing model. I show the basic structure of the argument here. Using the envelope condition $u_C = J_W$, and the normality implied by continuous time diffusions, Breeden saw that

$$Cov(u_c, r) = Cov(Jw, r) = E(J_{ww})Cov(W, r) + \sum_s E(J_{ws})Cov(S, r), \tag{13.19}$$

that is, the optimal consumption captures all state variables and hedging terms in Merton's model. Thus:

$$E(r) = Cov(r, -u_c(C_{t+1}))/E(J_w) = Cov(r, -u_c(C_{t+1}))/u_c$$
$$\Rightarrow E(r) = E(-u_{cc}/u_c)Cov(r, C_{t+1}), \tag{13.20}$$

where the second line uses Stein's lemma to express the covariances of returns with the marginal utility of consumption in terms of the covariances with consumption.

13.2.5   Optimal Portfolio Weights and Hedge Portfolios

Equation (13.18) can be solved to find the optimal portfolio weight x*. The trick is to substitute for wealth in terms of the optimal weight in the Cov(r, W) term. Since $W_{t+1} = (W_t - C_t)[R_{ft} + x'r_{t+1}]$, it follows that $Cov(\underline{r}, W_{t+1}) = (W_t - C_t)\Sigma x^*$, where $\Sigma = Cov(r)$. Plugging this expression into equation (13.18) and rearranging, we find

$$x^* = \{-E(J_w)/(W-C)E(J_{ww})\}\Sigma^{-1}E(r) + \Sigma_s E(J_{ws})/\{(W-C)E(J_{ww})\}\Sigma^{-1}Cov(\underline{r}, S). \tag{13.21}$$

In equation (13.21), the first term is a mean-variance efficient portfolio weight vector multiplied by a risk tolerance stated in terms of the indirect utility of wealth. The remaining terms, one for each state variable, are the additional "hedging demands" in the Merton model. When there are K state variables, the model displays "$K+2$" fund separation, meaning that the agent optimally chooses a combination of the mean variance excess return, the K hedge portfolios, and puts the remainder in (or borrows at) the risk-free rate.

As special cases, if $J_{ws} = 0$ or Cov(r, S) = 0 for all the state variables, the model reduces to the CAPM, and the investor holds a mean variance efficient portfolio. If either of these conditions holds for a given state variable, then there is no hedging demand for that state variable. The first case means that the agent does not care about the state variable in the sense that it does not influence her marginal utility of wealth. The second case means that the agent can't hedge the state variable in asset markets, so it can't influence the portfolio choice (or the equilibrium expected returns of assets). In general, the amount to hedge depends on the elasticity of J_w with respect to S. Campbell and Viceira (1999) and Lynch (2000) calibrate the importance of the hedging demands in equation (13.21) empirically, when there is predictability in asset returns.

Note that the hedge portfolio weights are proportional to $\Sigma^{-1}Cov(r, S)$, the regression betas of the state variable S on the vector of excess returns, r. Thus, the hedge portfolio weights are maximum (squared) correlation portfolios for S. Fama (1996) notes that the solution for the optimal portfolio weights in the Merton model is multifactor minimum variance efficient. That is to say, the optimal weight solves

$$Min_x \; x'\Sigma x - 2\lambda_1[x'E(r) - \mu_p] - 2\Sigma_s \lambda_s[x'Cov(\underline{r}, S) - C_s], \tag{13.22}$$

where μ_p and C_s are target expected returns and covariances. Merton's agents behave as if they were trying to minimize the portfolio variance, subject to a target for the mean return and a target covariance for each state variable. We will come back to this chapter 14.

Although we used Stein's lemma to derive the multibeta model, we could just as well have called it a first-order Taylor series for J(W, S). If we went to a second-order Taylor series, we would have a model in which co-skewness with the wealth and the state variables are priced, generalizing the single-period skewness preference model discussed in chapter 5. Chiang (2016) provides an example and evaluates the model empirically on bond returns, finding that models with conditional skewness and co-skewness help explain bond returns.

13.2.6   The Signs of the Risk Premiums

It should be clear that the model implies a positive risk premium for the covariance or beta on wealth, as the indirect value function is assumed to be concave and the risk aversion term in front of the covariance with future wealth is positive. The signs of the risk premiums or prices of risk for the other state variables may not be so obvious, but they are of the opposite sign as J_{ws}. Thus, if changes in a state variable are associated with higher marginal utility, the risk premium for sensitivity to that state variable should be negative. An asset that pays off more when the state variable is high, pays off more when the value of the payoffs is high, and investors are willing to pay for that with a higher price (or lower risk premium).

Maio and Santa Clara (2012) attempt to tie the risk premium for a state variable to its predictive ability. They argue that a state variable should predict either the future market return or its variance. If a state variable is positively related to wealth, it should have a positive risk premium, and if it is positively related to volatility, it should have a negative risk premium. They empirically examine some popular empirical multibeta models to see whether they satisfy this logic and find that not all of them do.

Applying the model to the market portfolio, we get an expression for the risk premium on the market portfolio:

$$E(r_m) = \{-E(J_{ww})/E(J_w)\}Var(r_m) + \sum_s \{-E(J_{ws})/E(J_w)\}Cov(r_m, S). \tag{13.23}$$

In the special case where $J_{ws} = 0$, such as for a log utility function, there are no state variables, and we have the Merton (1980) version of the model—essentially, the classical CAPM. That model links the expected market return to the variance of the market return.

Equation (13.23) shows how the asset pricing theory links the ability of a variable to predict the market returns with its ability to predict second moments, if we interpret the expectations in (13.23) as the conditional expectations given public information at time t about returns and state variables at time $t + 1$. Studies have examined the predictability of the market return variance and tried to relate market variance predictability to market excess return predictability through the Merton (1980) model without the hedging terms. The early empirical results were mixed. French, Schwert, and Stambaugh (1987) and Ghysels, Santa-Clara, and Valkanov (2004) were able to coax the data to reveal a positive relation between the conditional market expected return and the conditional market return variance. Scruggs (1998) shows that these two kinds of predictability line up better when a bond risk premium is in the model, consistent with a Merton (1973) hedging demand. Chang and Ferson (2018) emphasize that the risk premium on a state variable is determined by $\{-E(J_{ws})/E(J_w)\}$, while a state variable's contribution to the risk premium on the market portfolio depends also on $Cov(r_m, S)$. They evaluate a number of lagged predictor variables in the literature in terms of their ability to predict $Cov(r_m, S)$, the innovation market covariance. Chang and Ferson find that this term usually has more predictability than the expected market return.

Campbell (1996) expands the model, approximating the budget constraint log linearly, to express consumption growth approximately as a combination of the market return and a discounted sum of the future market returns. With the lower case denoting logs, Campbell finds that $c_{t+1} \approx a$ constant $+ r_{mt+1} + (1 - EIS)E_{t+1}\{\sum_{j>1} \rho^j r_{m,t+j}\}$ where EIS is the elasticity of intertemporal substitution. Substituting in for consumption, we get back to a model with a market beta and other factors, as in Merton's model, but now the factors are identified as the discounted future market returns. Campbell's version of the model states (in a simplified form):

$$E(r) \approx A \, Cov(r, r_{mt+1}) + A(1 - EIS)Cov(r, E_{t+1}\{\sum_{j \geq 1} \rho^j r_{m,t+j}\}). \tag{13.24}$$

So we have consumption-based asset pricing without consumption data, which Campbell motivates in terms of the problems of noisy and inaccurate consumption data. The signs of the risk premiums in this model are nailed down by the risk aversion, $A > 0$, and by whether the EIS is above or below 1.0.

14 Multibeta Models with Conditioning Information

This chapter concludes the material on the theory of empirical asset pricing. It revisits multibeta pricing and generalizes some of the results to consider explicit conditioning information. We also revisit the three paradigms of empirical asset pricing, now with conditioning information. Finally, we discuss a framework for testing asset pricing models with conditioning information, developed by Ferson and Siegel (2009).

14.1 Conditional and Unconditional Multifactor Efficiency

Recall the multibeta asset pricing or expected excess return expression from chapter 13:

$$E(r) = \{-(W-C)E(J_{ww})/E(J_w)\}\Sigma x^* + \Sigma_F\{-E(J_{wF})/E(J_w)\}Cov(r,F). \tag{14.1}$$

And the optimal weight vector x^* is

$$x^* = -E(J_w)/\{(W-C)E(J_{ww})\}\Sigma^{-1}E(r) + \Sigma_F E(J_{wF})/\{(W-C)E(J_{ww})\}\Sigma^{-1}Cov(r,F), \tag{14.2}$$

where subscripts on the indirect value function $J(\cdot)$ denote partial derivatives. Σ_F denotes the summation over the factors or state variables. The first term of (14.2) is a mean-variance efficient portfolio weight, and the remaining terms are hedging terms for each of the factors F that impact the indirect value function. The optimal weight is *multifactor minimum variance efficient* (MMV; Fama 1996); that is, for some target values μ_p and $\{C_F\}_F$, it solves the problem

$$Min_x x'\Sigma x - 2\lambda_1[x'E(r)-\mu_p] - 2\Sigma_F \lambda_F[x'Cov(r,F)-C_F]. \tag{14.3}$$

Merton (1973) and Fama (1996) had in mind expectations and covariances that are conditioned on the information Ω_t, available in the market at time t, and "fully reflected" in asset market prices. We now make the conditioning information explicit and consider an alternative kind of multifactor efficiency when there is conditioning information. A problem immediately arises, however, on the assumption that we cannot measure all the information reflected in asset market prices. Instead, we can only work with a subset Z.

14.1.1   Conditional and Unconditional Multifactor Efficient Portfolios

Ferson, Siegel, and Xu (2006) develop concepts of multifactor efficiency with respect to conditioning information, analogous to the results of Hansen and Richard (1987) and Ferson and Siegel (2001) for minimum variance efficiency. First, some definitions: A portfolio is *conditionally multifactor minimum variance efficient (CMMV) with respect to Z* iff the weights solve

$$\text{Min}_{x(Z)}\ \text{Var}[x(Z)'R|Z],$$
$$\text{s.t. } E[x(Z)'R|Z]=\mu_p(Z),\ E[Fx(Z)'R|Z]=C(Z),\ x(Z)'1=1, \tag{14.4}$$

where $\mu_p(Z)$ and $C(Z)$ are known functions of Z. Note that with $\mu_p(Z)$ fixed, we could just as well constrain $\text{Cov}\{F, x(Z)'R|Z\}$ instead of $E[Fx(Z)'R|Z]$, and obtain the same solution. This is what Merton (1973) and Fama (1996) had in mind, except conditioning on Ω instead of Z. It turns out that there is a version of the Hansen-Richard critique, discussed in chapter 9, for multifactor models, just like there is for mean variance models, which applies because we can't observe the full public information set. For now, we will define the conditioning in terms of the observable Z that we can work with. A portfolio is *unconditionally multifactor minimum variance efficient (UMMV) with respect to the information Z* iff the weights x(Z) solve

$$\text{Min}_{x(Z)}\ \text{Var}[x(Z)'R],$$
$$\text{s.t. } E[x(Z)'R]=\mu_p,\ E[Fx(Z)'R]=C,\ x(Z)'1=1, \tag{14.5}$$

where μ_p and C are constants. This generalizes the Hansen and Richard (1987) and Ferson and Siegel (2001) unconditional minimum variance efficiency to the case of multifactor efficient portfolios.

14.1.2   Explicit Solutions

Ferson, Siegel, and Xu (2006) derive UMMV portfolios using the calculus of variations. The approach is similar to that illustrated in chapter 9 when deriving the UE portfolio solution. The closed-form solutions are as follows. Let $\Delta(Z)=E\{RR'|Z\}^{-1}$ and $\Omega(Z)=[\Delta-\Delta 11'\Delta/(1'\Delta 1)]$. The CMMV solution is

$$x(Z)=\Delta 1/(1'\Delta 1)-\lambda_1(Z)\Omega(Z)\mu(Z)-\Omega(Z)E(RF'|Z)\lambda_2(Z), \tag{14.6}$$

where $\lambda_1(Z)$ and $\lambda_2(Z)$ are Lagrange multipliers that depend on Z. $\lambda_1(Z)$ is a scalar-valued function, and if there are K factors, $\lambda_2(Z)$ is a K vector-valued function. Ferson, Siegel, and Xu solve for the multipliers by plugging in the constraints. I am sure you could do that now, so I won't bother with the details. The CMMV weight exhibits $K+2$ fund separation. The first term of equation (14.6) is the conditional global minimum second-moment portfolio weight. The second term is a mean variance efficient excess return, and the third term encompasses K hedging terms for the K risk factors. Note that $1'\Omega(Z)=0$, so all but the first term are excess returns where the portfolio weights sum to zero.

The CMMV solution has several interesting special cases. One of them is that if the Lagrange multipliers λ_1 and λ_2 are constants, Ferson, Siegel, and Xu (2006) show that the solution is UMMV, as defined above. This implies that UMMV portfolios are a special case and thus a subset of CMMV portfolios. This generalizes the big result in Hansen and Richard (1987) about conditionally and unconditionally minimum variance efficient portfolios with respect to Z. In fact, if we set $\lambda_2 = 0$, we obtain the conditionally and unconditionally minimum variance efficient portfolios with respect to Z as special cases.

14.1.3   Representation

Let us now generalize the representation for minimum variance efficient portfolios to the case of UMMV and CMMV portfolios. Basically, by inspecting the explicit solutions in equation (14.6), we can see that any CMMV portfolio can be expressed as

$$R_0^* + w(Z)R_e^* + \sum_{j=1,\ldots,K} y_j(Z)R_{ej}^*, \tag{14.7}$$

where R_0^* is a CMMV portfolio with weights that sum to 1.0, and R_e^* and R_{ej}^* are CMMV excess returns (weights sum to 0.0). R_0^* and R_e^* are the portfolios used in the Hansen and Richard (1987) representation, and the R_{ej}^* are hedge portfolios for the factors F.

   Any UMMV portfolio can be expressed as

$$R_0^* + wR_e^* + \sum_{j=1,\ldots,K} y_j R_{ej}^*. \tag{14.8}$$

where R_0^*, R_e^*, and R_{ej}^* are the same portfolios, but w and the $\{y_j\}$ are constants. This again shows UMMV portfolios are special cases of CMMV portfolios.

   Finally, any return R_i can be expressed as its UMMV portfolio with the same mean plus noise:

$$R_i = R_0^* + w_i R_e^* + \sum_{j=1,\ldots,K} y_{ij} R_{ej}^* + \varepsilon_i, \tag{14.9}$$

where the weights w_i and $\{y_{ij}\}$ are now asset specific. This corresponds to forming combinations of the K + 2 funds to find a portfolio that lies on the UMMV boundary at the same mean return as the asset in question. The asset itself lies inside the boundary at a distance determined by the standard deviation of ε_i.

14.1.4   Relation to Asset Management

We discussed modern portfolio management in chapter 8. In this setup, managers try to earn abnormal returns or alphas relative to a benchmark. Chiang (2015) studies modern portfolio management with explicit conditioning information Z. Think of a problem where the manager has the Z, but the clients don't see it, so they care about the unconditional tracking error and alpha of the portfolio, defined in terms of the difference between the managed portfolio return and the benchmark return R_B. The problem is

$$\text{Min}_{\{x(z)'1=1\}} \text{Var}(R_p - R_B) \text{ s.t. } E(R_p - R_B) = \alpha_p, \tag{14.10}$$

where α_p is the target alpha, and the portfolio return is $R_p = x(Z)'R$. Apparently, defining the alpha and tracking error in this way assumes that the beta of the managed portfolio on the benchmark is 1.0. This is often done for simplicity and to avoid estimation error in the betas (Chiang considers other versions of the problem where this assumption is not imposed).

What does all this have to do with multibeta pricing? Since $Var(R_p - R_B) = Var(R_p) + Var(R_B) - 2Cov(R_p, R_B)$, and $Var(R_B)$ cannot be controlled by the manager, the above problem is equivalent to

$$\text{Min}_{\{x(z)'1=1\}} Var(R_p) \text{ s.t. } E(R_p) = \mu_p, Cov(R_p, R_B) = C, \qquad (14.11)$$

for some targets μ_p and C. This is the UMMV portfolio problem, where the benchmark return R_B is the factor. Modern portfolio managers, it seems, are induced to think of their benchmarks as risk factors that they would like to hedge.

14.1.5   Utility Maximization

Who would want to hold these UMMV portfolios, anyway? One answer suggested above is a modern portfolio manager who is evaluated relative to a benchmark and trades using information that her clients can't see. Ferson and Siegel (2001) show that unconditionally minimum variance efficient portfolios maximize a quadratic utility function when the agent has access to the conditioning information. So, quadratic agents in a single-period model would hold UE portfolios. Ferson, Siegel, and Xu (2006) show this generalizes to UMMV portfolios. Consider the case with a single factor and an agent with an indirect value function that depends on wealth and the factor J(W, F). It turns out that agents with particular forms of J($\cdot$) would find UMMV portfolios to be optimal:

$$\begin{aligned} &J(W, F) = W - aW^2 - bWF, \text{ or} \\ &J(W, F) = [1 - bF]W - aW^2. \end{aligned} \qquad (14.12)$$

Evidently, agents with quadratic indirect value functions, where the linear term depends on the factor, would find the UMMV portfolios to be optimal. I don't know any of these agents personally, but they might be around.

14.1.6   Hedge Portfolios and Maximum Correlation Portfolios

CMMV and UMMV portfolios minimize variance subject to target mean returns and target covariances with the factors. Suppose that we set the multiplier on the target mean to zero and consider a single factor for simplicity. In this special case, CMMV and UMMV portfolios may be said, by duality, to maximize the covariance with the factor, given a target for the portfolio variance. Such portfolios could be called *covariance-variance efficient*. This is just like mean-variance efficient, except that we replace the mean with the covariance. It turns out that the solutions to this problem are the hedge portfolios in the CMMV or UMMV solution depicted in equation (14.6).

Covariance-variance efficient portfolios are actually easy to analyze, and in fact you already know a great deal about them. Think of a mean-standard deviation bullet in the classical mean-variance problem of chapter 8. Replace the mean on the y-axis with covariance with the factor F. Since the covariance combines linearly in portfolios in exactly the same way that the mean does, virtually all the results for mean variance efficient portfolios and the classical bullet from chapter 8 can be applied to this problem. With these results, you know a great deal about the behavior of hedge portfolios!

Maximum correlation portfolios are a special case of covariance-variance efficient portfolios. Think of what is analogous to the Sharpe ratio on the new bullet: $\mathrm{Cov}(r_p, F)/\sigma(r_p) = \rho\sigma(F)$. Since we can't control $\sigma(F)$, maximizing the correlation ρ is equivalent to maximizing this ratio, the slope of a line drawn from the origin and tangent to a point on the new bullet. The portfolio at the tangent point is the maximum correlation portfolio for the factor F.

14.2 The Three Paradigms of Empirical Asset Pricing, Revisited

We described the three paradigms of empirical asset pricing in chapter 7. They are minimum variance efficiency, beta pricing, and of course, m-talk. The previous discussion waxed poetic about the beautiful three-way equivalence of the three paradigms (triality) but did not consider conditioning information. Now it's time to bring in the Z's! After we do that, we discuss a new framework for testing asset pricing models that relies on the relation between m-talk and minimum variance efficiency with conditioning information.

14.2.1 Conditional Version of the Three Paradigms

The simplest way, at least conceptually, to bring in conditioning information is to just reinterpret everything as applying to the conditional moments, given Z. Then chapter 7 can be applied directly. In particular, m-talk is equivalent to beta pricing, in the sense that m is conditionally linear in some factors if and only if those factors work for a conditional beta pricing model. By conditionally linear, I mean that the coefficients may be functions of Z:

$$m_{t+1} = a(Z_t) + \sum_j b_j(Z_t) F_{j,t+1}. \tag{14.13}$$

By analogy to the earlier case, it is OK to add an error term u_{t+1} to the above equation, provided it is conditionally uncorrelated with the asset returns: $E(u_{t+1}R_{t+1}|Z_t) = 0$.

To continue this version of the triality, conditional beta pricing is equivalent to conditional minimum variance efficiency in the sense that some combination of the factors (actually, their maximum conditional correlation portfolios) is a conditionally minimum variance portfolio. Finally, conditional minimum variance efficiency is equivalent to m-talk, in the sense that a maximum conditional correlation portfolio to m is a conditionally minimum variance efficient portfolio. Lots of conditioning here!

14.2.2   The Three Paradigms with Respect to Conditioning Information

I wonder: Did you see this coming? Now we want to work with unconditional means, variances, and covariances, but we want to consider conditioning information through the portfolio weights. Essentially, we expand the three paradigms to consider not just the given test asset returns R_{t+1} but also the infinite set of portfolios that can be formed using the information $x(Z_t)'R_{t+1}$. The three paradigms are all stated *with respect to Z*. This version of the three paradigms and their equivalence provides a framework for testing asset pricing models, coming up next.

Minimum variance efficiency in this framework is stated as in chapter 9, where we called it "unconditional efficiency" (UE) with respect to Z.

Minimum variance efficient w.r.t. Z:

$$\text{Min}_{x(Z)'1=1} \text{Var}(x(Z)'R) \text{ s.t. } E(x(Z)'R) = \mu_p. \tag{14.14}$$

We also need a version of beta pricing for the infinite set of portfolios.

Beta pricing w.r.t. Z:

$$E(x(Z)'r) = \sum_j [\text{Cov}(x(Z)'r, F_j)/\text{Var}(F_j)]\lambda_j, \text{ all } x(Z): x(Z)'1 = 1. \tag{14.15}$$

I defined the betas here as the simple regression betas, not multiple regression betas. However, given rotational indeterminacy as discussed in chapter 12, this is without any loss of generality. Note that the risk premiums in the beta pricing relation are constants over time and do not depend on Z.

Finally, we use a new version of *m*-talk. Maybe we should call this *m*-talk "with respect to Z:"

_m_-talk w.r.t. Z: $E(mx(Z)'R) = 1$, all $x(Z)'1 = 1$. $\tag{14.16}$

To complete the triality of the three paradigms with respect to Z, we need one more idea: the maximum (squared) correlation portfolio with respect to Z. A portfolio R_P is a *maximum correlation portfolio for a random variable m with respect to conditioning information Z*, if

$$\rho^2(R_P, m) \geq \rho^2[x'(Z)R, m] \quad \forall x(Z): x'(Z)1 = 1, \tag{14.17}$$

where $\rho^2(\cdot, \cdot)$ is the squared unconditional correlation coefficient, and we restrict the functions x to those for which the correlation exists.

With this machinery in place, we can state the equivalence between the three paradigms with respect to Z. First, *m*-talk with respect to Z is equivalent to minimum variance efficiency with respect to Z, in the sense that a portfolio that is maximally correlated to *m* with respect to Z is efficient with respect to Z (Ferson and Siegel 2009). Second, beta pricing with respect to Z is equivalent to *m*-talk with respect to Z, in the sense that you get beta pricing with respect to Z using a set of factors if and only if an *m* that works in

m-talk with respect to Z is linear in the factors. The linear relation of m to the factors may have time-varying weights that depend on Z. Finally, beta pricing with respect to Z is equivalent to minimum variance efficiency with respect to Z, in the sense that a combination of the factor portfolios must be efficient with respect to Z. The intuition is the same as the classical case with no conditioning information. The difference is that the set of returns is expanded to all $x'(Z)R$. The next section discusses how all this is used in asset pricing tests.

14.2.3 Testing Asset Pricing Models with Respect to Conditioning Information

This section presents selected results from Ferson and Siegel (2009). Most empirical work on "conditional asset pricing" actually uses unconditional moments, combined with ad hoc functional forms for the time-varying coefficients in the model. For example, "conditional CAPMs" are studied by Harvey (1989), Shanken (1990), Cochrane (1996), Jagannathan and Wang (1996), Lettau and Ludvigson (2001b), Lewellen and Nagel (2006), and many others. A typical example specifies

$$m = b(Z)'(1, R_m), \tag{14.18}$$

where $b(Z)$ is a linear function of Z, and R_m can be a vector of factor portfolios. Thus, m is linear in Z and $(R_m Z) : m = B_1' Z + B_2'(R_m Z)$. This is an example of a multiplicative approach to the conditioning information. In this example, we have an unconditional model where Z and $R_m Z$ are the factors, the coefficients, are constants, and unconditional moments are used.

More generally, the most common approach to testing an asset pricing model is to examine necessary conditions of equation (1.2). For example, multiplying both sides of equation (1.2) by the elements of Z_t and then taking the unconditional expectations leads to a multiplicative approach:

$$E\{m_{t+1}(R_{t+1} \otimes Z_t)\} = E\{\underline{1} \otimes Z_t\}. \tag{14.19}$$

Equation (14.19) uses the stochastic discount factor to price the dynamic strategy payoffs, $R_{t+1} \otimes Z_t$, on average (or "unconditionally"), where the $E\{\underline{1} \otimes Z_t\}$ are the average prices. However, the multiplicative approach captures only a portion of the information in equation (1.2). By using the right functions of Z_t, we can capture more of the information.

Of course, if the choice of Z excludes important information, this will result in a loss of power. Not to mention the Hansen-Richard critique, which states that we can't test the model at all if m depends on information Ω_t that we can't measure. The way out of this multivariate version of the Hansen-Richard critique is to assume that the representative agent has the indirect utility function specified in equation (14.12). Then the agent will optimally choose CMMV portfolios that are also UMMV, and we can test if the portfolio is UMMV. In this section we take the choice of Z as given and assume that the models for m depend at most on Z and other data that we can measure.

As discussed and proved in chapter 11, equation (1.2) is equivalent to the following equation holding for *all* bounded integrable functions $f(\cdot)$:

$$E\{m_{t+1}[R_{t+1} f(Z_t)]\} = E\{\underline{1} f(Z_t)\}. \tag{14.20}$$

The *m*-talk with respect to Z equation uses all portfolio weight functions $x(Z)$ in place of the general functions above, subject only to the restriction that the weights are bounded integrable functions that sum to 1.0. Using portfolio weight functions sacrifices some generality, but not much. The restriction that the weights sum to 1.0 for each realization of Z does not allow the solution to expand and contract the scale of the risky investment without borrowing or lending at the risk-free rate. With negative weights allowed, it does admit stylized short sales.

By referring to all portfolio weights in the *m*-talk with respect to Z equation (14.16), the tests can reject asset pricing models that previous methods would not reject. While many asset pricing models are rejected in the literature, Lewellen, Nagel, and Shanken (2010) argue, as discussed in chapter 12, that it may be too easy to find models that appear to explain some returns, like those of the Fama and French (1996) portfolios. Ferson and Siegel (2009) show that the *m*-talk with respect to Z approach appears to be powerful in that setting.

The Ferson and Siegel (2009) framework tests SDF models by finding the portfolio that a particular *m* says should be efficient with respect to Z and then testing the hypothesis that the indicated portfolio is indeed efficient with respect to Z. The tests involve the solution to portfolios that are efficient with respect to Z. Ferson and Siegel (2001) study the shape of the optimal weight function of such portfolios, as described in chapter 9. They argue that the portfolios are likely to be robust to extreme observations. The nonlinear shape makes them conservative in the face of extreme realizations of Z. Ferson and Siegel (2003) apply the expressions to the Hansen and Jagannathan (1991) bounds and find robustness in that setting. Ferson, Siegel, and Xu (2006) find evidence of robustness in the special case of maximum correlation portfolios with respect to Z.

The *m*-talk with respect to Z approach is inherently robust to misspecifying the conditional moments of returns. The argument is that with the wrong conditional moments, the expression used for the "optimal" $x(Z)$ that defines the efficient with respect to Z bullet is actually suboptimal. However, the solution still describes a valid portfolio strategy, and it should still expand the minimum variance boundary, if not to the maximum possible extent using Z. Thus, the tests may sacrifice power, but they remain valid with misspecified conditional moments.

The key to the *m*-talk with respect to Z approach is the relation between *m*-talk and minimum variance efficient portfolios with respect to Z. The three paradigms with respect to Z state that any model for *m* identifies a portfolio that should be minimum variance efficient with respect to Z. The Ferson-Siegel (FS) approach for any given model for *m* is to find the indicated portfolio and then test the hypothesis that it is efficient with respect to Z. This can be done by drawing the unconditionally efficient bullet with respect to Z, as described in chapter 9. Then we measure the distance of the tested portfolio from this minimum variance bullet and test whether it lies significantly inside the bullet, accounting for sampling error. Tests for the significance of the distance of a portfolio inside a bullet are described in more detail in chapter 21.

Ferson and Siegel (2009) develop an interesting application of their framework to *conditional* efficiency given Z. Previous studies test *conditional* efficiency given Z, where efficiency is defined in terms of the conditional means and variances given Z. Hansen and Hodrick (1983) and Gibbons and Ferson (1985) test versions of conditional efficiency given Z, assuming constant conditional betas. Campbell (1987) and Harvey (1989) test conditional efficiency restricting the form of a market price of risk, and Shanken (1990) and Ferson and Schadt (1996) restrict the form of time-varying conditional betas. Tests of conditional efficiency given Z may be handled as a special case of the FS framework.

If there is a (possibly time-varying) combination of the k benchmark returns R_B that is conditionally efficient, then there is an SDF, $m = A(Z) + B(Z)'R_B$. The coefficients are $A(Z) = R_f^{-1} - E(R_B|Z)'Cov(R_B|Z)^{-1}[\underline{1} - R_f^{-1}E(R_B|Z)]$ and $B(Z) = Cov(R_B|Z)^{-1}[\underline{1} - R_f^{-1}E(R_B|Z)]$. These coefficients are derived by requiring the model to price the benchmark returns. When $K = 1$, we have a single-factor model, as in the conditional CAPM. We can test conditional efficiency by constructing the maximum correlation portfolio for the indicated m with respect to Z. This portfolio, call it R_p^*, should be efficient with respect to Z. Note that R_p^* will be different from R_B when the coefficients A(Z) or B(Z) are time-varying functions of Z. Thus, for example, the conditional CAPM does not imply that the market portfolio is efficient with respect to Z. However, the conditional CAPM does identify a portfolio of the test assets that should be efficient with respect to Z, and its efficiency can be tested.

Ferson and Siegel (2009) present tests that show that the FS framework is powerful, in the sense that models can be rejected that would not be rejected using a multiplicative approach to the conditioning information. In some cases, models can be rejected that could not even be examined using a multiplicative approach, given the large number of implicit test assets in the multiplicative approach. Ferson and Siegel (2009) find that the Fama and French (1996) three-factor model is strongly rejected when using all the dynamic trading strategies and what at the time were standard lagged Z values, even when the model cannot be rejected using only the standard portfolios provided by Kenneth French.

To date, the main limitation of the FS framework is that there has been no sampling theory available for the test statistics. Ferson and Siegel (2009) resort to simulations to conduct statistical inference, and the simulations are somewhat complex. Stay tuned, however, as Ferson et al. (2018) provide and evaluate an asymptotic distribution theory that makes tests in the FS framework possible without the need for simulations.

IV EMPIRICAL ASSET PRICING TOOLS

15 Introduction to the Generalized Method of Moments

This chapter introduces the generalized method of moments (GMM) of Hansen (1982). It's Nobel prize winning work, and for good reason. I first motivate the approach from an asset pricing perspective and present a couple of simple examples that suggest why it is so general. Chapters 16–20, combined with this one, provide a fairly complete tool box for applying the GMM in applied problems. Appendix sections A.3 and A.4 provide some code for implementing the GMM.

15.1 Motivation and General Setup

Before Hansen's (1982) classic work, the main empirical methods in asset pricing were cross-sectional regressions and seemingly unrelated regression models, the latter often estimated under normal ML. Since all these things are special cases of the GMM, it makes sense to study the GMM first.

Normal ML assumes that the data—such as common stock returns—are normally distributed. But stock returns are not normally distributed. Their distributions are fat tailed and skewed (differently for individual stocks and portfolios), not homoskedastic and not serially independent. Fortunately, the GMM can handle all those messy aspects of financial market data. By this, I mean that the asymptotic distribution theory that gives us standard errors and confidence bands works under such conditions and can often be adjusted for the conditions at hand. The GMM also handles nonlinear models with aplomb. Many asset pricing models are nonlinear. For example, the simple consumption model, with $m_{t+1} = \beta(C_{t+1}/C_t)^{-\alpha}$, is nonlinear in its parameters (α, β).

Besides its generality and flexibility, the GMM paired with the m-talk equation (1.2) is a match made in heaven. Combining the two creates a virtual cookbook for the estimation of almost any asset pricing model. It should be no surprise that Lars Hansen shared the Nobel Prize in economics for his work in 2013.

Any asset pricing model is a statement about what the SDF m is. Suppose the model specifies $m_{t+1} = m(x_{t+1}, \theta)$; where x_{t+1} is data observed at time $t+1$. Define the error term

$u_{t+1} = m_{t+1}R_{t+1} - 1$. Then the *m*-talk equation (1.2), together with the informational efficiency assumption (that the observed instruments Z_t are fully reflected in market prices) implies $E(u_{t+1}|Z_t) = 0$. This says that u_{t+1} is orthogonal to Z_t, often referred to as an *orthogonality condition*. We saw in chapter 11 that this condition is equivalent to the statement that $E(u_{t+1}f(Z_t)) = 0$ for all reasonable functions $f(\cdot)$. However, with a few exceptions, a multiplicative approach is used, and only the condition $E(u_{t+1} \otimes Z_t) = 0$ is typically exploited. (The operator $\otimes$ is the Kronecker product, meaning each element of u is multiplied by the matrix or vector Z.)

Of course, there is nothing to say that the same instruments Z_t must be associated with each error term u_{t+1}. We can build up the orthogonality conditions without that assumption and the GMM will work just fine. For exposition, however, let's stick with the symmetric case.

Let $g_t = (u_{t+1} \otimes Z_t)$, and define $g = \sum_t g_t / T = (\sum_t u_{t+1} \otimes Z_t)/T = \text{Vec}(Z'u/T)$, where the Vec($\cdot$) operator stacks the columns of its argument to make a big column vector. Thus, g denotes the sample mean of the moment conditions. The goal in GMM estimation is to find the parameter vector $\varphi = \text{Vec}[\theta]$ that makes the sample mean of the moment condition as close to zero as possible. But first, some interpretation is in order.

In a typical asset pricing problem, the sample moment condition g is a "pricing error." To illustrate, let $Z = 1$, and observe that $E(g) = E(mR - 1)$ is the SDF alpha. This shows that if we find the GMM parameter estimates by minimizing a quadratic form in the vector g,

$$\text{Min}_\varphi \; g'Wg, \qquad\qquad\qquad (15.1)$$

it has the interpretation of minimizing the distance of the average pricing errors from zero. So, we fit the parameters of our asset pricing models, using the GMM, to make the pricing errors small. When we have lagged instruments and $g_t = (u_{t+1} \otimes Z_t)$, we are minimizing the pricing errors of the expanded set of dynamic trading strategies that are implied, in this case, by the multiplicative approach.

Results from the classic paper by Hansen (1982) justify the GMM approach. Hansen (1982, theorem 2.1) shows under technical assumptions that if the parameter estimates minimize $g'Wg$, for any W approaching a constant almost surely in large samples, then the GMM estimates are consistent and asymptotically normal. (The technical assumptions include the assumptions that the data $\{u_{t+1}, Z_t\}$ are strictly stationary and ergodic, that $\partial(Z'u/T)/\partial\varphi$ satisfies continuity conditions close to the true parameter values, and that the parameters lie in a compact set. It's much more carefully done and complicated in the original paper.) Hansen gives an expression for the covariance matrix of the parameter estimates in this general case, discussed below.

Hansen (1982, theorem 3.2) also shows that if we select W to be a consistent estimator for $\text{Acov}(g)^{-1}$ (defined below equation (15.2)), then the GMM estimates are efficient in the class of quadratic-form minimizing estimators (for a given choice of Z). In this case, we have "optimal GMM" estimation, and the parameter covariance matrix is asymptotically given by

$$\text{AVar}[\varphi] = [(\partial g/\partial\varphi)'W(\partial g/\partial\varphi)]^{-1}, \qquad\qquad (15.2)$$

where $\text{Avar}[\cdot]$ and $\text{Acov}(\cdot)$ denote the covariance matrix for the limiting distribution of $T^{1/2}$ multiplied by the estimated object in parentheses. We have to multiply by $T^{1/2}$ in order to obtain a nondegenerate random variable as T gets large, because as a consistent estimator approaches the true value of the coefficient in probability, the variance goes to zero.

Hansen's theorem also provides a goodness-of-fit statistic, which we discuss in chapter 18. This is known as the *J*-statistic: $Tg'Wg$, and it measures how close the sample moment conditions g are to zero. When the optimal weighting matrix W is used, Hansen shows that the *J*-statistic is asymptotically chi-square distributed.

15.2 Simple Examples

Almost every estimation scheme on the planet is a special case of the GMM. In chapter 20, we go through a bunch of examples. For now, let's start with two simple examples to illustrate the ideas. The first is ordinary least squares (OLS) and the second is standard method of moments.

15.2.1 Ordinary Least Squares

Let $u = y - z\beta$ be the $T \times 1$ vector of OLS regression residuals, y is $T \times 1$, z is $T \times K$, and β is the $K \times 1$ vector of coefficients to be estimated (z can include a leading column of ones if there is an intercept in the regression). If the data are not normal, we still have $E(u'z) = 0$ by the definition of the regression coefficient. This condition (along with the regularity assumptions described above) is good enough for the GMM. Normality not required. Let's show that the GMM estimator for this problem gives the OLS estimator. In this problem, $g = \text{Vec}(z'u/T) = z'y/T - (z'z/T)\beta$. This is an example of an exactly identified GMM problem, because the number of moment conditions, $\dim(g)$, is equal to the number of parameters, $\dim(\beta) = K$. This is an easy example, because we can minimize $g'Wg$ by just setting $g = 0$. Clearly, if W is positive semidefinite as assumed, then $g'Wg$ can't be smaller than zero, so $g = 0$ determines the minimum. Setting $g = 0$, we solve for the estimator $\beta = (z'z)^{-1}(z'y)$, which is our beloved OLS estimator.

Now, since $g = 0$ is the minimum for $g'Wg$ no matter which W is used, we can say that we used the optimal weighting matrix, and no one will ever tell. The asymptotic standard error is given by equation (15.2). In this example,

$$\text{Acov}[\beta] = [(\partial g/\partial \beta)' W (\partial g/\partial \beta)]^{-1} = [(-z'z/T)\text{Cov}(g)^{-1}(-z'z/T)]^{-1}$$
$$= (z'z/T)^{-1}\text{Cov}(g)(z'z/T)^{-1} = (z'z/T)^{-1}\,E(z'uu'z/T^2)(z'z/T)^{-1}.$$

A very important step above is between the first and second lines, where the inverse of the product of the three terms is written as the product of the three inverses. This can only be done if $(\partial g/\partial \beta)$ is invertible; in this case $K \times K$ and full rank. This is important in the implementation of exactly identified models, because it means that you don't have to invert $W^{-1} = \text{Cov}(g)$. In exactly identified models, you can't invert the sample $\text{Cov}(g)$, so it's a good thing that you don't have to in order to find the standard errors of your coefficients. You should beware of this case if you use standard GMM code from some high-level computer program, because some of them will try to invert the product in equation (15.2) when given an exactly identified problem, which can result in a poorly determined inverse and a big mess.

The preceding equation gives an expression for consistent standard errors when you use OLS to estimate the coefficients, but the usual OLS assumptions about the data are not satisfied. All we really had to assume to get this expression is that $E(u'z)=0$. For example, suppose that $E(uu'|z)=\Sigma$ is some general covariance matrix for the error terms of the regression. A consistent estimator for the covariance matrix of the OLS estimator of β is $(z'z)^{-1}E(z'\Sigma z)(z'z)^{-1}$.

In the special case where the errors are independent and identically distributed, then the GMM asymptotic variance gives the OLS standard errors. Assume $E(uu'|z)=\sigma^2 I$, and the standard error for the OLS estimator simplifies to $\mathrm{Acov}[\beta]=\sigma^2(z'z)^{-1}$.

15.2.2 Classical Method of Moments

Suppose $x_t \sim$ iid (μ, σ^2), and you want estimates of μ and σ^2. Let's take a GMM approach. Define $u1_t = x_t - \mu$; thus $E(u1_t)=0$. Letting $u2_t = \sigma^2 - [x_t - \mu]^2$, then $E(u2_t)=0$. Letting $Z_t = 1$, we can find the GMM estimates for μ and σ^2. While it is a starkly simple example, it illustrates a trick I like to use to help me formulate GMM problems and keep track of everything. I set up a table like table 15.1. North stands for the number of moment conditions and K stands for the number of parameters.

Adding up the number of parameters and moment conditions shows that we have two of each, so the model is exactly identified. Form $g_t = (u1_t, u2_t)'$ and $g = (1/T)\sum_t g_t$. Setting $g = 0$, we can solve for μ and σ^2, giving us expressions for the estimators that are equal to the usual sample mean and variance.

We can take this example further to illustrate the asymptotic standard errors. In this model, $(\partial g/\partial \varphi) = [(\partial g/\partial \mu), (\partial g/\partial \sigma^2)]$, which looks like a 2×2 identity matrix except with a -1 in the upper-left corner. Since the problem is exactly identified, we have

$$[(\partial g/\partial \varphi)' W (\partial g/\partial \varphi)]^{-1} = [(\partial g/\partial \varphi)^{-1} \mathrm{Cov}(g)(\partial g/\partial \varphi)'^{-1}]. \tag{15.3}$$

This is looking easy, because the derivative matrices are already their own inverses. We just need $\mathrm{Cov}(g)$. On the assumption that there is no serial dependence in x_t (generalizing

Table 15.1
GMM moment condition table.

Moment condition	Orthogonal to	North	K
$u1_t = x_t - \mu$	1	1	1
$u2_t = \sigma^2 - [x_t - \mu]^2$	1	1	1

Note: K is the number of parameters and north is the number of moment conditions.

later), we can estimate Cov(g) with $(1/T)\sum_t g_t g_t'$, where the moment conditions g_t are evaluated at the point estimates of the parameters. Since the point estimates are consistent, the resulting covariance matrix estimate will be consistent as well. This is a good exercise to try. When you are done, you have a 2×2 covariance matrix. The most complicated part is the asymptotic variance of the estimate of the variance, in the lower-right corner. Letting $x^* = (1/T)\sum_t x_t$, the solution is $(1/T)\sum_t[(x_t - x^*)^2 - (1/T)\sum_t(x_t - x^*)^2]^2$, the sampling variance of the sample variance! Pretty cool.

16 GMM Implementation

This chapter describes how to implement the GMM in practice. The approach used to implement the GMM differs depending on the type of problem, and it is useful to distinguish four basic cases. First, I provide an overview, and then describe how to solve the GMM estimation problem in practice, depending on which of four basic cases you have. This chapter finishes up with a first look at how to compute the GMM covariance matrix, a topic we then attack in detail in chapter 17.

16.1   Overview

The general idea is to pick ϕ to minimize $g'Wg$. Taking the first-order condition, we have

$$\text{Min } g'Wg \Rightarrow (\partial g/\partial \phi)'Wg(\phi) = 0. \tag{16.1}$$

It is said that nothing is ever really new in research. Maybe that is why they call it "research." We are always looking again—looking back at problems that others have looked at before. (Hopefully, we actually know that our problem has been looked at before!) In the case of Hansen (1982), the earlier work by Denis Sargan (1959) is cited. Sargan solved the method of moments problem when the condition was $Ag = 0$ for some fixed matrix A. Hansen's contribution was "only" to pick $A = [(\partial g/\partial \phi)'W]$.

Let's set up the dimensions of the (common asset pricing) problem, where $E(g) = E([mR - 1] \otimes Z) = 0$. Let $N = \dim(mR - 1) =$ number of assets, $L = \dim(Z) =$ number of lagged instruments, $\dim(g) =$ "North" = number of moment conditions, and $K = \dim(\phi) =$ number of parameters. The four basic cases of GMM problems are depicted in figure 16.1.

The solution method depends on whether the problem dimension is exactly identified or overidentified and whether the model is linear or nonlinear in the parameters. Each of the four cases justifies specialized code, and I keep four GMM programs in my toolbox: one for each of the above cases.

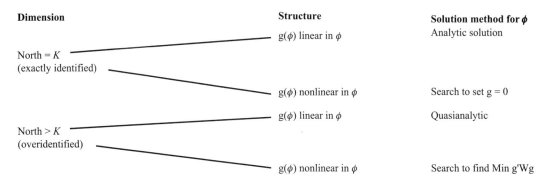

Figure 16.1
The four basic cases of GMM problems: two dimensions (exactly identified or overidentified)×two parameter types (linear or nonlinear).

16.2   Exactly Identified Models

In exactly identified models, $\dim(g) = \dim(\phi)$. This was the case in the OLS and sample method-of-moments problems that we started out with in chapter 15. In such models, we can simply set $g(\phi) = 0$ to find the parameter estimates algebraically. The estimates don't depend on the weighting matrix. Thus, to get the asymptotic standard errors, we pretend that we used the optimal GMM weighting matrix and employ the expression

$$\text{Var}(\phi) = [(\partial g/\partial \phi)' W (\partial g/\partial \phi)]^{-1} = (\partial g/\partial \phi)^{-1} \text{Cov}(g)(\partial g/\partial \phi)'^{-1}. \tag{16.2}$$

This is an important result. Sometimes, a generic program will try to use the expression on the left, essentially inverting twice. But if $\text{Cov}(g)$ is poorly conditioned, this approach might not work out well. Better to use the expression on the right. Otherwise, the first case in figure 16.1 is easy.

For the second case in figure 16.1, the model is exactly identified but nonlinear in the parameters. An example might be the simple power utility consumption model, where $m = \beta(C_{t+1}/C_t)^{-\alpha}$, which is nonlinear in the risk aversion parameter α. That model would be exactly identified if two assets and no conditioning information were used. But even in the exactly identified case, you can't find ϕ in closed form by setting $g(\phi) = 0$. A search routine has to be used.

Perhaps the simplest, workhorse search routine is the Newton-Raphson method, which can be described as follows. Start with an initial guess at the parameter vector, ϕ_0. Probably, we find that $g(\phi_0) \neq 0$. Then take an "educated" guess at how to try again, moving to ϕ_1, by using a Taylor series approximation and hoping for $g(\phi_1) = 0$:

$$0 - g(\phi_0) \approx [(\partial g/\partial \phi)|\phi_0](\phi_1 - \phi_0),$$
$$\phi_1 = \phi_0 - [(\partial g/\partial \phi)|\phi_0]^{-1} g(\phi_0). \tag{16.3}$$

If the model is well identified and ϕ_0 is not too far away from the true value, then $[(\partial g/\partial \phi)|\phi_0]$ should be invertible. The search can be iterated by replacing ϕ_0 on the right-hand side of the second line with ϕ_1, to find ϕ_2, and so on. The sequence should converge to the solution.

16.3   Linear Models

If the model is linear in the parameters, things are simpler, even if the model is over-identified. When the model is linear in the parameters, $g(\phi)$ is linear in φ: $g = A + B\varphi$, so $(\partial g/\partial \phi) = B$, and the matrices A and B do not depend on the model parameters. This means we can compute them once as a function of the data. We can solve $[(\partial g/\partial \phi)'W]g(\phi) = 0$ analytically for ϕ and W, even if the model is overidentified (i.e., even if $\dim(g) > \dim(\phi)$). For the linear model, the first-order condition (16.1) specializes to $B'W[A + B\phi] = 0$, which gives an expression for the parameter vector:

$$\phi = (B'WB)^{-1}(-B'WA). \tag{16.4}$$

However, the weighting matrix W depends on the parameter vector: $W(\phi) = [Cov(g)]^{-1}$. (This is why I called it "quasianalytic" in figure 16.1.) Start with an initial value for W, say, $W = I$. This provides consistent estimates for the parameters $\phi_0 = (B'B)^{-1}(-B'A)$. Given consistent estimates for the parameters, then Hansen's results ensure that $W(\phi_0)$ is a consistent estimate of $[Cov(g)]^{-1}$. It is possible to iterate back and forth between these two steps until the parameter estimates stop changing by any significant amount. In my experience, convergence occurs after only a few iterations.

16.4   Nonlinear, Overidentified Models

The fourth and final case is the most complicated, as we must search to minimize the nonlinear function $g'Wg$. There are, of course, many search algorithms out there that you can use. A version of the Gauss-Newton method is one of the simplest workhorse methods. This method is a special version of the Newton-Raphson approach described in section 16.2, designed for cases where the criterion is a sum of squared residuals ($g'Wg$ is a weighted sum of squared residuals if we think of g as the residual). As before, start with an initial guess at the parameter vector ϕ_0. When the model is overidentified, $g'Wg$ will be greater than zero at all parameter values. However, the first-order condition, $[(\partial g/\partial \phi)'W]g(\phi)$, will be zero for the parameter values that minimize $g'Wg$. To obtain the next guess for the parameter estimates, use a Taylor series to approximate $g(\phi_1)$, and write the first-order condition as

$$[(\partial g/\partial \phi)'W|\phi_0]\{g(\phi_0) + [(\partial g/\partial \phi)|\phi_0](\phi_1 - \phi_0)\} = 0,$$
$$\phi_1 = \phi_0 - [(\partial g/\partial \phi)'W(\partial g/\partial \phi)]^{-1} (\partial g/\partial \phi)'Wg(\phi_0), \tag{16.5}$$

where everything on the right-hand side is evaluated at ϕ_0.

The search can be iterated until convergence is achieved, as before. One of the cool things about this approach is that you "automatically" compute the asymptotic covariance matrix, $[(\partial g/\partial \phi)'W(\partial g/\partial \phi)]^{-1}$, as you go along. Note that if the model is exactly identified, the search reduces analytically to that for the exactly identified case.

16.5   One Step, Two Steps, N Steps, or More ...

When the GMM was new in the 1980s, finance researchers like me learned their GMM from the classical paper by Hansen and Singleton (1982). That paper uses the simple power utility consumption model and several lagged instruments for Z_t. It advocates a *two-step approach* to the GMM estimation. In the first step, set $W_1 = I$ and $\text{Min}\phi\ g'g$, producing the GMM estimates ϕ_1. According to Hansen (1982), these estimates are consistent but not efficient. In the second step, form a weighting matrix $W(\phi_1)$ using those parameter values. This matrix is a consistent estimate of $\text{Cov}(g)^{-1}$. Use this weighting matrix to find the second-step parameter estimates: $\text{Min}\phi\ g(\phi)'W(\phi_1)g(\phi) \Rightarrow \phi_2$. Now, the second-step estimates are asymptotically efficient in the class of quadratic form minimizing estimators. They can be used to compute the goodness-of-fit and other statistics described below. Also, you can compute the asymptotic covariance matrix from these using equation (15.2), which becomes $\text{Var}[\phi_2] = [(\partial g(\phi_2)/\partial[\phi_2])'W(\phi_2)(\partial g(\phi_2)/\partial[\phi_2])]^{-1}$.

Ferson and Foerster (1994) evaluate an iterated version of the GMM estimator, where at each step j, we compute

$$\phi_j = \text{Arg Min}\phi\ g(\phi)'W(\phi_{j-1})g(\phi),$$
$$J(\phi_j) = Tg(\phi_j)'W(\phi_j)g(\phi_j),$$
(16.6)

and if $[J(\phi_j) - J(\phi_{j-1})]$ is small enough, or the change in ϕ is small enough, we declare victory and stop. Ferson and Foerster find that the iterated approach has better sampling properties, in the sense that the finite sample properties discovered in simulations are closer to those predicted by the asymptotic formulas for the standard errors, and that the goodness-of-fit tests are more accurate than those using a two-step method.

Hansen, Heaton, and Yaron (1996) took this idea literally to the limit, developing a continuously updated GMM estimator, and showed that it is a limiting case of an iterated approach. They suggest that you solve

$$\phi = \text{Arg Min}_\varphi\ g(\phi)'W(\phi)g(\phi).$$
(16.7)

This is often a highly nonlinear problem, and it might sometimes be hard to solve, but they provide examples where it performs better in simulations than an iterated or two-step approach.

16.6   Search Problems

It's true that you can't always get what you want or even find what you want. In nonlinear problems, the search over the parameter space for the values that minimize $g'Wg$ might be difficult. Hansen's theorems justifying the asymptotic properties of GMM estimators assume that we have globally minimized $g'Wg$ over the whole parameter space. Yet most of the current search methods, such as Gauss-Newton and Newton-Raphson, are local search algorithms. If the criterion surface has multiple valleys, your search might get stuck in a valley that is not the lowest valley. If the criterion surface has a large flat spot, your search might just wander around the flat spot. Most people and some algorithms try multiple initial values (a *multistart* approach), hoping to find a starting point that leads the algorithm to the global minimum. But there is no obvious way to know whether that has worked. Another approach is to conduct a grid search over the space of economically reasonable initial values and select the parameter estimates that produce the smallest value of $g'Wg$.

A practical problem with the grid search is that if you have a lot of parameters, you must do a lot of calculations. Andrews (1997) gives an example with $K = 5$ parameters in the space $[0, L]^5$. You want to find the minimum of $g'Wg$ to within 1%. He computes that this takes 10^{20} function evaluations, which at one per second, requires 31.7 million years of computer time. Now, it is true that we can compute much faster than one function evaluation per second these days, but even at 100,000 times that speed, it takes over 30 years. They won't give you that much time to make tenure.

Andrews (1997) proposes a solution to this problem with a stopping rule. His procedure has several steps:

1. Run two-step GMM (or an initial grid search) to get the initial parameter value φ_0.

2. Check whether $T[g(\phi_0)'W(\phi_0)g(\phi_0)] \leq X^2(.05, \text{north-}K)$, the critical value of the chi-square distribution for degrees of freedom north-K, at the 5% significance level.

3. If the condition in step 2 is satisfied, run J Newton-Raphson steps using equation (16.5) to get the estimates $\phi_j, j = i, \ldots, J$.

4. The smallest value of $g'Wg$ taken over all parameter estimates encountered during the procedure determines the final parameter estimate.

Andrews argues that a test size of 5% should work well for the critical value in most cases, and that $J = 3$ should be large enough in practice in most cases.

Here comes the most remarkable part. Andrews (1997) shows that the GMM estimator found in this way is asymptotically equivalent to the efficient GMM estimator. He also shows that his procedure delivers values that approximate the global minimum value over the whole parameter space!

Now, what if you can't find parameter values that satisfy the second condition? This shows that either you would reject the null at the global minimum parameters or that the global minimization is intractable for a test of size 5%. You could try a larger test size. You could reformulate the problem, trying to find a more tractable specification of the moment conditions. Or you could reduce the number of parameters in the grid search by finding a nested model depending on a subset of the parameters, where the GMM works well for prespecified values of the troublesome parameter. Then, grid search over the troublesome parameter, holding it fixed in each of the well-behaved GMM subproblems. Good luck in your searches!

16.7   The Asymptotic Covariance Matrix: A First Cut

Recall that estimates of ϕ minimizing $g'Wg$ are consistent and asymptotically normal, for almost any fixed $W > 0$. If the W is the inverse of a consistent estimate of the asymptotic variance of $g = (\sum_t g_t)/T$, then the GMM estimates are efficient in the class of quadratic form minimizing estimators. In this case, Hansen (1982) shows that the GMM estimator converges in distribution:

$$T^{1/2}(\hat{\phi} - \phi) \rightarrow_d N(0, \text{Acov}(\hat{\phi})). \tag{16.8}$$

How do we estimate $\text{Acov}(\hat{\phi})$? It depends on the situation, and the details are fleshed out in chapter 17, but here is a first cut.

We want the sample counterpart to $[\text{E}(\partial g/\partial \phi)' W(\text{E}\partial g/\partial \phi)]^{-1}$, where $W = \text{Acov}(g)^{-1}$, and $g = (\sum_t g_t)/T$. The matrix of derivatives in $\text{E}(\partial g/\partial \phi)$ comes from the particular model, and we use the sample averages of the derivatives to estimate the expectations. The remaining trick is to find W. Note that because $T^{1/2}(g) \rightarrow_d N(0, \text{Acov}(g))$, with

$$\begin{aligned}
\text{Acov}(g) &= T \, \text{Var}(\sum_t g_t)/T) = (1/T) \, \text{Var}(\sum_t g_t) \\
&= (1/T)[\sum_t \text{Var}(g_t) + \sum \sum_{t \neq \tau} \text{Cov}(g_t, g_\tau)].
\end{aligned} \tag{16.9}$$

In general, the calculation will depend on the autocovariances of the g_t values. For this reason, we will spend a lot of time in chapter 17 discussing why the moment conditions might be autocorrelated and what to do about it. Of course, the simplest case is when there is no autocorrelation. Then we can estimate $\text{Acov}(g) = (1/T)[\sum_t \text{Var}(g_t)] = \text{Var}(g_t)$, where we assume stationarity so that the variance of each g_t is the same. This can be estimated using the time series of the sample moment conditions, evaluated at any consistent estimate of the model parameters as $\text{Acov}(g) = (1/T)\sum_t g_t g_t'$. The subtleties are coming up in the next chapter.

17 Covariance Matrices for the GMM

This chapter addresses the important problem of picking the weighting matrix in GMM problems. The trickiest part of this is dealing with serial dependence in the moment conditions. We first discuss how serial dependence in the moment conditions arises in asset pricing problems and in general problems. It turns out that asset pricing models give us some structure that can simplify the autocorrelations under the null hypothesis of mean-zero pricing errors, $E(g) = 0$. Then we talk about nonstandard cases and how to get consistent standard errors for your parameter estimates in nonstandard cases.

First, let's review the optimal covariance matrix in the GMM. The optimal choice is $W = \text{Acov}(g)^{-1}$, where $g = (1/T)\sum_t g_t$. Under H_0, $\sqrt{T}(g - 0) \to_d N(0, \text{Acov}(g))$. We can write $\text{Acov}(g)$ in a way that reveals its dependence on autocorrelation in the moment conditions. Assuming stationarity so that $\text{Cov}(g_t) = \text{Cov}(g_\tau)$, we have

$$
\begin{aligned}
\text{Acov}(g) &= \text{Var}(\sqrt{T}g) = T \ \text{Var}((1/T)\sum_t g_t) \\
&= (1/T)[\sum_t \text{Var}(g_t) + \sum\sum_{t \neq \tau} \text{Cov}(g_t, g_\tau')] \\
&= \text{Var}(g_t) + (1/T)\sum\sum_{t \neq \tau} E\{[g_t - E(g)][g_\tau - E(g)]'\}.
\end{aligned}
\tag{17.1}
$$

Evidently the asymptotic covariance of the parameter estimates depends crucially on the autocovariances of the moment conditions. To get good estimates of the standard errors for the parameter estimates, we have to get these autocovariances right. The first step is to describe the structure of GMM autocovariances for asset pricing problems and the reasons for serial dependence in asset pricing problems. These include overlapping returns, time aggregation in the data, and time-nonseparable models.

17.1 Autocorrelated Moment Conditions

There are some general principles for the autocorrelations of moment conditions in an asset pricing problem. The joint hypothesis implies $E_t(u_{t+1}) = 0$, where $u_{t+1} = m_{t+1}R_{t+1} - 1$, and $E_t(\cdot)$ denotes the conditional expectation at time t. Autocorrelation can also be present outside the usual asset pricing applications, when the theory does not say $E_t(u_{t+1}) = 0$ but only that $E(u_{t+1}Z_t) = 0$, a weaker condition.

In an asset pricing problem, the moment conditions will have no autocorrelation when u_t, u_{t-j}, $j \geq 1$, are in the information set at time t. An example is the simple power utility consumption model with $u_{t+1} = \beta(C_{t+1}/C_t)^{-\alpha} R_{t+1} - 1$, assuming that the consumption data are correctly measured (more on that below). Variables are observed when indicated by the time subscript. Because u_t is known at t and the model says $E_t(u_{t+1}) = 0$, it implies $E(u_t u_{t+1}) = E(E_t(u_{t+1})u_t) = 0$. The first equality uses the law of iterated expectations and the assumption that u_t is known at time t, so it can come outside of the conditional expectation $E_t(\cdot)$. The second equality follows from the asset pricing restriction that $E_t(u_{t+1}) = 0$. Thus, there should be no autocorrelation in the error terms of an asset pricing problem when u_t is known at time t.

We ultimately care about the autocovariances of the moment conditions, $g_{t+1} = u_{t+1} \otimes Z_t$. An important point is that, in an asset pricing problem, the autocovariance structure of $\{g_t\}$ is inherited directly from that of $\{u_t\}$. For example, if $E_t(u_{t+1}u_t') = 0$, then

$$\begin{aligned} \text{Cov}(g_{t+1}g_t') &= E[E_t\{(u_{t+1}u_t') \otimes (Z_tZ_{t+1}')\}] = E[E_t\{(u_{t+1}u_t')\} \otimes (Z_tZ_{t-1}')] \\ &= E[0 \otimes (Z_tZ_{t-1}')] = 0. \end{aligned} \tag{17.2}$$

What seems amazing about this result is that the autocovariance of $\{g_t\}$ does not depend on the serial correlation in the lagged instruments Z_t. In practice, the instruments are often highly persistent variables, like dividend yields and interest rates and spreads (whose auto-correlations can be 0.98 or so in monthly data). But the theory says that does not matter. Such is the power of the statement that $E_t\{u_{t+1}\} = 0$ in an asset pricing problem!

When the model says $E_t(u_{t+1}) = 0$, the orthogonality conditions inherit the moving average structure of the error terms. For example, the error terms are said to be a *moving average process* of order k, MA(k), if $E(u_tu_{t-j})$, $|j| \leq k$, can be nonzero, but $E(u_tu_{t-j}) = 0$ for $|j| > k$. This can occur in asset pricing problems if for some reason u_t is not observable at time t but is observable at time $t+k$. This happens if we have overlapping data or a non-time-separable model for m_{t+1}. Suppose that the errors were MA(k) because k lagged values of the error terms are not observable at time t. The error terms u_{t+1-j} are observable at time t for $j > k$. We find:

$$\begin{aligned} E &= (g_{t+1} g_{t+1-j}') = E[E_t\{g_{t+1} g_{t+1-j}'\}] \\ &= E[E_t\{u_{t+1}u_{t+1-j}'\} \otimes (Z_t Z_{t-j})'] \neq 0 \quad \text{for } j \leq k, \\ &= E[E_t\{u_{t+1}\}u_{t+1-j}' \otimes (Z_tZ_{t-j})'] = 0 \quad \text{for } j > k. \end{aligned} \tag{17.3}$$

While the serial dependence of the $\{g_t\}$ can depend on that of the instruments Z_t, the $\{g_t\}$ have the same MA(k) structure as the $\{u_t\}$. Such is the power of the statement that $E_t\{u_{t+1}\} = 0$ in an asset pricing problem!

17.1.1 Autocorrelated Moment Conditions Due to Overlapping Returns

Consider the example of a model of quarterly consumption optimization:

$$\text{Max } E\Sigma_t \beta^t U(C_t) \Rightarrow E_t\{[\beta^3 U'(C_{t+3})/U'(C_t)]R(t,t+3) - 1\} = 0, \tag{17.4}$$

where $R(t, t+3)$ denotes the quarterly return, from month t to month $t+3$. We might want to do this, because we take the view that agents optimize over longer than monthly horizons (e.g., Jagannathan and Wang 2007). We might also be interested in modeling the longer-horizon returns. Longer-horizon returns are known to differ from shorter-horizon returns in various respects. For example, they seem to be more predictable, and they have different skewness properties and different lead-lag dependencies across stocks than do shorter-horizon returns. If we have monthly data available, we might want to use it to form overlapping quarterly returns, and in this example, form quarterly consumption growth rates from the monthly National Income and Product Accounts (NIPA) consumption data. Suppose that we form a sample of monthly observations on quarterly overlapping growth rates and returns. For example, the first observation is for the first quarter of 2012, ending in March. The second observation starts in February and ends in April 2012, and so on. This overlap in the returns induces autocorrelated moment conditions.

Let $u(t, t+3) \equiv \beta^3[U'(C_{t+3})/U'(C_t)]R(t, t+3) - 1$. The model says $E_t\{u(t, t+3)\} = 0$, but the first lagged value of the error term that becomes known at time t is $u(t-3, t)$. Thus, $E_t\{u(t,t+3)u(t-3,t)\} = u(t-3,t)\,E_t\{u(t,t+3)\} = 0$, but we have $E_t\{u(t,t+3)u(t-1,t+2)\} \neq 0$, because $u(t-1, t+2)$ is not known at time t, and $E_t\{u(t, t+3)u(t-2, t+1)\} \neq 0$, because $u(t-2, t+1)$ is not known at time t. The error terms and therefore the moment conditions $\{g_t\}$ follow a moving average process of order two: an MA(2) process. In general the moving average structure is MA(k), with k equal to the number of periods (months, here) for which the long-horizon observations overlap.

17.1.2 Autocorrelated Moment Conditions Due to Time Aggregation

Time aggregation of the data creates spurious serial correlation and can induce one more moving average term than you would have in the model if the data were properly measured. For example, the aggregate consumption data we get refers to an average over the period. The annual consumption data is an average of consumption rates sampled at different points during the year and averaged together. The problem of time aggregation was famously studied by Working (1960) and in asset pricing by Breeden, Gibbons, and Litzenberger (1989). Here is a stylized, discretized example of the continuous-time models in those papers. It is assumed that the true log consumption process is the sum of stationary, independent increments Δ_j, as depicted in figure 17.1. Let us use c_t to refer to the natural logarithm of the consumption expenditures, so that the "true" consumption growth rate between two periods (say, 0 to t) is given by $c_t - c_0 = \sum_{j=1,\ldots,t} \Delta_j$

In this depiction, there are only two periods between each measurement interval for simplicity. If we had the true consumption, its growth rates would have no autocorrelation: $Cov(c_2 - c_1, \ c_1 - c_0) = Cov(\Delta_{1.5} + \Delta_2, \ \Delta_1 + \Delta_{0.5}) = 0$. For asset pricing, we are interested in consumption covariances. Consumption growth covariances with returns would also be accurately measured if there were no time aggregation, assuming that the returns are accurately measured.

Figure 17.1
Consumption as the sum of increments.

With time aggregation, the problem is that we don't measure the true consumption but only its time averages. Assume that we measure the average of the two periods between each measurement interval:

$$C_1 = (c_{0.5} + c_1)/2 = c_0 + \Delta_{0.5} + (1/2)\Delta_1,$$
$$C_2 = (c_{1.5} + c_2)/2 = c_0 + \Delta_{0.5} + \Delta_1 + \Delta_{1.5} + (1/2)\Delta_2,$$

then $\text{Cov}(C_2 - C_1, C_1 - C_0) = \text{Cov}\{(1/2)\Delta_1 + \Delta_{1.5} + (1/2)\Delta_2; \Delta_{0.5} + (1/2)\Delta_1\} = \text{Var}\{(1/2)\Delta_1\} = \sigma^2/4$. The measured consumption growth is autocorrelated, following an MA(1) structure, even when the true consumption growth is not.

In this example, the measured consumption growth is $C_1 - C_0 = \Delta_{0.5} + (1/2)\Delta_1$, and the true growth is $\Delta_{0.5} + \Delta_1$. The measured consumption growth puts less weight on the actual consumption innovations at the end of the period than it should. Thus, the Δ_1 term does not cancel out of the autocovariance. This can also impact measured consumption growth covariances with returns. Suppose we want $\text{Cov}_0(R_1, c_1) = \text{Cov}_0(R_1, \Delta_{0.5} + \Delta_1)$. We get $\text{Cov}_0(R_1, C_1 - C_0) = \text{Cov}_0(R_1, \Delta_{0.5} + (1.2)\Delta_1)$. The measured consumption covariances will suffer by downweighting the current information.

Zawadowski (2010) shows that if we use the average of the current and future period consumption growth rates in the presence of time aggregation, we can get better consumption covariances. If we use the two-period consumption growth, we get $\text{Cov}_0(R_1, C_2 - C_0) = \text{Cov}_0(R_1, \Delta_{0.5} + \Delta_1 + \Delta_{1.5} + (1/2)\Delta_2) = \text{Cov}_0(R_1, \Delta_{0.5} + \Delta_1) = \text{Cov}_0(R_1, c_1 - c_0)$, assuming that the returns don't predict future consumption. This approach is used by Chen, Roll, and Ross (1986).

Jagannathan and Wang (2007) argue that if annual consumption growth is desired, it helps to use the last-quarter to last-quarter data to measure the annual growth. Theirs is a behavioral story, where investors pay more attention to things in the fourth quarter, but it might reduce the impact of time aggregation as well. Both approaches assume that the returns are correctly measured and independent of consumption growth outside the return measurement interval. Fama (1981) argues that stock prices anticipate future economic growth, so this assumption may not be innocuous.

Time aggregation usually generates spurious autocorrelation. For example, if we put time-aggregated consumption growth on the left-hand side of a regression model, an MA term will be induced in the regression residuals. We would want to capture this effect to get consistent standard errors. The spurious autocorrelation can also show up as spurious growth predictability in a predictive regression.

17.1.3 Autocorrelated Moment Conditions in Time-Nonseparable Models

In models that are not time separable, m_{t+1} may depend on data not known at $t+1$, and this can induce autocorrelation in the moment conditions. An example follows Ferson and Constantinides (1991). This model assumes a single lag of internal habit persistence or durability, as discussed in chapter 5. The representative agent maximizes $E \sum_t \beta^t U(C_t - bC_{t-1})$. Let's ignore the time-aggregation problem for now and assume that C_t is the "true" consumption, so that we can isolate the effect of the time nonseparability. The SDF to use in the m-talk equation (1.1) is given in chapter 5 as

$$m_{t+1} = \{\beta U'(C_{t+1} - bC_t) - b\beta U'(C_{t+2} - bC_{t+1})\} / \{U'(C_t - bC_{t-1}) - b\beta E_t U'(C_{t+1} - bC_t)\}. \quad (17.5)$$

Note that $m_{t+1} = m_{t+1}(C_{t-1}, C_t, C_{t+1}, C_{t+2})$ depends on information that is not known at time $t+1$ but will be observable at time $t+2$. With $u_{t+1} \equiv m_{t+1}R_{t+1} - 1$, even though $E_t(u_{t+1}) = 0$, $E_t\{u_{t+1}u_t\} \neq 0$, so $E(g_{t+1}g_t) \neq 0$. However, u_{t-1} is known at t, so we have $E_t\{u_{t+1}u_{t-1}\} = 0$, and $E(g_{t+1}g_{t-1}) = 0$. The nonseparable model has an MA(1) error structure.

Another example follows Ferson and Harvey (1992). The representative agent maximizes $E \sum_t \beta^t U(C_t - bC_{t-4})$, which in quarterly data corresponds to seasonal habit persistence if $b > 0$. (The utility of my ski vacation is evaluated relative to what I did last skiing season.) If $b < 0$, there is seasonal durability (my skis only deliver consumption services when I use them during the skiing season). A perturbation reducing consumption at time t to earn a return at time $t+1$ will affect the subsistence levels that will be relevant at the future dates $t+4$ and $t+5$. The SDF to use in the m-talk equation (1.1) is

$$m_{t+1} = \{\beta U'(C_{t+1} - bC_{t-3}) - b\beta^5 U'(C_{t+5} - bC_{t+1})\} / \{U'(C_t - bC_{t-4}) - b\beta^4 E_t U'(C_{t+4} - bC_t)\}. \quad (17.6)$$

Note that $m_{t+1} = m_{t+1}(C_{t-4}, C_{t-3}, C_t, C_{t+1}, C_{t+4}, C_{t+5})$ depends on information that is not known at time $t+1$ but can be observed at time $t+5$. Thus, $u_{t+1} \equiv m_{t+1}R_{t+1} - 1$ is not known at $t+1$. As a result, $E_t\{u_{t+1}u_t\} \neq 0$, so $E(g_{t+1}g_t) \neq 0$. However, u_{t-4} is known at t, so the nonseparable seasonal model has an MA(4) error structure.

17.1.4 Autocorrelated Moment Conditions When $E_t(u_{t+1}) \neq 0$

An example is OLS, where $Y_t = Z_t'\beta + u_t$, with $E(u_t Z_t) = 0$, but we don't have the asset pricing condition, and $E_{t-1}(u_t)$ is not zero. Let's consider two popular ways to model the serial dependence in the error terms. The first example assumes an AR(1) model for the error terms: $u_t = \rho u_{t-1} + \epsilon_t$, where ϵ_t is iid. The second example assumes an MA(1) model: $u_t = \epsilon_t - \theta \epsilon_{t-1}$. In the OLS regression, the GMM moment condition is $g_t = Z_t u_t$ and

$$E(g_t g_{t-j}') = E(Z_t u_t u_{t-j}' Z_{t-j}') = E(Z_t E[u_t u_{t-j}' | Z] Z_{t-j}'), \quad (17.7)$$

where Z is the whole matrix of the Z_t values. The autocovariances of the moment conditions will in general now depend on that of the instruments because we do not impose the asset pricing restriction $E_{t-1}(u_t) = 0$.

In the AR(1) example for the residuals, $E[u_t u'_{t-j} | Z] = \rho^j \sigma_u^2$, and therefore $E(g_t g'_{t-j}) = \rho^j \sigma_u^2 E(Z_t Z'_{t-j})$. In the MA(1) example, we have $E[u_t u'_{t-j} | Z] = -\theta \sigma_\epsilon^2$ for $j = 1$, and zero otherwise. Therefore, the moment conditions will have $E(g_t g'_{t-j}) = -\theta \sigma_\epsilon^2 E(Z_t Z'_{t-j})$ for $j = 1$ and zero otherwise. In both cases, the autocovariance $E(g_t g'_{t-j})$ depends on $E(Z_t Z'_{t-j})$. Such is the cost of not imposing the asset pricing restriction $E_{t-1}(u_t) = 0$.

17.2   Optimal Weighting Matrices for Autocorrelated Moment Conditions

Hansen (1982) provides an optimal weighting matrix for when the moment conditions have autocorrelation:

$$W = \text{Acov}(g)^{-1} \text{Acov}(g) = [(1/T) \sum_t \sum_{j=-t,\dots,\tau} (g_t g'_{t-j})]. \tag{17.8}$$

To impress your nerd friends, you can call this the spectral density matrix at frequency zero. A special case is when $\tau = 1$, that is, the moment conditions are MA(1):

$$\text{Acov}(g) = (1/T) \sum_t (g_t g'_t) + (1/T) \sum_t \{(g_t g'_{t-1}) + (g_t g'_{t+1})\}. \tag{17.9}$$

This calculation allows the covariance of the current with the lag to be different from the covariance of the current with the lead, but with covariance stationarity, they should be the same. A problem is that this matrix is not always positive semidefinite, which can lead to negative variances, generally not a good thing.

Newey and West (1987a) provide the most popular solution for this problem. Their estimator is

$$\text{Acov}(g) = (1/T) \sum_t \sum_{j=-\tau,\dots,\tau} w(j, \tau)(g_t g'_{t-j}) \tag{17.10}$$

They propose $w(j, \tau) = [1 - |j|/(\tau+1)]$, the so-called Bartlett weights. These weights for the autocovariances of the moment conditions are tent shaped, putting the most weight on the near lags and the least on the distant lags. This makes sense in practice, because in a given finite sample, you have fewer observations with which to estimate more distant lagged autocovariances, so they will not be as precise. Newey and West show that if (1) $\tau = \tau(T) \to \infty$ in T, (2) $\tau(T)/T^{1/4} \to 0$ in T, and (3) $E\{|g_t|^{4+\delta}\} < \infty$ for small δ, then their estimator is positive semidefinite and consistent for Acov(g).

The choice of the lag length τ is sometimes determined by the structure of the model, as in an asset pricing problem with non-time-separable preferences or overlapping data, but sometimes it must be estimated using time-series methods. It is common to pick τ by looking at the sample autocorrelations of $\{g_t\}$, keeping only the lags with significant autocorrelations. For example, the approximate standard error of a sample autocorrelation

Weighting kernel	k(x)
Truncated	$k(x) = 1, |x| \le 1$
	$k(x) = 0, |x| > 1$
Bartlett	$k(x) = 1 - |x|, |x| < 1$
	$k(x) = 0, |x| > 1$
Tukey-Hanning	$k(x) = \{1 + \cos(\Pi x), |x| < 1\}$
	$k(x) = 0, |x| > 1$
Quadratic spectral	$k(x) = (25/12\Pi^2 x^2) \{\sin(6\Pi x/5)/(6\Pi x/5) - \cos(6\Pi x/5)\}$
Parzen	$k(j, \tau) = \{1 - 6\,(|j|^2/\tau^2) + 6\,(|j|^3/\tau^3)\},\qquad (j/\tau) \le 1/2$
	$k(j, \tau) = \{2\,(1 - |j|/\tau)^3\},\qquad\qquad (j/\tau) > 2$

Figure 17.2
Weighting kernels for GMM matrices covariance.

is $1/\sqrt{T}$, so a rule of thumb is to keep only those lagged autocorrelations larger than $2/\sqrt{T}$. Newey and West (1994) propose an "automatic" lag selection procedure along these lines.

17.3   A Catalog of Consistent Covariance Matrices

Many other weights $w(j, \tau)$ have been proposed and evaluated in the literature since the Newey and West (1987a) paper was published. Let's write the weights for a given number τ of lagged autocovariances in the estimator for Acov(g), as $k(x)$, where $x \equiv j/\tau$. A partial catalog of consistent covariance estimators can be described in terms of the weighting kernel functions $k(x)$, as shown in figure 17.2.

Wow! Of course, econometricians have been busy, and there are many more examples of consistent covariance matrices for the GMM, but these should provide a good start for most problems.

Another approach to estimating the GMM covariance matrix is to use parametric time-series models. For example, suppose that the model errors can be written in moving average form as an MA(k):

$$u_t = \epsilon_t + \Theta_1 \epsilon_{t-1} + \Theta_2 \epsilon_{t-2} + \cdots + \Theta_k \epsilon_{t-k}, \tag{17.11}$$

where the Θ_j are $n \times n$ matrices of parameters, and ϵ_{t-1} is an iid n-vector of shocks. The covariance matrix of the shocks, Σ_ϵ, can be estimated as $(1/T)\sum_t \epsilon_t \epsilon_t'$ from the autoregressive residuals. We can fill out the terms in the expression for Cov(g) using

$$\text{Var}(u_t) = (1 + \Theta_1 \Theta_1' + \cdots + \Theta_k \Theta_k')\Sigma_\epsilon. \tag{17.12}$$

Examples of studies using this approach include Andrews and Monahan (1992) and Anderson and Sorensen (1996). Simulation evidence in those studies reveals a tradeoff between bias and efficiency in using parametric time-series models. If the time-series structure is well specified they can improve efficiency, but they can also increase finite sample bias.

17.4   Alternative Non-optimal Weighting Matrices

Sometimes we want to use a weighting matrix W_0 to do the GMM, but it is not the optimal covariance matrix. There are several instances where this might be the right thing to do. Sometimes we set the weighting matrix equal to the identity matrix to weight the moment conditions equally. This is obviously the simplest GMM estimator, and sometimes it is robust compared with trying to estimate the covariances of the moment conditions. If the optimal weighting matrix is used but the covariances are poorly estimated, one would expect the finite sample properties of the parameter estimates to suffer. Sometimes in practice it might be hard to invert the optimal weighting matrix estimate, so the identity matrix is the best we can do. Several instances of this situation appear in published studies.

Hansen and Jagannathan (1997) advocate the weighting matrix $E(RR')^{-1}$, pointing out that it facilitates cross-model comparisons. For example, in an asset pricing problem with $E(g) = E(m(\varphi)R - 1)$, if we pick parameters that make m really noisy, one of the effects is to make $E(gg')$ large, even if the average pricing errors are not affected. In this case, the J-test, $Tg'Wg = Tg'Cov(g)^{-1}g$, is smaller, and we are less likely to reject the model using the J-test. This is really a matter of poor power. Hansen and Jagannathan point out that if we use $W_0 = E(RR')^{-1}$ as the weighting matrix, we don't face this issue. However, in this case the J-test statistic, $Tg'W_0g$, is not chi-squared. Jagannathan and Wang (1996) show in this case that $Tg'W_0g$ converges to a weighted sum of chi-squares, weighted by the eigenvalues of a particular matrix.

17.5   Asymptotic Variance for GMM When Non-optimal Weighting Matrices Are Used

When we use a non-optimal weighting matrix W_0, the asymptotic variance expression in equation (17.8) is invalid and produces an inconsistent estimate of the standard error. We can understand the situation and get a matrix that works by looking at the first-order condition:

$$\text{Min } g'W_0g \Rightarrow [(\partial g/\partial \varphi)'W_0]g(\varphi) = 0. \tag{17.13}$$

Think of the first-order condition as a moment condition, $h(\varphi) = 0$, for a K-dimensional problem. There are now K moment conditions and K parameters, so this problem is exactly identified. In an exactly identified problem, the parameter estimates do not depend on the weighting matrix, so we can assume that we used the optimal matrix and minimized $h'[E(hh')^{-1}]h$. Therefore, the asymptotic covariance matrix of the parameter estimates is given by

$$\text{Acov}(\varphi) = [(\partial h/\partial \phi)'\text{Cov}(h)^{-1}(\partial h/\partial \phi)]^{-1},$$
$$= (\partial h/\partial \phi)^{-1}\text{Cov}(h)(\partial h/\partial \phi)'^{-1}, \tag{17.14}$$

where the second line uses the fact that the transformed problem is exactly identified. Using (17.13) for $h(\phi)$, we have $(\partial h/\partial \phi) \approx [(\partial g/\partial \varphi)'W_0(\partial g/\partial \varphi)]$, and $E(hh') \approx [(\partial g/\partial \varphi)'W_0W^{-1}W_0(\partial g/\partial \varphi)]$, where $W^{-1} = E(gg')$ is the optimal GMM weighting matrix. Plugging this expression into equation (17.14) leads us to Hansen's (1982) theorem 3.1:

$$\text{Acov}(\varphi) = [(\partial g/\partial \varphi)'W_0(\partial g/\partial \varphi)]^{-1}[(\partial g/\partial \varphi)'W_0W^{-1}W_0(\partial g/\partial \varphi)] \, [(\partial g/\partial \varphi)'W_0(\partial g/\partial \varphi)]^{-1}. \tag{17.15}$$

So now we can find the covariance matrix of the parameter estimates for any weighting matrix W_0. Note that if $W_0 = W$, equation (17.15) reduces to the optimal GMM covariance matrix expression in (17.8).

18 GMM Tests

Armed with a tool kit for estimating the parameters of models using the GMM, the next thing to do is to test some hypotheses. Of course, we want to test hypotheses about the parameters, such as whether they are equal to zero. We might have hypotheses about combinations or other functions of the model parameters. Also, because the GMM moment conditions in an asset pricing setting can be interpreted as pricing errors, we are interested in testing hypotheses about the pricing errors. Here we go.

18.1 An Aside on Chi-Square Variables

The main result needed here is that if some random variables ε_i are distributed as normal, $N(0, 1)$, and independent, then $\sum_{i=1,\dots,k} \varepsilon_i^2$ is distributed as a chi-square with k degrees of freedom. Consider a k-vector x that is normal $N(\mu, \Sigma)$, and let $\Sigma^{-1/2}$ be such that $\Sigma^{-1/2} \Sigma \Sigma^{-1/2\prime} = I$. An example of finding $\Sigma^{-1/2}$ is through a Cholesky factorization. Now let the vector $\varepsilon = \Sigma^{-1/2}(x - \mu)$. Then $E(\varepsilon) = 0$, and

$$E(\varepsilon\varepsilon') = \Sigma^{-1/2} E\{(x - \mu)(x - \mu)'\} \Sigma^{-1/2\prime} = I, \tag{18.1}$$

so ε is distributed as an $N(0, I)$ vector. Therefore, $\varepsilon'\varepsilon = \sum_{i=1,\dots,k} \varepsilon_i^2$ is distributed as a chi-square with k degrees of freedom. This is the trick that we will now use repeatedly to get chi-square test statistics for GMM problems. We take some vector whose asymptotic distribution is known to be normal with some mean and covariance matrix, and we transform it like the vector x above to get something that has an asymptotic chi-squared distribution.

18.2 Hansen's J-Statistic

This is a famous goodness-of-fit test for overidentified models in the GMM. Let $x \equiv T^{1/2}g$, then $T^{1/2}g \to_d N(0, \text{Acov}(g))$, where $\to_d$ means converges in distribution. We apply the trick in the last section, letting $\varepsilon = T^{1/2} \text{Acov}(g)^{-1/2}g$. Then $E(\varepsilon) = 0$, and in large samples, $\varepsilon \sim N(0, I)$. So we have that

$$J \equiv (\varepsilon' \varepsilon) = T g' \text{Acov(g)}^{-1} g \tag{18.2}$$

is chi-square in large samples. Hansen (1982) shows this is so, with degrees of freedom = $NL - \dim(\varphi)$. The degrees of freedom are reduced by the number of parameters, because the estimated orthogonality conditions are not linearly independent. Recall that a number of linear combinations of moment conditions, k, equal to the number of parameters, is set to zero when choosing parameter values.

Be warned that we are relying on asymptotic distributions. Different choices made in implementing the GMM can be asymptotically equivalent, so relying on the asymptotic properties can't guide these choices. For example, whether we use two-step GMM or iterated GMM, as discussed in section 16.5, is a choice that must be made based on other criteria, such as finite sample performance. Here is another example of asymptotically equivalent choices that we must make. Under the null hypothesis, E(g) = 0. But at the sample estimates, unless the model is exactly identified, the sample mean g will not be zero. When computing the asymptotic covariances, we can de-mean the moment conditions, subtracting their sample averages, or we can just use the outer products of the sample moment conditions without subtracting the means. The two approaches are asymptotically equivalent and result in estimators with the same asymptotic distribution under the null hypothesis. But the two approaches will produce different results when working with actual finite samples. Hall (2000) shows that subtracting the sample means results in a goodness-of-fit statistic $T g' W g$ that has more power in finite samples. Intuitively, the sample second-moment matrix of g_t is at its smallest when the sample mean is subtracted, resulting in a smaller Cov(g) estimate and thus a larger W and a larger test statistic $T g' W g$.

18.3 Asymptotic Wald Tests and the Delta Method

Our chi-square trick applies to an estimator for φ, $\acute{\varphi}$, where $T^{1/2} (\acute{\varphi} - \varphi) \rightarrow_d N(0, \text{Acov}(\acute{\varphi}))$.

Now, $\varepsilon = T^{1/2} \text{Acov}(\acute{\varphi})^{-1/2} \acute{\varphi} \sim N(0, I)$, so $\varepsilon' \varepsilon = T \acute{\varphi}' \text{Acov}(\acute{\varphi})^{-1} \acute{\varphi} \sim$ chi-square(dim(ϕ)) under the null hypothesis that $\varphi = 0$. If we had a different point null hypothesis that the parameter $\varphi = \varphi_0$, we would just replace the estimated value in the quadratic form with its difference from the hypothesized value. We can use these results with any of the GMM asymptotic covariance matrices for the parameter estimates described in chapter 17.

A version of this chi-square trick can also be used for any hypothesis about the coefficients that can be expressed as a function H(φ) = 0, where H($\cdot$) is M-valued, M $\leq$ dim(φ), and may be a nonlinear function. To do this, let us look at the *delta method,* which is just a first-order Taylor series. We expand the function H($\cdot$):

$$\text{H}(\varphi) \approx \text{H}(\acute{\varphi}) + (\partial \text{H}/\partial \varphi)(\acute{\varphi} - \varphi) + \text{higher order terms.} \tag{18.3}$$

An amazing thing about the usual T-asymptotics that we are using here is that the higher-order terms do not affect the asymptotic variances and may be ignored. Thus, we may use the delta method to get the asymptotic variance of the function $H(\cdot)$ evaluated at the estimated parameters as

$$\text{Acov}\{H(\acute{o})\} = (\partial H/\partial \varphi) \, \text{Acov}(\acute{o})(\partial H/\partial \varphi).' \tag{18.4}$$

This is a very helpful result for computing asymptotic standard errors for functions of the model parameters. Using this result, we can also test hypotheses about $H(\varphi)$, such as that it is equal to zero. Let $\hat{H} = H(\acute{o})$, with its asymptotic variance $\text{Acov}\{\hat{H}\}$ given by the last equation. We define $\varepsilon = T^{1/2}\text{Acov}(\hat{H})^{-1/2} \, \hat{H}$, so that

$$\varepsilon'\varepsilon = T\hat{H}'[(\partial H/\partial \varphi) \, \text{Acov}(\acute{o})(\partial H/\partial \varphi)']^{-1}\hat{H} \sim \text{chi-square}, \tag{18.5}$$

with degrees-of-freedom $M = \dim(H)$.

18.4   Hypotheses about Pricing Errors

Recall that in a typical asset pricing problem, the sample average moment condition, g, can be interpreted as an average pricing error. So, we might want to draw inferences about the model's average pricing errors. However, we have a problem. The GMM picks the parameter estimates to make the pricing errors as close to zero as possible. In an exactly identified model, $g = 0$ evaluated at the parameter estimates. So, even a really terrible model will have zero average pricing errors if it is estimated in an exactly identified setup. This is sort of like saying that a regression R-squared will be 100% if the number of variables is equal to the number of observations. That doesn't mean that the regression model is any good. The more degrees of freedom that the model uses up in estimating the parameters, the closer to zero will the fitted pricing errors appear. We need to somehow adjust for this problem when drawing inferences about the expected pricing errors.

Hansen (1982), naturally, comes to the rescue. He derives a standard error for the estimated moment condition, $\hat{g}$, that accounts for the fact that the parameters have been estimated by the GMM. This is his lemma 4.1:

$$\text{Acov}(\hat{g}) = \text{Acov}(g) - (\partial g/\partial \varphi)\text{Acov}(\acute{o})(\partial g/\partial \varphi)'. \tag{18.6}$$

Note that the covariance matrix of the estimated moment condition is reduced to reflect the fact that the moment conditions will be "too close" to zero because of the parameter estimation. A special case is when we use $W = \text{Acov}(g)^{-1}$, the optimal W, to get the covariance matrix of $\acute{o}$. Then $\text{Acov}(\acute{o}) = \{(\partial g/\partial \varphi)'\text{Acov}(g)^{-1} \, (\partial g/\partial \varphi)\}^{-1}$, so in that case:

$$\text{Acov}(\hat{g}) = \text{Acov}(g) - (\partial g/\partial \varphi)\{(\partial g/\partial \varphi)'\text{Acov}(g)^{-1}(\partial g/\partial \varphi)\}^{-1}(\partial g/\partial \varphi)'. \tag{18.7}$$

Note what happens if the model is exactly identified.

In equation (18.7), $\text{Acov}(\hat{g})$ is generally singular, because when we estimate the model by GMM, k linear combinations of the sample moment conditions $\hat{g}$ are set to zero through the first-order condition: $(\partial\hat{g}/\partial\acute{\phi})W\hat{g}(\acute{\phi})=0$. So, we won't be able to invert Acov($\hat{g}$). But we can use it to compute standard errors and t-ratios for the average pricing errors and functions of the average pricing errors.

18.5   Conditional Moment Encompassing Tests

Wooldridge (1990) develops tests for comparing different model specifications in a nonlinear least squares (NLS) setting (which of course is a special case of the GMM). The moment condition for the null model can be written as $g_t(\phi)=y_t-m_t(\phi, z_{t-1})$, where the parameter vector $\phi \in R^K$. The null model is to be compared to an alternative model, where the moment condition is $g_t(b)=y_t-\mu_t(b, z_{t-1})$, with $b \in R^Q$. The models do not have to be nested. The approach derives from manipulating the first-order condition of a consistent estimator for b as follows:

$$
\begin{aligned}
0 &= (1/T)\sum_t(\partial\mu_t(b, z_{t-1})/\partial b)[y_t-\mu_t(b, z_{t-1})] \\
&= (1/T)\sum_t(\partial\mu_t(b, z_{t-1})/\partial b)[(y_t-m_t(\phi, z_{t-1}))-(y_t-\mu_t(b, z_{t-1}))], \\
&= (1/T)\sum_t(\partial\mu_t(b, z_{t-1})/\partial b)(y_t-m_t(\phi, z_{t-1})).
\end{aligned}
\tag{18.8}
$$

The last line follows from writing the minimized sum of squares in the NLS problem as $(y_t-\mu_t(b, z_{t-1}))^2=(y_t-m_t(\phi, z_{t-1}))^2+(m_t(\phi, z_{t-1})-\mu_t(b, z_{t-1}))^2$, noting that you can't control the first of these two squared terms when minimizing over b, then adding and subtracting y_t from the second term.

The tests are based on the third line of equation (18.8), which states that the residuals of the null model are orthogonal to the gradient of the alternative model, $\partial\mu_t(b, z_{t-1})/\partial b$. You could imagine testing this orthogonality condition by regressing $y_t-m_t(\phi, z_{t-1})$ on $m_t(\phi, z_{t-1})$ and $\partial\mu_t(b, z_{t-1})/\partial b$ and testing for the significance of the coefficient on the second term. In fact, if the residuals $y_t-m_t(\phi, z_{t-1})$ are homoskedastic, TR^2 from this regression is asymptotically a chi-square (Q). This idea works even under heteroskedasticity. Wooldridge shows the way in his procedure 3.1:

1. Get consistent estimates of ϕ and b from their respective models. Estimate the residuals $\varepsilon_t=y_t-m_t(\phi, z_{t-1})$. Estimate $\partial\mu_t(b, z_{t-1})/\partial b$ and $\partial m_t(\phi, z_{t-1})/\partial\phi$.

2. Run a regression of $\partial\mu_t(b, z_{t-1})/\partial b$ on $\partial m_t(\phi, z_{t-1})/\partial\phi$, and compute the residuals v_t.

3. Run a regression of a vector of ones on the products $\{\varepsilon_t v_t\}$, and compute the R^2. TR^2 from this regression is asymptotically a chi-square (Q).

Wooldridge (1990) shows that procedure 3.1 is consistent under heteroskedasticity and, if the residuals are homoskedastic, is efficient. If some parts of $\partial\mu_t(b, z_{t-1})/\partial b$ are redundant with $\partial m_t(\phi, z_{t-1})/\partial\phi$, throw them out and reduce the degrees of freedom. An example is regressions models with overlapping regressors.

18.6   Other Tests

Newey and West (1987b) evaluate four alternative test statistics for the general non-linear hypothesis H(φ) = 0. These are based on two versions of the GMM estimator. The first is the one we have been using so far, the unrestricted GMM estimator ó, with J-test $J(\acute{o}) = \text{Min}_\varphi\ g'Wg$. The second estimator is called the "minimum chi-square restricted estimator." This estimator, φ^*, satisfies

$$\varphi^* = \text{Arg Min}_\varphi\{(\acute{o} - \varphi)'\text{Acov}(\acute{o})^{-1}\ (\acute{o} - \varphi)\}\ \text{s.t. } H(\varphi) = 0. \tag{18.9}$$

The equation suggests the name. We want to get as close as possible to the unrestricted estimator, as measured by the quadratic form, subject to satisfying the null hypothesis. This is akin to a restricted ML estimator.

In this setup, Newey and West study four test statistics for the null hypothesis that H(φ) = 0:

$$\text{Wald} = T\hat{H}'[(\partial H/\partial\varphi)\ \text{ACov}(\acute{o})(\partial H/\partial\varphi)']^{-1}\hat{H},$$

$$D = T\ [g(\varphi^*)'Wg(\varphi^*) - g(\acute{o})'Wg(\acute{o})],$$

$$\text{LM} = T\ [g(\varphi^*)'\text{Cov}(g)^{-1}(\partial g/\partial\varphi)]\{\text{Acov}(\varphi^*)\}^{-1}[(\partial g/\partial\varphi)'\text{Cov}(g)^{-1}g(\varphi^*)],$$

$$\text{MC} = T\ (\acute{o} - \varphi^*)'\{\text{Acov}(\acute{o})\}^{-1}(\acute{o} - \varphi^*). \tag{18.10}$$

The important thing about the comparison is that each test uses the same $W = \text{Acov}(g)^{-1}$. The first test, the Wald test, has the advantage that you don't have to compute the restricted estimator. The second test, the Newey-West D test, is perhaps the most highly cited and used in asset pricing problems, after the Wald test. It is akin to a likelihood ratio test, in that you are comparing the minimum value of a criterion function with and without the constraint on the coefficient. The LM test examines the hypothesis that the first-order condition to the unrestricted GMM problem, $(\partial g/\partial\varphi)'\text{Cov}(g)^{-1}g(\varphi) = 0$, holds at the restricted parameter values.

Newey and West provide some results about the various test statistics, illustrating that a lot of things are asymptotically equivalent. All four test statistics are asymptotically chi-squared distributed. When the GMM problem is exactly identified, the D and LM test statistics are equal in the probability limit. When g(φ) is linear in the parameters, then D = LM = MC. Finally, when g(φ) is linear and H(φ) is linear, all four tests are equivalent.

19 Advanced GMM

This chapter briefly reviews some advanced tricks and applications of the GMM. The choice of instruments is a problem that we all must confront. Unfortunately, there is no clean prescription for how to make the choice. Perhaps too many studies just follow previous studies, hoping the referee will let them get away with it. Missing data is another common problem that studies often fail to treat with due respect. Hopefully, you can do better in your work. Warning: Some of these tricks are powerful and should be used with caution!

19.1 Scaling Factors in GMM

The use of scaling factors is one of the dirty little secrets of the GMM. If $E[g_t] = 0$, then for "nice" functions $f(\varphi) \neq 0$, $E[g_t f(\varphi)] = 0$. Scaled sample GMM moment conditions are still valid orthogonality conditions. Similarly in conditional models, if $E[g_t | Z_{t-1}] = 0$, then $E[g_t f(\varphi, Z_{t-1}) | Z_{t-1}] = 0$ for nice nonzero functions of the model parameters and information. This scaling offers the researcher many degrees of freedom, so be careful!

The literature encounters situations in which scaling like this is convenient. Consider the time-nonseparable power utility consumption model with internal habit persistence in Ferson and Constantinides (1991). The unconditional Euler equation can be written as

$$
g = E[\beta \; mru_{t+1} \; R_{t+1} - mru_t] = 0,
$$
$$
mru_t = (C_t + bC_{t-1})^{-\alpha} + \beta b E_t (C_{t+1} + bC_t)^{-\alpha}.
$$

(19.1)

There is a trivial solution to the Euler equation when we set $\alpha = 0$, $(1 + b\beta) = 0$. Ferson and Constantinides use scaling factors to "trick" the numerical search and keep it from choosing those values. They divide the Euler equation by $(C_t + bC_{t-1})^{-\alpha} (1 + b\beta)$, so that the criterion function explodes when the parameters approach the unwanted configurations.

If $f(\varphi)$ is a scalar and the optimal GMM weighting matrix is used, then the minimized quadratic form and Hansen's J-test are invariant to scaling at unique optimal parameter values. Let the scaled moment condition be $h_t = g_t \, f(\varphi)$, $h = (1/T) \sum_t h_t$, and pick the parameters to minimize

$$h'Cov(h)^{-1} h = g'f(\varphi)Cov[g\ f(\varphi)]^{-1}gf(\varphi) = g'Cov(g)^{-1}g. \tag{19.2}$$

Thus, Hansen's (1982) J-test is invariant to the scaling, and the same parameter values maximize the rescaled and original problems.

Scaling as described here is just a trick to get to the optimal parameter values during the numerical search when using an iterated approach that holds W fixed at some point while trying to minimize g'Wg. The GMM parameter estimates are typically affected by scaling. Consider the first-order conditions for the scaled GMM problem:

$$\text{Min } h'Cov(h)^{-1}h \Rightarrow [(\partial h/\partial \phi)'Cov(h)^{-1}]h(\phi)$$
$$= (\partial g/\partial \phi)'Wg(\phi) + [(\partial f/\partial \phi)'/f(\varphi)]g'Wg = 0, \tag{19.3}$$

where $W = Cov(g)^{-1}$. Equation (19.3) shows that in overidentified problems, the first-order conditions for the scaled and the original problem differ.

Scaling factors generally extract a cost in terms of efficiency. It is not hard to show that the optimal asymptotic covariance matrix of the original GMM problem is smaller than the covariance matrix for the scaled problem: $Acov_s(\varphi) = [(\partial h/\partial \phi)'Cov(h)^{-1}(\partial h/\partial \phi)]^{-1}$. With $h = gf(\varphi)$, we obtain $Acov_s(\varphi) = \{[(\partial g/\partial \phi)'f(\phi) + g(\partial f/\partial \phi)]'(1/f(\varphi))^{\wedge 2}\ W\ [(\partial g/\partial \phi)'f(\phi) + g(\partial f/\partial \phi)]\}$. Using the first-order condition for the scaled problem, this simplifies to $Acov_s(\varphi) = \{[(\partial h/\partial \phi)'Cov(h)^{-1}(\partial h/\partial \phi)] - C\}^{-1}$, where C is a positive semidefinite matrix. Thus, $Acov_s(\varphi) \geq Acov(\varphi)$, and the scaled model is less efficient.

Ni (1997) evaluates the loss of efficiency with several scaling factors using simulation, for the nonseparable consumption model described at the beginning of this section: (1) $(C_t)^{-\alpha}$, (2) $[(C_t + bC_{t-1})^{-\alpha}]$, (3) $[(C_t + bC_{t-1})^{-\alpha}](1 + b\beta)$, (4) $[(C_t + bC_{t-1})^{-\alpha}](1 + b\beta)(1 + e^{5 + 10\alpha})$, and (5) $(C_t)^{-\alpha}(1 + b\beta)$. Scaling factor 3 is used by Ferson and Constantinides (1991), and scaling factor 4 is used by Braun, Constantinides, and Ferson (1993). Ni finds that version 3 performs better than versions 4 or 5.

19.2 Optimal Instrument Choice

The optimal choice for your instruments is too complex a problem to cleanly formulate, as many issues will be involved for a particular study. This is a subtle part of the experimental design. The choice may depend on the number and nature of the moment conditions, the number of parameters, the available data, the structure of the model, etc. It can also depend on how your work is related to previous studies and what those studies did. We can ask two questions: How many instruments? Which specific ones?

As to the first question, the asymptotics provide a clear answer. Using more instruments leads to more moment conditions, and having more moment conditions always produces a (weakly) smaller asymptotic variance for the parameters. A more efficient estimator typically delivers more statistical power. The improved asymptotic efficiency with more moment conditions can be seen by partitioning a big vector of moment conditions into two smaller ones as $g' = (g'_1, g'_2)'$, plugging this into the asymptotic variance matrix expression (15.2) and comparing the covariance using only g_1 with the matrix using both moment conditions. Life would be simpler if only the asymptotic efficiency of the parameter estimator were at issue! However, in finite samples, we can be hurt by more moment conditions. The accuracy of the asymptotics will likely be worse, finite sample biases might be worse, and the finite sample efficiency might actually be worse with too many moment conditions.

As for the specific choice of instruments, sadly, there is no clear answer. Weak instruments, discussed below, are best avoided. Studies often seem to follow the lead of earlier studies in selecting their instruments, although that is not very creative. In asset pricing problems, recall that the instruments imply dynamic trading strategies, so we can ask whether we want to confront the model with the particular strategies implied. The GMM weighting matrix determines how the various returns or trading strategies are to be weighted when estimating the parameters, and sometimes the economics of the weighting can be used to determine the design.

Several papers have explored special cases of the problem of optimal instrument selection, but only in very limited special cases. Examples include Tauchen (1986), Hansen and Singleton (1996), and Nagel and Singleton (2010). What follows is a basic overview of the general problem. Perhaps naturally, it comes from Hansen in a 1985 paper.

Suppose that the full set of potential instruments is denoted by $\Omega_{T \times L}$, and express the L-vector of the GMM moment condition as $g = \Omega' u / T$, where u is the Euler equation error for a single asset. The first-order condition for the GMM optimization is $(\partial g / \partial \varphi)' W \Omega' u / T = 0$. This has an interesting interpretation. Let $Z^* = \Omega W (\partial g / \partial \varphi)$ be a $T \times K$ matrix of "optimal" instruments, and the first-order condition is $Z^{*\prime} u / T = 0$. We have transformed the first-order condition to an exactly identified problem. The matrix of instruments Z^* is optimal in the sense that no other $T \times K$ matrix of instruments delivers a smaller asymptotic variance matrix for the parameter estimates. Hansen (1985) further simplifies the expression for Z^*:

$$Z^* = E(uu' \mid \Omega)^{-1} \, \partial E(u \mid \Omega) / \partial \varphi. \tag{19.4}$$

Even this result is not very useful in practice, especially if we interpret Ω as the unobservable public information set.

When we specialize to a GMM estimation based on the *m*-talk equation (1.1), we can simplify equation (19.4) further, but we can't get rid of those expectations conditional on Ω. If we are willing to assume that the instruments that can be measured comprise the full public information set, then (19.4) can be attacked, assuming enough time-series structure on the problem. This is the approach used in Hansen and Singleton (1996) and Nagel and Singleton (2010).

19.3   Weak Instruments

Instruments are weak when they have little correlation with the variables in the model. Weak instruments can lead to all sorts of problems, like inconsistent estimates of the model parameters, and they can cause the standard asymptotics to break down.

Weak instruments are easily explained in a special case of the GMM: linear instrumental variables. Consider two time-series regressions:

$$Y = X\beta + u,$$
$$Z = \Pi X + v.$$
(19.5)

The first equation corresponds to our GMM moment condition $g = E(Z'u/T) = 0$, also called the *exclusion condition* for the instrument, Z. The second equation of (19.5) is an auxiliary regression to describe the strength of the relation between Z and X. The condition that $\Pi \neq 0$ is called the *inclusion condition*. We require $\dim(Z) = L \geq \dim(\beta) = K$.

If $\Pi = 0$, the instruments Z are worthless. If Π is small, the instruments are weak. This definition is obviously not rigorous. For asymptotic theory for weak instruments and more precise definitions, see, for example, Stock, Wright, and Yogo (2002), Wright (2003), or Andrews and Stock (2007).

Of course, we don't want weak instruments, we want strong ones! One intuition for strong instruments comes from thinking about what we are trying to use the instruments Z_t to predict in an asset pricing problem: $u_{t+1} = m_{t+1} R_{t+1} - 1$. The exclusion restriction or GMM moment condition states that u_{t+1} is not predictable by any valid Z_t. If we pick instruments that do a good job of predicting m_{t+1} and R_{t+1}, testing that their product is not predictable is likely a powerful test.

How can you tell if you have weak instruments? There are several practical tip-offs. First, if the instruments don't predict the variables in the model, they might be weak. One could examine the F-statistics or R-squares of regressions of the model variables on instruments to assess this. However, if you did sort through instruments to find the ones that seem to deliver large R-squares in your sample, it would complicate the inference, because you would have to account for the randomness in your search. If two-stage, iterated, and continuously updated GMMs lead to vastly different results, the instruments may be weak. Finally, there are formal tests for weak instruments, devised for the special case of linear instrumental variables (see Stock, Wright, and Yogo [2002]).

What are the implications of weak instruments? The standard asymptotics may not be valid. For example, consider the linear instrumental variables case with $\Pi = 0$. The GMM estimator converges to the true parameter value plus the limiting distribution of $[Z'u/Z'v]$, which might be a ratio of normals (leading to a Cauchy distribution). The instrumental variables estimator is inconsistent. Also, with weak instruments, quadratic form tests like Hansen's J-test will not follow asymptotic chi-squared distributions. Bootstrapping to get a finite sample distribution might not help if the bootstrap depends on the choice of a value for Π. For a fairly complete discussion of these issues, see Andrews and Stock (2007).

19.4 Missing Data

Lynch and Wachter (2013) examine the problem of having incomplete data in GMM problems. They consider the common situation where some of the data series start later in the sample than others. This happens so frequently in practice that it is worth discussing.

Suppose that the sample spans the time period $\{1, \ldots, T_0, \ldots, T\}$. The data X_1 are available for the whole period, but observations for X_2 are only available from T_0 to T. To do the asymptotics, Lynch and Wachter parameterize $T - T_0 = \lambda T$ and derive results for fixed λ as $T \rightarrow \infty$. Partition the GMM moment condition g into two parts, where one part depends on X_1 only and the other subset of moments depends on X_2: $g = (g_1(X_1, \varphi)', g_2(X_2, \varphi)')'$. The weighting matrix W is conformably partitioned.

Four different approaches to the missing data are studied.

1. The short estimator just ignores the data from the $\{1, T_0\}$ period and minimizes $g'Wg$ using the rest of the data.

2. The long estimator uses all the data to Min $g'Wg$, but the upper left block of the weighting matrix W_{12} uses $[T_0, T]$ only.

3. The adjusted moment estimator solves: Min $g^{*\prime}Wg^*$, $g^* = (g_1', g_2^{*\prime})'$, and $g_2^* = g_2 + \text{Cov}(g_1)^{-1}\text{Cov}(g_2, g_1)[g_1(1:T) - g_1(T_0:T)]$, where the notation like $(T_0:T)$ indicates the range of the data that is used. This estimator essentially regresses the short estimator on the long estimator over the period when they both have data, and it uses the fitted values of the regression to fill in the short estimator. A similar idea is used by Breeden, Gibbons, and Litzenberger (1989) to fill in consumption data.

4. The overidentified estimator minimizes $g^{\#\prime}Wg^\#$, with $g^\# = [g_1(1:T_0), g_1(T_0:T), g_2(T_0:T)]$.

Lynch and Wachter (2013) show that all four estimators are consistent and asymptotically normal under their conventions, which implies that both T_0 and $T - T_0$ go to infinity. They show that, the estimators 3 and 4 have the same asymptotic variance and are more efficient than the simpler estimators 1 and 2. Simulations confirm these results, using international predictive regressions, with $T_0 = 30$ and $T = 124$.

19.5 Simulated Method of Moments

Duffie and Singleton (1993), Gallant and Tauchen (1996), and others develop the simulated method of moments (SMM), which is a version of the GMM, for cases where you can't solve for the moment conditions in closed form but you can simulate data from the model for given parameter values. For given parameters Φ, suppose you can simulate stationary data $\{X_t^s(\Phi)\}_t$ from the model in some way. Consider an "auxiliary" model for the actual data, $F(X_t|X_{t-j}, \theta)$. The auxiliary model is used to evaluate the fit of the simulated data, using particular parameter values Φ, to the actual data. For example, you might form the auxiliary model for the actual data by assuming that the variables are a vector AR(1) with normal disturbances, but that is not required. Estimate the auxiliary model by quasi-ML:

$$\text{Max}_\theta (1/T)\Sigma_t \ln f(X_t|X_{t-j}, \theta) \Rightarrow \hat{\theta}. \tag{19.6}$$

Let the first-order condition to this problem be a GMM moment condition:

$$g = (1/T)\Sigma_t[\partial \ln f(X_t|X_{t-j}, \theta)/\partial\theta] = 0. \tag{19.7}$$

This condition can be evaluated, substituting the simulated data in for the actual data. The first-order condition (19.7) won't be exactly zero for the simulated data, but if the null hypothesis is correct, it should be close to zero. The idea is to find Φ by GMM to make the scores in equation (19.7) close to zero in the simulated data. That is, we pick the parameters Φ to minimize $g(X^s(\Phi))'Wg(X^s(\Phi))$, where the weighting matrix W is the measure of how close to zero $g(\cdot)$ is: $W = (1/T)\Sigma_t[\partial \ln f(X_t|X_{t-j}, \theta)/\partial\theta][\partial \ln f(X_t|X_{t-j}, \theta)/\partial\theta]'$.

Duffie and Singleton (1993) show that the SMM estimator for Φ is consistent and asymptotically normal for time-homogeneous Markov processes. Foster, Richardson, and Smith (1993) examine the small sample properties of the estimator and find that for their simulations, the properties are as good as those for the classical GMM estimator. With this approach, you need $\dim(\theta) > \dim(\Phi)$ to identify Φ. The approach also relies on a good auxiliary model for the data.

19.6   Testing Inequality Restrictions

An example of this last topic comes from Boudoukh, Richardson, and Smith (1993). The goal is to test the hypothesis that the market's expected excess return must be positive, as in the CAPM with risk-averse agents. The variable r_{mt+1} is the market excess return. The first step is to find some strong instruments that are themselves nonnegative. For example, $z_t+ = \text{Max}(z_t, 0)$. The model says:

$$E[\{r_{mt+1} \otimes z_t+\} - \theta+] = 0, \text{ for } \theta+ \geq 0. \tag{19.8}$$

Boudoukh, Richardson, and Smith propose to first compute an unrestricted estimate of $\theta+$ as

$$\hat{\theta} = (1/T)\Sigma_t\{r_{mt+1} \otimes z_t+\}. \tag{19.9}$$

This is unrestricted because it does not have to be nonnegative. In a second step, estimate $\theta+$ from a quadratic program with constraints

$$\theta+ = \text{Arg Min}_\theta \; T(\hat{\theta}-\theta)'\text{Cov}(\theta+)^{-1}(\hat{\theta}-\theta), \;\; \text{s.t. } \theta \geq 0. \tag{19.10}$$

The minimized value serves as a test statistic and is distributed as a weighted sum of chi-square variables.

20 GMM Examples

In my ideal world, Econometrics 101 starts with the GMM, and then everything else is seen as a special case. This chapter starts with some examples to illustrate. Then we move on to formulations of asset pricing models. The chapter not only shows you how to formulate problems, it also provides a bit of literature review along the way.

20.1 Examples Illustrating the Generality of the GMM

It has been claimed that every estimation method on the planet is a special case of the GMM. It's pretty general, this generalized method of moments. We previously went through examples of special cases including OLS regression and the classical method of moments. Chapter 19 used NLS and linear instrumental variables regressions. A few more examples are presented here to show just how general the GMM really is.

20.1.1 OLS with White's Standard Errors

Let's revisit the OLS problem from chapter 15. We initially assumed that the error term satisfies $E(u_{t+1}|Z_t) = 0$ and that u_t is uncorrelated over time. If $E(uu'|z) \neq \sigma^2 I$, so that $Cov(u_t^2 z_t)$ can be nonzero, then the asymptotic covariance of equation (15.2) specializes to: $A\,cov(\beta) = (z'z/T)^{-1}\{(1/T)\sum_t z_t u_{t+1} u'_{t+1} z'_t\}(z'z/T)^{-1}$. This produces the famous White (1980) standard errors, consistent in the face of general conditional heteroskedasticity of the residuals.

20.1.2 Linear Instrumental Variables

We used part of this example in the discussion of weak instruments in chapter 19. Consider the regression

$$Y = X\beta + u, \tag{20.1}$$

where there are K parameters in β. However, we have the problem that $E(u'X) \neq 0$, perhaps due to endogeneity. Let's assume we can find L instruments Z that are orthogonal to the error terms (i.e., satisfy the exclusion restriction), and let $g = E(Z'u/T) = 0$ be the GMM moment conditions. There are several interesting special cases.

If $L = K$, the GMM problem is exactly identified, and we can find the solution by setting the sample mean $g = 0$:

$$\beta_{IV} = [(Z'X/T)]^{-1}(Z'Y/T), \tag{20.2}$$

leading to an estimator that is sometimes called "indirect least squares." In this case, we can use the exactly identified version of the asymptotic covariance matrix of the estimator:

$$A \operatorname{cov}(\beta_{IV}) = (X'Z/T)^{-1}\{(1/T)\sum_t z_t u_t u'_t z'_t\}(Z'X/T)^{-1}. \tag{20.3}$$

This delivers consistent standard errors for instrumental variables estimation, allowing for conditional heteroskedasticity in the error terms, but it assumes that there is no time-series dependence.

A second example occurs when $L = \dim(Z) > K$, and the model is overidentified. The first-order condition to the GMM problem is

$$(\partial g/\partial \phi)' Wg(\phi) = (-Z'X/T)' W[(Z'Y/T) - (Z'X/T)\beta] = 0, \tag{20.4}$$

leading to the instrumental variables estimator:

$$\beta_{IV} = [(Z'X/T)'W(Z'X/T)]^{-1}(Z'X/T)'W(Z'Y/T). \tag{20.5}$$

If we assume that the error term is iid, so that W is proportional to $(Z'Z/T)^{-1}$, we can compute the slope coefficient in two steps. First, regress X on Z using OLS and take the fitted values: $Z(Z'Z)^{-1}Z'X$. Then regressing Y on the fitted values delivers β_{IV}. This is sometimes called two-stage least squares.

20.1.3   Maximum Likelihood

Yes, even ML can be thought of as a special case of the GMM. This may seem strange, recalling that an ML estimator attains the Cramer-Rao lower bound (meaning that you can't get more efficient than that). GMM cannot be more efficient than ML, but it can match the efficiency of ML when the two coincide. It is in the computation of the ML estimators that we can see ML as a special case of the GMM. Along the way, we get to talk about quasimaximum likelihood. Consider the first-order condition for the ML estimator:

$$(1/T)\sum_t \partial \ln L_t(\varphi)/\partial \varphi = 0, \tag{20.6}$$

where $\ln L_t(\varphi)$ is the log likelihood for the data at time t, and the example assumes that the data are independent over time. Letting $g_t = \partial \ln L_t(\varphi)/\partial \varphi$, and $g = (1/T)\sum_t g_t \equiv \partial \ln L(\varphi)/\partial \varphi$, we have a GMM problem. We can apply the standard GMM optimal covariance matrix, because the ML first-order condition produces an exactly identified GMM problem, and we can say that we used the optimal weighting matrix and minimized g′Wg, with $W = E(g_t g_t')^{-1}$. Thus,

$$
\begin{aligned}
\text{Acov}[\varphi] &= [(\partial g/\partial \varphi)' W (\partial g/\partial \varphi)]^{-1} \\
&= [E(\partial^2 \ln L_t(\varphi)/\partial \varphi \partial \varphi')' \{E(\partial \ln L(\varphi)/\partial \varphi)(\partial \ln L(\varphi)/\partial \varphi)'\}^{-1} \\
&\quad (E\partial^2 \ln L_t(\varphi)/\partial \varphi \partial \varphi')]^{-1}.
\end{aligned}
\tag{20.7}
$$

This expression provides a consistent estimator for the parameter covariance matrix when the ML criterion function is used, but it does not require that the data actually be distributed according to the density function $L(\cdot)$. All we have to do is make sure the first-order condition meets the conditions of a GMM moment condition. This is called "quasimaximum likelihood." Under normality, equation (20.7) simplifies, because we can plug in the "information matrix identity" in that case, $E\{(\partial \ln L(\varphi)/\partial \varphi)(\partial \ln L(\varphi)/\partial \varphi)'\} = E(-\partial^2 \ln L_t(\varphi)/\partial \varphi \partial \varphi')$, and two of the three terms in the second line of (20.7) cancel out.

20.2 Formulating Asset Pricing Models with GMM

Starting in the early 1980s, studies used the GMM to set up and estimate asset pricing models. Since this is a book on empirical asset pricing, it should be fun to review a few historical examples from the literature.

20.2.1 Conditional CAPMs

The first conditional asset pricing model was developed by Hansen and Hodrick (1983) and Gibbons and Ferson (1985), who assumed that conditional betas were constant, while conditional expected risk premiums, given an observable set of instruments Z, are time-varying as a function of Z. Papers by Campbell (1987), Ferson, Foerster, and Keim (1993), and others refined this model, which became known as a "latent variable" model for expected asset returns. The model was given this name because, as Gibbons and Ferson pointed out, under the assumptions about what is time varying and what isn't, you could estimate and test the model without actually observing the market portfolio return. Thus, it was thought, one could avoid the famous Roll (1977) critique, which said that you can't test the CAPM if you can't measure the true market portfolio. Wheatley (1989) criticized this logic, pointing out that since there is always a conditionally mean variance efficient portfolio, if the conditional betas on that portfolio are approximately constant, then the latent variable model would appear to work even if the CAPM was false.

An empirical formulation of the model is as follows. Let $r_i = R_i - R_f$, the excess return of asset i over that of the risk-free asset. The model states:

$$
\begin{aligned}
E(r_{it+1} | Z_t) &= \beta_{im} E(r_{mt+1} | Z_t), \\
E(r_{mt+1} | Z_t) &= \delta_m' Z_t,
\end{aligned}
\tag{20.8}
$$

where β_{im} is a constant parameter.

The trick in the latent variable model is to substitute out for the expected market portfolio return. Pick some reference asset excess return, say, asset 1, and write:

$$E(r_{it+1}|Z_t) = (\beta_{im}/\beta_{1m}) \, E(r_{1t+1}|Z_t). \tag{20.9}$$

Because the betas are assumed to be constant, we can express expected returns for all the assets in terms of the coefficients $C_i = (\beta_{im}/\beta_{1m})$. A formulation of the model for estimation by the GMM is given in table 20.1, using the tabular representation that I find so helpful, introduced in chapter 15. Note that for each row, table 20.1 counts only the number of new parameters that the moment condition introduces. Since the L-dimensional vector δ is counted in the first row, it is not counted again in the second row. One of the amazing things about this model is that it can be tested with only two assets, provided there are at least $L = 2$ lagged instruments available to predict the excess returns.

Harvey (1989) developed a time-varying covariance model, making the assumption that the ratio of the expected excess return to the conditional covariance of return with the market portfolio (the market price of risk) is a constant parameter. At the same time, he allowed the conditional covariance of asset returns with the market return to be time-varying with no restrictions on its functional form. He argued that this set of assumptions is natural, because in the Merton (1980) continuous-time CAPM, the ratio of expected excess return to covariance should be equal to a relative risk aversion coefficient, and this is plausibly a fixed parameter over time. Harvey's model can be written as

$$
\begin{aligned}
E(r_{it+1}|Z_t) &= \lambda \mathrm{Cov}(r_{i,t+1}, r_{mt+1}|Z_t), \\
E(r_{mt+1}|Z_t) &= \delta'_m Z_t,
\end{aligned}
\tag{20.10}
$$

where λ is a constant parameter, the market reward-to-risk ratio. A GMM formulation is given in table 20.2.

Table 20.1
A latent variable model.

Moment condition	Orthogonal to	North	K
$u1_{t+1} = r_{1t+1} - \delta'Z_t,$	Z_t	L	L
$u2_{t+1} = r_{it+1} - C_i \, (\delta'Z_t)$	Z_t	L	1

Table 20.2
A conditional CAPM with time-varying covariances.

Moment condition	Orthogonal to	North	K
$u1_{t+1} = r_{mt+1} - \delta'Z_t,$	Z_t	L	L
$u2_{t+1} = r_{it+1} - \lambda(u1_{t+1} \, r_{it+1})$	Z_t	L	1

The trick here is to recall that $\text{Cov}(r_{i,t+1}, r_{mt+1}|Z_t) = E\{(r_{i,t+1} - E(r_{i,t+1}|Z_t))(r_{mt+1} - E(r_{m,t+1}|Z_t))\} = E\{r_{i,t+1}(r_{mt+1} - E(r_{m,t+1}|Z_t))\}$. The last expression is used in the moment condition in the bottom row of table 20.2. This model has the same degrees of freedom as the previous latent variable model.

20.2.2 Multibeta Asset Pricing Models

Ferson, Foerster, and Keim (1993) developed and tested a multibeta version of the latent variables model. Define the $T \times N$ excess return matrix r, with typical element $r_{it} = R_{it} - R_{0t}$, $i = 1, \ldots, N$, and $t = 1, \ldots, T$, where R_{0t} is the return of an arbitrarily chosen zeroth asset. Define the $T \times K$ matrix of the risk premiums $\lambda_j(Z_{t-1})$ as $\lambda(Z)$, where K is the number of latent variables. Define the $T \times L$ matrix of Z_{t-1} values as Z. Assuming fixed betas, the multibeta model for expected excess returns is

$$E(r|Z) = \lambda(Z)\beta, \tag{20.11}$$

where β is the $K \times N$ matrix of fixed betas for the excess returns, and $E(r|Z)$ is the $T \times N$ matrix of the $E(r_{it}|Z_{t-1})$, for $t = 1, \ldots, T$ and $i = 1, \ldots, N$. Partition the excess returns as $r = (r_1 \; r_2)$, where r_1 is a $T \times K$ matrix of *reference assets*, and r_2 is a $T \times (N-K)$ matrix of *test assets*. Partition the matrix of betas conformably as $\beta = (\beta_1 \beta_2)$. The reference assets are chosen so that the $K \times K$ matrix β_1 is nonsingular. From the partitioned equation, solve for the risk premiums and substitute back $\lambda(Z) = E(r_1|Z)\beta_1^{-1}$, to obtain the following restrictions:

$$E(r_2|Z) = E(r_1|Z)C, \tag{20.12}$$

where $C = \beta_1^{-1}\beta_2$ is a $K \times (N-K)$ matrix. Following Gibbons and Ferson (1985), assume that the conditional expected excess returns of the reference assets are linear regression functions of the instruments. The GMM formulation is given in table 20.3.

This model is overidentified, provided $L > K$. The latent variable model implies that if there are K common factors that describe expected excess returns over time, then linear combinations of the regression functions that predict the excess returns of K reference assets are sufficient to capture the predictable variation in all the returns.

Dittmar (2002) proposed a kurtosis-based asset pricing model, which was mentioned in chapter 5. This model may be stated as

Table 20.3
A K-factor expected return model.

Moment condition	Orthogonal to	North	K
$u1_t = r_{1t} - \delta'Z_{t-1}$,	Z_{t-1}	KL	KL
$u2_t = r_{2t+1} - C(\delta'Z_{t-1})$	Z_{t-1}	$(N-K)L$	$K(N-K)$

$$E(r_i) = \beta_i \lambda_1 + S_i \lambda_2 + K_i \lambda_3,$$
$$\beta_i = \text{Cov}(r_i, r_m)/E\{[r_m - E(r_m)]^2\},$$
$$S_i = \text{Cov}(r_i, [r_m - E(r_m)]^2)/E\{[r_m - E(r_m)]^3\}, \qquad (20.13)$$
$$K_i = \text{Cov}(r_i, [r_m - E(r_m)]^3)/E\{[r_m - E(r_m)]^4\},$$

where β_i is the usual market beta, S_i is the systematic co-skewness, and K_i is the systematic co-kurtosis parameter. The λ_1, λ_2, and λ_3 are parameters to be estimated, representing unconditional expected risk premiums. The parameters β_i, S_i, K_i, u_m, σ_m, S_m, λ_1, λ_2, and λ_3 are constant over time. A GMM system to estimate and test this model is given in table 20.4.

This model is exactly identified with three assets and is overidentified for $N > 3$. It is a good illustration of using GMM moment conditions to identify co-moments used in asset pricing models.

Here is a beta pricing example from Ferson and Korajczyk (1995) that illustrates using the GMM to analyze predictable variation in asset returns. Consider a conditional beta pricing model stated in terms of excess returns and the measured instruments Z:

$$E(R_{it}|Z_{t-1}) = \lambda_o(Z_{t-1}) + \sum_{j=1,\ldots,K} b_{ij} \lambda_j(Z_{t-1}). \qquad (20.14)$$

The conditional betas are the b_{ij} (assumed constant here for simplicity). The conditional expected risk premiums are $\lambda_j(Z_{t-1}) = E(F_{jt}|Z_{t-1})$, where the F_{jt} are traded factor portfolios. Ferson and Korajczyk assume that the expected risk premiums are given by linear regressions of the traded factor-mimicking excess returns F on the lagged Z. A conditional model should explain the predictability of returns, and the goal is to measure empirically what fraction of the predictable variation the model actually does capture. Writing for excess returns,

$$r_{it} = E(r_{it}|Z_{t-1}) + u_{it}, \qquad (20.15)$$

then u_{it} should be unpredictable using Z_{t-1} and the model-fitted $E(r_{it}|Z_{t-1})$, a function of the betas and risk premiums, should capture the predictability of the returns if the model works perfectly. The approach is to decompose the predictability in returns into the part associated with betas and risk premiums, and a residual part unrelated to these, and to see how

Table 20.4
A model with co-skewness and co-kurtosis.

Moment condition	Orthogonal to	North	K
$u_{mt} = r_{mt} - \mu_m$	1	1	1
$u_{1t} = \beta_i u_{mt}^2 - r_{it} u_{mt}$, $i = 1, \ldots, N$	1	N	N
$u_{2t} = S_i u_{mt}^3 - r_{it}[u_{mt}^2 - \sigma_m^2]$, $i = 1, \ldots, N$	1	N	N
$u_{3t} = \sigma_m^2 - u_{mt}^2$	1	1	1
$u_{4t} = s_m - u_{mt}^3$	1	1	1
$u_{5t} = K_i u_{mt}^4 - r_{it}[u_{mt}^3 - S_m]$, $I = 1, \ldots, N$	1	N	N
$u_{6t} = r_{it} - \beta_i \lambda_1 - S_i \lambda_2 - K_i \lambda_3$, $i = 1, \ldots, N$	1	N	3

much of the predictability is associated with each part. Each part of the decomposition is modeled using linear regressions:

$$E(r_{it}|Z_{t-1}) = \sum_j \beta_{ij} E(F_{jt}|Z_{t-1}) + E(u_{it}|Z_{t-1}),$$

$$\delta_i' Z_{t-1} = \sum_j \beta_{ij} \Gamma_j' Z_{t-1} + E(u_{it}|Z_{t-1}). \tag{20.16}$$

The predictable variation is studied using the following variance ratios:

$$VR1 = \text{var}\{E(\text{model-fitted expected return}|Z)\}/\text{var}\{E(r|Z)\} \tag{20.17}$$

$$= \text{var}\{\beta\Gamma'Z\}/\text{var}\{\delta'Z\}.$$

We might call this an "ANOPVA" (analysis of predictable variance), by analogy to ANOVA (analysis of variance). The GMM moment conditions to estimate this model for a given asset i are shown in table 20.5.

The lettered rows in table 20.5 are interpreted as follows. Row (a) says $E(r_{it}|Z_{t-1}) = Z_{t-1}'\delta_i$, (b) says $E(F_t|Z_{t-1}) = Z_{t-1}'\gamma$, (c) says $\beta_i = \text{Cov}[(F_tF_t')|Z_{t-1}]^{-1}\text{Cov}[(F_t r_{it})|Z_{t-1}]$, (d) says $\theta_i = E\{E(r_{it}|Z_{t-1})\} = E\{r_i\}$, (e) says $E\{r_i\} = \alpha_i + E\{\beta_i' E[F_t|Z_{t-1}]\}$, and (f) says $VR1_i = \text{Var}\{\beta_i' E(F_t|Z_{t-1})\}/\text{Var}\{E(r_{it}|Z_{t-1})\} = \text{Var}\{\beta_i'\gamma'Z_{t-1}\}/\text{Var}\{Z_{t-1}'\delta_i\}$.

20.2.3 Long-Run Risk Models

The long-run risk model based on Bansal and Yaron (2004) has been a phenomenal success, useful for the equity premium puzzle, size and book-to-market effects, momentum, long-term return reversals, risk premiums in bond markets, real exchange rate movements, and more. Here we formulate a stationary version of a long-run risk model, following Bansal and Yaron (2004). Chapter 31 expands the analysis and review. Here is the stationary model:

$$\Delta c_t = x_{t-1} + \sigma_{t-1}\varepsilon_{ct},$$

$$x_t = \mu + \rho_x x_{t-1} + \phi\sigma_{t-1}\varepsilon_{xt}, \tag{20.18}$$

$$\sigma_t^2 = \underline{\sigma} + \rho_\sigma(\sigma_{t-1}^2 - \underline{\sigma}) + \varepsilon_{\sigma t},$$

Table 20.5
Analysis of predictable variance.

	Moment condition	Orthogonal to	Number of moment conditions
(a)	$u1_{it} = (r_{it} - Z_{t-1}'\delta_i)$	Z_{t-1}	L
(b)	$u2_t = (F_t' - Z_{t-1}'\gamma)'$	Z_{t-1}	LK
(c)	$u3_{it} = [(u_{2t}u_{2t}')\beta_i - (F_t u1_{it})]$	Z_{t-1}	LK
(d)	$u4_{it} = (Z_{t-1}'\delta_i - \theta_i)$	1	1
(e)	$u5_{it} = (Z_{t-1}'\gamma_i\beta_i - \theta_i + \alpha_i)$	1	1
(f)	$u6_{it} = (u4_{it}^2)VR1_i - u5_{it}^2$	1	1

Note: See text for interpretation of lettered rows in the table.

where c_t is the natural logarithm of aggregate consumption expenditures, Δ is the first difference, and x_{t-1} is the conditional mean of consumption growth. The conditional mean is assumed to be a stationary but persistent stochastic process, with ρ_x less than but close to 1.0. (In Bansal and Yaron 2004, $\rho_x = 0.98$.) Consumption growth is conditionally heteroskedastic, with conditional volatility σ_{t-1}, given information at time $t-1$. The conditional volatility follows an autoregressive process. The shocks $\{\varepsilon_{ct}, \varepsilon_{xt}, \varepsilon_{\sigma t}\}$ are assumed to be homoskedastic and independent over time, although they might be correlated.

Bansal and Yaron (2004) show that in this model, the innovations in the log of the stochastic discount factor are, to good approximation, linear in the three-vector of shocks $u_t = [\sigma_{t-1}\varepsilon_{ct}, \phi \, \sigma_{t-1}\varepsilon_{xt}, \varepsilon_{\sigma t}]$. The linear function has constant coefficients. Because the coefficients in the stochastic discount factor are constant, unconditional expected returns are approximately linear functions of the unconditional covariances of returns with the shocks.

For the stationary model, Ferson, Nallareddy, and Xie (2013) estimate versions of the moment conditions shown in table 20.6.

In this table, r_t is an N-vector of asset returns in excess of a proxy for the risk-free rate, and β is an $N \times 3$ matrix of the betas of the N excess returns in r_t with the priced shocks $\{u_{1t}, u_{3t}, u_{4t}\}$. The last equation in table 20.6 identifies the three unconditional risk premiums λ. These are the expected excess returns on mimicking portfolios for the shocks to the state variables. The system is exactly identified. This model illustrates that the error terms from one equation in the system can serve as instruments in other equations.

The state vector in this model is the risk-free rate and aggregate price/dividend ratio: $s_t = \{rf_t, \ln(P/D)_t\}$. The formulation reflects the fact, as shown by Bansal, Kiku, and Yaron (2010) and Constantinides and Ghosh (2011), that the conditional mean of consumption growth and the conditional variance can be identified as affine functions of these state variables.

Table 20.6
A stationary long-run risk model.

Moment condition	Orthogonal to	North	K
$u_{1t} = \Delta c_t - [a_0 + a_1 \, rf_{t-1} + a_2 \, \ln(P/D)_{t-1}$ $\equiv \Delta c_t - x_{t-1}$	$(1, rf_{t-1}, \ln(P/D)_{t-1})$	3	3
$u_{2t} = u_{1t}^2 - [b_0 + b_1 \, rf_{t-1} + b_2 \, \ln(P/D)_{t-1}]$ $\equiv u_{1t}^2 - \sigma_{t-1}^2$	$(1, rf_{t-1}, \ln(P/D)_{t-1})$	3	3
$u_{3t} = x_t - d - \rho_x x_{t-1}$	$(1, x_{t-1})$	2	2
$u_{4t} = \sigma_t^2 - k - \rho_\sigma \sigma_{t-1}^2$	$(1, \sigma_{t-1}^2)$	2	2
$u_{5t} = r_t - \mu - \beta(u_{1t}, u_{3t}, u_{4t})$	$(1, u_{1t}, u_{3t}, u_{4t})$	4	4
$u_{6t} = \lambda - [\beta'\beta]^{-1}\beta r_t,$	1	3	3

20.2.4 Returns-Based Fund Performance Measurement

Here we follow the approaches used in Farnsworth et al. (2002) and Ferson, Henry, and Kisgen (2006). We allow that a mutual fund, with gross return $R_{p,t+1}$, may not be priced exactly by the SDF. Its SDF alpha is defined as $\alpha_{pt} \equiv E_t(m_{t+1}R_{p,t+1} - 1)$. The conditional performance of a fund and the parameters of the SDF model can be estimated simultaneously using the following system of moment conditions:

$$E\{[m(\varphi)_{t+1} R_{t+1} - \underline{1}] \otimes Z_t\} = 0,$$
$$E\{[m(\varphi)_{t+1} R_{pt+1} - 1 - \alpha'_p Z_t] \otimes Z_t\} = 0. \tag{20.19}$$

The first equation states that the SDF prices a set of primitive returns R_{t+1}. The conditional alpha of the fund is $\alpha_{pt} = \alpha'_p Z_t$, which is assumed to be a linear function of Z as in Christopherson, Ferson, and Glassman (1998). By now, you can construct the table of moment conditions on your own. Farnsworth et al. (2002) show that estimating a system like this for one fund at a time produces the same point estimates and standard errors for alpha as a system where the second equation is stacked up for many funds. This is convenient, as the number of available funds exceeds the number of monthly time series, and joint estimation with all funds is not feasible. Efficient GMM parameter estimates can be obtained using any subset of funds, and the individual standard errors are numerically equivalent to those in the full system.

20.2.5 Holdings-Based Fund Performance Measurement

Ferson and Khang (2002) present a conditional version of holdings-based performance measures, estimated by the GMM. An example is:

$$CWM_t = E\left\{ \sum_{j=1}^{N} (\tilde{w}_{jt} - w_{bjtk})(\tilde{r}_{jt+1} - E(\tilde{r}_{jt+1}|Z_t)) \mid Z_t \right\}, \tag{20.20}$$

where $w_{bjtk} = w_{jt-k} \prod_{\tau=t-k+1}^{t} \left(\dfrac{1+\tilde{r}_{j\tau}}{1+\tilde{r}_{p\tau}} \right)$ is the benchmark weight of asset j at time t, based on a lag of k periods, updated assuming a buy-and-hold strategy.

To estimate CWM, Ferson and Kang estimate the conditional expected returns of the individual assets as regression functions: $E(\tilde{r}_{jt+1}|Z_t) = b'_j Z_t$, where Z_t is an L-vector of information variables, including a constant. The conditional covariance between the weight changes and the future abnormal returns is approximated by a linear function of the conditioning information. This assumption follows Christopherson, Ferson, and Glassman (1998) and can be motivated by a Taylor series approximation: $CWM_t = CWM + \gamma' z_t$, where the instruments are de-meaned, and $z_t = Z_t - E(Z)$, excluding the constant. In this equation, CWM is the average conditional covariance, $E(CWM_t)$.

Here is a GMM system to carry out the estimation:

$$\tilde{e}_{jt+1} = \tilde{r}_{jt+1} - b'_j Z_t,$$

$$\tilde{e}_{cwmt+1} = \sum_{j=1}^{N} (w_{jt} - w_{bjtk})(\tilde{r}_{jt+1} - b'_j Z_t) - \gamma' z_t - CWM,$$

$$\tilde{e}^u_{jt+1} = \tilde{r}_{jt+1} - E(\tilde{r}_j),$$

$$\tilde{e}_{uwmt+1} = \sum_{j=1}^{N} (w_{jt} - w_{bjtk})(\tilde{r}_{jt+1} - E(\tilde{r}_j)) - UWM.$$

(20.21)

The parameters of the model are b, γ, CWM, UWM, and $E(\tilde{r})$, where b is the stacked b_j values, and $E(\tilde{r})$ is the stacked $E(\tilde{r}_j)$ values. Each of the moment conditions is orthogonal to the lagged Z_t. The vector of sample moment conditions g is an $(N+1)(L+1) \times 1$ vector. The GMM system is exactly identified with $(N+1)(L+1)$ parameters and $(N+1)(L+1)$ orthogonality conditions.

Ferson and Mo (2016) develop holdings-based performance measures using m-talk. Let x_t be the vector of holdings at time t, and let $r_{pt+1} = x'_t r_{t+1}$ be the "hypothetical" fund excess return based on the future excess returns r_{t+1} for the securities held at time t. Ferson and Mo show a fund's SDF alpha is

$$\alpha_p = Cov(x', mr) \equiv \sum_i Cov(x_{it}, m_{t+1}r_{it+1}).$$

(20.22)

Equation (20.22) says that the right measure of performance with holdings is the sum of the covariances between the portfolio weights and the "risk-adjusted" abnormal returns populating the vector mr. Previous holdings-based measures, such as the classical Grinblatt and Titman (1993) holdings-based measure is an estimate of $Cov(x', r)$ for the fund portfolio x. This leaves out the risk adjustment, m. Ferson (2013) provides conditions under which the two approaches are equivalent. These conditions are reviewed in chapter 27, but they are stringent and unlikely to be met in practice.

Ferson and Mo (2016) assume that the SDF is given by a linear factor model:

$$m = a - b'r_B,$$

(20.23)

where r_B is a vector of the K benchmark portfolio excess returns, and a and b are marketwide parameters to be estimated. This example includes a CAPM, where $K=1$, and a broad stock market index is the benchmark, as well as the Fama and French (1996) and Carhart (1997) factor models.

Consider a factor model regression for the excess returns of the N underlying securities:

$$r = a_0 + \beta r_B + u,$$

(20.24)

where β is the $N \times K$ matrix of regression betas, and $E(ur_B) = E(u) = 0$. Let the vector of idiosyncratic returns be the sum of the intercepts plus residuals: $v = a_0 + u$. A fund forms a portfolio of the N assets in r using the predetermined weights x: $r_p = x'r = (x'\beta) r_B + x'v$.

Table 20.7
Holdings-based performance evaluation.

Moment condition	Orthogonal to	Number of moments	Number of parameters
$\varepsilon_1 = (a - b'r_B)r_B$	1	K	$K+1$
$\varepsilon_2 = (a - b'r_B)R_f - 1$	1	1	0
$\varepsilon_3 = r_B - \mu_B$	1	1	1
$\varepsilon_4 = \alpha_m - a(r_B - \mu_B)'w$	1	1	1
$\varepsilon_5 = \alpha_\sigma + b'(r_B r_B')w - a\mu_B'w$	1	1	1
$\varepsilon_6 = \alpha_S - [(a - b'r_B)\,v'x]$	1	$\underline{1}$	$\underline{1}$

Note: Total number of moments and of parameters is $K+5$.

In this formulation, the "cash" position invested in the short-term Treasury security is $1 - \underline{1}'x$, where $\underline{1}$ is an N-vector of ones. Define $w' = x'\beta$ as the "asset allocation" weights. The approach is to estimate these using a "bottom-up" method and daily data for the underlying asset returns and benchmarks, similar to Jiang, Yao, and Yu (2007) and Elton, Gruber, and Blake (2011). Unfortunately, this means that the sampling error in the daily betas will not be captured in the GMM specification.

Substituting for r_p in the definition of alpha, we obtain

$$\alpha_p = a\text{Cov}(w'r_B) - b'E\{[r_B r_B' - E(r_B r_B')]w\} + E\{(a - b'r_B)x'v\}. \tag{20.25}$$

Ferson and Mo (2016) estimate the marketwide parameters a and b through the short-term Treasury return R_f and the excess returns of the benchmarks. For each fund, the market-level timing ability in alpha is denoted α_m, a volatility timing component is denoted by α_σ, and a selectivity component is α_S. The total SDF alpha for each fund is $\alpha_p = \alpha_m + \alpha_\sigma + \alpha_S$. Table 20.7 shows the GMM formulation.

The GMM system is exactly identified and has a block diagonal structure with respect to the fund-specific performance parameters. This allows fund-by-fund estimation based on the results in Farnsworth et al. (2002). The fifth equation in the table uses the condition $0 = aE(r_B)' - b'E(r_B r_B')$ to avoid the need to estimate the parameters of the matrix $E(r_B r_B')$. The fourth and fifth equations use the asset allocation weights w.

We are at the end of our journey through the GMM. Now you should be ready to use it to tackle almost any problem!

21 Multivariate Regression Methods

This and the next chapter review regression models for tests of asset pricing relations. This chapter is about multivariate regression models, and the next is about cross-sectional regressions. The order is historically reversed, as the cross-sectional regressions appeared in the literature first, in the 1960s. The multivariate regression approach was introduced to asset pricing, I believe, by Chang and Lee (1977) and James D. MacBeth (1979), and it was further developed in contemporaneous Chicago dissertations by Michael Gibbons (1982) and Robert Stambaugh (1982). In those days, the focus was testing the CAPM, but the approach works with multibeta models. When the risk factors are not traded assets, then mimicking portfolios must first be formed. The cross-sectional regression approach described in chapter 22 automatically forms mimicking portfolios. The multivariate regression approach is best when the number of time series observations, T, is much greater than the number of test assets, N. Most of the asymptotic properties are taken as T goes to infinity. In contrast, the cross-sectional regression approach described in chapter 22 works best when $N \gg T$. The asymptotic properties have N going to infinity, although in some cases, also assume T is large.

21.1 Regression Model Restrictions

The multivariate regression approach starts with time series regressions, where the test assets are regressed on the factor portfolios returns. The regressions are stacked up across the N test assets, resulting in a multivariate regression model:

$$R_{it} = a_i + \sum_j b_{ij} R_{jt} + u_{it}, t = 1, \dots, T, i = 1, \dots, N, \tag{21.1}$$

where the right-hand side $\{R_{jt}\}$ are the factor returns, like the market portfolio return for the CAPM or factor-mimicking portfolios in the multiple-beta case. The beta pricing model is tested by examining the restrictions that it imposes on the parameters in the system of regressions. The system (21.1) is a simple GMM problem, exactly identified and linear in the parameters. There are $N(K+1)$ moments and $N(K+1)$ parameters in a K-factor model. Since the variables on the right-hand side are the same for each test asset, the regression is a "seemingly unrelated" regression model, and the equation-by-equation OLS estimates are the same as the GMM estimates of the full system. The returns can be measured either in raw return form or in excess return form, and the asset pricing restrictions differ in the two cases.

The multivariate regression approach is related to the panel regression approach discussed in chapter 23. The main difference is that in the regressions here, the coefficients differ across assets, while in the panel setup, the coefficients are typically the same across assets.

Let's develop the asset pricing restrictions, first in the raw returns case. The multibeta expected return model says:

$$E(R_i) = \gamma_0[1 - \textstyle\sum_j b_{ij}] + \textstyle\sum_j b_{ij}E(R_j) \tag{21.2}$$

which implies that $a_i = \gamma_0[1 - \sum_j b_{ij}$ for some γ_0. In the unrestricted model, we have N intercepts, which the restriction asserts can be represented with only one new parameter γ_0, thus reducing the degrees of freedom by $N - 1$. The task is to impose this nonlinear restriction, estimate the zero-beta rate, γ_0, and test the model.

Things are easier when returns are measured in excess return form:

$$r_{it} = \alpha_i + \textstyle\sum_j \beta_{ij} r_{jt} + u_{it}, \; t = 1, \ldots, T, \; i = 1, \ldots, N, \tag{21.3}$$

where the r_{it} are excess returns. Now we don't have to estimate the zero-beta rate as a parameter. (The returns do not have to be in excess of a risk-free rate. But if not, the betas must be the excess return betas.) Multi-beta pricing implies

$$E(r_i) = \textstyle\sum_j \beta_{ij}E(r_j). \tag{21.4}$$

so that $\alpha_i = 0; \; i = 1, \ldots, N$. In the excess return case, we have N linear restrictions, setting the α_i parameters equal to 0 in the regression system.

21.1.1   Mean Variance Spanning and Intersection

The restrictions on the multivariate regression system are related to the concepts of mean variance spanning and intersection, as explored by Huberman and Kandel (1987). These ideas are precursors, historically, to our current understanding of the triality among the three paradigms of empirical asset pricing. Basically, any asset pricing model implies that some portfolio should be minimum variance efficient. In particular, a K-beta asset pricing model says that some combination of the traded factors (or their mimicking portfolios) is a minimum variance efficient portfolio.

Huberman and Kandel consider two minimum variance boundaries. The inner boundary is formed from combinations of the factor-mimicking portfolios. The outer boundary is formed from the test assets, including the mimicking portfolios. The intercept restriction on equation (21.1) implied by the K-beta pricing model using raw returns is

$$a_i = \gamma_0[1 - \sum_j b_{ij}]. \tag{21.5}$$

Huberman and Kandel show that this is equivalent to *mean variance intersection*, which means that the outer minimum variance boundary of all asset returns and the inner minimum variance boundary, of only the mimicking portfolios, coincide at one point. The portfolio at the point where the two boundaries coincide has the zero-beta rate γ_0. I think I know why they didn't call it the mean variance "tangency," which would seem to be more descriptive. When we graph the *variance* instead of the standard deviation on the x-axis of the minimum variance bullet, the point where the two boundaries coincide is a point of intersection. Students of the late and great Stephen A. Ross were trained to think that way, and these authors were students of Steve Ross.

There are three possibilities for the relation between the inner boundary formed from the factor portfolios and the outer boundary formed from all test assets. The boundaries can lie one inside the other, can have one point in common, or intersect at more than one point. If they intersect at more than one point, then the two boundaries must coincide, because of two-fund separation, which Huberman and Kandel denote as *mean variance spanning*. They show that the following restrictions on the regression system (21.1) for raw returns are equivalent to mean variance spanning:

$$\alpha_i = 0, \sum_j b_{ij} = 1, \text{ for all } i \tag{21.6}$$

Note that this is a stronger restriction than implied by an asset pricing model. You can think of it as saying that we have multibeta pricing using these factors for all zero-beta rates. If $a_i = \gamma_0[1 - \sum_j b_{ij}$ for all zero-beta rates, it implies the condition above.

Mean variance spanning has been used in the literature to ask whether certain assets expand the investment opportunity set formed from some smaller set of assets. For examples, De Santis (1993) asks: Can access to international equities expand the opportunities for a mean variance investor, given the ability to trade domestic equities? Does trading in emerging markets (De Roon, Nijman, and Werker 2001) add to the opportunities given that we can trade US stocks? Do closed-end funds (Bekaert and Urias 1996) expand the boundary, given that we can trade the underlying returns the funds hold? Do swaps (Aragon 2005) expand the bullet given trading in the underlying swapped assets? A review of mean variance spanning is provided by De Roon and Nijman (2001).

21.1.2   Spanning and Intersection in *m*-Talk

Everything is a special case of *m*-talk! Spanning and intersection can therefore be expressed in *m*-talk. We saw in chapter 7 that a portfolio R_p is minimum variance efficient if and only if it has the maximum squared correlation to *m* among all portfolios of the test asset returns. Multiple beta pricing says that an efficient portfolio is a combination of the K mimicking portfolios. Therefore, a combination of the factor-mimicking portfolios must have the maximum correlation to *m*. Therefore, if we run a regression of *m* on the vector of all returns R, only the factor portfolio returns have nonzero coefficients. This turns out to be equivalent to the characterization of mean variance intersection in Huberman and Kandel (1987), as shown by Ferson (1995).

Let R be the $T \times N$ matrix of asset returns where a subset of K returns, R_1, is hypothesized to be the mimicking portfolios that can be used to price the remaining assets in R_2. That is, a minimum variance efficient portfolio for all of R can be formed using only the returns in R_1. Partition the vector of expected returns as $E(R) = (\mu_1, \mu_2)$, and partition the covariance matrix $Cov(R) = \Sigma$ conformably. Consider a regression of m on the full set of asset returns:

$$m = E(m) + [R - \mu]\,\beta + \varepsilon, \tag{21.7}$$

where $\beta = \Sigma^{-1} Cov(R, m)$ is the regression coefficient. The fitted values are $m^* = E(m) + [R - \mu]\beta$, satisfying $E(m^*R) = 1$. If we can represent m^* using only R_1, it should be possible to set $\beta_2 = 0$ in the above regression, where $\beta = (\beta_1', \beta_2')'$. Cochrane (1996) and De Santis (1993) test the restriction that $\beta_2 = 0$ in the regression for m. Setting $\beta_2 = 0$, using m-talk to write $Cov(R, m) = 1 - \mu E(m)$, and using a standard expression for the inverse of a partitioned matrix, the expression $\beta_2 = 0$ states:

$$0 = (\Sigma_{22} - \Sigma_{21}\Sigma_{11}^{-1}\Sigma_{12})^{-1}\Sigma_{21}\Sigma_{11}^{-1}(\underline{1} - \mu_1 E(m)) + (\Sigma_{22} - \Sigma_{21}\Sigma_{11}^{-1}\Sigma_{12})^{-1}(\underline{1} - \mu_2 E(m)), \tag{21.8}$$

which implies that

$$\Sigma_{21}\Sigma_{11}^{-1}(\underline{1} - \mu_1 E(m)) = (\underline{1} - \mu_2 E(m)). \tag{21.9}$$

Mean variance intersection says that there is only one value of $E(m) = \lambda_0^{-1}$ at which the two frontiers coincide. Substituting $E(m) = \lambda_0^{-1}$ in equation (21.8), it can be seen that the restriction is equivalent to beta pricing for the returns in R_2 using the returns in R_1 as the factors, which is the condition for intersection proposed by Huberman and Kandel (1987).

Mean variance spanning says that the two minimum variance boundaries are coincident, which means that equation (21.9) holds for all possible values of $E(m)$. This is true only if

$$\Sigma_{21}\Sigma_{11}^{-1}\underline{1} = \underline{1}, \text{ and } \Sigma_{21}\Sigma_{11}^{-1}\mu_1 = \mu_2, \tag{21.10}$$

which are the restrictions for mean variance spanning proposed by Huberman and Kandel (1987). Ferson, Foerster, and Keim (1993) conduct GMM tests of both conditional and unconditional mean variance spanning using these conditions and the GMM.

21.2   Testing Portfolio Efficiency in the Multivariate Regression Model

Let's start with a single-factor model, the CAPM. This is how the literature started, although Gibbons, Ross, and Shanken (1989) recognize that asset pricing models are always about what portfolio is minimum variance efficient, and it need not be the market portfolio. We still ignore conditioning information for now.

21.2.1 The Setup

We start with excess returns for simplicity. The null hypothesis is that R_{pt} is a minimum variance efficient portfolio, which from Gonzales-Gaverra (1973), Fama (1973), and Roll (1977) is equivalent to

$$E(r_t) = \beta E(r_{pt}); \ \beta \equiv \text{Cov}(r_t; r_{pt})/\text{Var}(r_{pt}), \tag{21.11}$$

where the r_t are excess returns. The multivariate regression model is

$$r_t = \alpha + \beta r_{pt} + \varepsilon_t, \ \varepsilon_t \sim (0, \Omega). \tag{21.12}$$

The restrictions on the multivariate regression implied by the null hypothesis are that the N-vector $\alpha = \underline{0}$. The alternative hypothesis is that $\alpha \neq 0$. This setup generalizes easily, as shown by Gibbons, Ross, and Shanken (1989), to the case of a multibeta model, where r_{pt} is a vector and β is an $N \times K$ matrix.

21.2.2 Using the GMM to Test Efficiency

MacKinlay and Richardson (1991) were among the first to use the GMM in this setup. This is basically a stack of time series regression models, made easy by the GMM. The set of moment conditions are based on the error terms:

$$u_t = r_t - \alpha - \beta r_{pt} E(u_t) = 0 = E(u_t r_{pt}), \tag{21.13}$$

so the instruments are $Z'_t = (1, r_{pt})$, and they are not lagged relative to the returns. The vector of moment conditions is $g = \sum_t (u_t \otimes Z_t)/T$, and the parameter vector is $\varphi = (\alpha', \beta')'$, with $N(K+1)$ moments and $N(K+1)$ parameters ($K = 1$ in the CAPM). Without the parameter restrictions, this is an exactly identified regression model, and the parameter estimates are the same as the OLS estimates. Since this is a seemingly unrelated regression model, the OLS estimates for the stacked system are the same as the estimates for each equation taken separately (which is easily checked by solving for the GMM estimates). The null hypothesis can be tested using a standard Wald Test: $T\alpha' \text{Acov}(\alpha)^{-1}\alpha$, asymptotically a chi-square (N). Alternatively, the null hypothesis can be imposed by suppressing the intercept vector. Now we have an overidentified linear model, and the null hypothesis can be tested using Hansen's (1982) J-test. This test is asymptotically equivalent to the first, also distributed as a chi-square (N).

There is a complication here if the tested portfolio r_p is a linear combination of the left-hand-side assets. Then the residual covariance matrix is singular, and $\text{Acov}(\alpha)$ may not be invertible. We can randomly remove some of the test assets to avoid this problem.

Do not decide which ones to remove to get results that you like. That would be specification mining, one of the deadly sins of empirical work.

We can also use the GMM to test efficiency using raw returns. The system of moment conditions is

$$u_t = R_t - \gamma_0(1 - \beta) - \beta R_{pt}, \tag{21.14}$$

and the instruments are $Z_t' = (1, R_{pt})$. In this formulation, we have imposed the asset pricing restriction on the intercepts, and the model is overidentified. The model is also nonlinear in the parameters because of the product $\gamma_0(1 - \beta)$. The restrictions can be tested with Hansen's J-test, which is asymptotically chi-square $(N - 1)$.

Harvey and Kirby (1996) propose a GMM formulation for nontraded factors F and raw returns. The system of moment conditions for a one-factor model with no conditioning information is

$$
\begin{aligned}
u_{1t} &= R_t - \mu, \\
u_{2t} &= F_t - \mu_F, \\
u_{3t} &= (F_t - \mu_F)^2 - \sigma_F^2, \\
u_{4t} &= R_t - \lambda_0 1 - \lambda_1[(R_t - \mu)(F_t - \mu_F)/\sigma_F^2],
\end{aligned} \tag{21.15}
$$

where λ_1 is the K-vector of market prices of risk. The degrees of freedom with N assets and K factors is $N - 2$. I think that we could remove the first moment condition and remove μ from the fourth, and it would still work.

As an alternative to the restricted regression, you could estimate the unrestricted regression system for raw returns using the GMM and test the nonlinear hypothesis that $H(\varphi) = a - \gamma_0(1 - \beta) = 0$, where a in the N-vector of intercepts replaces $\gamma_0(1 - \beta)$ in equation (21.14). The unrestricted system does not produce an estimate of γ_0, so you have to get the zero-beta rate some other way. If you wish, you can impose that the zero-beta rate is the average of some risk-free rate proxy in your data. Black, Jensen, and Scholes (1972) suggested estimating the zero-beta rate in the raw return model as $\sum_i h_i[a_i/(1 - \beta_i)]$, with the weights h_i chosen to minimize the mean squared error. This is essentially a cross-sectional regression of the intercepts on $(1 - \beta_i)$ with the intercept suppressed. Shanken (1986) developed a zero-beta rate estimator based on minimizing over the choice of γ_0, a quadratic form in the average residuals. Shanken and Zhou (2007) compare the properties of several zero-beta rate estimators using simulations.

21.2.3   Quadratic Form Tests

Recall the general approach to χ^2 tests that we used in chapter 18. Suppose that we have an estimator of alpha, $\hat{\alpha}$, for which we know $\sqrt{T}[\hat{\alpha} - \alpha] \to_d N(0, \text{Acov}(\hat{\alpha}))$. The null hypothesis is that alpha is zero. We form $\varepsilon = \sqrt{T} \, \text{Acov}(\hat{\alpha})^{-1/2} \to_d N(0, I)$, so that $\varepsilon'\varepsilon = T\hat{\alpha}' \text{Acov}(\hat{\alpha})^{-1}\hat{\alpha} \to_d \chi^2(N)$. This results in the standard Wald test for the zero alpha null. Under GMM or normal regression,

$$\text{Acov}(\hat{\alpha}) = (1 + S^2(r_p)) \, \Omega, \tag{21.16}$$

where $S^2(r_p)$ is the sample squared Sharpe ratio of r_p, and Ω is the $N \times N$ covariance matrix of the regression model residuals. With this substitution, we can write the Wald test as

$$\text{Wald} = T(1 + S^2(r_p))^{-1}\hat{\alpha}'\Omega^{-1}\hat{\alpha}. \tag{21.17}$$

This result turns out to have some fascinating—in a nerdy sort of way—economic interpretations.

21.3   Economic Interpretation

To quickly recap, we are testing for the mean variance efficiency of a portfolio r_p (excess return), using the stacked regression $r = \alpha + \beta_p r_p + u$, $E(uu') = \Omega$, where α is the N-vector of pricing errors, and minimum variance efficiency states that $\alpha = 0$. The Wald test is

$$\text{Wald} = T(1 + S^2(r_p))^{-1} \ \hat{\alpha}'\Omega^{-1}\hat{\alpha} = T\hat{\alpha}'\text{Acov}(\hat{\alpha})^{-1}\hat{\alpha}, \tag{21.18}$$

where $\text{Acov}(\hat{\alpha}) = (1 + S^2(r_p)) \ \Omega$.

21.3.1   Facts about Sharpe Ratios and Alphas

We need a few facts to display the economic intuition of the tests. These are developed in a series of lemmas. The first one characterizes the maximum squared Sharpe ratio.

Lemma 21.1. $\text{Max}_x\{(x'E(r))^2/x'\Omega x\} = E(r)'\Omega^{-1}E(r).$ \hfill (21.19)

Proof. Doing the calculus, the optimal weight is $x = c\Omega^{-1}E(r)$; plug this result back in to find the maximized value. QED.

Define $S^2(r)$ as the maximum squared Sharpe ratio for a portfolio of the excess returns on the test assets, r, with $V = \text{Cov}(r)$. By lemma 21.1, $S^2(r) = E(r)'V^{-1}E(r)$. Define $S^2(r_p)$ as the squared Sharpe ratio for the tested portfolio, r_p. On the assumption that the tested portfolio is a fixed combination of the test assets, its squared Sharpe ratio may be written as $S^2(r_p) = (x_p'E(r))^2/x_p'Vx_p$, where the x_p are the weights that define the portfolio r_p. Finally, define $S^2(\alpha)$ as the largest abnormal excess return per unit of nonsystematic risk of any portfolio, where nonsystematic risk is defined relative to the benchmark r_p. That is, $S^2(\alpha) = \text{Max}_x\{(x'\alpha)^2/x'\Omega x\} = \alpha'\Omega^{-1} \alpha$, by lemma 21.1. The solution to this problem, $x_\alpha = c\Omega^{-1} \alpha$, is the *optimal orthogonal portfolio*.

21.3.2   The Optimal Orthogonal Portfolio

The optimal orthogonal portfolio defined above (the OOP) is the key to the economic interpretation of the tests for mean variance efficiency in multivariate regression models. We met the OOP in chapter 8 when we discussed modern portfolio management. It has several interesting properties. First, it is the "most mispriced" by the benchmark r_p, as suggested by its definition. That is the "optimal" part in its name, and its main attraction for modern portfolio managers. The second word in its name is "orthogonal." This refers to the fact that the portfolio has the property that $x_a'\beta = 0$. That is, the portfolio is orthogonal to r_p.

The last word is "portfolio," but I guess that is pretty obvious. To establish the orthogonal part, we need another lemma.

Lemma 21.2. In the stacked regression $r = \alpha + \beta r_p + u$, $E(uu') = \Omega$, we have $\alpha' \Omega^{-1} \beta = 0$ and $\alpha' V^{-1} \beta = 0$.

Proof. For $\alpha' V^{-1} \beta = 0$, we use the fact that r_p is a combination of the tested assets: $r_p = W'r$, so $\mu_p = W'\mu$. Then we have $\beta' = (W'VW)^{-1}W'V$, and $\alpha' V^{-1} \beta = [\mu - VW(W'VW)^{-1}\mu_p]'V^{-1}[VW (W'VW)^{-1}] = \mu'W(W'VW)^{-1} - \mu'W(W'VW)^{-1} = 0$. For $\alpha' \Omega^{-1} \beta = 0$, write $V = \sigma_p^2 \beta\beta' + \Omega$, and apply the Sherman Morrison formula to write $\Omega^{-1} = V^{-1} + V^{-1}(\sigma_p^2 \beta\beta')V^{-1}/(1 + \sigma_p^2 \beta'V^{-1}\beta)$. We see that if $\alpha' V^{-1} \beta = 0$, then $\alpha' \Omega^{-1} \beta = 0$. QED.

Since the weight of the OOP portfolio is $x_\alpha = c \, \Omega^{-1}\alpha$, lemma 21.2 implies that the OOP is orthogonal to r_p. We express the optimal orthogonal portfolio in terms of the residual covariance matrix, but it can also be written as $x^\alpha = cV^{-1}\alpha$. To see this, use $V = \sigma_p^2\beta\beta' + \Omega$. Orthogonality then implies that $x_\alpha' \Omega x_\alpha = x_\alpha'[V - \sigma_p^2 \, \beta\beta']x_\alpha = x_\alpha' V x_\alpha$. Thus, the variance in the denominator of the maximization that defines the optimal orthogonal portfolio can use either matrix.

The OOP is called the "active" portfolio by Gibbons, Ross, and Shanken (1989), which makes sense, given its definition as getting the largest alpha per unit of residual risk. It is a strong tenet of active portfolio management that you must trade off abnormal performance, or alpha, against the nonsystematic risk that you take to get alpha. But the OOP has another interesting property that justifies calling it the "active" portfolio. Suppose that a mean variance efficient portfolio is desired, but the benchmark portfolio, r_p, is not efficient. This is typically the case. It turns out that a combination of the OOP and its benchmark, r_p, will be efficient. So, the OOP is the thing you want to trade to move from an inefficient benchmark to the mean variance frontier. This is the next lemma.

Lemma 21.3. A combination of the portfolio, x_α, and the benchmark r_p, is minimum-variance efficient.

Proof. Consider $r^* = wr_\alpha + (1 - w)r_p$, where $r_\alpha = x_\alpha'r$ is the OOP. We want to show that the N-vector a is zero in the regression $r = a + br^* + \varepsilon$, for some choice of w. This will be true if the weights of r^* are $\propto V^{-1}E(r)$ for some w. Call the weights x^*. Then if x_p are the weights of r_p, $x^* = wx_\alpha + (1 - w)x_p = cwV^{-1}\alpha + (1 - w)x_p = cwV^{-1}[\alpha + \{(1 - w)/cw\}Vx_p]$ for any w. Now pick w so that $(1 - w)/cw = E(r_p)/Var(r_p)$, then $x^* = cwV^{-1}[\alpha + \{(Vx_p)/Var(r_p)\} E(r_p)] = cwV^{-1}[\alpha + \beta \, E(r_p)] = cwV^{-1}E(r)$. Thus, the weights x^* are proportional to those of a minimum variance efficient portfolio, $V^{-1}E(r)$. QED.

This proof uses the population moments, but everything can be stated in term of sample moments. Gibbons, Ross, and Shanken (1989, table VII) provide an illustration. In the example, the OOP does not have positive weights in the positive alpha small stocks. It shows that the OOP weights can be more subtle than simply buying the positive alpha assets.

21.3.3 Graphical Interpretation

The OOP drives a graphical interpretation of quadratic form tests in the multivariate regression model. This requires another lemma.

Lemma 21.4. The law of conservation of squared Sharpe ratios states that

$$S^2(r) = \alpha' \Omega^{-1} \alpha + S^2(r_p). \tag{21.20}$$

Proof. By lemma 21.1,

$$
\begin{aligned}
S^2(r) &= E(r)' V^{-1} E(r) \\
&= [\alpha + \beta\, E(r_p)]' V^{-1} [\alpha + \beta\, E(r_p)] \\
&= \alpha' V^{-1} \alpha + 2E(r_p)\alpha' V^{-1}\beta + E(r_p)^2 \beta' V^{-1}\beta \\
&= \alpha' V^{-1} \alpha + 0 + E(r_p)^2 \beta' V^{-1}\beta, \text{ by lemma 21.2. Now, substituting } \beta = (x_p' V x_p)^{-1} V x_p
\end{aligned}
$$

$S^2(r) = \alpha' V^{-1} \alpha + S^2(r_p)$. QED.

Now we can use the law of conservation of squared Sharpe ratios to graphically interpret the quadratic form tests: $W = T(1 + S^2(r_p))^{-1} \alpha' \Omega^{-1} \alpha = T(1 + S^2(r_p))^{-1} [S^2(r) - S^2(r_p)]$. Use your imagination for the graph. The graph is the sample mean-standard deviation bullet. The tested portfolio r_p must lie somewhere inside the sample boundary. Its Sharpe ratio $S(r_p)$ is the slope of a line drawn from the average risk-free rate on the y-axis through the tested portfolio inside the boundary. The Sharpe ratio of the boundary portfolio is $S(r)$. This is the slope of a line drawn from the average risk-free rate tangent to the bullet. Thus, the quadratic form measures the distance between the tested portfolio and the boundary, through the difference in their squared Sharpe ratios, or the two slopes in the graph. This difference is proportional to the squared Sharpe ratio of the OOP. When the two slopes are close together, the test statistic is small, and we do not reject efficiency. When the two slopes are far apart, we are more likely to reject the efficiency of the tested portfolio.

21.3.4 Efficiency with Respect to Z, Again

Ferson and Siegel (2015) develop optimal orthogonal portfolios with conditioning information and relate them to efficiency with respect to information Z, as discussed in chapter 9. They define the OOP in this case as follows. The tested portfolio is R_p. For a given zero-beta rate, γ_0, the *most mispriced* (or *optimal orthogonal*) portfolio, R_c, with respect to R_p and conditioning information Z, uses the conditioning information to maximize α_c^2/σ_c^2, where σ_c^2 is the unconditional variance of R_c, and α_c is the unconditional alpha of R_c with respect to R_p.

Ferson and Siegel present several results for the OOP with respect to Z. Among these, a law of conservation of squared Sharpe ratios is found to apply. That is,

$$S_s^2 \equiv \text{Max}_{\{x(Z):x(Z)'\underline{1}=1\}}\{E[x(Z)'R - \gamma_0]\}^2 / \text{Var}\{x(Z)'R\} = S_p^2 + S_c^2, \tag{21.21}$$

where S_p^2 and S_c^2 are the squared unconditional Sharpe ratios of R_p and R_c. They show that the efficient portfolio with respect to Z can be written as a combination of the tested portfolio and its OOP with respect to Z. Finally, orthogonality holds in the sense that $\text{Cov}(R_c, R_p) = 0$.

21.4 Some Results from Gibbons, Ross, and Shanken

21.4.1 A Review of Standard Tests under Normality

The log likelihood function under normality for the multivariate regression model in excess returns can be written as

$$\ln(L) = (NT/2)\ln(2\pi) - (T/2)\ln|\Omega| - (1/2)\Sigma_t(r_t - \alpha - \beta r_{pt})'\Omega^{-1}(r_t - \alpha - \beta r_{pt}). \tag{21.22}$$

Gibbons, Ross, and Shanken (1989) compare three standard tests based on the log likelihood. The first is the Wald test (W), stated here for the general nonlinear hypothesis that $H(\theta) = 0$:

Wald Test $T\,H(\theta)'\{(\partial H(\theta)/\partial\theta)\text{Cov}(\theta)\,(\partial H(\theta)/\partial\theta)'\}^{-1}H(\theta) \sim$ chi-square (dim $H(\cdot)$).

The next is the likelihood ratio test (LRT) of the hypothesis that $\theta = \theta_0$, which was used by Gibbons (1982):

Likelihood Ratio Test $2[\text{Max}_\theta \ln L(\theta) - \ln L(\theta_0)] \sim$ chi-square (number of restrictions by null).

The third test is the Lagrange multiplier test (LMT), studied by Stambaugh (1982):

Lagrange Multiplier Test $T[\partial \ln L(\theta_0)/\partial\theta]\text{Cov}(\theta)(\partial \ln L(\theta_0)/\partial\theta)' \sim$ chi-square (number of restrictions).

For the LMT, $\text{Cov}(\theta)$ may be estimated as a special case of the GMM, as shown in chapter 20, where the moment condition is $g = E(\partial \ln L(\theta_0)/\partial\theta)$.

Under normality, the Wald test is proportional to an exact F-test, for which the Gibbons, Ross, and Shaken (1989) paper is widely cited:

$$[(T - N - 1)/N](1 + S_p^2)^{-1}\hat\alpha'\Omega^{-1}\hat\alpha \sim F_{N,T-N-1}. \tag{21.23}$$

With K mimicking portfolio excess returns, the generalization is

$$[(T - N - K)/N][1 + E_T(f)'V(f)^{-1}E_T(f)]^{-1}\hat\alpha'\Omega^{-1}\hat\alpha \sim F_{N,T-N-K}, \tag{21.24}$$

where $E_T(f)$ denotes the sample mean of the factor portfolio excess returns and $V(f)$ is their sample covariance matrix. Sample moments are used, of course, for all terms in this expression.

21.4.2 A Comparison of Wald, LMT, and LRT Tests

Buse (1982) showed that in any given sample, the W, LRT, and LMT test statistics are transformations of each other. In any finite sample, $W > LRT > LMT$, but they have the same asymptotic distribution. Gibbons, Ross, and Shanken (1989) use these results to compare the sampling properties of the various statistics in finite samples under the assumption of normality. In this case, the exact F-test would deliver the "true" p-value of the test. Defining $W = (1 + S_p^2)^{-1} \hat{\alpha}' \Omega^{-1} \hat{\alpha}$, then $[(T-N-1)/N]W \sim F$. Given N, T, and a "true" p-value, this determines the value of the Wald test statistic, W. Then, using W, they compute what the other test statistic values would have been: $LRT = T \ln(1+W)$ and $LMT = TW/(1+W)$. Given the values that the other tests would have, their asymptotic distributions are referenced to see what p-values they would report. These values can be compared with the true p-value initially selected. (See table 21.1.)

The results of this exercise shed light on what had been a puzzling set of results in the literature that tested the CAPM with multivariate regression models. Gibbons (1982), using the Wald test, had strongly rejected the CAPM, while Stambaugh (1982), using the LMT, had not. Table 21.1 suggests that the finite-sample properties of the tests likely had something to do with it: The stated p-values of the LRT are too small in finite samples, while those of the LMT are too large.

21.5 MacKinlay's Power Analysis and the Fama-French Factors

The power of a test is the probability that it rejects when the alternative ($\alpha \neq 0$) hypothesis is true. If the alternative hypothesis in an asset pricing test is true, then the quadratic form in the tests is $\hat{\alpha}' \Omega^{-1} \hat{\alpha} = S^2(r) - S^2(r_p) > 0$. We saw in chapter 8 that the correlation of an arbitrarily chosen portfolio with an efficient portfolio is the ratio of the two Sharpe ratios. Let the correlation be ρ^*, and write $S^2(r_p) = \rho^{*2} S^2(r)$. Then the Wald test can be written as $T(1 - \rho^{*2})/[(1/S^2(r) + \rho^{*2}]$. Intuitively, a larger test statistic means more power, so we see that the power is higher when $S^2(r)$ is higher and the bullet has a steeper face, and it is lower when ρ^{*2} is higher. Gibbons, Ross, and Shanken (1989) graph some power functions based on this relation.

MacKinlay (1995) uses power to analyze tests of the three-factor model of Fama and French (1993). That model adds two factors to the market portfolio of the CAPM: the excess returns of small less those of large firms (SMB) and the excess returns of high book/market less those of low book/market stocks (HML). Under normality, the exact F-test of Gibbons, Ross, and Shanken (1989) follows a central F distribution in finite samples when the null hypothesis is true. If the zero alpha null is false, the test is noncentral, $F_{N,T-N-1}(\delta)$, where the noncentrality parameter $\delta = T(1 + S_p^2)^{-1} \hat{\alpha}' \Omega^{-1} \hat{\alpha} \leq T \hat{\alpha}' \Omega^{-1} \hat{\alpha}$. A larger δ shifts the distribution to the right and spreads it out. MacKinlay uses this characterization of the alternative to examine rejections of the CAPM in favor of the Fama and

Table 21.1
Comparison of four asymptotically equivalent tests of ex ante efficiency of a given portfolio.

			p-values using asymptotic approximations		
N	*T*	*p*-value using exact distribution of W	Wald	LRT	LMT
10	60	0.05	0.008	0.027	0.071
20	60	0.05	0.000	0.007	0.094
40	60	0.05	0.000	0.000	0.173
58	60	0.05	0.000	0.000	0.403
10	120	0.05	0.023	0.038	0.060
20	120	0.05	0.005	0.023	0.070
40	120	0.05	0.000	0.003	0.094
58	120	0.05	0.000	0.000	0.122
118	120	0.05	0.000	0.000	0.431
10	240	0.05	0.035	0.044	0.055
20	240	0.05	0.109	0.035	0.059
40	240	0.05	0.003	0.017	0.069
58	240	0.05	0.000	0.006	0.079
118	240	0.05	0.000	0.000	0.123
238	240	0.05	0.000	0.000	0.451
10	60	0.10	0.025	0.061	0.122
20	60	0.10	0.000	0.019	0.146
40	60	0.10	0.000	0.000	0.216
58	60	0.10	0.000	0.000	0.404
10	120	0.10	0.056	0.081	0.111
20	120	0.10	0.017	0.053	0.122
40	120	0.10	0.000	0.010	0.147
58	120	0.10	0.000	0.000	0.175
118	120	0.10	0.000	0.000	0.432
10	240	0.10	0.076	0.090	0.106
20	240	0.10	0.048	0.075	0.111
40	240	0.10	0.009	0.041	0.122
58	240	0.10	0.001	0.018	0.133
118	240	0.10	0.000	0.000	0.178
238	240	0.10	0.000	0.000	0.452

Source: Gibbons, Ross, and Shanken (1989), table 1.
Note: The W statistic is distributed as a transform of a central *F* distribution in finite samples. The Wald test, the likelihood ratio test (LRT), and the Lagrange multiplier test (LMT) are monotone transforms of W, and each is distributed as chi-square with *N* degrees of freedom as *T* approaches infinity. *N* is the number of assets used together with portfolio *p* to construct the ex post frontier, and *T* is the number of time series observations.

French (1993) factors. He argues that if the Fama-French factors are based on risks that are missing in the CAPM, as suggested by Fama and French (1996), then the missing factor(s) bound δ by "reasonable" values of the squared Sharpe ratio of the OOP. For example, MacKinlay argues the squared Sharpe ratio should be ≤ 0.02 in monthly data, so $\delta = 7.1$. However, data snooping, market irrationality, trading costs, and other nonrisk-based alternatives perhaps allow the Sharpe ratio of the OOP to be arbitrarily large. MacKinlay finds that for the Fama and French (1993) sample, the F-test statistic is in the far right tail under a risk-based alternative. So, the HML and SMB factors, added to the CAPM, appear more consistent with nonrisk-based alternatives, as suggested by Lakonishok, Shleifer, and Vishny (1994) than with the hypothesis that they are risk factors missing from the CAPM.

21.6 Finite-Sample Results

Most of the tests discussed above rely on asymptotic distributions. The Gibbons, Ross, and Shanken (1989) test is an exact F-test under normality, when we use excess returns, and has a known distribution in finite samples. It is central F under the null hypothesis that the pricing errors are mean zero, and noncentral F under the alternative hypothesis for the pricing errors. The literature provides only a few further results on the finite-sample distributions of tests for multivariate regression models. For most of these results, normality of the returns is assumed. The case of raw return tests under normality is studied by Shanken (1986), Zhou (1991), and Velu and Zhou (1999). They provide upper and lower bounds and simulation methods to obtain the finite-sample distributions when the zero-beta rate must be estimated. Most of what we know about the finite-sample properties of multivariate regression tests of asset pricing models comes from simulation studies, and there are many. Lewellen, Nagel, and Shanken (2010) review some of the classics and present their own evidence using simulations.

22 Cross-Sectional Regression

Cross-sectional regression (CSR) methods have been important in empirical asset pricing since the beginning of the field, when researchers started testing the CAPM in the 1960s. These methods remain important to this day, and I will bet that in the next issue of the *Journal of Finance*, *Journal of Financial Economics*, or *Review of Financial Studies*, you will find a CSR analysis. This chapter provides most of what you need to use these methods in your work. There are some interesting interpretations and some pitfalls to be avoided. Research on these methods continues, so I include a discussion of some recent work.

22.1 CSR Tests

The multivariate regression approach discussed in chapter 21 is based on N time series regressions of the tested asset returns on the priced factors, with T observations. We estimate the betas and intercepts, and we look for zero alphas (if portfolio excess returns are on the right-hand side) or restricted intercepts (if raw returns are used). CSRs use the N test assets' returns or expected returns as the left-hand side observations, and the right-hand side variables are the betas or other characteristics of the test assets. The slope coefficients are risk premiums, and we can estimate a time series of risk premiums and zero-beta rates. (If excess returns are used, the expected zero-beta rates should be zero.)

CSRs are natural in asset pricing problems, because the number of assets, N, and thus the number of observations in the regression can be large. For example, even though the number of listed stocks on US exchanges has declined in recent years, you can get more than 18,000 stocks' returns data from the Center for Research in Security Prices for 2013. The CSR approach also makes it easy to handle randomly missing data for subsets of the asset returns by just leaving out those stocks when their returns data are missing. Another advantage of CSRs, compared with multivariate time series regression models, is that CSRs can handle nontraded factors. We can estimate betas on any old factors and put them on the right-hand side of the regressions. CSR delivers "mimicking portfolios," whose mean excess returns are the risk premiums. Therefore, the asset pricing model can be tested by comparing the average asset returns with the returns predicted by the products of the betas and risk premiums.

In contrast, nontraded factors "mess up" time series regression intercepts. To see this, consider the system of time series regressions on a nontraded factor, F_t:

$$r_t = a + \beta F_t + u_t, \; t = 1, \dots, T. \tag{22.1}$$

The asset pricing model allowing for a nonzero alpha implies $E(r) = \alpha + \beta\lambda$, with risk premium λ, which implies that the regression intercept is $a = \alpha + \beta[\lambda - E(F_t)]$. If $E(F) = \lambda$, then $a = \alpha$, and we can test the null hypothesis of zero alphas by looking at the time series regression intercepts, a. The problem is that the expected value of a nontraded factor, $E(F)$, may differ from the expected risk premium λ. If the factor is not traded, its scale could be almost anything. This invalidates the regression intercept as a measure of abnormal return. Using the intercepts in time series regressions with nontraded factors as if they were alphas is an all too common mistake.

22.1.1   The Big Picture

The basic idea of the CSR approach is very simple. We wish to run a regression across assets or portfolios i, of returns on their betas with respect to some factor(s):

$$r_i = \lambda_0 + \lambda_1\beta_i + e_i, \; i = 1, \dots, N, \tag{22.2}$$

where the r_i may be excess returns, as shown in equation (22.2), or raw returns, described shortly. The CSR delivers estimates of the coefficients $\{\lambda_0, \lambda_1\}$.

Consider the CAPM using excess returns and betas, which implies $E(r_i) = \beta_i E(r_m)$, so that the asset pricing hypothesis states that the slope of the fitted line in figure 22.1 is $\lambda_1 = E(r_m)$, and the intercept, $\lambda_0 = 0$. We can use excess returns on the left-hand side as illustrated, or we can use raw returns. If we use raw returns R_i, and if we had the betas of the raw returns on the x-axis, then the Black (1972) zero-beta version of the CAPM would imply $E(R_i) = \lambda_0 + \beta_i E[R_m - \lambda_0]$, so that the slope of the line is $\lambda_1 = E(R_m - \lambda_0)$, and the intercept λ_0 is the expected zero-beta rate. With raw returns and betas, the Sharpe (1964) CAPM implies that λ_0 is the average risk-free rate.

22.1.2   A Two-Step Approach in Various Flavors

CSRs follow a two-step approach. In the first step, we get estimates of betas from time series regressions. For example, if r_m is the factor, then

$$r_{it} = \alpha_i + \beta_i r_{mt} + \varepsilon_{it}, \; t = 1, \dots, T \text{ for each } i. \tag{22.3}$$

These are the separate time series regressions that comprise the multivariate regression model examined in chapter 21. For now, we ignore the estimation error in the estimates of beta, which are addressed in section 22.5.

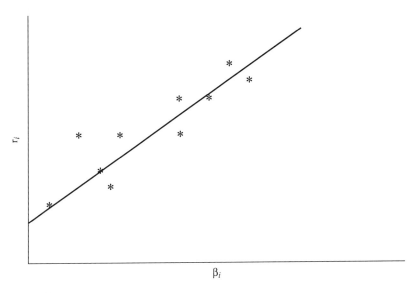

Figure 22.1
Cross-sectional regression of returns on betas.

In the second step, the beta estimates are used as regressors in a CSR of returns on betas, possibly for each period, t:

$$r_{it} = \lambda_{0t} + \lambda_t' \beta_i + u_{it}, \quad i = 1, \ldots, N. \tag{22.4}$$

There can be K betas in the second-step regression, as the notation suggests. A couple of points deserve mention. First, we could use excess returns (as shown above), or we could use raw returns on the left-hand side of the regressions. The interpretation of the CSR coefficients is different in the two cases. Importantly, the "betas" on the right-hand side of the second-step CSR do not even have to be betas. They could be other characteristics of the firms' stocks, such as book-to-market ratios, firm size, or dividend yield. Use your imagination. We can obviously avoid the first step for those variables.

The literature has used the two-pass CSR method for many decades, and in several flavors. Black, Jensen, and Scholes (1972) hold the betas constant over the full sample period and run one CSR with sample mean returns on the left-hand side. Fama and MacBeth (1973) use rolling time series regressions (with 60-month windows) to get the betas and then run a CSR for next month's returns each month, rolling the whole thing forward month by month. This approach remains popular to this day. When applied in the context of, say, mutual fund or hedge fund evaluation, where the lengths of the time series can be limited, a 36-month (or even 24-month) window is often used. With daily returns data, a 240-trading-day window is popular for the beta estimation. Some studies use a fixed beta for each stock or portfolio but still run CSRs for each period. This is useful for theoretical analyses of the sampling properties.

Early studies, including Fama and MacBeth (1973), sorted stocks into portfolios based on past beta estimates. The idea was to reduce measurement error in the betas (as the betas of portfolios are more accurately estimated than the betas of individual stocks), while still preserving cross-sectional variation in the betas, which is important for the accuracy of the CSRs. Some studies use portfolio-level betas and assign the portfolio beta to be the beta of the individual stock, in an attempt to reduce estimation error in the individual stock betas (e.g., Fama and French, 1993). Ang, Liu, and Schwartz (2017) show that this is equivalent to using the portfolios instead of the individual stocks as the cross section of assets.

Starting in about the 1980s, studies began to sort stocks on firm characteristics that are found to be correlated with average returns in the cross section, but not necessarily with betas. Often these patterns in average returns are anomalous relative to a beta pricing model. This approach may deliver better power to reject the beta pricing model, and it results in less explanatory power for the betas. If the characteristic is the result of data mining, it can bias the regression results (e.g., Lo and MacKinlay 1990). Some recent studies have advocated the use of large cross sections of individual stocks instead of portfolio formation (e.g., Kim 1995; Gagliardini, Ossola, and Scaillet 2016; Ang, Liu, and Schwartz 2017). Such an approach allows samples with large N and avoids the loss of information that portfolio formation implies, but it requires a solution to the errors-in-variables problem in the individual stock betas, as described below.

22.1.3 A Portfolio Interpretation and Mimicking Portfolios in CSRs

The first part follows Fama (1976b), in a classic but out-of-print book. Consider the CSR model with $K = 1$:

$$R_{it} = \lambda_{0t} + \lambda_{1t}\beta_i + u_{it}, \ i = 1, \ldots, N. \tag{22.5}$$

Using raw returns R_{it}, then the asset pricing model (e.g., CAPM) implies H_0: $E(\lambda_{1t}) = E(R_{mt} - R_{0t})$, and $E(\lambda_{0t}) = E(R_{0t})$ is the expected zero-beta return. We know the OLS estimator of the slope, $\hat{\lambda}_{1t}$, is a best linear unbiased estimator, meaning it solves:

Min $\sum_i u_{it}^2$, subject to
Unbiased: $E(\hat{\lambda}_{1t}) = \lambda_{1t}$, $\qquad\qquad\qquad\qquad\qquad\qquad\qquad\qquad$ (22.6)
Linear: $\hat{\lambda}_{1t} = \sum_i w_i R_{it}$.

Thus, the estimator for λ_{1t} is a "portfolio" of the returns. What kind of portfolio? Substituting, we see that

$$\hat{\lambda}_{1t} = \sum_i w_i R_{it} = \sum_i w_i(\lambda_{0t} + \lambda_{1t}\beta_i + u_{it}) = (\sum_i w_i)\lambda_{0t} + \lambda_{1t}(\sum_i w_i\beta_i) + (\sum_i w_i u_{it}). \tag{22.7}$$

Now unbiasedness, $E(\hat{\lambda}_{1t}) = \lambda_{1t}$, implies

$(\sum_i w_i) = 0:$ a zero net investment portfolio;
$(\sum_i w_i \beta_i) = 1:$ beta = 1. $\qquad\qquad\qquad\qquad\qquad\qquad\qquad\qquad\qquad$ (22.8)

Given that the average market beta of stocks is about 1.0 and the portfolio weights sum to 0.0, the first condition of equation (22.8) says that the portfolio must have both long and short positions. It has to long the higher-beta and short the lower-beta assets.

Suppose that we used a characteristic like the book/market of the portfolio in place of its beta on the right-hand side. The same conditions apply. The portfolio is a "long-short" portfolio, long the high book/market stocks and short the low book/market stocks. This is like the famous HML factor of Fama and French (1996), except that its average book/market is normalized to 1.0.

The intercept or zero-beta rate estimator also has a portfolio interpretation, with different weights. Equation (22.7) applies. Unbiasedness, $E(\hat{\lambda}_{0t}) = \lambda_{0t}$, implies

$(\sum_i w_i) = 1:$ a fully invested portfolio,
$(\sum_i w_i \beta_i) = 0:$ a zero-beta portfolio. $\qquad\qquad\qquad\qquad\qquad\qquad\qquad$ (22.9)

This generalizes for K factors. In general the portfolio weights of the zero-beta estimator, like that of the slope coefficient estimator, will have both negative ("short") and positive values. For example, if all the asset betas were positive, it would have to go both short and long to have a portfolio beta of zero.

Note that the mimicking portfolio weights that are automatically constructed by CSR are "well diversified" in the large market APT sense. Think of the regressors as the $N \times K$ matrix of loadings, β. Then using OLS to run the regressions, the weights in the portfolios are $w = (\beta'\beta)^{-1}\beta'$ and therefore, $w'w = (\beta'\beta)^{-1}$, which goes to zero as N gets large, under the assumption that the limit of $(\beta'\beta/N)^{-1}$ exits.

If we run a GLS CSR of excess returns on their betas using the covariance matrix $V(r)$, we get $\hat{\lambda} = [\beta'V(r)^{-1}\beta]^{-1}\beta'V(r)^{-1}r$, and the mimicking portfolio weights are $w' = V(r)^{-1}\beta[\beta'V(r)^{-1}\beta]^{-1}$.

An interesting side note here is that betas measured on the mimicking portfolios are the same as the betas measured on the risk factors that we started with, as was first shown by Breeden (1979). Letting r_p be the mimicking portfolios, we have

$$\begin{aligned} \text{Cov}(r, r_p)V(r_p)^{-1} &= \text{Cov}(r, r'V(r)^{-1}\beta[\beta'V(r)^{-1}\beta]^{-1})V(r_p)^{-1} \\ &= V(r)V(r)^{-1}\beta[\beta'V(r)^{-1}\beta]^{-1}\{[\beta'V(r)^{-1}\beta]^{-1}\beta'V(r)^{-1}V(r)V(r)^{-1}\beta \\ &\quad [\beta'V(r)^{-1}\beta]^{-1}\}^{-1} \\ &= \beta. \text{ Remarkable!} \end{aligned}$$
$\qquad\qquad\qquad\qquad\qquad\qquad\qquad\qquad\qquad\qquad\qquad\qquad\qquad$ (22.10)

22.1.4 Traded Factors in CSR

With traded factors f_t and excess returns, Shanken (1992) shows that the ML (or GLS) risk premium estimator of λ is the sample mean of the traded factor excess return. To see this:

Table 22.1
CSR is GMM.

	North	Parameter
$u1_t = R_t - \gamma_0 - \beta f_t$	N	$\gamma_0, \beta = NK + 1$
$u2_t = \text{vec}([R_t - \gamma_0 - \beta f_t] * f_t'$	NK	0
$u3_t = R_t - \gamma_0 - \beta \lambda$	N	$\lambda = K$
$g_t = (u1_t, u2_t, u3_t),$	$NK + 2N$	$NK + K + 1$

Note: The vec($\cdot$) operator stacks the columns of its matrix argument.

$$\hat{\lambda} = [\beta' V(r)^{-1}\beta]^{-1}\beta' V(r)^{-1}r = [\beta' V(r)^{-1}\beta]^{-1}\beta' V(r)^{-1}(\beta f + u)$$
$$= f + [\beta' V(r)^{-1}\beta]^- \beta' V(r)^{-1}u, \tag{22.11}$$

and $\beta' V(r)^{-1}u = 0$ in a GLS CSR. The most efficient estimator for the risk premium on a traded factor is its average excess return.

22.2   Everything, Including CSR, Is a Special Case of GMM

I have asserted this so many times in the GMM chapters that I should make good on it here. The GMM problem in table 22.1 illustrates my point.

This formulation allows for beta estimation simultaneously with the other model parameters, thereby automatically solving the errors-in-variables problem that occurs when estimating the betas in a two-step method. However, with the large number of moment conditions, the GMM problem here is only feasible for small numbers of assets. CSR methods can use arbitrarily large numbers of assets. Indeed, this system is overidentified when $N > (K + 1)/2$, so only small numbers of assets are needed to test asset pricing models. This formulation is used in Ferson, Henry, and Kisgen (2006).

22.3   Testing Unconditional Models with CSR Methods

22.3.1   Why the Usual Standard Errors Are Wrong

Consider the CSR model applied each month, say, to monthly returns data. The excess returns in a given month t are on the left-hand side of the regression $r_t = B\lambda_t + u_t$. (Or include a column of ones with the betas on the right-hand side and estimate the zero beta rate as the coefficient on the ones.) Let the covariance matrix of the residuals be the $N \times N$ matrix $E(u_t u_t') = \Sigma_u$. The $N \times K$ matrix of betas on the right-hand side of the regression is B. We take the B values as given for now, but the effects of their estimation error, which creates an errors-in-variables problem, is considered below. The OLS CSR estimator of

$\{\lambda_{t,}\}$, $t=1,\ldots,T$, is given for month t as $\hat{\lambda}_t = (B'B)^{-1}B'r_t$. Let's figure out its standard error:

$$\hat{\lambda}_t = (B'B)^{-1}B'(B\,\lambda_t + u_t) = \lambda_t + (B'B)^{-1}B'u_t. \tag{22.12}$$

Under the null hypothesis, the regression error is mean zero, which implies

$$\begin{aligned}\operatorname{Var}(\hat{\lambda}_t - \lambda_t) &= E\{(\hat{\lambda}_t - \lambda_t)(\hat{\lambda}_t - \lambda_t)'\} = (B'B)^{-1}B'E\{u_t u_t'\}B(B'B)^{-1}\\ &= (B'B)^{-1}B'\Sigma_u B(B'B)^{-1}.\end{aligned} \tag{22.13}$$

When the hypothesis to be tested is $E(\lambda_t) = \lambda$, some constant (e.g., zero), then we can write $\operatorname{Var}(\hat{\lambda}_t - \lambda_t) = \operatorname{Var}(\hat{\lambda}_t - \lambda + \lambda - \lambda_t)$, with $\operatorname{Cov}(\hat{\lambda}_t, \lambda_t) = \operatorname{Var}(\lambda_t)$, so that

$$\operatorname{Var}(\hat{\lambda}_t - \lambda) = \operatorname{Var}(\lambda_t) + (B'B)^{-1}B'\Sigma_u B(B'B)^{-1}. \tag{22.14}$$

This equation says that the usual OLS standard errors are wrong. The first term is the variance of the population risk premium over time, and it would be zero if λ_t is a constant parameter. The second term is $\operatorname{Var}(\hat{\lambda}_t - \lambda_t)$. You only get the OLS standard errors for this term when Σ_u is proportional to an identity matrix. With strong cross-sectional correlation and heteroskedasticity in stock returns, this is not a good approximation. The moral of the story is not to use the usual OLS standard errors in CSR models for stock returns. Instead, use equation (22.14). Better yet and, amazingly, equivalent, use the standard errors suggested by Fama and MacBeth (1973).

22.3.2 The Fama-MacBeth Standard Errors

Fama and MacBeth (1973) suggest a simple standard error calculation for the CSR. When the null hypothesis is that $E(\lambda_t) = 0$, they suggest a t-ratio of the following form:

$$\begin{aligned}T\text{-stat}: &\ (1/T)\textstyle\sum_t \hat{\lambda}_t / \operatorname{Var}((1/T)\sum_t \hat{\lambda}_t)^{1/2}, \text{ with } \operatorname{Var}(\hat{\lambda}_t)\\ &= (1/T)\textstyle\sum_t[\hat{\lambda}_t - (1/T)\sum_t \hat{\lambda}_t]^2.\end{aligned} \tag{22.15}$$

Thus, they suggest that you simply compute the standard error of the time-series average, taking the monthly CSR coefficient estimators as if they were data. Let's see what this delivers. If returns and the slope estimators are iid over time, $\operatorname{Var}((1/T)\sum_t \hat{\lambda}_t) = (1/T^2)$ $\operatorname{Var}(\sum_t \hat{\lambda}_t) = (1/T)\operatorname{Var}(\hat{\lambda}_t)$. Substituting in $\hat{\lambda}_t = \lambda_t + (B'B)^{-1}B'u_t$, we see that the Fama-MacBeth estimator of $\operatorname{Var}(\hat{\lambda}_t)$ converges, as T gets large, to $\operatorname{Var}(\lambda_t) + (B'B)^{-1}B'\Sigma_u B(B'B)^{-1}$. Holy mackerel, they got it right! Taking the time series of the CSR coefficients and simply computing the standard error of the mean as if they were just data ends up capturing the estimation error in the CSR slopes and the impact of correlation across the assets on the standard errors. In particular, there is no need to estimate Σ_u, which can be a challenge if the cross section is large. Amazing!

22.3.3 Accounting for Serial Dependence

With monthly CSRs for stock returns, the returns have little serial correlation in most cases, so the lack of autocorrelation assumed in the Fama-MacBeth estimators is a good approximation. However, with overlapping data or serially dependent returns, the CSR estimates for each period might be dependent over time. Studies have used methods similar to those described for consistent covariance matrix estimation in GMM. For example, we can use

$$\text{Var}[(1/T)\Sigma_t \hat{\lambda}_t] = (1/T)\text{Var}[\hat{\lambda}_t] + (1/T)^2 \Sigma \Sigma_{i \neq j} \text{Cov}[\hat{\lambda}_i \, \hat{\lambda}_j] \tag{22.16}$$

The first term of this estimator is the Fama and MacBeth estimate, while the second term accounts for serial dependence. We can use Hansen's (1982) estimator or the Newey and West (1987a) estimator for this term, for example. This estimator is evaluated by simulation in panel regressions by Petersen (2009), discussed in chapter 23.

22.3.4 Goodness of Fit in CSRs

There are several measures for the goodness of fit in CSRs for asset returns. This section briefly describes some of the popular measures. The first three are based on the average pricing errors from the CSR at the second step, defined as the sample estimate of $\acute{\alpha}_i = \Sigma_t (\lambda_{0t} + u_{it})/T$. when the null hypothesis implies that $E(\lambda_{0t}) = 0$. When λ_{0t} is a zero-beta rate with a nonzero return, under the null hypothesis, use only the residuals. When the same beta estimates are used in each CSR, the sample mean value of these alphas is the same as if you had regressed sample mean returns, $\Sigma r_{it}/T$, on those betas, and taken the constants $(\lambda_0 + u_i)$ from that regression to measure the average alphas. Alternatively, you could suppress the intercept in the CSRs for excess returns and use the time series sample mean of the residual as the average pricing error.

Perhaps the most natural goodness-of-fit statistic at this point in our discussion is the classical Wald test for the hypothesis that the vector of alphas is zero. MacBeth (1975) might have been the first to propose a version of this in CSR. Suppose $\sqrt{T}(\acute{\alpha} - \alpha) \rightarrow_d N(0, \text{Acov}(\acute{\alpha}))$, then $\varepsilon = \sqrt{T}\text{Acov}(\acute{\alpha})^{-1/2} \acute{\alpha} \rightarrow_d N(0, I)$, so $\varepsilon'\varepsilon = T\acute{\alpha}'\text{Acov}(\acute{\alpha})^{-1}\acute{\alpha} \rightarrow \chi^2_{N-K}$. This is the Shanken (1985a) T^2 test. The $\text{Acov}(\acute{\alpha})$ can be estimated Fama-MacBeth style. Let $\acute{\alpha}_t = (\hat{\lambda}_{0t} + \hat{u}_t)$ from monthly CSRs, $\acute{\alpha} = \Sigma_t \acute{\alpha}_t/T$, then estimate $\text{Acov}(\acute{\alpha}) = \Sigma_t (\acute{\alpha}_t - \acute{\alpha})$ $(\acute{\alpha}_t - \acute{\alpha})'/T$. This can be adjusted for serial dependence as before. It may have invertibility problems, if linear combinations of the test asset returns are equal to traded factor returns in the regressions, as the residuals will be linearly dependent. You might need to drop a test asset.

Cochrane (1996) popularized a simple graphical depiction of the fit for asset pricing models. On the y-axis are the average returns or excess returns of the test assets. On the x-axis are the model predicted returns, equal to the betas multiplied by the lambdas. If the fit is perfect, all the dots lie on the 45-degree line in the graph. This makes it easy to visualize which assets have positive and which have negative pricing errors.

The average of the CSR R-squares is a commonly reported measure of fit. But the sample R-square in a CSR can be unreliable. Lewellen, Nagel, and Shanken (2010) evaluate the average cross-sectional R-square using simulation and find that it has horrible finite-sample properties. In a three-factor model with zero true R-squares, there is a 50% chance that your sample R-square is more than 50%! GLS R-squares perform better, so let's talk about them.

22.3.5 GLS in CSR Models

In theory, GLS CSRs are better in the sense of being more efficient. But that assumes that you have a good estimate of the $N \times N$ covariance matrix of asset returns or their residuals. The efficiency result is asymptotic for a consistent covariance matrix estimate. Shanken (1985a) shows that when GLS is used, the second stage of a two-pass CSR coefficient estimator is asymptotically equivalent to an ML estimate and is thus efficient as T gets large.

Define the GLS R-square as

$$\text{GLS } R^2 = 1 - \{\alpha' V^{-1} \alpha / (\mu' V^{-1} \mu)\}, \tag{22.17}$$

where μ is the vector of mean returns. Recall that we can transform a GLS regression and implement it as an OLS regression on the transformed variables. In particular, we can regress $\Sigma_u^{-1/2} R_t$ on $(\Sigma_u^{-1/2} \beta)$ using OLS. It turns out that this is the same as regressing $V^{-1/2} R_t$ on $(V^{-1/2} \beta)$ using OLS, as noted by Litzenberger and Ramaswamy (1979). Kandel and Stambaugh (1995) show that the GLS CSR R^2 is proportional to the distance of the tested portfolio from the mean variance boundary. The OLS CSR R^2 has no such interpretation. Roll and Ross (1994) showed that the relation between the OLS cross-sectional R-square and the proximity of the tested portfolio to the bullet can be arbitrary. The quadratic form in the GLS CSR R-square, $\alpha' V^{-1} \alpha$, is proportional to the distance between the tested portfolio and the sample minimum variance bullet, as we saw in chapter 21.

In practice, full GLS is usually not feasible if the number of test assets is reasonably large. In such cases, weighted least squares might be attractive. The following analysis, based on Ferson and Harvey (1999), shows how weighted least squares is related to the classical Fama and MacBeth regressions. This is a good warm-up for panel regressions in the next chapter. Consider the pooled time series and CSR:

$$Y = X\gamma + u, \ E(uu') = \Omega, \tag{22.18}$$

where the dimensions are $Y_{TN \times 1}$, $X_{TN \times k}$. Assume $\Omega = \text{Diag}\{\Omega_t\}_t$. That might be a good approximation if the returns have little serial dependence. The GLS estimator is

$$\gamma_{\text{gls}} = (X' \Omega^{-1} X)^{-1} X' \Omega^{-1} Y = (\sum_t X_t' \Omega_t^{-1} X_t)^{-1} (\sum_t X_t' \Omega_t^{-1} Y_t). \tag{22.19}$$

For comparison, a GLS Fama-MacBeth estimator for month t is given by

$$\gamma_{\text{fm},t} = (X_t' \Omega_t^{-1} X_t)^{-1} (X_t' \Omega_t^{-1} Y_t). \tag{22.20}$$

Thus, the GLS (really, weighted least squares with the diagonal covariance matrix assumption) estimator is a weighted average of the time series of the Fama-MacBeth estimators:

$$\gamma_{gls} = \Sigma_t w_t \gamma_{fm,t}, \text{ with } w_t = (\Sigma_t X_t' \Omega_t^{-1} X_t)^{-1} (X_t' \Omega_t^{-1} Y_t). \tag{22.21}$$

Standard errors can be obtained by replacing $\gamma_{fm,t}$ with $\{w_t \gamma_{fm,t}\}$ and using the Fama-MacBeth calculation like in equation (22.15). Shanken and Zhou (2007) evaluate the tradeoff between small sample bias and efficiency with this approach using simulations.

22.4   Errors in the Betas

CSRs have some attractive features for asset pricing research. As we have seen, they automatically create mimicking portfolios if we start with the betas on nontraded factors or if we use characteristics on the right-hand side. They can handle large numbers of assets, N. However, we have so far ignored the fact that the betas computed in the first step have an estimation error in them. This error creates an errors-in-variables bias in the second-step CSRs.

22.4.1   Shanken's Adjustment for Fama-MacBeth Standard Errors

Much of the analysis of the sampling properties of CSRs assumes that the same betas are used in each cross-sectional month. If so, then the same estimated beta (and its estimation error) appear in each CSR month. This creates time series dependence in the estimates. Shanken (1992) shows how to correct Fama-MacBeth standard errors for this fact. His adjustment adds the term in curly brackets to the Fama-MacBeth standard errors:

$$\text{Var}(\hat{\lambda}_t) \approx \text{Var}(f_t) + (B'B)^{-1}B'\Sigma_u B(B'B)^{-1}\{1 + E(f_t)'\text{Cov}(f_t)^{-1}E(f_t)\} \tag{22.22}$$

Under the null hypothesis that $\lambda = E(f_t) = 0$, there is no need to adjust. In practice, the adjustment often has a small impact. The larger the squared Sharpe ratio of the factors is, the larger the adjustment will be. Importantly, this does not correct the coefficient estimates for the errors-in-variables problem, just the standard errors. (Under the null hypothesis that the coefficients are zero, that is OK.) Shanken (1992), discussed in the next section, does address the errors-in-variables bias on the CSR coefficients due to estimated betas.

22.4.2   Theil's Adjustment to CSR Coefficients for Errors in Betas

An early version of a correction for errors-in-variables bias in the CSR coefficients is proposed by Theil (1971). Consider the CSR model with no errors in betas:

$$R_t = \lambda_0 + B \lambda_t + u_t, \text{ where } \text{Cov}(u_t, B) = 0. \tag{22.23}$$

We don't see the true B; instead, we have the estimates $B^* = B + v = \text{true} + \text{noise}$, and $\text{Cov}(v, B) = 0$ is typically assumed. A time series GMM (or multivariate regression model) gives estimates of $\text{Cov}(v)$, the $K \times K$ cross-sectional covariance matrix of the errors in betas. We run the CSR:

$R_t = \lambda_0 + B^* \lambda_t^* + e_t$, then

$$\begin{aligned}
\lambda_t^* &\to_p \text{Cov}(B^*)^{-1} \text{Cov}(B^*, R_t) \\
&= \text{Cov}(B^*)^{-1} \text{Cov}(B + v, B\lambda_t + u_t) \\
&= \text{Cov}(B^*)^{-1} \text{Cov}(B)\lambda_t.
\end{aligned} \tag{22.24}$$

From this, Theil (1971) shows that the bias-adjusted estimator is

$$\lambda_t^{**} \equiv [B^{*\prime} B^* / N - \text{Cov}(v)]^{-1} [B^{*\prime} B^* / N] \, \lambda_t^{**} \to_p \lambda_t. \tag{22.25}$$

Versions of this approach to adjusting the coefficients in CSR for errors in the betas are used by Black and Scholes (1974), Litzenberger and Ramaswamy (1979, 1982), and others. Wang (2014) evaluates the sampling properties of the Theil estimator with simulations and finds that it works pretty well.

Shanken (1992) derives a version of the bias-adjusted estimator that relates $\text{Cov}(v)$ to the variance matrix of the factor realizations: $\text{Cov}(v) = \sigma_u^2 (F'F)^{-1}$, where σ_u^2 is a degrees-of-freedom adjusted mean of the trace of the covariance matrix of the pricing errors, and F is the $T \times K$ matrix of the de-meaned factors. Gagliardini, Ossola, and Scaillet (2016) derive a more general version of the bias adjustment for models with time-varying betas. The bias adjustment in all these cases goes to zero as T gets large, because the time series regression estimates converge to their limiting constants as T gets large.

Kim (1995) models the errors-in-betas problem in CSR coefficients by appending the lagged errors in the betas from a time series regression up to time $t - 1$, with the CSR of time t returns on the lagged betas and characteristics. He allows that the betas have measurement error while the characteristics do not. Building up the joint likelihood for the returns and the lagged beta errors, he solves for the ML estimates, which automatically correct the slopes for the errors in the betas. Lee and Chen (2013) review alternative approaches for errors-in-variables correction.

22.4.3 Instrumental Variables Solutions

Wang (2014) suggests an instrumental variables approach to the errors-in-betas problem. He proposes using lagged betas, estimated from an earlier period, as instruments for the noisy current-period betas. On the assumption that returns have little serial correlation, this approach produces consistent estimates of the CSR coefficients, and simulations suggest that it works pretty well. Jegadeesh et al. (2017) evaluate an approach where the odd months are used to estimate betas and the even months are used to run the CSRs (or vice versa). They show with simulations that this approach works pretty well.

22.5   Asymptotic Distributions for CSR Coefficients

Several papers derive asymptotic distribution results for CSR coefficients under various assumptions. As previously described, Shanken (1985a) shows that a GLS CSR is asymptotically equivalent to ML as T gets large. Shanken (1992) develops asymptotic distributions for the CSR estimators under fixed betas, highlighting the role of the errors in variables in the analysis. Kan, Robotti, and Shanken (2013) develop further asymptotic results for CSR with estimation error in the betas. Let $X = (1, \beta)$ and $\lambda^* = \sum \lambda_t / T$. They show that $\sqrt{T}(\lambda^* - \lambda) \to N(0, \text{Acov})$, $\text{Acov} = E(\sum_j h_t h_{t+j})$ and provide the expression for h_t:

$$
\begin{aligned}
h_t = (\lambda_t - \lambda^*) && \text{FM term} \\
+ \{[\lambda_{0t}, (\lambda_t - F_t)']' - [\lambda_0, (\lambda - \mu)']'\}\lambda' V(F)^{-1}(F_t - \mu_F) && \text{Errors-in-betas term} && (22.26) \\
+ (X'V^{-1}X)^{-1}[0, (F_t - \mu_F)'V^{-1}]'\alpha' V(F)^{-1}(R_t - \mu_F)1. && \text{Misspecification term}
\end{aligned}
$$

Kan, Robotti, and Shanken (2013) provide asymptotic distribution results for CSR estimators in some special cases. They also provide asymptotic distributions for the CSR R^2. They show that TR^2 is distributed as a weighted sum of $\chi^2(1)$ if the true $R^2 = 0$ or 1, and that $\sqrt{T}R^2$ is normal, with a covariance expression they provide, if the true R^2 is between 0 and 1.

Most of the asymptotic results so far require large T and fixed N. This assumption seems unfortunate in asset pricing problems, where N can be large relative to T. Kim and Skoulakis (2014) derive an N-consistent estimator by estimating the first- and second-pass regressions with nonoverlapping subperiods. The idea is similar to the instrumental variables solutions to the errors-in-betas problem, relying on serial independence in the asset returns. Bai and Zhou (2015) investigate N and T for CSR estimators. Ramponi, Robotti, and Zaffaroni (2016) provide asymptotic distribution theory for CSRs when N goes to infinity but T can be small. They handle the errors-in-betas problem for finite T by using Shanken's (1992) bias-adjusted estimator, which they find is asymptotically unbiased as N gets large. They also provide a goodness-of-fit test based on the sum of squared estimated model average pricing errors, and they derive its limiting distribution as N goes to infinity. Most of these asymptotic results apply to the case where the betas are assumed to be fixed over time. In applications of CSRs to conditional asset pricing, this will not be the case.

22.6   Conditional Asset Pricing with CSR

Suppose that we start with a conditional multiple-beta model:

$$
E(R_{it}|Z_{t-1}) = \gamma_0(Z_{t-1}) + \sum_{j=1,\dots,k} b_{ij,t-1}\,\gamma_j(Z_{t-1}), \tag{22.27}
$$

where the $b_{ij,t-1}$ are conditional betas (multiple regression coefficients) of the R_{it} on K factors, $j = 1, \dots, K$, conditional on Z_{t-1}. The risk premiums $\gamma_j(Z_{t-1})$, $j = 0, \dots, K$ are the conditional zero-beta rate and risk premiums. If we run CSRs for each month t, excess returns on the conditional betas are

$$r_{it} = \lambda_{0t} + \sum_{j=1,\ldots,k} \lambda_{jt} \, \beta_{ij,t-1} + e_{it}; \; i = 1,\ldots,N, \tag{22.28}$$

where the $\beta_{ij,t-1} = b_{ij,t-1} - b_{fj,t-1}$ are proxies for the conditional betas of the excess returns. Many studies use rolling regression estimates to proxy for the conditional betas (e.g., Ferson and Harvey 1991; Braun, Nelson, and Sunier 1995; Lewellen and Nagel 2006). The CSR slope coefficient estimate λ_{jt}, $j = 1, \ldots, K$ is a "mimicking portfolio" excess return whose conditional expected value is an estimate of the conditional expected risk premium, or price of beta, $\gamma_j(Z_t)$, for factor j.

The CSR (22.28) provides a decomposition into two parts of the excess return each month. The first part is $\sum_{j=1,\ldots,k} \lambda_{jt} \, \beta_{ij,t-1}$, the part of the return i that is related to the conditional measures of risk in the cross section. The second part is $\lambda_{0t} + e_{it}$, the part of the asset return that is uncorrelated with the measures of risk. If excess returns are used, a "perfect" model implies that $E(e_{it}|Z_{t-1}) = 0$ and $E(\lambda_{0t}|Z_{t-1}) = 0$, but the model is not going to be perfect. Ferson and Harvey (1991) estimate this decomposition for some standard risk factors and portfolios, and they find that most of the predictability in returns resides in the part of the returns that is related to risk, while little of it resides in the part that is not related to risk. Related analysis based on APT is provided in Ferson and Korajczyk (1995), using the GMM and assuming that conditional betas are linear in lagged instruments.

Gagliardini, Ossola, and Scaillet (2016) provide an analysis of conditional two-pass CSRs where the risk-premium coefficients and the conditional betas are both assumed to be linear functions of lagged instruments Z_{t-1}. They prove consistency and asymptotic normality as both T and N get large. They show that a quadratic form in the average pricing errors is asymptotically distributed as a weighted sum of chi-square variates.

22.7   CSR versus Spread Portfolios

Since Fama and French (1993, 1996), and with the convenience of Kenneth French's data library, many studies use "long-short" or spread portfolios, such as SMB and HML, in asset pricing studies. These portfolios go long the high book/market or small firms and short the low book/market or large firms, with a net investment of zero. This seems a lot like how we described CSR slopes as portfolios. If we use a characteristic like book/market on the right-hand side and run a univariate regression, the CSR slopes are versions of "high minus low" or spread-portfolio returns. Multiple sorts are more like running a multiple regression. The regression picks the weights to minimize the sum of squares, while the spread portfolios typically weight the assets equally or by market value but avoid the need to explicitly estimate anything, when they are formed based on observable characteristics.

CSRs have some advantages over sorting. Multiple regressions can be used to tease out marginal effects. You can check the residuals for functional form problems. You can usually put a lot more variables on the right-hand side of a CSR before running out of degrees of freedom, compared with the number of dimensions along which you can sort assets into portfolios before you run out of assets in some of the bins. Since the OLS approach is equal weighted, regressions can be dominated by small stocks in a cross section. A value-weighted CSR could be used to address this difficulty. CSRs can work on individual stocks, across portfolios, or even within portfolios (e.g., Fama and French 2008).

In contrast, HML-type spread portfolios likely obtain a larger spread on the sorting criteria. Characteristics like book/market can be measured with less error than can betas, which can enhance the power of the tests. The portfolios can be weighted to avoid being dominated by small stocks (but value-weighted versions may be dominated by big stocks).

One serious concern is that HML-type spread portfolios can hide anomalies as "factors," even if the anomaly in returns bears no actual relation to risk, as shown by Ferson, Sarkissian, and Simin (1999). The basic idea is as follows. The cross-sectional variation in average returns that is not related to risk is cross-sectional variation in alpha. Sorting on alpha produces HML spread portfolios that approximate the optimal orthogonal portfolio. By construction, combining the test asset returns (say, the market portfolio) with the optimal orthogonal spread portfolio provides a factor model where a combination of the factors is minimum variance efficient and thus appears to price the test assets.

Reducto ad absurdum is to find a portfolio with weights $w_{(N \times 1)}$ so that the covariances of all the assets with $r_p = w'r$ are proportional to the vector of alphas. This portfolio can pretend to be the perfect new factor for "explaining" the anomaly behind the alpha. We can easily find the portfolio weights, as the covariances are $\Sigma w = c\alpha \Rightarrow w = c\Sigma^{-1}\alpha$. This is the optimal orthogonal portfolio with respect to the factor(s) in the model that produced α. Chapter 21 shows that the optimal orthogonal portfolio is the same, using as Σ either the covariance matrix of returns or the residual covariance matrix of returns. Including the optimal orthogonal portfolio in the model would drive the alphas to zero, because a combination of the original factor(s) and the optimal orthogonal portfolio is minimum variance efficient. We find the optimal orthogonal portfolio if we regress returns on alphas, as the slope coefficient for that regression is $\mathrm{Cov}(\alpha)^{-1}\alpha'R$. In the multivariate regression model we saw that $\mathrm{Cov}(\alpha)$, the covariance of the N-vector of alphas, is proportional to the covariance matrix of the asset returns, so the weights are proportional to $\Sigma^{-1}\alpha$.

23 Introduction to Panel Methods

This chapter introduces panel regression methods. Panel methods are increasingly used in economics and finance research. Their use in finance first became popular in corporate finance problems and now is becoming popular in asset pricing, to the point where you can't really be an empirical asset pricer if you don't know something about panels. This chapter is a basic introduction, skewed toward asset pricing. The chapter spends some time on predictive panel regressions, since in asset pricing we are often interested in predicting things like stock returns. It finishes up with a brief discussion of causal inference in panel regressions. I point out that the classical "event" study is an early version of a difference-in-differences approach. Our colleagues in corporate finance apply causal inference methods and have become quite sensitive to causality issues—perhaps you have heard of the "causality police." But I predict that the causality police will come to asset pricing town soon, so you had best be ready.

23.1 Concepts and Terminology

The simplest setup for panel regressions is a pooled OLS regression:

$$y_{it} = X'_{it}\beta + \varepsilon_{it}, \tag{23.1}$$

where y_{it} is the dependent variable, the X_{it} are K explanatory variables, the time subscripts are $t = 1, 2, \ldots, T$. There are $i = 1, 2, \ldots, N$ individuals (units, firms, stocks, states, jurisdictions, agents, etc.). If there are T observations for each firm, the panel is considered to be "balanced." Think of the $T \times N$ data matrix X, suspended from a string in the middle. If $t = 1, 2, \ldots, T_i$ is not the same for each firm i, the panel is "unbalanced."

OLS is consistent in equation (23.1) under the usual assumption that $E\{X_{it}\varepsilon_{it}\} = 0$. Sometimes we use asymptotics where NT goes to infinity, but traditionally, panels are designed for datasets in which $N \gg T$; that is, the $T \times N$ data matrix is short and wide. Sometimes we use asymptotics that allow for a fixed, finite T but assume that N goes to infinity. Sometimes the estimators assume strict exogeneity, where the error term is uncorrelated with all past and future values of the X variables. Sometimes, this assumption fails.

Some of the methods we have studied in earlier chapters are almost special cases of panel regressions. For example, the multivariate regression model of chapter 21 combines time series regressions for a cross section of stock returns. That model allows the slope coefficients, alphas and betas, to be different for each stock. A panel regression more typically assumes that the slope coefficient is the same for all stocks, but it allows the predictors X_{it} to be different for different stocks. Fama-MacBeth style CSRs, discussed in chapter 22, combine stocks for each period in a time series of CSRs. That model allows the CSR coefficients to be different for each period. A panel regression typically assumes that the coefficients are fixed over time. Of course, all these approaches are special cases of the GMM, so I show you how to do panel regressions in the GMM later in this chapter.

23.2   Firm Effects and Time Effects

The regression errors ε_{it} in equation (23.1) may contain effects, or shocks that are common across firms or over time. Simple firm effects can be expressed in the residual as $\varepsilon_{it} = \gamma_i + u_{it}$, where u_{it} is independently distributed. The term γ_i represents the unobserved firm effect. This is unobserved heterogeneity across firms. Examples that seem like firm effects include a firm's leverage, dividend policy, CEO, a fund's family structure or incentive compensation program, and the like. It is common to assume that firm effects are uncorrelated across firms, so that $\mathrm{Cov}(\gamma_i, \gamma_j) = 0$ when $i \neq j$. Firm effects are often assumed to be fixed over time and are commonly referred to as "fixed" effects. Fixed effects mean that the γ_i are fixed parameters that do not vary over time, but they can vary in the cross section of firms. The common fixed effects assumption is that $\mathrm{Cov}(\gamma_i, X_{it}) \neq 0$. It is hard to imagine the examples above being uncorrelated with the firm-level characteristics that are typically measured. This implies that OLS is not consistent, so we will discuss estimating the regression with fixed effects, using firm dummies or by differencing the equation across time.

There are also models with random firm effects, where the γ_i are assumed to be drawn from some distribution with a smaller number (less than the N firms) of parameters. The usual assumption with random firm effects is that $\mathrm{Cov}(\gamma_i, X_{it}) = 0$. This assumes that OLS is a consistent estimator, which seems more like an assumption of convenience than of descriptive realism.

With simple time effects, the regression residual may be modeled as $\varepsilon_{it} = \delta_t + u_{it}$, where δ_t is the unobserved time effect. Examples that seem like time effects include monetary policy shocks, political shocks, market panics, and changes in regulations. Panel econometrics has been an active and productive field since at least the early 1980s, and there are models in which the various effects die out slowly over time or across firms. In this introduction, we will stick with the simpler examples.

The two most important things to remember about firm and time effects are

Firm Effects Cause Time Series Dependence:

$$t \neq \tau; \; \mathrm{Cov}(\varepsilon_{it}, \varepsilon_{i\tau}) = \mathrm{Cov}(\gamma_i + \mathrm{u}_{it}, \gamma_i + \mathrm{u}_{i\tau}) = \mathrm{Var}(\gamma_i). \tag{23.2}$$

Time Effects Cause Cross-Sectional Dependence:

$$i \neq j; \; \mathrm{Cov}(\varepsilon_{it}, \varepsilon_{jt}) = \mathrm{Cov}(\delta_t + \mathrm{u}_{it}, \delta_t + \mathrm{u}_{jt}) = \mathrm{Var}(\delta_t). \tag{23.3}$$

There is a common loose use of language among researchers, where people refer to firm dummies as "firm fixed effects," or to time dummies as "time fixed effects." This usage is imprecise, because fixed effects are not dummy variables, they are statistical assumptions about the model. Dummy variables can be used to estimate fixed effects. It can be useful to keep in mind the distinction between the theoretical model for the data and the method used to estimate the model, for example, when thinking about misspecification.

23.3   Standard Errors and Clustering

Given consistent estimators, the next concern in panel regression models is getting the standard errors of the coefficients right. This requires capturing the dependence across firms or time in the regression residuals, as would be induced by time or firm effects. Let's start out with a simple model in which OLS is consistent and is used to estimate the coefficients, so we can focus on getting the standard errors right. We assume that $E(\varepsilon_{it}) = 0$, and $E(X_{it}\varepsilon_{it}) = 0$, so that OLS is consistent. This is the standard assumption. However, the OLS standard errors are typically not consistent. To further simplify, let $E(X_{it}) = 0$, $\dim(\beta) = K = 1$. (Extending to $K > 1$ is easy. The mean of the X_{it} is addressed below.) The most common method for getting consistent standard errors is called "clustering." The pooled OLS estimator of β is

$$\beta_{\mathrm{OLS}} = (\textstyle\sum_i \sum_t X_{it} Y_{it}) / (\sum_i \sum_t X_{it}^2) = \beta + (\sum_i \sum_t X_{it}\varepsilon_{it}) / (\sum_i \sum_t X_{it}^2), \tag{23.4}$$

with

$$\mathrm{Var}(\beta_{\mathrm{OLS}} - \beta) = E\{(\beta_{\mathrm{OLS}} - \beta)(\beta_{\mathrm{OLS}} - \beta)'\} = E\{[\textstyle\sum_i \sum_t X_{it}^2]^{-1}(\sum_i \sum_t X_{it}\varepsilon_{it})^2 [\sum_i \sum_t X_{it}^2]^{-1}\}. \tag{23.5}$$

Let's take X fixed, in the classical regression sense, or condition on X and use asymptotic approximations, assuming $[(1/NT)\sum\sum X_{it}^2]^{-1} \approx (\sigma_x^2)^{-1}$. This allows us to focus the analysis on the central term in equation (23.5), which is where all the action is for clustering:

$$V = E(\textstyle\sum_i \sum_t X_{it}\varepsilon_{it})^2 = E((E(\sum_i \sum_t X_{it}\varepsilon_{it})^2 | X)). \tag{23.6}$$

Five special cases of the central V term illustrate the impact of the different kinds of effects and how we adjust the standard errors, by clustering, to handle them.

The baseline and simplest special case is homoskedasticity and independence, when there are no firm or time effects. Then $\varepsilon_{it} \sim (0, \sigma_\varepsilon^2)$, ε is stationary and independent of X, and

$$V = E(\sum_i \sum_t X_{it} \varepsilon_{it})^2 = \sum_i \sum_t X_{it}^2 \, E(\varepsilon_{it}^2) \to_p \sigma_\varepsilon^2 NT \, \sigma_x^2, \tag{23.7}$$

where the notation indicates convergence in probability (formally, after dividing by NT). In this case, we would estimate V using the sample variance of the regression residuals taken across both time and firms: $(1/NT)\sum_i \sum_t \varepsilon_{it}^2$, and estimate the variance of X by its sample variance in the same fashion. Then we have

$$\text{Var}(\beta_{\text{OLS}} - \beta) = (NT\sigma_x^2)^{-1}(NT\sigma_x^2\sigma_\varepsilon^2)(NT\sigma_x^2)^{-1} = (\sigma_\varepsilon^2)/(NT\sigma_x^2). \tag{23.8}$$

Stating the asymptotics a little more precisely, $(NT)^{1/2}(\beta_{\text{OLS}} - \beta) \to_d N(0,(\sigma_\varepsilon^2/\sigma_x^2))$, which emphasizes that the number of observations is NT. With that under our belts, we have more special cases to consider.

The second special case, where we have heteroskedasticity but no firm or time effects, leads to White's (1980) heteroskedasticity-consistent standard errors for panels. Assume $E(\varepsilon_{it}\varepsilon_{i\tau}) = 0$, $t \neq \tau$ (no firm effects), which implies $V = E(\sum_i \sum_t X_{it} \varepsilon_{it})^2 = \sum_t E(\sum_i X_{it}\varepsilon_{it})^2$. Assuming also that $E(\varepsilon_{it}\varepsilon_{jt}) = 0$, $i \neq j$ (no time effects), then we have $\sum_t E(\sum_i X_{it} \, \varepsilon_{it})^2 = \sum_t \sum_i E(X_{it}^2\varepsilon_{it}^2)$. By clustering together X_{it}^2 and ε_{it}^2 in this expression, we allow that $E(\varepsilon_{it}^2|X_{it})$ may depend on X_{it}. We don't have to specify the form of the dependence of the conditional variance on X_{it}. In this case, we can estimate V with

$$1/(NT)\sum_i \sum_t X_{it}^2\varepsilon_{it}^2 \equiv V_{\text{W}}, \tag{23.9}$$

and thus

$$\text{Var}(\beta_{\text{OLS}} - \beta) = (NT\sigma_x^2)^{-1} \, (\sum_i \sum_t X_{it}^2\varepsilon_{it}^2 / NT)(NT\sigma_x^2)^{-1}. \tag{23.10}$$

Now we should be ready to do some clustering to address time and firm effects. In the third special case, we have firm effects but no time effects. We handle the firm effects in the standard errors by clustering by firm. Firm effects induce time series dependence in the regression error terms: $\text{Cov}(\varepsilon_{it}\varepsilon_{i\tau}) \neq 0$ for $t \neq \tau$, but for now, we assume $\text{Cov}(\varepsilon_{it}\varepsilon_{j\tau}) = 0$ $i \neq j$ for all t, τ, that is, cross-sectional independence (no time effects). Then we have

$$V = E(\sum_i \sum_t X_{it}\varepsilon_{it})^2 = E\sum_i (\sum_t X_{it}\varepsilon_{it})^2. \tag{23.11}$$

The term $(\sum_t X_{it}\varepsilon_{it})^2$ is a "cluster" for firm i. The cluster keeps together the squared terms for firm i over time and thus preserves the time dependence for firm i. To see this, expand the sum of squares in the cluster as $(\sum_t X_{it}\varepsilon_{it})^2 = (\sum_t X_{it}^2\varepsilon_{it}^2 + \sum \sum_{t \neq \tau} X_{it}\varepsilon_{it}X_{i\tau}\varepsilon_{i\tau})$. Taking the expectation of this term shows that the cluster captures the autocovariances of the $X_{it}\varepsilon_{it}$ in its second term. Using the assumed cross-sectional independence, we can estimate V with

$$(1/N)\sum_i (\sum_t X_{it}\varepsilon_{it})^2 \equiv V_{\text{F}}. \tag{23.12}$$

To cluster by firms, you need enough firms in your sample to get good estimates of the average across the firms. These are the Rogers (1993) standard errors. They are the result of using the "cluster" switch in Stata^{TM}, for example.

The fourth special case is clustering by time to capture time effects. Time effects induce cross-sectional correlation in the regression errors: $\text{Cov}(\varepsilon_{it}\varepsilon_{jt}) \neq 0$, $i \neq j$. Now assume no time dependence: $\text{Cov}(\varepsilon_{it}\varepsilon_{j\tau}) = 0$, $t \neq \tau$, all i, j. Then we can write $V = \text{E}(\sum_i \sum_t X_{it}\varepsilon_{it})^2 = \text{E}\sum_t (\sum_i X_{it}\varepsilon_{it})^2$. The expression $(\sum_i X_{it}\varepsilon_{it})^2$ is a cluster for time t. The cluster preserves the cross-sectional dependence over firms at time t. To see this, expand the sum of squares as $(\sum_i X_{it}\varepsilon_{it})^2 = (\sum_i X_{it}^2 \varepsilon_{it}^2 + \sum\sum_{i \neq j} X_{it}\varepsilon_{it} X_{jt}\varepsilon_{jt})$. Taking the expectation of this term shows how the cluster for time t captures, in its second term, the dependence across firms. Using the assumed time series independence, we can estimate V with

$$(1/T)\sum_t (\sum_i X_{it}\varepsilon_{it})^2 \equiv V_T. \tag{23.13}$$

When clustering by time, you need enough time series observations in your sample to get good estimates of the time series averages. You are estimating the average of the cross-sectional dependence over time. And, you are ignoring any time series dependence.

Finally, we can cluster by both firm and time, if we have enough data. In the fifth special case, we allow both cross-sectional and time series dependence in the regression residuals: $\text{Cov}(\varepsilon_{it}\varepsilon_{j\tau}) \neq 0$ (if $i \neq j$ and $t = \tau$ or $i = j$ and $t \neq \tau$). However, we assume that there is no correlation across firms at different points in time: $\text{Cov}(\varepsilon_{it}\varepsilon_{j\tau}) = 0$ {if $i \neq j$ and $t \neq \tau$}. In this case, we expand V as

$$\begin{aligned} V &= \text{E}(\sum_i \sum_t X_{it}\varepsilon_{it})^2 \\ &= \text{E}(\sum_t \sum_i X_{it}^2 \varepsilon_{it}^2) + \text{E}\sum_i (\sum\sum_{t \neq \tau} X_{it}\varepsilon_{it} X_{i\tau}\varepsilon_{i\tau}) + \text{E}\sum(\sum\sum_{i \neq j} X_{it}\varepsilon_{it} X_{jt}\varepsilon_{jt}). \end{aligned} \tag{23.14}$$

Note that there are three terms on the right-hand side of equation (23.14). The first delivers the estimator under OLS, allowing for heteroskedasticity, and is equal to the White (1980) estimator V_W (defined in equation (23.9)). The second term reflects clusters by firm, allowing for dependence across time for a given firm, like V_F. The third term reflects clusters by time, allowing dependence across firms at the same point in time, like V_T. Note that both V_F and V_T include the contemporaneous parts that define V_W. To avoid double counting, therefore, we can compute the double-clustered standard errors using $V = V_F + V_T - V_W$. To cluster in both dimensions, you need enough firms to get good estimates of the averages in the second term and enough time to get good estimates of the averages in the third term.

23.4   Panels in GMM

It is said that almost everything is a special case of the GMM, and so it is! Let's think about panel regressions as a GMM problem. In a pooled panel, we have $g_{it} = (X_{it}\varepsilon_{it})$ and $g = (1/(NT))\sum_i \sum_t g_{it}$. With K predictors in X_{it}, then $\dim(g) = K$. The problem is exactly identified. Setting the sample mean $g = 0$, the coefficient estimator β_{GMM} is the pooled OLS panel estimator of β from equation (23.4). Note that $\partial g/\partial \beta = -\text{E}[(1/(NT))\sum_i \sum_t X_{it}X_{it}']^{-1}$, and the GMM asymptotic covariance matrix for the exactly identified problem is $[\partial g/\partial \beta]^{-1}E(gg')[\partial g/\partial \beta]^{-1}$, which is the same as equation (23.5) with the middle term $E(gg') = V = E(1/(NT)(\sum_i \sum_t X_{it}\varepsilon_{it}\varepsilon'_{it}X'_{it})$.

Special cases of the GMM version of panel regressions naturally emerge. Let $g_t = (1/N)\sum_i g_{it}$, and run the "time series" GMM on the cross-sectional averages. Thus, $g = (1/T)\sum_t g_t$ is the same as before, resulting in the same pooled panel estimator. Also $\partial g/\partial \beta = -E[(1/(NT))\sum_i \sum_t X_{it}X'_{it}]^{-1}$, the same as before. But now we have

$$E(gg') = E[((1/T)\sum_t g_t)((1/T)\sum_t g_t)']$$
$$= E\{((1/T)^2 \sum_t g_t g'_t + ((1/T)^2 \sum_t \sum_{\tau \neq t} g_t g'_\tau)\}, \tag{23.15}$$

which appeared earlier in our discussion of the GMM. If we assume the error structure is MA(0), the GMM problem turns out to be equivalent to clustering by time in the pooled panel. Working with the cross-sectional averages is similar to keeping the cross-sectional average in a cluster by time. The second term in equation (23.15) is zero in this case, and the first term becomes

$$E(gg') = E\{((1/T)^2 \sum_t g_t g'_t)\} = E\{(1/T)^2 \sum_t ((1/N)\sum_i g_{it})((1/N)\sum_i g_{it})'\}$$
$$= E\{(1/T)^2 \sum_t ((1/N)^2 \sum_i g_{it}g'_{it} + (1/N)^2 \sum_{i \neq j} g_{it}g'_{jt})\}. \tag{23.16}$$

The first term in the second line of equation (23.16) is V_W, or White's (1980) standard errors, and second term captures clustering by time. This result is due to Driscoll and Kraay (1998).

We can swap the *i*s and *t*s and get a version of cross-sectional GMM. Let $g_i = (1/T)\sum_t g_{it}$, and run cross-sectional GMM on the time series averages. Thus, $g = (1/N)\sum_i g_i$ is the same as before, resulting in the same pooled panel estimator. Also $\partial g/\partial \beta = -E[(1/NT)\sum_i \sum_t X_{it}X'_{it}]^{-1}$, the same as before. But now

$$E(gg') = E[((1/N)\sum_i g_i)((1/N)\sum_i g_i)']$$
$$= E\{(1/N)^2 \sum_i g_i g'_i + (1/N)^2 \sum_i \sum_{j \neq i} g_i g'_j\}. \tag{23.17}$$

If there is no cross-sectional dependence of the $\{g_i\}_i$, then the second term in equation (23.17) is zero. The result is equivalent to clustering by firm in the pooled panel. The first term of (23.17) becomes V_F from equation (23.12):

$$E(gg') = E((1/N)^2 \sum_i g_i g'_i) = E[(1/N)^2((1/T)\sum_t g_{it})((1/T)\sum_t g_{it})']$$
$$= E\{((1/N)^2 \sum_i ((1/T))^2 \sum_t g_{it}g'_{it}) + ((1/T)^2 \sum_{t \neq \tau} g_{it}g'_{i\tau})\}. \tag{23.18}$$

The first term, once again, is V_W, or White's (1980) standard errors. The second term captures clustering by firm. Arellano and Bond (1991) use cross-sectional GMM on a panel regression with lagged dependent variables and fixed effects, and they provide some specification tests and simulation evidence for their finite-sample properties.

23.5   How Dummy Variables Work in Panels

Now we move beyond the focus on standard error estimation and examples where OLS is consistent. If there are unobserved firm or time effects in the model that may be correlated with the X variables in your regression, then the regression error terms will be correlated with X, and OLS will not be consistent. This can be thought of as a "missing variables" bias in the regression. If the unobserved effects are fixed, we can estimate them using dummy variables and recover consistent estimates for the regression coefficient β. Let's first consider a model with fixed firm effects.

23.5.1   Fixed Firm Effects and Firm Dummies

With fixed firm effects, we model the regression residual as $\varepsilon_{it} = \gamma_i + u_{it}$, where γ_i is the fixed firm effect. It is randomly drawn once and hits firm i at all times t. If it is correlated with the right-hand side variables X_{it}, then OLS will be inconsistent. Write the theoretical model with the expanded error term as

$$y_{it} = X'_{it}\beta + \gamma_i + u_{it}, \tag{23.19}$$

where u_{it} is mean zero and is uncorrelated with both X_{it} and γ_i. For estimation of the model, write the $T \times N$ data matrix for the dependent variable as y, and let $Y = \mathrm{Vec}(y)$ be the $NT \times 1$ column vector that results from stacking the columns of y. Similarly, let X be the $NT \times K$ matrix of the stacked up right-hand side variables, where each column of X contains one of the K predictors for each firm. Introduce a data matrix of firm dummy variables, $D = (d_1, \ldots, d_N)$, which is $NT \times N$. Each d_i is an $NT \times 1$ vector of dummy variables for firm i. The dummy variable d_i is equal to 1.0 if the y_{it} observation is for firm i; otherwise, it is equal to zero. In matrix notation, $D = I_N \otimes \underline{1}_T$, where I_N is the $N \times N$ identity matrix, $\otimes$ is the Kronecker product, and $\underline{1}_T$ is a T-vector of ones. Note that the use of firm dummies in a panel regression leads to large data matrices, as D is of dimension $TN \times N$, which can be pretty large.

With this setup, we can write the stacked regression equation as

$$Y = X\beta + D\gamma + u, \tag{23.20}$$

where the parameter γ is the N-vector of the γ_i, and u is the $TN \times 1$ vector of the $\{u_{it}\}$. Now OLS on this model delivers a consistent estimate for β, and all is right with the world. But if we dig a little deeper into this model, we find some interesting interpretations. This merits an aside.

23.5.2   The Frisch-Waugh Theorem and Firm Dummies

We were all exposed to the famous theorem from Frisch and Waugh (1933) when we studied "Regression 101." In that context, we first studied univariate regressions of some variable y on X. Then we encountered multivariate regression of y on both X and Z.

We learned that the coefficient on X in the multivariate regression was the same as the one we get using a series of univariate regressions. First, we regress y on Z and take the fitted residuals. Then we regress X on Z and take the fitted residuals. Finally, we run the y residuals on the X residuals, and this regression delivers the multivariate regression coefficient of y on X, when Z is also in the regression. In our dummy variable problem, the theorem applies when Z is the dummy variable matrix, D. In short, to get the OLS estimate of β in the multivariate model, we regress the residuals from the y-on-D regression on the residuals from the X-on-D regression. That is, we regress $Y - D(D'D)^{-1}D'Y$ on $X - D(D'D)^{-1}D'X$. Defining the matrix $M = [I_{NT} - D(D'D)^{-1}D']$, we note that M is idempotent ($M'M = M$), and the regression that delivers the estimate of β is that of MY on MX. Thus, the OLS estimator of β in the stacked regression is $(X'MX)^{-1}(X'My)$.

This use of the Frisch-Waugh theorem provides a nice interpretation of regression models with firm dummies to handle fixed firm effects. Using the definitions of M and D, simple matrix calculations show that $M = I_N \otimes [I_T - (\underline{1}_T \underline{1}'_T)/T]$. This matrix M operates on X or Y to remove the time series mean. The regression of MY on MX is therefore the regression:

$$y^*_{it} = X^{*\prime}_{it}\beta + \eta_{it}, \tag{23.21}$$

where y^*_{it} and $X^{*\prime}_{it}$ are the values of the dependent and independent variables, respectively, after each variable's time series sample mean has been subtracted. Running OLS on equation (23.21) delivers what is called the "within" estimator, or the "within-group" estimator. You have to think of i as indexing a "group." For simplicity and to conform to a common finance context, a firm is a group, and the within-group variation is the time series variation for the firm. This estimator uses the time series variation to identify the coefficient. The firm dummy variables soak up the cross-sectional information. This is the regression chosen using "AREG" within the Stata package, for example.

Starting with the simple statistical model $y_{it} = X'_{it}\beta + \gamma_i + u_{it}$, we can see that de-meaning the variables has removed the firm fixed effects from the model. (Write the model for the time series averages and subtract.) This suggests that we might also proceed by just first differencing the model, estimating $\Delta y_{it} = \Delta X'_{it}\beta + \Delta u_{it}$. This is OLS in first differences, and it is a consistent estimator, given a simple fixed firm effect. If the error term is highly autocorrelated, by approximating it as a unit root process, the difference estimator can be efficient. But it requires dealing with the error term Δu_{it}. If the original u_{it} was iid, for example, its first difference is autocorrelated, with an MA(1) structure. This requires more time series information and may not be desired if your T is small.

Sometimes empirical studies in finance have de-meaned only the left-hand side variables. The observations for firm i might be measured net of the industry averages. Returns on the left-hand side may be adjusted with risk factors to get "abnormal" returns. Taking the simple fixed firm effects model seriously, these approaches are inconsistent, as emphasized by Gormley and Matsa (2013). The "missing" de-meaned X_{it} becomes a correlated missing variable. Gormley and Matsa also criticize the practice of putting group averages (e.g., industry averages) of the y_{it} on the right-hand side of the regression, as this also measures the properly de-meaned variable with error.

23.5.3 Time Dummies and Fixed Time Effects

The analysis here is fairly similar to that of the previous case, except we swap firms and times. With fixed time effects, we modeled the regression residual as $\varepsilon_{it} = \delta_t + u_{it}$, where δ_t is the time effect. This is drawn randomly at each date t and is assumed to hit all firms at time t. If $\mathrm{Cov}(\delta_t, X_{it}) \neq 0$, OLS is inconsistent, but we can pull out the time effect with dummy variables to fix the problem. Write the theoretical model with the expanded error term as

$$y_{it} = X'_{it}\beta + \delta_t + u_{it}. \tag{23.22}$$

As before, we write the $T \times N$ data matrix for the dependent variable as y and let $Y = \mathrm{Vec}(y)$ be the $NT \times 1$ column vector that results from stacking the columns of y. Here X is the $NT \times K$ matrix of the stacked-up right-hand side variables, where the columns are the K predictors for each firm.

For time dummies, we introduce a different data matrix of dummy variables, $D = (d_1, \ldots, d_T)$, which is $NT \times T$, where each d_t is an $NT \times 1$ vector of dummy variables for time t. Each of these dummy variables is equal to 1.0 if the y_{it} observation is for time t; otherwise it is equal to zero. In matrix notation, $D = \underline{1}_N \otimes I_T$, where $\underline{1}_N$ is an N-vector of ones.

With this setup, we can write the stacked regression equation as

$$Y = X\beta + D\delta + u, \tag{23.23}$$

where δ is the T-vector of the δ_t. Now as NT gets large, OLS delivers a consistent estimate for β, and all is right with the world once again.

We can use the Frisch-Waugh theorem on this regression to obtain an interpretation in terms of de-meaned variables. In this case, the variables are cross-sectionally de-meaned; that is, the cross-sectional averages are subtracted from each variable. There is also a first-differences approach, call it OLS in "firm differences." It seems a little weird to me to run regressions on the differences between firms, especially if the way we order them in the dataset has no inherent meaning, but it would be consistent, given a fixed time effect. We could just as well pick some reference firm and run regressions on the excess returns and X-variables over that of the reference firm.

Of course, we don't have to limit ourselves to firm effects and time effects. We can have models with other group effects, such as industries, states, families, and so on. We can think of models with time effects by day, month, quarter, year, or decade. In section 23.9, we consider dummy variables indicating before versus after some "treatment," dipping our toes into the realm of causal inference. Dummy variables can be used for all these things and more. The Frisch-Waugh theorem works for all these cases.

23.5.4 Time Series versus Cross-Sectional Information

It is interesting to compare regressions with firm dummies and time dummies, in terms of the information that is being used to estimate the regression coefficient. Consider an example with only a single regressor, where the dependent variable is a stock excess return and where we have both firm and time fixed effects:

$$r_{it+1} = \gamma_i + \delta_t + bx_{it} + u_{it+1}. \tag{23.24}$$

To estimate the coefficient b, you could average the model over time or across firms before de-meaning:

$$(1/T)\sum_t r_{it+1} = \gamma_i + (1/T)\sum \delta_t + b(1/T)\sum x_{it} + (1/T)\sum u_{it+1}, \tag{23.25}$$

$$(1/N)\sum_i r_{it+1} = (1/N)\sum \gamma_i + \delta_t + b(1/N)\sum x_{it} + (1/N)\sum u_{it+1}. \tag{23.26}$$

Now the estimator for b using firm dummies works on (23.24)–(23.25), which differences out the firm fixed effects. The estimator using only time dummies works with (23.24)–(23.26), differencing out the time effects. Let's examine the OLS estimators for these two cases:

Within Estimator (Assuming $\delta_t = 0$, Firm Dummies Only):

$$\begin{aligned}\hat{b} &= \sum_i \sum_t (r_{it+1} - (1/T)\sum_t r_{it+1})(x_{it} - (1/T)\sum_t x_{it}) / \sum_i \sum_t (x_{it} - (1/T)\sum_t x_{it})^2 \\ &= (1/N)\sum_i \text{TS Cov}(r_{it+1}, x_{it}) / (1/N)\sum_i \text{TS Var}(x_{it}).\end{aligned} \tag{23.27}$$

Estimator with Only Time Dummies (Assuming $\gamma_i = 0$):

$$\begin{aligned}\hat{b} &= \sum_i \sum_t (r_{it+1} - (1/N)\sum_i r_{it+1})(x_{it} - (1/N)\sum_t x_{it}) / \sum_i \sum_t (x_{it} - (1/N)\sum_i x_{it})^2 \\ &= (1/T)\sum_t \text{CS Cov}(r_{it+1}, x_{it}) / (1/T)\sum_t \text{CS Var}(x_{it}).\end{aligned} \tag{23.28}$$

The terms "TS" and "CS" refer to time series and cross-sectional, respectively, covariances and variances. Expression (23.27) shows with firm dummies (the within estimator) that time series covariability is what determines the slope coefficients. Its average across the firms is reflected in the coefficient estimate. In contrast, using time dummies, we identify the coefficient from the cross-sectional relation between the left- and right-hand variables, and we report its average over time as the coefficient. If we have both firm and time dummies, we identify b from the covariance between y and X at different dates across different firms.

We used cross-sectional covariances to identify the coefficients in the Fama-MacBeth CSRs studied in chapter 22, where we could report the averages of the cross-sectional coefficients over time. This sounds suspiciously similar to the panel estimator with time dummies, and the two are indeed related. We can write the time series average of the Fama-MacBeth estimator in the model above as $(1/T)\sum_t [\text{CS Cov}(r_{it+1}, x_{it})/(\text{CS Var}(x_{it}))]$. The expected Fama-MacBeth estimator is like $E(x/y)$, and the pooled estimator with time dummies is like $E(x)/E(y)$.

23.5.5 Sum-of-Squares Decomposition

We have seen that using firm dummies, the within estimator identifies the regression coefficients from the averaged time series covariation. Using time dummies, the OLS estimator identifies the regression coefficients from the cross-sectional covariation. Without any dummies, the pooled OLS estimator uses both the time series and the cross-sectional covariation.

The relative impact of the two kinds of variation on the pooled OLS estimator depends on how much of which kind of variation is in the data. This can be expressed using sum-of-squares decompositions. Let $\underline{X}_i = (1/T)\sum_t X_{it}$. and $\underline{\underline{X}} = (1/N)\sum_i X_{i\cdot}$ Then we can define

$$
\begin{aligned}
SS_{xx}^{Total} &= \sum_i \sum_t (X_{it} - \underline{\underline{X}})^2 = \sum_i \sum_t [(X_{it} - \underline{X}_i) + (\underline{X}_i - \underline{\underline{X}})]^2 \\
&= \sum_i \sum_t [(X_{it} - \underline{X}_i)^2 + (\underline{X}_i - \underline{\underline{X}})^2 + 2(X_{it} - \underline{X}_i)(\underline{X}_i - \underline{\underline{X}})],
\end{aligned}
\tag{23.29}
$$

where the third term sums to zero. The first term is the sum of squares within group or firm i, SS_{xx}^{within}. This is the variation used by the within estimator. The second term is the sum of squares between groups, $SS_{xx}^{between}$. We have $SS_{xx}^{Total} = SS_{xx}^{within} + SS_{xx}^{between}$.

We can also decompose the sum of squared products in the numerator of the OLS estimator. Write $SS_{xy}^{total} = \sum_i \sum_t (X_{it} - \underline{\underline{X}})(Y_{it} - \underline{\underline{Y}}) = SS_{xy}^{within} + SS_{xy}^{between}$. Now, the within estimator can be expressed in terms of these sums of squares as

$$
\beta_{within} = (SS_{xx}^{within})^{-1}(SS_{xy}^{within}).
\tag{23.30}
$$

We can also define the between estimator as

$$
\beta_{between} = (SS_{xx}^{between})^{-1}(SS_{xy}^{between}).
\tag{23.31}
$$

The between estimator is similar to Fama-MacBeth using average returns, as in Black, Jensen, and Scholes (1972). For comparison, the pooled OLS estimator is

$$
\begin{aligned}
\beta_{OLS} &= (SS_{xx}^{total})^{-1}(SS_{xy}^{total}) \\
&= (SS_{xx}^{within} + SS_{xx}^{between})^{-1}(SS_{xy}^{within} + SS_{xy}^{between}) \\
&= \alpha \beta^{within} + (I - \alpha)\beta^{between}, \quad \alpha = (SS_{xx}^{total})^{-1}(SS_{xx}^{within}).
\end{aligned}
\tag{23.32}
$$

Thus, the pooled OLS estimator is a matrix-weighted average of the within and between estimators, where the matrix captures how much of the total sum of squares comes from within versus between the groups. If most of the variation is within the groups, then OLS is close to the within estimator. If most of the variation is across groups, then OLS is close to the between estimator.

23.6 Predictive Panel Regressions

Sometimes we want to predict the future values of variables, like stock returns, in a panel using predictors lagged in time. From empirical corporate finance, panel models like this expanded into mutual fund research, perhaps starting with new money flows (Sirri and Tufano 1998), and now include studies of fund performance and other outcome variables of interest to asset pricers. Predictive panel regressions introduce a couple of biases that we have to deal with.

23.6.1 The Stambaugh Bias

The lagged stochastic regressor bias was first studied in a single-variable time series context by Stambaugh (1999) and is sometimes called the "Stambaugh bias." Stambaugh considered simple time series regressions of stock market returns on a lagged dividend yield, $x_t = D_t/P_t$:

$$r_{t+1} = \alpha + bx_t + \varepsilon_{t+1}. \tag{23.33}$$

The OLS slope coefficient estimator is

$$b_{OLS} = (1/T)\sum_t r_{t+1}(x_t - (1/T)\sum_t x_t)/[(1/T)\sum_t (x_t - (1/T)\sum_t x_t)^2], \tag{23.34}$$

so that

$$b_{OLS} - b = \sum_t \varepsilon_{t+1}(x_{it} - (1/T)\sum_t x_{it})/[\sum_t (x_t - (1/T)\sum_t x_t)^2] \tag{23.35}$$

$$\Rightarrow E(b_{OLS} - b) = -E((1/T)\sum_t \varepsilon_{t+1} \sum_{\tau \geq t+1} x_{i\tau})/[\sum_t (x_t - (1/T)\sum_t x_t)^2]. \tag{23.36}$$

The expected bias is not zero if the future values of the $x_{i\tau}$ for $\tau > t$ can be correlated with the residual ε_{t+1}. Intuitively, a positive shock to market returns at time $t+1$ increases prices and thus lowers dividend yields at all future dates, so $E[\varepsilon_{t+1}x_{i\tau}] < 0$ for $\tau \geq t+1$. Because of the negative sign in equation (23.36), the expected bias is positive. The extent of the bias depends on how long the return shocks persist in the future dividend yields. Stambaugh (1999) shows how this is related to a bias in the sample autocorrelation of x_t. Assuming a first-order autoregression for x_t,

$$x_{t+1} = a + \rho x_t + u_{t+1}, \ t = 1, \ldots, T, \tag{23.37}$$

the OLS slope coefficient estimator ρ_{OLS} has

$$\rho_{OLS} - \rho = \sum_t u_{t+1}(x_{it} - (1/T)\sum_t x_{it})/[\sum_t (x_t - (1/T)\sum_t x_t)^2], \tag{23.38}$$

which is the same as equation (23.36) with u_{t+1} replacing E_{t+1}. Regress ε_t on u_t to get $\varepsilon_t = \gamma u_t + v_t$, with u_t and v_t uncorrelated. Plug this result into expression (23.36) for the bias in b_{OLS} to see that

$$E(b_{OLS} - b) = \gamma E(\rho_{OLS} - \rho). \tag{23.39}$$

Kendal (1954) showed that $E(\rho_{OLS} - \rho) \approx -(1 + 3\rho)/T$. Thus, the bias in b_{OLS} is approximately $E(b_{OLS} - b) \approx -\gamma(1 + 3\rho)/T$, where $\gamma = Cov(u, \varepsilon)/Var(u)$. Stambaugh (1999) provides the approximate correction:

$$b^* = b_{OLS} + \gamma(1 + 3\rho)/T. \tag{23.40}$$

The Stambaugh bias is a finite-sample bias that goes to zero as T gets large. It would be zero under the strict exogeneity assumption $E(\varepsilon_{t+1}|X) = 0$, where X refers to the full time series of the X values, as can be seen from equation (23.36).

The Stambaugh bias can also be thought of as a missing variables bias. The original predictive regression is "missing" the future values of the innovations in the predictor variable on its right-hand side, which biases the regression coefficients, because they are correlated with the error in the regression. The missing variables approach is used by Amihud and Hurvich (2004) and Amihud, Hurvich, and Wang (2008, 2010). Consider a two-step approach where the autoregression for x_t is first run to get the fitted residuals, u_t. Putting those on the right-hand side of the predictive regression should deliver

$$r_{t+1} = \alpha + bx_t + \gamma u_{t+1} + v_{t+1} \tag{23.41}$$

and an approximately unbiased estimate of b.

23.6.2 Lagged Stochastic Regressor Bias in Panels

Now we are ready to talk about lagged stochastic regressor bias in predictive panel regressions. Such panel regressions are found in several asset pricing studies. Sirri and Tufano (1998) use panels of mutual funds' new money flows regressed on lagged fund characteristics and performance. Pástor, Stambaugh, and Taylor (2015) regress measures of fund performance on fund and industry size. Hjalmarsson (2008, 2010) studies the problem of lagged stochastic regressor bias in panels. Anderson and Hsiao (1982) study dynamic panel models (which include lagged independent variables on the right-hand side) and derive consistency results for various specifications of firm effects.

Ferson and Wang (2018) regress stock returns on the lagged portfolio weights of the fund; this discussion follows their paper. This panel regression problem inherits properties of a dynamic panel regression, because the lagged predictor variables have a feedback effect. If there is a positive shock to a stock return, a fund's portfolio weight in the stock automatically increases in the future unless the fund takes action to reverse the effect.

For each fund, assuming that the fund portfolio contains N stocks and exists for T periods, the panel regression is

$$r_{t+1}^i = \beta w_t^i + \varepsilon_{t+1}^i, \ i = 1, \ldots, N; \ t = 1, \ldots, T. \tag{23.42}$$

In this regression, r_{t+1}^i is the excess return of stock i at time $t+1$, and w_t^i is the weight of the stock held by the fund at time t. The slope coefficient β captures the ability of a manager's weights to predict future excess stock returns; thus, the estimated β should be positive when the manager is informed. The null hypothesis is that the coefficient is zero.

The OLS slope coefficient estimator in the panel regression is

$$\hat{\beta} = (1/T)\sum_i \sum_t (r_{t+1}^i w_t^i)/[(1/T)\sum_i \sum_t (w_t^i w_t^i)]. \tag{23.43}$$

The numerator of the slope coefficient for a particular fund corresponds to the classical holdings-based performance measures discussed in chapter 25. In those examples, either the returns r_{t+1}^i or the portfolio weights w_t^i are replaced with benchmark-adjusted variables.

The regression (23.42) is unrealistic for stock returns, because under the null hypothesis that $\beta = 0$, if the residuals are mean zero, it implies that all the stocks have the same expected return equal to zero. This shortcoming is addressed by introducing stock fixed effects, and the model becomes

$$r_{t+1}^i = a^i + \beta w_t^i + \varepsilon_{t+1}^i, \tag{23.44}$$

where the coefficient of a dummy variable for stock i is a^i, the fixed effect of stock i. Under the null hypothesis that $\beta = 0$, the fixed effect is the expected return of the stock. The fixed effects are likely to be correlated with the right-hand side variables if funds' holdings respond to the expected returns of the stocks. Now, introducing the firm dummy variables means that the OLS estimator is equivalent to a panel regression with the returns and weights measured net of their time series averages. Because the lagged portfolio weights have a feedback effect, the dummy variables or first difference estimators will be inconsistent.

Funds' portfolio weights are autocorrelated. Assume they follow an AR(1) model:

$$w_{t+1}^i = \gamma^i + \rho w_t^i + v_{t+1}^i \tag{23.45}$$

This equation and the regression (23.45) for returns form a panel "predictive system" (e.g., Pástor and Stambaugh 2009). The lagged stochastic regressor bias can be understood as follows. Consider the numerator of the least squares dummy variable (the within) estimator, written with only the weights de-meaned (without loss of generality):

$$\hat{\beta}_{w,num} = (1/T) \sum_i \sum_t r_{t+1}^i (w_t^i - (1/T) \sum_t w_t^i). \tag{23.46}$$

Substituting the panel regression (23.44) model for the returns yields

$$E(\hat{\beta}_{w,num} - \beta_{num}) = -E[(1/T) \sum_i \sum_t \varepsilon_{t+1}^i (1/T) \sum_{\tau \geq t+1} w_\tau^i]. \tag{23.47}$$

This shows that the lagged stochastic regressor bias arises if ε_{t+1}^i is correlated with $\sum_{\tau \geq t+1} w_\tau^i$. Under a buy-and-hold strategy for the fund, for example, we would expect a positive correlation between ε_{t+1}^i and w_{t+1}^i, as a positive return shock increases the portfolio weight, and therefore a negative bias is expected in the slope coefficient. The larger the serial correlation in the weights is, the larger will be the expected bias.

The feedback effect in the portfolio weights operates similarly to a lagged independent variable in a dynamic panel regression. Nickell (1981) considers dynamic panel regressions with fixed effects and derives expressions for the asymptotic bias as N gets large. These depend on the autocorrelations of the independent variables. In the current problem, the bias depends on the autocorrelations of the portfolio weights.

Hjalmarsson (2008) proposes a parametric correction for the lagged stochastic regressor bias in panels. He assumes a local-to-unity structure for the autoregressive parameter in equation (23.45): $\rho = 1 - c/T$. He also makes the strong assumption that the errors in (23.44) are independent across stocks. He uses sequential limit theory (first, fix N and let T go to infinity, then let N go to infinity) to obtain several results. First, when there are no fixed effects in the model, there is no Stambaugh bias and the pooled OLS estimator of the panel regression for returns is consistent. However, with fixed effects, the estimator with the stock dummy variables is biased. Letting $R_{t+1}^i = r_{t+1}^i - \overline{r^i}$ and $W_t^i = w_t^i - \overline{w^i}$, the corrected estimator is

$$\hat{\beta}_c = \sum_i \sum_t (R_{t+1}^i W_t^i - NT \, \mathrm{Cov}(\varepsilon, v)\theta(c))/[\sum_i \sum_t (W_t^i \, W_{t'}^i)], \tag{23.48}$$

where $\mathrm{Cov}(\varepsilon, v) = (1/N)\sum_i \mathrm{Cov}(\varepsilon_{it}, v_{it})$ is consistently estimated from the OLS residuals of the predictive panel regression system. Hjalmarsson (2008) finds that the estimator is relatively insensitive to how the residuals are estimated. The crucial parameter is c in $\theta(c) = -(e^c - c - 1)/c^2$. Hjalmarsson proposes to estimate the parameter c as $T(1 - \rho_{\mathrm{pool}})$, where ρ_{pool} is the pooled estimator of ρ with no intercept in the regression (23.45).

Hjalmarsson (2010) proposes a forward recursively de-meaned estimator for the fixed effects case in panels. Let $w_{it}^{bd} = w_{it} - [1/(t-1)]\sum_{s=1}^{t-1} w_{is}$ and $r_{it}^{fd} = r_{it} - [1/(T-t-1)]\sum_{s=t+1}^{T} r_{is}$ be the backward de-meaned weights and the forward de-meaned returns, respectively. The estimator is

$$\hat{\beta}_H = \sum_i \sum_t (r_{it+1}^{fd} w_{it}^{bd})/[\sum_i \sum_t (w_{it}^{bd} w_{it'}^{fd})]. \tag{23.49}$$

The backward de-meaned weight is used as an instrument. The intuition is that the forward de-meaned returns should be independent of the lagged weights. The recursive de-meaning induces a moving average term in the errors of the de-meaned model, which needs to be handled by including the autocovariances in the standard error computations. While the recursive de-meaning gives up some data and thus some efficiency, Hjalmarsson (2010) finds that it effectively controls the lagged stochastic regressor bias, and Ferson and Wang (2018) confirm this in simulations for mutual funds. Pástor, Stambaugh, and Taylor (2015) use a similar approach with forward and backward de-meaning.

A differenced estimator takes the difference between equation (23.44) and the same equation τ periods before. The fixed effects cancel out:

$$r_{t+1}^i - r_{t+1-\tau}^i = \beta(w_t^i - w_{t-\tau}^i) + \varepsilon_{t+1}^i - \varepsilon_{t+1-\tau}^i. \tag{23.50}$$

Ferson and Wang (2018) show that the classical difference estimator based on (23.50) suffers a Stambaugh bias for $\tau \geq 2$. This result is similar to one for dynamic panels shown by Anderson and Hsiao (1981). An unbiased estimator is the *differenced IV estimator*:

$$\hat{\beta}^{\text{DiffIV}} = \frac{\sum\limits_{t}\sum\limits_{i}((w_t^i - w_{t-\tau}^i)(r_{t+1}^i - r_{t+1+\tau}^i))}{\sum\limits_{t}\sum\limits_{i}((w_t^i - w_{t-\tau}^i)(w_t^i - w_{t+\tau}^i))}. \tag{23.51}$$

In this estimator, the lagged first difference in the portfolio weights is used as an instrument and the stock return is forward differenced. Ferson and Wang (2018) find that the Hjalmarsson (2010) and differenced IV estimators perform well in removing the Stambaugh bias in examples for mutual funds.

23.7   Random Firm Effects and Efficient Estimation

We have discussed fixed firm or time effects that are unobserved and may be correlated with the X variables in the regression. In this case, OLS will not be consistent, but if the effects are fixed, we can estimate them using dummy variables, and we can still recover consistent estimates for the regression coefficient. But we might have random firm effects that are not correlated with the X variables, so that the dummies are not needed. In fact, the usual random effects assumption is that $\text{Cov}(\gamma_i, X_{it}) = 0$. In this case, OLS will be consistent but will not be efficient, because a random firm effect will lead us to violate the OLS assumption that the residuals in the regression are iid with variance matrix $\sigma^2 I$. In this case, GLS is efficient, but we need to model the covariance matrix. This is the task to which we now turn.

23.7.1   An Error Components Model and GLS Estimation

With a fixed firm effect, we modeled the regression residual as $\varepsilon_{it} = \gamma_i + u_{it}$, where γ_i is the unobserved effect that hits firm i at all times t. Let's consider the case where $\text{Cov}(\gamma_i, X_i) = 0$, so OLS will be consistent but not efficient, and we can focus on efficiency. The model is $y_{it} = X_{it}'\beta + \gamma_i + \eta_{it}$, and we model the covariance matrix using an "error components" model. The model assumes $E(\gamma_i^2) = \sigma_\gamma^2$, $E(\eta_{it}^2) = \sigma_\eta^2$ for all i, t, and $E(\gamma_i \eta_{it}) = 0$ for all i, j, t.

To figure out a GLS estimator, let's stack up the observations as Y_i, a $T \times 1$ column vector for each i, and X_i, a $T \times K$ matrix of the X_{it} variables for firm i. Writing the regression model as

$$Y_i = X_i\beta + \varepsilon_i \tag{23.52}$$

reveals that $E(\varepsilon_i \varepsilon_i') = \sigma_\eta^2 I_T + \sigma_\gamma^2 (1_T 1_T') \equiv \Omega$, where I_T is the $T \times T$ identity matrix and is a T-vector of ones. The first term is what we would get under the usual OLS assumptions, while the second term captures the time dependence caused by the fixed firm effect. The GLS estimator regresses $\Omega^{-1/2}Y_i$ on $\Omega^{-1/2}X_i$, and the transformed residual $\Omega^{-1/2}\varepsilon_i$ has the identity matrix for its covariance matrix.

A feasible version of the GLS estimator uses consistent estimates for the parameters in the covariance matrix Ω and obtains asymptotic efficiency. This can be done by starting with OLS estimation to get the fitted $\varepsilon_{it} = \gamma_i + u_{it}$. Then use the sample autocovariance of ε_{it} as an estimate of $E(\gamma_i^2) = \sigma_\gamma^2$. Using all the data, you could estimate σ_γ^2 by $(\sum_i \sum_t \sum_{\tau \neq t} \varepsilon_{it} \varepsilon_{i\tau})/[NT(T-1)]$. Then the time series variance of $\text{Var}(\varepsilon_{it}) = \text{Var}(\gamma) + \text{Var}(\eta)$ delivers σ_η^2. Using these estimates, a consistent estimate of Ω can be calculated, inverted, and used to form the GLS estimator. The special structure of the error components model allows us to find $\Omega^{-1/2} = I_T - (1/T)\theta(1_T 1_T')$, where $\theta = (1 - \sigma_\eta/(T\sigma_\eta^2 + \sigma_\gamma^2)^{1/2})$. This is called the Balestra-Nerlove estimator.

This GLS approach is not used very often in finance papers. Perhaps the homogeneity of variances assumed in the error components model seems unrealistic when applied to stock returns. But other versions of GLS can be used in finance studies. One that seems compelling to me for stock returns is to model correlation across firms but assume no time series dependence of the NT vector of residuals in an equation like (23.20) or (23.23). Then $E(uu') = \Omega = \sum \otimes I_T$, where $\sum$ is the $N \times N$ covariance matrix. If N is too large, a diagonal matrix with the standard errors can be used to capture heteroskedasticity.

23.7.2   Consistent OLS Estimation with Standard Error Adjustment

Under the conditions imagined in the last section, GLS is efficient. However, its efficiency assumes that the covariance matrix estimate Ω contributes no variance. For asymptotic efficiency, we need a consistent estimate of Ω. In finite samples, we need the estimators to be at their probability limits. This is a lot to ask, so studies often use the OLS coefficient estimators, sacrificing theoretical efficiency for robustness. But now the OLS estimate of the standard errors is inconsistent. Of course, we can cluster by firm, as discussed above. An alternative is called the Moulton (1987) correction to the standard errors. Here we use the estimate for Ω and compute $\text{Cov}(\beta_{OLS}) = (X'X)^{-1}X'\Omega X(X'X)^{-1}$. This standard error is consistent for OLS, given a consistent estimate for Ω, when the standard OLS assumptions fail.

23.7.3   Summary

To summarize, if $\text{Cov}(\gamma_i, X_{it}) = 0$, we have observed that GLS is efficient and consistent, while OLS with firm dummies is consistent but inefficient. OLS in first differences is less efficient than using firm dummies. If the fixed effects are correlated with the X variables, $\text{Cov}(\gamma_i, X_{it}) \neq 0$, then firm dummies and differences are consistent, but OLS and GLS (without dummies) are inconsistent.

23.8   Some Results from Petersen (2009)

Petersen (2009) was one of the most frequently cited papers in the *Review of Financial Studies* in recent times. This section reviews some of the main results from that paper. The paper surveys how panel methods had been applied across preceding published finance papers. It focuses on standard error estimation, allowing for firm effects and time effects in simulation experiments designed to be representative of empirical studies in finance. It focuses in particular on OLS and clustered standard errors, the White (1980) standard errors, Newey and West (1987a) standard errors for panels, the Fama-MacBeth (1973) standard errors and adjustments to these for autocorrelation.

Assuming mean zero regressors, the Fama-MacBeth (1973) estimator is characterized by Petersen as $\beta_{FM} = (1/T)\sum_t \hat{\beta}_t; \hat{\beta}_t = (\sum_i X_{it} Y_{it})/(\sum_i X_{it}^2)$, and the squared Fama-MacBeth standard error is $S^2(\beta_{FM}) = (1/T)\sum_t (\hat{\beta}_t - \beta_{FM})^2/(T-1)$. Petersen's simulations show that this estimator performs well, given simple fixed time effects. This is not surprising, as this is the environment that the estimator was designed for. Petersen finds that the Fama-MacBeth estimator performs poorly with a firm fixed effect, or when highly autocorrelated firm-specific variables are on the left-hand side. Simulations show that the bias in the standard errors can be similar to that of pooled OLS. The trap here is that you run pooled OLS and Fama-MacBeth, and you get similar results. Don't conclude that your results are "robust." They could both be horribly biased.

Petersen considers "adjusted" Fama-MacBeth estimators that attempt to deal with serial dependence as: $\text{Var}(\beta_{FM}) = T^{-2}\sum_t \text{Var}(\hat{\beta}_t) + T^{-2}\sum_\tau \sum_t \text{Cov}(\hat{\beta}_t, \hat{\beta}_\tau)$, and with different methods to get the $\text{Cov}(\cdot, \cdot)$ values for adjustment. The first is to simply use the first-order sample autocovariance of the $\hat{\beta}_t$, an approach described in chapter 22. A second approach is to model the $\hat{\beta}_t$ as a first-order autoregressive (AR(1)) process. The autocovariance can be approximated with a finite number of lags, or the solution for an AR(1) with infinite lags can be used. Finally, we can model the coefficients as a moving average process of some order, MA(k). Petersen finds (in a working-paper version of the published paper) that in the presence of a fixed firm effect, the first method performs the worst, while the other methods perform better and similarly to each other. This is because with a simple fixed firm effect, the time series dependence persists instead of dying out after a few lags.

Petersen examines the performance of the Newey and West (1987a) estimator for panels: $\text{Var}(\beta_{OLS} - \beta) = E\{[\sum_i \sum_t X_{it}^2]^{-1} V [\sum_i \sum_t X_{it}^2]^{-1}\}$. If we cluster by firm and apply the Newey-West weighting scheme to the autocovariances, we get

$$V = E\sum_i \{\sum_t X_{it} X_{it}' \varepsilon_{it}^2 + \sum\sum_{t \neq \tau}[(T - |\tau - 1|)/T] X_{it} \varepsilon_{it} X_{i\tau}' \varepsilon_{i\tau}\}. \tag{23.53}$$

Petersen finds that this estimator works better if firm effects die out over time. If the effects are long lived, the simpler Hansen (1982) unweighted estimator, where the term in brackets $[\cdot]$ is set to 1.0, performs better.

There are many more results in this classic paper and some good practical advice for standard error calculation in panel regressions.

23.9   Differences in Differences

Recent empirical work in economics and corporate finance has centered on causal inference. The variable y_{it} is some outcome variable; for example, a stock return, a measure of firm operating performance, an individual's taxable income or employment status, or a student's grades. We are interested in measuring the causal effect of some treatment on the outcome variable. The treatment could be the change in a trading venue or regulation, the enactment of a minimum wage law, a job search assistance program, a student's classroom size, a natural disaster, and so forth. We want to estimate the expected, or average, effect of the treatment on the y_{it}. We are talking about so-called observational studies, where no controlled experiments are conducted. The problem with observational studies is that each agent is either treated or is not treated, and this may not be random, so the causal effect of treatment cannot be cleanly observed like it can in a controlled experiment, where treatment can be assigned at random.

The difference-in-differences (diff-in-diff) approach addresses the causal inference problem by estimating the change in outcomes for a treated group before versus after treatment, relative to the changes in outcomes for a control group, before versus after treatment of the treated group. The assumption is that the change in the control group is a good estimate for what the change in the treatment group would be if it had not been treated. This is called the "parallel trends" assumption. This setup is depicted in figure 23.1.

The bottom curve in the figure is the control group's observations, and the top one is the treated group. The causal effect of the treatment is measured as the change in the top curve after the event or treatment date less the change in the bottom curve after the treatment date. The assumption that the two curves track each other, except for the effect of the treatment, is the parallel trends assumption.

Now, one might think that corporate finance scholars get all the kudos here for incorporating causal inference into their research programs far earlier than did the asset pricers,

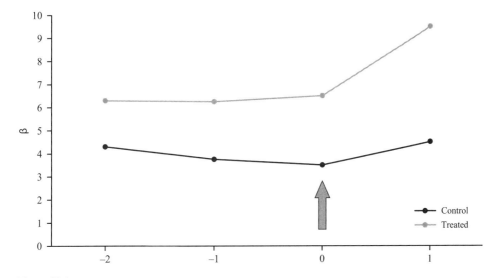

Figure 23.1
Diff-in-diff.

and to some extent this is true. However, it turns out that asset pricers (and corporate finance scholars) have been conducting so-called event studies since Fama et al. (1969), and these studies are also concerned with causal inference. I argue below that an event study is a special case of a diff-in-diff regression.

To set this up in a panel regression, define two dummy variables. The variable D_{it}^1 is equal to one if time t is after the treatment occurs for stock i and equal to zero before the treatment date. The dummy variable D_{it}^2 is equal to one if agent i is in the treated group and equal to zero if in the control group. The approach can allow for agents' receiving treatment at different points in time. If each treated agent i is paired with a member of the control group $j(i)$, then the dummy variable D_{it}^1 can turn on for both i and $j(i)$ when agent i is treated.

Consider the panel regression:

$$y_{it} = \beta_0 + \beta_1 D_{it}^1 + \beta_2 D_{it}^2 + \beta_3 (D_{it}^1 \times D_{it}^2) + \varepsilon_{it}. \tag{23.54}$$

Obviously, the product or interaction dummy turns on only if both conditions are true. Taking the conditional expected values and assuming that the error terms are independent of the dummies, we see that

$$\begin{aligned} \beta_0 &= E(y_{it}|\text{pretreatment, control}), \\ \beta_0 + \beta_1 &= E(y_{it}|\text{posttreatment, control}), \\ \beta_0 + \beta_2 &= E(y_{it}|\text{pretreatment, treated}), \\ \beta_0 + \beta_1 + \beta_2 + \beta_3 &= E(y_{it}|\text{posttreatment, treated}). \end{aligned} \tag{23.55}$$

Rearranging, we see that the coefficient β_3 captures the diff-in-diff treatment effect, because

$$\begin{aligned} \beta_3 &= E(y_{it}|\text{posttreatment, treated}) - (\beta_0 + \beta_1 + \beta_2) \\ &= E(y_{it}|\text{posttreatment, treated}) - (\beta_0 + \beta_2) - [(\beta_0 + \beta_1) - \beta_0] \\ &= E(y_{it}|\text{posttreatment, treated}) - E(y_{it}|\text{pretreatment, treated}) \\ &\quad - [E(y_{it}|\text{posttreatment, control}) - E(y_{it}|\text{pretreatment, control})]. \end{aligned} \tag{23.56}$$

Often studies will include firm fixed effects in the model and replace the common intercept β_0 in equation (23.54) with the fixed effects γ_i. In this case, it is not possible to identify the fixed effects as well as β_1 and β_2. Studies might include only the fixed effects and the interaction dummy $(D_{it}^1 \times D_{it}^2)$, and its coefficient β_3 is used to proxy for the local average treatment effect. In this approach, $\beta_3 = E(y_{it}|\text{posttreatment, treated}) - E(y_{it}|\text{pretreatment or control})$.

What problems should we watch out for in diff-in-diff regressions? The error terms of equation (23.45) could be correlated with the treatment dummies. If agents can choose to be treated or not, the treatment is endogenous and may be correlated with various characteristics of the agents. The treatment and control groups may differ for reasons unrelated to the treatment, violating the assumption that the change in the control group is equal to what the change in the treated group would have been if they had not been treated.

Now, if we can find a set of control variables X_{it}, we can throw them into the panel regression to solve some of these problems. Think about the Frisch-Waugh theorem again, where the coefficients on the dummies in the regression with the controls are the coefficients for the residuals, once we regress both outcomes and assignment variables on the controls. The assumption that, conditional on the variables X_{it}, there are no more unobserved factors correlated with both the assignment to treatment (the D values) and the outcomes is called "unconfoundedness."

Studies work hard to justify the parallel trends assumption. Are the measured relevant characteristics of the treated and control groups similar before treatment? Sometimes you can design a "placebo" test, for example, pretending that treatment occurs at some random time and hoping to find no effect, thus making the case that your results are not spurious.

Perhaps not surprisingly, there are many extensions and refinements of diff-in-diff style analysis. After all, this is a technique that started out in labor economics (e.g., Ashenfelter and Card 1986), a field which tends to be econometrically sophisticated. You can have multiple treatment periods and different treatment intensities, and you can combine untreated firms to create optimized controls. There may be another dimension (e.g., state of incorporation), for which treatment occurs or does not occur. We could take the difference of the outcome variable (e.g., firm sales) across states where the treatment did or did not occur. The difference between the treated before versus after treatment, less the controls before versus after treatment, less the states with or without treatment, comprise a diff-in-diff-in-diff, or triple-diff design. This makes my head hurt.

Versions of the event study go back as far as Dolley (1933), who examined the impact of stock splits on stock prices. The "modern" version of the event study was set out by Ball and Brown (1968) and Fama et al. (1969). The idea is that corporate events, such as dividend or earnings announcements, stock splits, or the surprise death of a CEO, may occur at different times for different firms and cause abnormal returns. Like stars in the universe, there must be billions and billions of event studies in the finance literature! Abnormal returns were traditionally defined as the stocks' residuals in a market model regression, although Brown and Warner (1980) showed using simulations that with daily returns data, the choice of the benchmark does not matter much in practice. The event study takes measures of the abnormal returns of firms experiencing the event and averages them across firms, where different firms may experience the event at different times, and computes an average abnormal return, a proxy for a local average treatment effect.

This approach can be expressed in a multivariate regression model with firm fixed effects, following the lead of Izan (1978), Gibbons (1980), Schipper and Thompson (1983), and Binder (1985):

$$r_{it} = \gamma_i + \beta_i r_{mt} + \delta_i D_{it} + u_{it}, \tag{23.57}$$

where $D_{it} = (D_{it}^1 \times D_{it}^2)$ is the dummy variable that turns on if firm i is treated to the event at time t. The fixed effect γ_i allows for nonzero alphas with respect to the CAPM. The event dummy can turn on for an event window around the event. Different parts of the event window can be examined by including multiple dummies, like the pre-event run up and postevent performance.

To see this as a diff-in-diff, the relevant expectations are:

$$E(r_{it}|\text{posttreatment, control}) = \gamma_i + \beta_i E(r_{mt}|D_{it}=0),$$
$$E(r_{it}|\text{posttreatment, treated}) = \gamma_i + \beta_i E(r_{mt}|D_{it}=1) + \delta_i. \tag{23.58}$$

The diff-in-diff measure of the causal effect is $E(r_{it}|\text{posttreatment, treated}) - E(r_{it}|\text{posttreatment, control}) = \delta_i + \beta_i\{E(r_{mt}|D_{it}=1) - E(r_{mt}|D_{it}=0)\}$. If there is no market impact of the event, the last two terms cancel, and δ_i measures the causal effect.

When the events occur at widely dispersed times for the treated firms, it might be reasonable to assume that the market index returns are not affected by the event. If the events cluster in time, this might not be such a good assumption. If the event changes the firm's beta, we can replace β_i with $(\beta_{i0} + \beta_{i1} D_{it})$, resulting in an interaction term $D_{it} r_{mt}$, which allows the beta to change.

The main statistic in the event study is the average abnormal return, $AR = (1/N)\sum_i \delta_i$, a version of a local average treatment effect. Since the elements of the cross-sectional average capture abnormal returns at different times for different firms, they are unlikely to have much cross-sectional dependence. This independence simplifies the calculation of standard errors, as the variance of the sum is roughly the sum of the variances. It is common to put additional dummy variables to indicate a window in time before and after the event. If the event is at $\tau = 0$, we have an abnormal return AR_0 in the specification above. With dummies that turn on for firm i at time $\tau > 0$ after the event, we have $\{AR_\tau\}$. The cumulative abnormal return over the window is then defined as $CAR = \sum_\tau AR_\tau$.

23.10   Regression Discontinuity

In an ideal experiment, causality can be established when units are randomly assigned to treated and control groups. Regression discontinuity (RD) tries to approximate random assignment when there is a treatment that occurs at some mechanical, exogenous threshold. If the units very close to the threshold are randomly chosen to be just above or just below the threshold, the assignment to treatment is essentially random, and the two groups can be compared to measure the causal effect of the treatment at the threshold.

An example in finance is membership in the Russell 1000 index. On the last day of May each year, Russell ranks US stocks on the basis of their market capitalization and the largest 1,000 are in the index for the next year. The stock ranked 1,001 is in the Russell 2000, but this is a different situation. Institutional investors, such as mutual funds that track the Russell 1000, hold the stocks in the 1000 index, and studies have observed that institutional holdings jump up by about 10 percent when a stock just makes the Russell 1000.

The stock ranked 1,001 is a small component of the Russell 2000, and there does not seem to be any significant change in institutional holdings for these stocks. Thus, new membership in the Russell 1000 is an instrument for—and can be used in—an RD design to estimate the causal effects of institutional ownership. Studies have found stock price effects (Chang, Hong, and Liskovich 2014), increases in dividend payouts (Crane, Michenaud, and Weston 2016), and greater CEO turnover (Mullins 2014) attributed to the greater institutional ownership associated with entry into the Russell 1000 index. Because corporate performance might cause institutional investors to be interested in a stock, institutional ownership is endogenous to measures of firm performance. The assumption that breaks the endogeneity is that small random changes in a stock's price near the end of May determine on which side of the threshold it ends up. (If institutional owners could somehow bid up the price at the end of May just to push a favorite stock into the 1000, there would be a bias.)

Let y_{it} be the outcome variable. D_{it} is a dummy variable indicating treatment of unit i at time t. There is assumed to be a continuous forcing variable X_{it}, which determines treatment according to a threshold value X_0, but the forcing variable itself is continuous at the threshold (e.g., market capitalization in the Russell 1000 example). RD can be *sharp* or *fuzzy*. It is sharp when treatment at the threshold is deterministic: $\text{Prob}(D_{it} = 1|X_{it} \geq X_0) = 1$, and $\text{Prob}(D_{it} = 1|X_{it} < X_0) = 0$. It is important that the agents cannot choose to be on one side of the boundary or the other. This is assumed to be the case in the Russell 1000 example. Unconfoundedness is always satisfied in a sharp RD design, because after conditioning on X, there is no variation left in D_{it}, so it cannot be correlated with any other factors. RD is fuzzy when the probability of treatment jumps discretely at the boundary: $\text{Lim}_{\{x^*x0\}} \text{Prob}(D_{it} = 1|x) \neq \text{Lim}_{\{x^*x0\}} \text{Prob}(D_{it} = 1|x)$. The earliest examples of RD studies estimated a local average treatment effect, $E(y|x \geq X_0) - E(y|x < X_0)$ from the sample means of the y_{it} over a bandwidth on either side of the threshold (Thistlethwaite and Campbell 1960). Nonlinear regression is more common these days.

Consider a linear regression example:

$$y_{it} = a + bD_{it} + \delta X_{it} + \epsilon_{it}, \tag{23.59}$$

where D_{it} is the treatment indicator, and b is the true causal effect of treatment. Treatment, however, is correlated with X_{it}. The idea is to approximate b by comparing cases close to and on either side of the threshold value for x. Assuming that the regression errors are independent, we have $E(y_{it}|D_{it} = 1) - E(y_{it}|D_{it} = 0) = b + \delta[E(X_{it}|X_{it} \geq X_0) - E(X_{it}|X_{it} < X_0)]$. On the assumption that X is continuous near the threshold, the second term should be small for values of X near the threshold, and we should get a good estimate of the local average treatment effect, b. One of the tricks here is to pick a bandwidth near the threshold, wide enough so that there are enough observations for good estimates but not so wide as to bring in data that do not follow the underlying relation.

When the outcome is not linear in the X variables, RD studies have to pay serious attention to functional form. A quadratic relation, for example, could be mistaken for a discontinuity if a linear regression is used. Polynomials have been popular outside the finance literature, for example, replacing the previous regression with

$$y_{it} = a + bD_{it} + \sum_{j=1,\ldots,4} \delta_{0j}(X_{it})^j + \sum_{j=1,\ldots,4} \delta_{1j}(D_{it}X_{it})^j + \epsilon_{it}. \tag{23.60}$$

RD designs lend themselves easily to placebo tests, by simply making up a fake threshold and hoping to find nothing. It is also easy to check for continuity of the X variables at the threshold by running the RD model with an X variable on the left-hand side. The covariates should be relatively balanced on either side of the threshold.

The RD approach is increasingly popular in finance. Some of the designs are cleaner examples than others. Here are a few examples of recent studies. Kerr, Lerner, and Schoar (2010) examine the effect on a startup firm's probability of survival relative to obtaining angel financing. Bronzoni and Iachini (2014) find an increase in investment spending associated with applications for an investment subsidy in Italy, based on a scoring threshold for the application. Flammer (2015) finds improved accounting performance is caused by corporate social responsibility measures, based on the discontinuity of close votes for the measures. Cuñat, Gine, and Guadalupe (2012) measure higher stock valuations and better long-term performance for firms with corporate governance provisions that respect shareholding rights, similarly using close votes for those provisions. Rauh (2006) finds impact of internal corporate financing sources on investment and R&D expenditures, using discrete boundaries in required contributions to pension plans. Malenko and Shen (2016) measure an impact of negative recommendations from Institutional Shareholder Services on shareholder voting, using a discontinuity in the guidelines used to produce the recommendations. Roberts and Sufi (2009) find lower net debt issuance when firms violate a discrete debt covenant. Bakke, Jens, and Whited (2012) find lower firm employment, investment, and cash when a stock crosses the threshold for delisting on the NASDAQ. Most of the examples I can find are in the corporate finance literature. But hold on, asset pricers: Causal inference is coming!

24 Bootstrapping Methods and Multiple Comparisons

As computing power gets more impressive and cheaper, more studies are using bootstrap methods to conduct statistical inference. This chapter starts with a general discussion of the bootstrap and then moves to illustrations from the literature using mutual funds and hedge funds. This is a natural setting for wanting to make inferences in a large cross section, and so issues of multiple comparisons, false discovery rates, Bayes factors, and confusion naturally arise. Hopefully there will be little confusion when reading this chapter.

24.1 Bootstrap 101

24.1.1 The Basics

The parametric bootstrap is a special case of the simple, or nonparametric, bootstrap, itself an example of a resampling scheme. Introduced by Efron (1987), the bootstrap is useful when we wish to conduct statistical inferences but don't have an analytical formula for the sampling variation of a statistic or don't wish to assume normality (or some other convenient distribution that allows for an analytical formula), or when we have a sample that is too small to trust the asymptotic distribution theory. The basic idea is to build a sampling distribution by resampling from the data at hand. In the simplest example, we have some statistic that has been estimated from a sample, and we want to know its sampling distribution. We resample from the original data, randomly with replacement, to generate an artificial sample of the same size, and compute the statistic on the artificial sample. Repeating this many times, the histogram of the statistics from the artificial samples is an estimate of the sampling distribution for the original statistic. This distribution can be used to estimate standard errors, confidence intervals, and the like. We can think of the bootstrap samples as being related to the original sample as the original sample is to the population. There are many variations on the bootstrap, and a good overview is provided by Efron and Tibshirani (1994).

In the *simple*, or *nonparametric bootstrap*, no assumptions are made about the form of the distribution. It is assumed, however, that the sample accurately reflects the underlying population distribution, which is critical for reliable inferences. This assumption may not always hold. For example, suppose that the true distribution is uniform on $[0, M]$. In a sample drawn from this distribution, the maximum value is likely to be smaller than M, so that the bootstrap, based on this sample, will likely understate the true variability in the population. For another example, suppose that the true distribution has infinite variance. Then the bootstrap distribution for the sample mean is inconsistent (e.g., Athreya 1987), and it will obviously underestimate the variance. This kind of problem is worse if the original sample has fewer observations. In contrast, if the data are contaminated with measurement errors, then the extent of the true variability in the population can be overstated.

Here is a simple example of resampling to test the hypothesis that the mean of X is zero. The data are $X = \{X_t\}_t$, and the sample mean is the statistic of interest, $\hat{\mu} = (1/T)\sum_t X_t$. In a simple bootstrap, we follow several steps. To create artificial samples, we do the following:

1. Draw T rows from X, randomly with replacement. (Or we could randomly reorder the original rows; see Noreen 1989.)

2. Estimate the sample mean on each artificial sample.

3. Do this many times, and draw a frequency distribution of the sample mean estimates.

4. Use this distribution to make standard errors, confidence intervals, and so forth.

One of the most popular statistics generated by simulation is the *empirical p-value*. Suppose that we compute a simple *t*-ratio in the sample for the hypothesis that the sample mean is zero. We can resample from the original data after subtracting the sample mean from each observation, so that we are drawing from a "population" that actually has mean zero. Then, in each artificial sample, we compute the simple *t*-ratio. The empirical *p*-value for the null hypothesis that the mean is zero is the fraction of the artificial samples in which the *t*-statistic is larger in absolute value (for a two-tailed test) than the one we computed in the original sample. If the empirical *p*-value is small, it is unlikely that the original data are drawn from a population with mean equal to zero, and we reject the null hypothesis.

When bootstrapping statistics, it is best to use pivotal statistics. A *pivotal statistic* is one whose distribution (perhaps under the null hypothesis) does not depend on any parameters. In the example above, the sample mean is not a pivotal statistic, because its distribution depends on the population mean μ. The population is centered around μ, but the sample is centered around the sample mean $\hat{\mu}$. Because the two can differ, the resampled distribution may not accurately capture the population distribution. An example of a pivotal statistic is the *t*-ratio, which, under the null hypothesis, converges to a normal distribution with mean zero and variance one. This distribution does not depend on any parameters, so the bootstrap will produce more accurate results for the *t*-ratio than it will for the sample mean.

24.1.2 Serial Dependence

Several approaches in the simple bootstrap can be used to handle serial dependence in the data. The most popular are versions of the block bootstrap. Here the basic idea is to resample, not a single observation at a time, but a block of time series observations long enough to capture the time series dependence in the block. The *regular block bootstrap* does this with a fixed block length. The *stationary bootstrap* uses random block lengths, where the length may be drawn from an auxiliary distribution. One of the problems with the block bootstrap is that it tends to undersample the beginning and the end of the original data sample. This can be addressed with a "circular" block, which wraps the blocks around, so that a block at the end of the sample can be completed using data from the beginning of the sample. For a more detailed discussion of these and other methods, see Efron and Tibshirani (1994).

24.1.3 The Parametric Bootstrap

With a parametric bootstrap, we can sometimes do better than with a nonparametric boot-strap, where "do better" means, for example, obtaining more accurate confidence intervals (e.g., Andrews 2005). The idea of the *parametric bootstrap* is to model some of the structure of the data, conditioning on the estimated parameters of the model for the data. This might be as simple as assuming the form of the probability distribution. For example, assuming that the data are independent and normally distributed, we can draw artificial samples from a normal distribution (a Monte Carlo simulation), using the original sample mean and variance as the parameters of the normal distribution. This is not exactly the right thing to do, because we should be sampling from a population with the true parameter values, not the estimated values. But if the estimates of the mean and variance are good enough, we should be able to obtain reliable inferences.

Here is an example to illustrate and further suggest the flexibility of the parametric bootstrap. Suppose that X_t is serially dependent, and you want to test H_0: $E(X) = 0$. Because of the serial dependence, the usual standard error calculation for the sample mean is biased. To address this problem, we can model the serial dependence parametrically. For example, suppose we use a first-order autoregression:

$$X_t = a + \rho X_{t-1} + e_t. \tag{24.1}$$

We estimate this model on the original data and keep the estimated parameters $\{\hat{a}, \hat{\rho}\}$ as "true" parameters of the simulation. We also keep estimated residuals, $\{e_t\}$, and resample from the $\{e_t\}$, as above. We can then build each artificial sample recursively. Start by drawing one of the $\{X_t\}_t$ at random, and use it as the initial value of an artificial sample X_0. Then we form:

$$X_1 = a + \hat{\rho} X_0 + e_1,$$
$$X_2 = a + \hat{\rho} X_1 + e_2, \ldots. \tag{24.2}$$

Now we can use the histogram of the estimated sample means to test the null that $E(X) = 0$, and the sampling variability in those estimates allows for the serial dependence in X. Of course, it would be better to use a t-ratio to test the null hypothesis, because the t-ratio is pivotal.

There are at least two approaches in the previous example. One would be to compute the t-ratio using the usual but biased sample standard deviation of X, $\sigma(X)$, assuming that the standard error of the mean is $\sigma(X)/\sqrt{T}$. This version of the t-statistic is biased, but it is also biased when computed on the artificial samples, because of the serial dependence that we capture in the bootstrap. We may still be able to conduct valid inferences by comparing the sample statistic with the bootstrap resampling distribution. However, this approach is not quite right, because the bias in the standard deviation estimate depends on the serial dependence in the data, so the biased t-ratio is not actually pivotal. A second approach would be to compute a t-statistic, where the standard error is adjusted for serial dependence in the original data. For example, with a first-order autoregression, the variance of the mean is a function of the autocorrelation, and an adjusted variance can be used in the t-ratio. If we replicate this approach in each artificial sample, estimating the autocorrelation in that sample and using it to adjust the standard error for the t-ratio, we should be able to obtain valid inferences by comparing the sample statistic with the bootstrap resampled distribution. There is a trade-off between the two approaches. Using the biased statistic, the variance of the t-ratio is too small, so the t-ratio does not approach a normal (0, 1) variable in large samples and is no longer pivotal. Using the adjusted statistic, however, there is additional noise introduced by estimating the sample autocorrelation, which can reduce the power. (We could take this further, recognizing and trying to adjust for the bias in the sample first-order autocorrelation, but I hope you see the idea.)

Another issue with serial dependence arises when there are missing values in the data series. These typically occur in blocks, such as at the beginning of the sample when new data enter, or at the end of the sample when some of the data series die off (e.g., firms or funds cease to exist). When the bootstrap resampling is done by picking one time period at random, then in the simulated samples, the missing values will be randomly scattered throughout the artificial sample. If the original data have no serial dependence, this need not be an issue, but with serial dependence, it can create an inconsistency in the bootstrap.

24.1.4   Application to Mutual Funds

Mutual funds are a great context to illustrate bootstrap resampling methods, especially their application to cross sections. We have a large number of mutual funds, and we want to be able to describe and draw inferences about them. For example, consider mutual fund alphas from a regression of fund excess returns on some factors, such as the Fama and French (1993) three-factor model or the Carhart (1997) four-factor model, which adds a factor representing momentum. The large cross section presents a problem in empirical mutual fund studies. Imagine that you have estimated the alphas for, say, 4,000 mutual funds. Now what do you do with them? How do you summarize such a large cross section of test statistics, and how do you draw inferences about them?

The particular factors described above remain controversial in asset pricing, as they originate from empirical observations of patterns in the data, as opposed to theory. However, the controversy is less intense when these models are used to generate mutual fund alphas than it is for asset pricing problems. The philosophy in this application is that we are comparing fund returns to that of a benchmark. Even if we don't really know why there are momentum returns, we know that historically high momentum returns have been observed, and we might want to avoid attributing abnormal performance to a fund that simply follows a momentum strategy, earning the well-known anomaly returns. Using the momentum strategy excess return as one of the benchmark factors is an attempt to abstract from the part of the funds' performance that comes from following a simple momentum strategy.

One of the first studies to apply bootstrap simulation methods to an entire cross section of mutual fund returns was Kowsowski et al. (2006). Their approach was to estimate for each fund the factor model regression

$$r_{it} = \alpha_i + \beta_i r_{mt} + c_i \text{SMB}_t + d_i \text{HML}_t + e_i \text{MOM}_t + \varepsilon_{it}, \tag{24.3}$$

where r_{mt} is the excess return on a stock market index, and SMB, HML, and MOM are long-short portfolios formed on the basis of market capitalizations, book/market ratios, and lagged relative returns. Kowsowski et al. (2006) keep the estimates of $\{\alpha_i, \beta_i, c_i, d_i, e_i\}$ as true parameters of the parametric bootstrap, and they keep the fitted regression residuals $\{\varepsilon_{it}\}$. For each simulation trial, they draw for each fund from the $\{\varepsilon_{it}\}$, randomly with replacement, and append the resampled residual to the original factors and coefficients, leaving off the intercept to generate artificial fund returns that satisfy the null hypothesis that $\alpha_i = 0$. They estimate the t-statistics for the alphas for each fund on each artificial sample, and repeating this many times, produce an empirical distribution for the t-ratios.

24.2 Multiple Comparisons

Perhaps the most important innovation in Kowsowski et al. (2006) is to conduct inferences about the entire cross section of the mutual funds' alphas. The inferences are based on N individual tests for the significance of the N mutual fund alphas. This is the *multiple comparisons problem*. As described below, their bootstrap methods are designed to draw inferences that account for the number of funds that are tested.

You could just test each fund separately for significance. If you set the size of your tests to, say, $\gamma = 5\%$, then even if all the alphas are really zero, you might expect 200 or so of the 4,000 estimates to have t-ratios larger than 2.0, and so to be falsely rejected if $N = 4,000$. These false detections or false discoveries are a problem when multiple comparisons are made.

24.2.1 Elevating the Issue

Our jobs as finance researchers is to discover new truths about financial economics and to communicate them to other researchers, students, policy makers, and other interested parties. For empirical work, I like the way Merton Miller once described it. He said the data have in them some truth, and your job as an empiricist is to figure out what that truth is and to bring it to light. Sounds noble. Use all your skills and intuition to do it. Recently, the finance profession, like many other fields of scientific and empirical research, has begun to question the extent to which our published research actually accomplishes this goal. Scholars have raised doubts about the external validity of the results in published papers. Meta studies have examined how the findings of published papers fare in samples outside those of the original work, and often find a lack of external validity. Ferson, Sarkissian, and Simin (2003) find that much of the earlier evidence for stock market predictability is consistent with a spurious mining process, where the interaction of data mining and systematic statistical bias due to spurious regression produces a false impression of predictability. Welch and Goyal (2007) examine several predictor variables proposed in previous studies and find their out-of-sample forecasting ability to be lacking. McLean and Pontiff (2016) examine anomaly factors found by previous studies and find that the variables' explanatory power for the cross section of stock returns is vastly reduced beyond the samples of the original studies. Jones and Mo (2016) find a similar pattern among variables that were previously found to be associated with superior performance in mutual funds. Linnainmaa and Roberts (2018) examine a large set of factors using data prior to the published studies' samples and also find a reduction in explanatory power. Harvey, Liu, and Zhu (2016) examine a large number of factors for the cross section of returns, question their statistical significance using standard criteria, and suggest some alternative criteria.

Technological changes in the way we do empirical research in finance, combined with adverse incentives in the profession, seem to be conspiring against the external validity of our research findings. There is a publication bias in our journals. Results that appear "significant" are given more attention, and those that do not appear significant are typically screened out. Journal editors and researchers compete for citations, and it's hard to publish papers with t-ratios smaller than 2. Knowing that, researchers might stop working on a project if they aren't "finding" anything. This is a form of selection bias. Harvey, Liu, and Zhu (2016) report the distribution of reported t-statistics from factor model studies, and it appears to be truncated at the left, with very few t-ratios less than 2 reported. Invalid studies can also be published by chance. When a significance level is set, say to 5%, it means that if there is no relation, then one will be found about 5% of the time. Suppose that such a 5% paper becomes influential, and 100 scholars set out to work on it. We expect five of them to find something. If the five papers are published, and the other 95 stay in someone's file drawer, a whole stream of literature can emerge that is fundamentally wrong.

With the strong incentives to find an interesting dissertation and get tenure, some researchers unfortunately might fall prey to these adverse incentives. If you ran 100 independent tests and only reported the expected five with t-ratios larger than 2.0, this would be outright academic fraud. I am not sure what the penalty is for this, but you don't want to find out. Even short of fraud, the temptations for p-hacking are huge. "P-hacking" refers to cherry picking the empirical results that you like without reporting the others. We make so many choices in our empirical research, some version of this could be common and even hard to avoid. We might look at several definitions of the main variables. We might try various statistical approaches or regression methods on our problem. There are different ways to standardize the variables. There are different subperiods to focus on. The reason that technology changes have made these things worse is that it is increasingly easy to run different versions of our empirical analysis. So many databases are available at the touch of a keystroke that "hand collected" data has come to mean the researcher's hands were on a computer keyboard. Canned software packages are like weapons of mass research destruction. There are eight ways to cluster the standard errors sitting at our fingertips. Robustness checks invite p-hacking. We might have a research assistant who does some of this without our knowledge to produce a result the professor will like. When the results of only some of the choices are shown, and it skews the results in some particular direction, this is p-hacking. Try not to do it! Merton Miller once suggested that an empiricist should be conservative and not report the strongest results, so when someone else tries to replicate the findings, the results will be even stronger.

There are several potential solutions to the problem. One idea would be to spearhead a sea change in the culture of finance research in favor of replication. Our field seems to have a bias against replication studies. I am not holding my breath for this one. Some journals have started to require that publishing authors share their data and programs, and there is some talk of using preregistered analysis plans. PhD students typically love the idea of getting free data and code. Some scholars have argued that the p-value is misunderstood by many researchers and have proposed alternative statistical criteria, such as odds ratios (discussed in section 24.2.5). Some scholars have argued for simple changes in the conventional standards for statistical significance. If more stringent tests with smaller sizes are used, the rate of false discoveries is reduced. Benjamin et al. (2018) argue that we should replace 5% with 0.5% as our conventional significance level. Harvey, Liu, and Zhu (2016) famously suggest that we should require t-ratios larger than 3 for new factors in asset pricing. Another approach is to try to adjust more carefully for multiple comparisons. This is the topic to which we now turn.

24.2.2 Multiple Comparisons: Classical Solutions

A couple of simple techniques have been popular in the empirical literature to deal with multiple hypothesis tests. One is the *Bonferroni inequality*. This relation follows from the fact that for N probabilistic "events" e_i, $\text{Prob}\{U_i e_i\} = \Sigma_i \text{Pr}\{e_i\} - \text{Pr}(\text{intersection events}) \leq \Sigma_i \text{Pr}\{e_i\}$ where U denotes the union. In words, the probability that any of the events occurs is the sum of the probabilities that the separate events occur, less the probability that more than one occurs. Suppose that you estimated the t-ratios for the alphas of 4,000 mutual funds and found that 300 of them had t-ratios larger than 2.0 and that the largest absolute t-ratio was 3.5. Is there any evidence that funds have alphas different from zero? The events e_i in question are the rejection of the null hypothesis of zero alpha for fund i. The Bonferroni inequality notes that $\Pr\{\cup_i e_i\} \leq \sum_i \Pr\{e_i\}$. Suppose that we want a critical value, so that the type-I error rate (if the null is true) is 5%. The Bonferroni adjustment is to test each of the N cases against a critical value that corresponds to a size of .05/N. Loosely interpreting the p-value for fund i as $\Pr\{e_i\}$, we can take the smallest p-value that we find and multiply it by the number of tests, N. This provides a conservative upper bound on the p-value for the joint hypothesis that all the alphas are zero. This is conservative in that, if we reject the null based on this p-value, we would have found an even smaller p-value if we could have considered the dependence across the funds. In our mutual fund example, the Bonferrroni p-value is $0.0002 \cdot (4,000) = 0.8$, so we do not reject the null that all the funds have zero alphas.

The classical Bonferroni test controls the *familywise error rate* (FWER), which is the probability of even one false rejection in the N mutual funds. The power of the Bonferroni test can be low, since it relies on the information in only the most extreme value. In our example, this is the mutual fund alpha with a t-ratio of 3.5. When we have many funds, we might be willing to accept a few false positives to get a more powerful test. This trade-off could be appropriate if we think there are a lot of funds with nonzero alphas that produce modest t-ratios.

In the mutual fund example, 300/4,000 or 7.5% had t-ratios bigger than 2. Is this statistically significant? We can model the situation using a binomial probability model. Let's use a t-ratio of 2 as the critical value and assume that the probability of seeing this event under the zero-alpha null hypothesis is 5%. The event indicator, e_i, takes the value 0.0 with probability 0.95, and it takes the value 1.0 with probability 0.05 under the null. The fraction of N tests that produce p-values greater than 2 is distributed under the null as $(1/N)\sum_i e_i$, which has mean equal to 0.05 and variance equal to $(.05)(.95)[N\{1 + (N-1)\rho\}]$, where ρ is the average of the correlations of the tests. A t-ratio can be formed for the difference between the observed and expected fractions. In our mutual fund example, the t-ratio for the null of zero alphas when we find 300 cases out of 4,000 with t-ratios above 2 is $(0.075 - 0.05)/[(.05)(.95)/(4,000\{1 + (N-1)\rho\})]^{0.5}$. If $\rho = 0$, the binomial t-ratio is 7.25, which looks pretty significant. However, mutual fund returns and even their factor model regression residuals are correlated, so the tests are likely correlated. As the preceding expression suggests, the binomial t-statistic is highly sensitive to the correlation. For example, if the average correlation of the tests is 0.05, the t-ratio becomes 0.51, which no longer looks very significant.

24.2.3   False Discovery Rates

A persistent problem in the mutual fund performance literature is separating skill from luck. Funds that have no skill but produce t-ratios above the critical value for rejecting the null, are just lucky "false discoveries." The *false discovery rate* (FDR) is the expected ratio of rejected nulls (discoveries) that actually have zero alphas, to the total number of rejected funds. The FDR is a natural extension of the idea of the size of a test to a setting like this, where we are testing for performance in a cross section of funds.

An illustration of the FDR follows Harvey (2017). Consider first the detection of breast cancer, which appears in about 1% of women aged 40–50. Let $(1 - \pi_0) = .01$ be the fraction of the population where the null hypothesis of no cancer is false. Suppose that if we run a test at the 5% size, then the mammogram has a power of 90%, $\beta = .90$. The probability of rejecting the null when there really is cancer is 90%. Suppose that we run 100 tests. Then the FDR is $99(.05)/[99(.05) + 1(.90)] = 5/6$. The false discovery rate seems pretty high. Now let's carry the example over to high blood pressure. A recent article cited the American Heart Association as finding that almost half of the adults in the United States have high blood pressure, using their revised standard reading of 130/80 or above. Let $(1 - \pi_0) = .50$ be the fraction of the population where the null hypothesis of no high blood pressure is false. Suppose that if we run a test at the 5% size, then the blood pressure test has a power of 90%, $\beta = .90$. Suppose that we run 100 tests. Then the expected FDR is $50(.05)/[50(.05) + 50(.90)] = 5.3\%$. We can make the FDR smaller if the size of the test is smaller or if the power of the test is larger, but for most test sizes, it strongly depends on the "prior" probability π_0 that the null is true in the population.

Suppose the population of mutual funds consists of a fraction π_0 with zero alphas, while $(1 - \pi_0)N$ funds have nonzero alphas. The FDR is related to the FWER by a scale factor. Suppose that the zero-alpha null is rejected for fraction F of the N funds. Then FWER = FDR · F. Barras, Scaillet, and Wermers (2010) apply the FDR to a mutual fund problem. They compute the FDR as $= \pi_0 \gamma/F$. Controlling the FDR involves using simulations, searching for the value of γ (the size of the individual tests) that delivers the desired value of the FDR, where both π_0 and F depend on the chosen γ.

24.2.4   Step-Down *p*-Values for Multiple Comparisons

Benjamini and Hochberg (1995) propose a stepwise approach to Bonferroni type tests that control the FDR. This uses the result that, under the null hypothesis, the *p*-value of the test for the null hypothesis is uniformly distributed on [0, 1]. It also assumes that the tests are independent, although they show how to generalize to allow dependence. It works for any value of π_0. Given the *p*-values for N tests, order them from smallest to largest. Reject the null for all funds where the *i*th from the smallest *p*-values is below $(i/N)\gamma$. This procedure controls the FDR at level γ. That is, the expected fraction of false to total discoveries is $\leq \gamma\pi_0$, and thus for any value of π_0, the expected fraction is below γ. This is an example of a step-down procedure, where larger *p*-values are evaluated with larger cutoffs. The classical Bonferroni approach described above just compares the smallest *p*-value to $\gamma(1/N)$.

Lehmann and Romano (2005) provide some generalizations of step-down approaches directed at controlling the FWER and also the FDR. They distinguish between the *false discovery proportion* (FDP), which they define as the number of false rejections to the total number of rejections, and the FDR, which they define as the expected value of the FDP. They develop an approach that guarantees that $\text{Prob}\{\text{FDP} > \gamma\} \leq \alpha$, where the α here is a generalization of the size of a test, not the performance of a mutual fund (sorry). A more general step-down method evaluates the ordered p-values with an increasing sequence of cutoff values, $\alpha_i = (\gamma i + 1)\alpha/[N + \gamma i + 1 - i]$. They also consider step-down methods to control the k-FWER, which is the probability of k or more false rejections. This method uses different cutoff values above and below the kth ordered p-value. For $I \leq k$, the cutoff is $\alpha_i = k\alpha/N$, and for $I > k$, it is $\alpha_i = k\alpha/[N + k - i]$. They show that you can't increase the cutoffs any further than in this sequence without violating the k-FWER.

24.2.5 Bayes Factors

Recent commentators have argued that many scholars misunderstand and misuse the p-value. We know the p-value is NOT the probability that the null hypothesis is true. It is the probability that a test statistic of the observed magnitude or larger would be found if the null hypothesis were true. We saw above that the FDR depends on the prior probability that the null hypothesis is true in the population. As any good Bayesian knows, prior beliefs are unavoidable when evaluating research results, so why not just be explicit about it?

Basic probability theory states:

$$\text{Prob}(H_1|\text{data}) = \text{Prob}(H_1, \text{data})\text{f}(\text{data}) = \text{f}(\text{data}|H_1)\text{Prob}(H_1)/\text{f}(\text{data}),$$
$$\text{Prob}(H_0|\text{data}) = \text{Prob}(H_0, \text{data})\text{f}(\text{data}) = \text{f}(\text{data}|H_0)\text{Prob}(H_0)/\text{f}(\text{data}). \tag{24.4}$$

So we can infer that the $\text{Prob}(H_1|\text{data}) > \text{Prob}(H_0|\text{data}) \Leftrightarrow \text{f}(\text{data}|H_1)\text{Prob}(H_1) > \text{f}(\text{data}|H_0)\text{Prob}(H_0)$, or

$$\text{f}(\text{data}|H_1)/\text{f}(\text{data}|H_0) > \text{Prob}(H_0)/\text{Prob}(H_1). \tag{24.5}$$

The term on the left is the Bayes factor for the alternative, H_1, and the term on the right is the prior odds ratio for the null. After we look at the data, the alternative is more likely than the null if the Bayes factor is larger than our prior odds in favor of the null.

24.3 Bootstrap 201: Inferences in Cross Sections

Suppose that in the original data, the funds' alpha estimates (or better, t-ratios) are ranked. We would like to ask questions like: Do the top 5% of the mutual funds have alphas significantly different from zero? A simulation approach is called for!

In each artificial sample, we can rank the t-ratios in the cross section of funds just as we did in the original data. The hypothesis is that the alpha value in the top 5% of the actual fund data is drawn from a population with true alphas equal to zero. We can bootstrap the cross sections of fund returns and factors, imposing the null hypothesis that all the alphas are zero. For example, Fama and French (2010) draw random rows from $\{(r_{pt} - \alpha_{pt}), f_t\}_t$, from equation (24.3), thus capturing randomness in the factors. An empirical p-value for a one-sided test of the 5% hypothesis is computed as the frequency with which the estimated alpha in the 5% right tail in the simulations is larger than the value at the 5% right tail estimated in the data. If this p-value is small, we reject the null in favor of the alternative hypothesis that funds at the 5% right tail in the sample have positive alphas.

This simulation approach can accommodate non-normality, heteroskedasticity, and other features of the data. Provided that we resample the entire N-vector of funds together, the artificial samples will preserve the correlation across the funds. If the funds and the factors are sampled jointly, the artificial data will also reflect conditional heteroskedasticity, correlated with the factors, that may be present in the original data.

Kowsowski et al. (2006) use a resampling approach for inferences on a cross section of mutual fund returns. They find evidence that there are mutual funds with positive alphas. In their table II, panel B, where the t-ratio is used, almost all the simulation trials deliver t-ratios for alpha in the Carhart (1997) four-factor model, whose values at the 5% right tail of the cross section are smaller than in the original data. Thus, the null hypothesis that the funds in the right 5% tail of the cross section are drawn from a population where all the alphas are zero is rejected. In addition, there is evidence of significant negative alphas in the left tails.

Fama and French (2010) also present a bootstrap analysis of the cross section of mutual fund returns and factor model alphas. They find little evidence for positive alpha funds, but they do find evidence of negative alpha funds. Several differences in sample and design potentially explain the different results. Fama and French use more recent data, and mutual fund performance might be weaker in the more recent sample. In their baseline results, Kowsowski et al. (2006) resample funds' residuals but reuse the factors. This can understate the sampling variability. Kowsowski et al. use a 60-month survival criterion in selecting the funds. Fama and French prefer an 8-month selection criterion to reduce survivorship bias. If funds die when their performance is poor, then selecting on funds that survive for 5 years is likely to result in a sample of funds with relatively good performance, compared to the underlying population. Fama and French (2010) present an analysis of the effect of some of these differences on the result and find that the exclusion rules make a difference.

With more structure on the problem, we can use simulations that reveal the power of the tests to draw more refined inferences in a cross section. Imagine that the population of fund managers consists of three subpopulations, described in terms of their abnormal performance, or alphas. A fraction of fund managers, π_0, has zero alphas, while a fraction π_g of "good" managers has positive alphas centered at $\alpha_g > 0$, and a fraction π_b are "bad" managers with negative alphas, centered at $\alpha_b < 0$. Barras, Scaillet, and Wermers (BSW; 2010) consider this model and estimate it by simulating the cross section of funds' returns. Chen and Ferson (2017) develop the model further and estimate it on samples of mutual

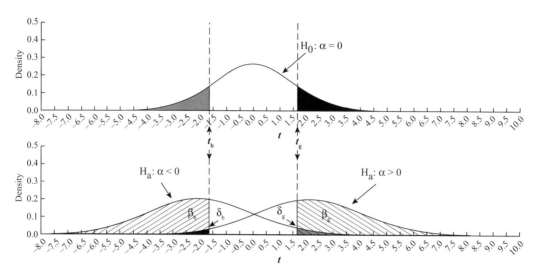

Figure 24.1
A mixture of alpha distributions.

funds and hedge funds. Chen, Cliff, and Zhao (2017) refine the estimation and apply the model to hedge funds, and Harvey and Liu (2016) offer further refinements applied to mutual funds.

For investment funds, groups associated with zero, negative, or positive alphas seem natural. That some funds should have zero alphas is predicted by Berk and Green (2004). Under decreasing returns to scale, new money should flow to positive-ability managers until the performance left for investors is zero. There are also many reasons to think that some funds may have negative or positive alphas. Costs keep investors from quickly pulling their money out of bad funds or sending money quickly into good funds. These costs include taxes, imperfect information, agency costs, and human imperfections such as the disposition effect, where for psychological reasons, investors hold on to their losing funds for longer than the economics would suggest.

The basic probability model behind BSW is laid out by Chen and Ferson (2017) as follows and is illustrated in figure 24.1. They use three simulations to implement the approach. First, set the size of the tests, say, to 10%. Simulating the cross section of mutual fund alphas under the null hypothesis that all the alphas are zero produces two critical values for t-statistics, t_g and t_b. The simulations at this step are very similar to those of Fama and French (2010). The critical value t_g is the t-ratio above which 10% of the simulated t-statistics lie when the null of zero alphas is true. The critical value t_b is the value below which 10% of the simulated t-statistics lie.

The second step is to simulate the cross section of alphas under the alternative hypothesis that managers are good, that is, the alphas are centered not at zero but around a value

$\alpha_g > 0$. The fraction of the simulated t-ratios above t_g is the power of the test for good managers. This power is denoted in figure 24.1 as β_g. The fraction of the simulated t-ratios below t_b is an empirical estimate of the probability of rejecting the null in favor of a bad manager when the manager is actually good. This *confusion parameter* is denoted in the figure as δ_b.

A third simulation is based on the alternative hypothesis that managers are bad, that is, the alphas are centered around a value $\alpha_b < 0$. The fraction of the simulated t-ratios below t_b is the power of the test for bad managers, β_b. The fraction of the simulated t-ratios above t_g is an empirical estimate of the probability of rejecting the null in favor of a good manager when the alternative of a bad manager is true. This confusion parameter is denoted as δ_g.

Chen and Ferson (2017) combine the estimates as follows. Let F_b and F_g be the fractions of rejections of the null hypothesis in the actual fund data using the simulation-generated critical values, t_b and t_g, respectively. The probability model implies that

$$
\begin{aligned}
E(F_g) &= P(\text{reject at } t_g | H_0)\pi_0 + P(\text{reject at } t_g | \text{Bad})\pi_b + P(\text{reject at } t_g | \text{Good})\pi_g \\
&= 0.10\pi_0 + \delta_g\pi_b + \beta_g\pi_g,
\end{aligned}
\tag{24.6}
$$

and similarly,

$$
\begin{aligned}
E(F_b) &= P(\text{reject at } t_b | H_0)\pi_0 + P(\text{reject at } t_b | \text{Bad})\pi_b + P(\text{reject at } t_b | \text{Good})\pi_g \\
&= 0.10\pi_0 + \beta_b\pi_b + \delta_b\pi_g.
\end{aligned}
$$

The second lines of (24.6) present two equations in the two unknowns, π_b and π_g, (since $\pi_0 = 1 - \pi_b - \pi_g$), which can be solved given values of $\{\delta_g, \beta_g, \delta_b, \beta_b, F_g, F_b\}$. The solution to this problem is found numerically by minimizing the squared errors of the two equations subject to the constraints that $\pi_b \geq 0$, $\pi_g \geq 0$, and $\pi_b + \pi_g \leq 1$. $E(F_g)$ and $E(F_b)$ are replaced by the fractions rejected in the actual data at the simulation-generated critical values, and the parameters $\{\delta_g, \beta_g, \delta_b, \beta_b\}$ are from the three simulations. The estimator of π_0 derived above reduces to the estimator used in BSW, when $\beta g = \beta b = 1$ and $\delta g = \delta b = 0$. BSW estimate π_0 following Storey (2002), as:

$$
\hat{\pi}_0 = [1 - (F_b + F_g)]/(1 - \gamma),
\tag{24.7}
$$

where γ is the size of the test. This estimator assumes that the fraction of zero-alpha funds in the population, multiplied by the probability that the test will not reject the null of zero alpha when it is true, is equal to the fraction of funds for which the null of zero alpha is not rejected in the sample. But the test also will not reject the null in the sample for cases where alpha is not zero, if the power of the tests is below 100%. Consider the case of a two-tailed test with size γ and power $\beta < 1$, and let $F = (F_b + F_g)$. Then we have

$$
\begin{aligned}
1 - E(F) &= P(\text{don't reject} | H_o) \, \pi_0 + P(\text{don't reject} | H_a)(1 - \pi_0) \\
&= (1 - \gamma) \, \pi_0 + (1 - \beta)(1 - \pi_0).
\end{aligned}
\tag{24.8}
$$

Solving this expression for π_0 gives a generalization of Storey's (2002) estimator:

$$\pi_0^* = [\beta - E(F)]/(\beta - \gamma), \tag{24.9}$$

where Storey's estimator is a special case when $\beta = 1$. Storey's assumption that the test has 100% power is justified by choosing the size of the tests to be large enough. He motivates this as a "conservative" choice.

Storey's estimator of π_0 has two offsetting biases. Setting the β parameters to 1.0 creates an upward bias in the estimator, while setting the δ parameters to 0.0 creates a downward bias in the estimator. The net effect is an empirical question, which Chen and Ferson (2017) address in bootstrap experiments, finding that the upward bias from assuming perfect power dominates.

In this model, a false discovery rate considers not just the "lucky" funds with zero alphas but also the expected fraction of "extremely lucky" bad funds, where the test is confused and indicates that the fund is good. The fraction of unlucky funds, as a ratio to the total number of funds that the tests find to have alphas below α_b, is the false discovery rate for bad funds:

$$FDR_b = [(\gamma/2)\pi_0 + \delta_b\pi_g]/F_b. \tag{24.10}$$

The false discovery rate for good funds takes the following form:

$$FDR_g = [(\gamma/2)\pi_0 + \delta_g\pi_b]/F_g. \tag{24.11}$$

The estimates of the π values that result from the last two equations are conditioned on the values of $\{\alpha_g, \alpha_b\}$. Because the estimates of the population fractions are sensitive to the values of the alphas assumed, Chen and Ferson (2017) simultaneously estimate the π fractions and the alpha values, $\alpha = (\alpha_g, \alpha_b)$.

The information from the cross section of funds and the estimated parameters of the probability model can be used to draw inferences about individual funds. Given a fund's alpha estimate α_p, we can use Bayes' rule to compute

$$P(\alpha > 0|\alpha_p) = f(\alpha_p|\alpha > 0)\pi_g/f(\alpha_p), \tag{24.12}$$

$$f(\alpha_p) = f(\alpha_p|\alpha = 0)\pi_0 + f(\alpha_p|\alpha > 0)\pi_g + f(\alpha_p|\alpha < 0)\pi_b. \tag{24.13}$$

The densities $f(\alpha_p|\alpha = 0)$, $f(\alpha_p|\alpha > 0)$, and $f(\alpha_p|\alpha < 0)$ can be modeled using the empirical conditional densities of the alpha estimates from the three subpopulations of funds, fit using a standard kernel density estimator for $f(\cdot)$, as described in Chen and Ferson (2017). Chen, Cliff, and Zhao (2017) and Harvey and Liu (2016) also use information from the cross section of funds to form posterior inferences for a particular fund. Earlier Bayesian analyses of fund return performance include Baks, Metrick, and Wachter (2001), Pástor and Stambaugh (2002), Avramov and Wermers (2006), Busse and Irvine (2006), Jones and Shanken (2005), and Kosowski, Naik, and Teo (2007).

The inference for a given fund in equation (24.12) reflects the estimated fractions of funds in each subpopulation. For example, as the prior probability π_g that there are good funds approaches zero, the posterior probability that a particular fund has a positive alpha, given its point estimate α_p, approaches zero. This captures the idea that if there are not many good funds, a particular fund with a positive alpha estimate was likely to have been lucky. The inference also reflects the locations of the groups through the likelihood that the fund's alpha estimate could be drawn from the subpopulation of good funds. If the likelihood $f(\alpha_p | \alpha > 0)$ is small, then the fund is less likely to have a positive alpha than if $f(\alpha_p | \alpha > 0)$ is large. This uses information about the position of the fund's alpha estimate relative to the other funds in a category.

Chen, Cliff, and Zhao (2017) refine the estimation of the group model and apply it to a sample of hedge fund returns for 1994–2011. Harvey and Liu (2016) further refine ML estimation of the group model for mutual funds, maximizing a joint likelihood that depends on both the alpha and beta parameters, which more fully incorporates the impact of estimation error in the alphas into the analysis. Unlike Chen and Ferson (2017), Harvey and Liu find that about 10% of the mutual funds have positive alphas. They attribute this to a substantial fraction of funds (almost 30% in their sample) that have positive alphas with small t-ratios, which might have been counted as zero-alpha funds in the previous methods. This seems like an area of research that has more room to run.

V INVESTMENT PERFORMANCE EVALUATION

25 Classical Investment Performance Evaluation

This chapter provides an overview of mutual funds and the classical measures of risk-adjusted investment portfolio performance. It borrows heavily from work with George Aragon (Aragon and Ferson 2006). Thanks, George! It also borrows from work with other authors (thanks, other authors!). I first describe the general logic that lies behind all classical investment performance measures and then define the common measures and discuss their properties. Then I look at empirical estimation of the measures on managed portfolios and review the empirical evidence.

25.1   The Classical Performance Measures

The main idea behind most classical measures of investment performance is to compare the return of a managed portfolio over some evaluation period to the return of a benchmark portfolio. The benchmark portfolio should represent a feasible investment alternative to the managed portfolio being evaluated. If the objective is to evaluate the investment ability of the portfolio manager or management company, as has typically been the case, the benchmark should represent an investment alternative that is equivalent, in all return-relevant aspects, to the managed portfolio being evaluated, except that it should not reflect the investment ability of the manager or firm. Aragon and Ferson (2006) call such a portfolio an "otherwise equivalent" (OE) benchmark portfolio. An OE benchmark portfolio requires that all portfolio characteristics that imply differences in required expected returns are the same for the fund being evaluated and for the benchmark.

To operationalize the OE benchmark, it is necessary to have some model for what aspects of a portfolio should lead to higher or lower expected returns. That is, some asset pricing model is required. While a few studies have claimed the ability to bypass the need for an asset pricing model, I would argue that some benchmark is always implied. For this reason, portfolio performance measures and asset pricing models are inextricably linked, and the development of portfolio performance measures in the literature mirrors the development of empirical asset pricing models.

Historically speaking, the earliest asset pricing models made relatively simple predictions about what it means for a benchmark to be OE to a managed portfolio. The CAPM of Sharpe (1964) suggests that an OE portfolio is a broadly diversified portfolio, combined with safe assets or cash, mixed to have the same market risk exposure, or beta coefficient, as the fund being evaluated. This is the logic of Jensen's (1968) alpha, which remains one of the most widely used measures of risk-adjusted performance. If alpha is positive, the manager earns an abnormal return relative to the alternative of holding the benchmark portfolio strategy.

Following the CAPM, empirical asset pricing in the 1970s began to explore models in which exposure to more than a single market risk factor determines expected returns. In the Merton (1973) and Long (1974) models, investors should not simply hold a broad market index and cash but should also invest in "hedge portfolios" for other economically relevant risks, like interest rate changes and commodity price inflation. Some investors may care more about inflation or interest rate changes than others, so they should adjust their portfolios in different ways to address these concerns. For example, an older investor whose anticipated lifetime of labor income is relatively short, and who is concerned about future cost-of-living risks, may want a portfolio that will pay out better if inflation accelerates in the future. A younger investor, less concerned about these issues, may not be satisfied with the lower expected return that inflation hedging implies. It follows from a model with additional hedge portfolios that an OE portfolio for performance evaluation should have the same exposures as the managed portfolio to be evaluated, not just with respect to the overall market but also with respect to the other relevant risk factors.

The papers of Long and Merton suggested interest rates and inflation as risk factors, but their models did not fully specify what all the hedge portfolios should be or how many there should be. APT specifies the factors only in a loose statistical sense. Thus, it is left to empirical research to identify the risk factors or hedge portfolios. Chen, Roll, and Ross (1986) empirically evaluate several likely economic factors, and Chen, Copeland, and Mayers (1987) use these factors in an evaluation of equity mutual funds. Connor and Korajczyk (1988) show how to extract statistical factors from stock returns in a fashion theoretically consistent with the APT (Ross 1977), as described in chapter 12. Connor and Korajczyk (1986) and Lehmann and Modest (1987) use these factors in models for mutual fund performance evaluation.

Current practice typically assumes that the OE portfolio is defined by the fund manager's investment style. Roughly, "style" refers to a subset of the investment universe in which a manager is constrained to operate, such as small capitalization stocks versus large stocks, or value versus growth firms. The style constraint may be a self-declared specialization, or it may be imposed on the manager by the firm. This leads to the idea of "style exposures," analogous to the risk exposures implied by the multiple-beta asset pricing models. In this approach, the OE portfolio has the same style exposures as the portfolio to be evaluated. The style-based approach is reflected prominently in academic studies following Fama and French (1996), such as Carhart (1997), who evaluates mutual funds using four factors derived from stocks' characteristics previously observed to be related to average returns (a market index plus size, book-to-market, and momentum factors).

Daniel et al. (1997) refine style-based performance measures by examining the actual holdings of mutual funds and measuring the characteristics of the stocks held by the fund. The characteristics include the market capitalization or size, a measure of value (the book-to-market ratio), and the return over the previous year. For a given fund, the OE portfolio is formed by matching the characteristics of the portfolio held by the fund with "passive" portfolios constructed to have matching characteristics.

In some cases, the style of a fund is captured using the returns of other managed portfolios in the same market sector. The OE portfolio is then a combination of a manager's peers' portfolios. With this approach, the measured performance is a zero-sum game, as the average performance measured in the peer group must be zero. This approach can make it easy to control for costs and risks, to the extent that the portfolio and its peers are similar in these dimensions. In such cases, the performance differential can be a relatively clean measure of value added. However, finding truly comparable peer funds may be a challenge.

25.1.1   The Role of Costs

Performance measures have often been crude in their treatment of investment costs and fees, but recent studies are moving toward a more careful consideration of costs. Mutual fund companies charge an expense ratio that gets deducted from the net assets of the funds and sometimes additional "load fees," such as those paid to selling brokers. Transactions fees have been charged to compensate the other shareholders for the costs of buying and selling the underlying assets when new money comes into or leaves a fund.

The fund expense ratio is the most commonly available measure of fund costs and has been used in empirical work to proxy for a variety of things. The expense ratio includes the management fee, which might account for 40–70% of the expense ratio (at the 25th–75th percentile for 1995–2014, according to data from Morningstar Direct. The expense ratio also includes custodial and operating costs as well as some of the expenses associated with marketing funds to investors. An example of the latter is the 12b-1 fee, named after a rule established by the SEC in 1980 that allowed these fees to be used to compensate fund sellers, such as brokers. The 12b-1 fee for a long time could be 0.25% and up to even 1% for some funds. According to the Investment Company Institute (2016), the average 12b-1 fee was 0.17% per year in 2000, diminishing to 0.08% in 2016. The decline in popularity of the 12b-1 fee has been associated with a Labor Department rule established in 2016 that required advisors on retirement accounts to serve the fiduciary duty of putting the clients' best interest before their own compensation. This rule remains under debate by politicians at this writing.

Funds' trading costs represent a drain from the net assets of the fund. Trading costs are not reported by funds and are hard to estimate precisely given currently available data, but they can be roughly approximated based on the reported turnover of the fund and estimates for the trading costs of the types of assets held by it. This calculation won't capture funds' abilities to negotiate trading fees or the income from securities lending, among other aspects of their trading activities.

A manager may be able to generate higher returns than an OE benchmark before costs, and yet after costs, investors' returns may be below the benchmark. If a fund can beat the OE benchmark on an after-cost basis, we say that the fund *adds value* for investors, to distinguish this situation from one where the manager has investment ability but either extracts the rents from this ability in the form of fees and expenses or dissipates it through trading costs. We say that a manager has *investment ability* if the managed portfolio outperforms the OE portfolio on a before-cost basis. Berk and Green (2004) assume that investors compete with their fund purchases and sales to find managers with investment ability, which is a scarce resource. They argue that managers with investment ability will attract flows of new money until decreasing returns to scale removes any value added for investors. Pástor, Stambaugh, and Taylor (2015) find empirical evidence of decreasing returns to scale in gross performance measures at the fund industry level.

25.1.2   The Sharpe Ratio

Perhaps the simplest risk-adjusted performance measure is the Sharpe ratio, used by Sharpe (1966) to evaluate mutual fund performance. We studied applications of the Sharpe ratio in portfolio management in chapters 8 and 9 and in tests of asset pricing models in chapters 21 and 22. The Sharpe ratio for a portfolio p is defined as

$$SR_p = E(r_p)/\sigma(r_p), \tag{25.1}$$

where $r_p \equiv R_p - R_f$ is the return of the portfolio p, net of the return to a safe asset or cash, R_f, and $\sigma(r_p)$ is the standard deviation or volatility of the portfolio excess return. Sharpe (1966) referred to the measure as the "reward-to-variability" ratio. The Sharpe ratio measures the degree to which a portfolio is able to yield a return in excess of the risk-free return to cash, per unit of risk. If we were to graph expected excess return against risk, measured by the volatility, or standard deviation of the excess return, the Sharpe ratio would be the slope of a line from the origin through the point for portfolio p. Any fixed portfolio that combines the fund with cash holdings would plot on the same line as the portfolio R_p itself. We saw how this works in chapter 8 on the classical mean variance analysis. As a performance measure, the Sharpe ratio of the fund is compared with the Sharpe ratio of the OE benchmark. If the ratio is higher for the fund, it performs better than the benchmark.

The Sharpe ratio is traditionally thought to make the most sense when applied to an investor's total portfolio, as opposed to any particular fund that represents only a portion of the investor's portfolio. The assumption is that what the investor really cares about is the volatility of his or her total portfolio, and various components of the portfolio combine via diversification, depending on the correlations among the various components. Chapter 28 presents a somewhat heretical, alternative view.

25.1.3 Jensen's Alpha

Alpha is perhaps the best-known of the classical measures of investment performance. Using the market portfolio of the CAPM to form the benchmark, Jensen (1968) advocated the original version, or Jensen's alpha. The most convenient way to define Jensen's alpha is as the intercept α_p^J in the following time series regression:

$$r_{p,t+1} = \alpha_p^J + \beta_p r_{m,t+1} + u_{pt+1}, \tag{25.2}$$

where $r_{p,t+1}$ is the return of the fund in excess of a short-term "cash" instrument. Using the fact that the expected value of the regression error is zero, we see how Jensen's alpha represents the difference between the expected return of the fund and that of its OE portfolio strategy:

$$\alpha_p^J = E\{r_p\} - \beta_p E\{r_m\} = E\{R_p\} - [\beta_p E\{R_m\} + (1 - \beta_p)R_f]. \tag{25.3}$$

The second line expresses the alpha explicitly in terms of the expected returns and portfolio weights that define the OE portfolio strategy. The OE portfolio has weight β_p in the market index with return R_m, and it has weight $(1 - \beta_p)$ in the risk-free asset or cash with return R_f.

Versions of Jensen's alpha remain the most popular performance measures in academic studies. However, the measure has some disadvantages. For example, alpha does not control for nonsystematic sources of risk that could matter to investors (e.g., Fama 1972).

25.1.4 The Treynor Ratio

One problem with Jensen's alpha arises when $\beta_p > 1$, as might occur for aggressive funds, such as small capitalization stock funds, aggressive growth funds, or technology funds. In these cases, the OE portfolio strategy involves a negative weight on the safe asset. This implies borrowing at the same rate as the risk-free return. In empirical practice, a short-term Treasury bill rate is often used to represent the safe asset's return, and few investors can borrow as cheaply as the US Treasury, so the implied OE benchmark strategy is typically not feasible.

Treynor (1965) proposed a measure that penalizes the portfolio in proportion to the amount of leverage employed:

$$T_p = E\{r_p\}/\beta_p. \tag{25.4}$$

Like the Sharpe ratio, a higher value of the Treynor ratio suggests better performance. Unlike the Sharpe ratio, the excess return is normalized relative to the systematic risk or beta, not the total risk or volatility. The Treynor ratio and Jensen's alpha are related by $T_p = \alpha_p^J/\beta_p + E\{r_m\}$. Thus, if $\alpha_p^J = 0$, $T_p = E\{r_m\}$. For a portfolio with a nonzero alpha, $T_p - E\{r_m\} = \alpha_p^J/\beta_p$ is invariant to the amount of borrowing or lending at the rate R_f.

25.1.5 The Treynor-Black Appraisal Ratio

Treynor and Black (1973) studied a situation where security selection ability implies expectations of nonzero Jensen's alphas for individual securities, or equivalently, the benchmark market return index is not mean variance efficient given those expectations. They derive the mean variance optimal portfolio in this case and show that the optimal deviations from the benchmark holdings for each security depend on the appraisal ratio:

$$AR_i = [\alpha_i/\sigma(u_i)]^2, \tag{25.5}$$

where $\sigma(u_i)$ is the standard deviation of the residual for security i in the factor model regression. This was introduced in chapter 8 in a discussion of modern portfolio management. Taking the deviations from benchmark for all the securities defines the "active portfolio," and Treynor and Black show how the active portfolio's appraisal ratio depends on the ratios of the individual stocks. They note that the portfolio's appraisal ratio, unlike Jensen's alpha, is invariant both to the amount of benchmark risk and the amount of leverage used in the portfolio. Market timing ability may lead to changes in the market risk and leverage used. Treynor and Black argue that this invariance recommends the appraisal ratio as a measure of a portfolio manager's skill in gathering and using information specific to individual securities.

25.1.6 The Merton-Henriksson Market Timing Measure

Classical models of market timing use convexity in the relation between the fund's return and the market return to indicate timing ability. In these models, the manager observes a private signal about the future performance of the market and adjusts the market exposure (or beta) of the portfolio at the beginning of the period. Successful timing implies higher betas when the market subsequently goes up, or lower betas when it goes down, leading to the convex relation. In the model of Merton and Henriksson (1981), the manager shifts the portfolio weights discretely, and the resulting convexity can be modeled with put or call options. The Merton-Henriksson market timing regression is

$$r_{pt+1} = a_p + b_p r_{mt+1} + \Lambda_p \text{Max}(r_{mt+1}, 0) + u_{t+1}. \tag{25.6}$$

The coefficient Λ_p measures the market timing ability. If $\Lambda_p = 0$, the regression reduces to the market model regression used to measure Jensen's alpha. Thus, under the null hypothesis that there is no timing ability, the intercept measures performance as in the CAPM. However, if there is timing ability and Λ_p is not zero, the interpretation of the intercept a_p is different. The OE portfolio for this model and the interpretation of the intercept is more complex, as discussed below.

25.1.7   The Treynor-Mazuy Market-Timing Measure

The Treynor and Mazuy (1966) market-timing model is a quadratic regression:

$$r_{pt+1} = a_p + b_p r_{mt+1} + \Lambda_p r_{mt+1}^2 + v_{t+1}. \tag{25.7}$$

Treynor and Mazuy argue that $\Lambda_p > 0$ indicates market-timing ability. When the market is up, the fund will be up by a disproportionate amount, and when the market is down, the fund will be down by a lesser amount. Admati et al. (1986) formalize the model, showing how it can be derived from a timer's optimal portfolio weight, assuming normal distributions and managers with exponential utility functions. They show that the timing coefficient Λ_p is proportional to the product of the manager's risk tolerance and the precision of the signal about the future market returns. Admati et al. (1986) show how to separate the effects of risk aversion and signal quality by estimating regression (25.7) together with a regression for the squared residuals of (25.7), on the market excess return. Coggin, Fabozzi, and Rahman (1993) implement this approach on equity mutual fund data.

25.1.8   Multibeta Models

Multibeta models arise when investors optimally hold combinations of a mean variance efficient portfolio and hedge portfolios for the other relevant risks. In this case, the OE portfolios are combinations of a mean variance efficient portfolio and the relevant hedge portfolios. The simplest performance measures implied by multibeta models are straighforward generalizations of Jensen's alpha, estimated as the intercept in a multiple regression:

$$r_p = \alpha_p^M + \sum_{j=1,\dots,K} \beta_{pj} r_j + v_p, \tag{25.8}$$

where the r_j, $j = 1, \dots, K$, are the excess returns on the K hedge portfolios ($j = 1$ can be a market index). The performance measure α_p^M is the difference between the average excess return of the fund and the OE portfolio:

$$\alpha_p^M = E\{r_p\} - \sum_j \beta_{pj} E\{r_j\} = E\{R_p\} - [\sum_j \beta_{pj} E\{R_j\} + (1 - \sum_j \beta_{pj}) R_f]. \tag{25.9}$$

Multibeta models using style-related factors, such as the Carhart (1997) model, have been popular funds in recent research.

25.1.9   Classical Holdings-Based Measures

In a returns-based measure, the return earned by the fund is compared with the return on the OE benchmark over the evaluation period. The OE benchmark is designed to control for risk, style, investment constraints, and other factors, and the manager who returns more than the benchmark has a positive alpha. One strength of returns-based methodologies is their minimal information requirements. One needs only returns on the managed portfolio and the OE benchmark. However, this minimal requirement ignores potentially useful information that is sometimes available in practice: the composition of the managed portfolio. In a weight-based or holdings-based measure, the manager's choices are directly analyzed for evidence of superior ability. A manager who increases the fund's exposure to a security or asset class before it performs well, or who anticipates and avoids losers, is seen to have investment ability.

Cornell (1979) was among the first to propose using portfolio weights to measure the performance of trading strategies. Copeland and Mayers (1982) modify Cornell's measure and use it to analyze Value Line rankings. Grinblatt and Titman (1993) develop weight-based measures of mutual fund performance. Applications to mutual funds include Grinblatt and Titman (1989b), Grinblatt, Titman, and Wermers (1995), Daniel et al. (1997), Zheng (1999), Ferson and Mo (2016), and others. The intuition behind weight-based performance measures can be motivated, following Grinblatt and Titman (1989a), with a single-period model where an investor maximizes the expected utility of terminal wealth:

$$\text{Max}_x \ E\{U(W_0[R_f + x'r]) \,|\, \Omega\}, \tag{25.10}$$

where R_f is the risk-free rate, r is the vector of risky asset returns in excess of the risk-free rate, W_0 is the initial wealth, x is the vector of portfolio weights on the risky assets, and Ω is the manager's private information signal, available at time 0. Private information, by definition, is correlated with the future excess returns, r. If returns are conditionally normal given Ω, and the investor has nonincreasing absolute risk aversion, the first- and second-order conditions for the maximization imply (see Grinblatt and Titman 1993) that

$$E\{x(\Omega)'[r - E(r)]\} = \text{Cov}\{x(\Omega)'r\} > 0, \tag{25.11}$$

where $x(\Omega)$ is the optimal weight vector, and $r - E(r)$ is the unexpected (or abnormal) returns, from the perspective of an observer without the signal. Equation (25.11) states that the sum of the covariances between the weights of a manager with private information, Ω, and the abnormal returns for the securities in a portfolio is positive. If the manager has no private information, then the covariance is zero.

From the definition of covariance, we can implement the inequality (25.11) by de-meaning the weights or the returns: $\text{Cov}\{x(\Omega)'r\} = E\{[x(\Omega) - E(x(\Omega))]r\} = E\{x(\Omega)[r - E(r)]\}$. Copeland and Mayers (1982) and Daniel et al. (1997) de-mean returns; Grinblatt and Titman (1993) de-mean the weights; and Ferson and Khang (2002) de-mean both.

Using a set of benchmark weights x_B to de-mean the weights, the holdings-based measure is $\text{Cov}\{x(\Omega)'r\} = E\{[x(\Omega) - x_B]'r\}$. The OE portfolio implied by the measure is $r_B = x_B'r$, and $E\{[x(\Omega) - x_B]'r\}$ is the average difference between the hypothetical returns of the fund and those of the benchmark portfolio. If the benchmark contains no special information about future returns, then $\text{Cov}\{x_B'r\} = 0$, and in theory, it does not change the measure to introduce the benchmark weights, although it will in practice if the mean of the weight difference is not zero (see Ferson and Wang 2017 for a recent analysis).

In weight-based measures, the return of the fund is a "hypothetical" return, since it is constructed using a snapshot of the fund's actual weights at the end of a period (usually at the end of a quarter or half-year) and the subsequent returns of the securities. These hypothetical returns reflect no trading in the quarter and no trading costs or management fees. Typically, the benchmark also pays no costs. Holdings-based performance measures are thus fairly clean measures of managerial ability before costs, but they ignore the effects of interim trading between the reporting dates, and they may be subject to biases, such as "window dressing," where managers move to more favorable-looking positions before holdings reporting dates.

Studies use fund holdings data for more than weight-based performance measures. A simple example is the "return gap" measure of Kacperczyk, Sialm, and Zheng (2008), defined as $r_p - x(\Omega)'r$, where r_p is the return reported by the fund. The return gap reflects trading during the quarter, trading costs, and funds' expenses. Cremers and Petajisto (2009) propose a measure of active portfolio management, the "active share": $(1/2)|x - x_B|'1$, which is the mean absolute difference between the fund's portfolio weights and those of its benchmark. Doshi, Elkamhi, and Simutin (2015) propose a refinement called "active weight" that simply compares the fund's holdings to that of a value-weighted index; they find that it is simpler and works better than active share. Recent research has brought in the wide range of available data on the characteristics of the securities held by funds, allowing many creative analyses. This is a leading edge in fund performance research.

Daniel et al. (1997) combine holdings, returns, and characteristics of the stocks held by the fund. For a given fund, a benchmark is formed by matching the characteristics of each stock i in the portfolio held by the fund with benchmark portfolios R_t^{bi} constructed *for each stock* to have matching characteristics. This is similar to the matching-firm approach used in some corporate finance studies. Specifically, each security in the fund's portfolio is assigned to one of 125 groups, depending on its size, book-to-market ratio, and lagged return. The benchmark for each stock is a value-weighted portfolio of the stocks in the characteristic group. Daniel et al. (1997) start with the Grinblatt and Titman (1993) performance measure, introducing a set of benchmark weights equal to a fund's actual holdings reported k periods before: $x_{i,t-k}$, to obtain:

$$DGTW_{t+1} = \sum_i x_{it}(R_{i,t+1} - R_{t+1}^{bi}) + \sum_i (x_{it}R_{t+1}^{bi} - x_{i,t-k}R_{t+1}^{bi(t-k)}) + \sum_i x_{i,t-k}R_{t+1}^{bi(t-k)}, \qquad (25.12)$$

where $R_{t+1}^{bi(t-k)}$ is the benchmark return associated with security i at time $t-k$. The assumption is that matching the characteristics of each stock defines the OE benchmark. The first term in equation (25.12) is interpreted as "selectivity," the second term as "characteristic timing," and the third as the return attributed to style exposure. While the sum of the components is equal to—and thus has the same theoretical justification as—the Grinblatt and Titman (1993) measure, the individual terms are ad hoc, and no theoretical justification for their interpretation is known.

25.2   Properties of the Classical Measures

25.2.1   Selectivity versus Timing

In the model of Merton and Henriksson (1981), the OE portfolio holds a weight of b_p in the market index returning R_m, a weight of $\Lambda_p P_0$ in a call option with beginning-of-period price P_0 and return $\{\text{Max}(R_m - R_f, 0)/P_0 - 1\}$ at the end of the period, and a weight of $(1 - b_p - \Lambda_p P_0)$ in the safe asset returning R_f. Call option is a claim on the relative value of the market index, $V_m/V_0 = 1 + R_m$, with strike price equal to $1 + R_f$. Given a measure of the option price P_0, it is possible to construct the fund's return in excess of its OE portfolio.

The excess return of a fund over the OE portfolio is not the same as the intercept in the timing model regression, so don't make the mistake that some authors have made of calling the intercept timing "adjusted abnormal performance" or "timing-adjusted selectivity." Unless the manager has perfect clairvoyance, the intercept reflects the imperfect timing ability and the effects of security selection, and the two are hard to disentangle. For example, a manager with timing ability but poor security selection ability may be hard to distinguish from a manager with no ability who buys options at the market price (e.g., Jagannathan and Korajczyk 1986; Glosten and Jagannathan 1994). Indeed, without an estimate of the option price P_0, the intercept in the timing regression has no clean interpretation in general. In the Treynor-Mazuy or Merton-Henriksson models, the intercept does not capture the return in excess of an OE portfolio, because r_m^2 or the option payoff are not portfolio returns.

25.2.2   Differential Information

The existence of investment ability presumes there is differential information: The portfolio manager usually needs to know more about future security returns than the average investor. Mayers and Rice (1979) study the question of whether a manager who knows more than "the market" as a whole would deliver a positive Jensen's alpha. They assume that the manager with superior information has a small enough amount of capital that he will have no effect on market prices. Thus, the information reflected in the market as a whole does not include the superior information. This is a strong assumption, perhaps more tenable at the time of their paper than today, given that institutional trading represents a larger fraction of trading volumes in recent years. Mayers and Rice provide assumptions under which the alpha of a manager with superior information is positive, as assessed by the investor without the information. They show that the manager, who better knows which states of the world are more likely, will invest more money in securities that pay off in the more likely states, while investing smaller amounts in the less likely states, thereby generating a higher average return than would be expected given the risk as perceived by the uninformed investor. However, Mayers and Rice were not able to show that a positive alpha would be found for an informed manager under general assumptions.

Verrecchia (1980) shows that a positive alpha could be expected in the model of Mayers and Rice, if the fund manager maximized a utility function with constant relative or absolute risk aversion, but he also presented an example where a manager with a quadratic utility function would not deliver a positive alpha. A manager with quadratic utility will optimally choose a mean variance efficient strategy, conditional on his information. Dybvig and Ross (1985a) assume that the informed manager uses her information to form a mean variance efficient portfolio (conditionally efficient). They show that when the manager's returns are viewed by an investor without the information, the returns would not generally appear to be efficient (unconditionally mean variance inefficient). Hansen and Richard (1987) showed in general that conditionally mean variance efficient portfolios need not be unconditionally mean variance efficient, as we discussed in chapter 9.

If a manager with superior information knows that he will be evaluated based on the unconditional mean and variance of the portfolio return, he may be induced to form a portfolio that uses his information to maximize the unconditional mean relative to the unconditional variance.

While there are many studies of differential information, the literature has not fully resolved the effects of differential information in the problem of measuring portfolio performance. Given an OE portfolio that reflects a sensible alternative strategy to the portfolio being evaluated and controls for all the managed portfolio's return-relevant characteristics, the alpha relative to the OE portfolio captures the extra return that the managed portfolio delivers. At a deeper level, however, it is disconcerting to think that a manager with investment ability in the form of superior information could be using her information to maximize essentially the same objective as the investor/client and yet be seen as delivering an inferior return from the client's perspective. There is more work to be done in this area.

25.2.3   Nonlinear Payoffs

When managers can trade assets whose payoffs are nonlinearly related to the benchmark return, this may generate convexity or concavity in the fund returns. For example, Jagannathan and Korajczyk (1986) argue that a fund holding call options can appear to be a good market timer, based on the convexity of the fund's returns. Managers trading options may also affect the fund's Sharpe ratio and other measures of performance. For example, a strategy of selling out-of-the-money put options has little risk given small stock price changes, but it has exaggerated risk given large stock price movements. Goetzmann et al. (2007) solve for the maximal Sharpe ratio obtainable from trading options on the market index and find that such strategies can generate large Sharpe ratios. Strategies that involve taking on default risk, liquidity risk, or other forms of catastrophic risk may also generate upwardly biased Sharpe ratios.

Nonlinearities may be more important for some kinds of investors and strategies than for others. Some mutual funds face explicit constraints on derivatives use, whereas private investment firms such as hedge funds are not typically subject to such constraints. The problem of nonlinear payoffs for performance evaluation is more important for portfolios that have broad investment mandates, as these mandates afford more opportunity to trade dynamically across asset classes. Dynamic trading is akin to the use of options or other derivatives in its ability to generate nonlinear payoffs. For example, in many option pricing models, a derivative can be replicated by dynamic trading in the underlying assets.

There are two cases to consider when addressing the problem of nonlinear payoffs in a managed portfolio. In the first case, it may be possible to replicate the nonlinearity in the funds' returns by trading other assets whose returns are measured at the same frequency as the funds' returns. In this case, the OE portfolio includes the replicating securities, and the funds' ability can be measured as described above.

In the second case, it may not be possible to replicate the funds' nonlinearities with other security returns measured at the same frequency. The market may be "incomplete" with respect to the funds' payoffs: The patterns generated by the fund are just not available elsewhere. In this situation, it is necessary to use an asset pricing model based on an assumption about investors' utility functions. For example, Leland (1999) argues that a model featuring power utility is useful in such a context, as the nonlinearity of the marginal utility function captures the prices of nonlinear portfolio payoffs. Possibly, the funds' nonlinearities can be replicated by trading at a higher frequency than is used to measure the funds' returns. This situation leads to the problem of interim trading bias.

25.2.4   Interim Trading Bias

A potential cause of nonlinearity is "interim trading," which refers to a situation where fund managers trade or cash flows in and out of the fund during the period over which returns on the fund are measured. With monthly fund return data, a common choice in the literature, interim trading definitely occurs, as flows and trades occur typically each day. The problems posed by interim trading were perhaps first studied by Fama (1972), and later by Goetzmann, Ingersoll, and Ivkovic (2000), Ferson and Khang (2002), and Ferson, Henry, and Kisgen (2006).

Consider an example where returns are measured over two periods, but the manager of a balanced fund trades each period. The manager has neutral performance, but the portfolio weights for the second period can react to public information at the middle date. By assumption, merely reacting to public information at the middle date does not require superior ability.

Suppose that a terrorist event at the middle date increases market volatility in the second period, and the manager responds by moving part of the portfolio into cash at the middle date. The higher volatility might indicate that the expected return-to-risk trade-off for stocks has become less favorable for the second period, so the optimal portfolio is now more conservative. If only two-period returns can be measured and evaluated, the manager's strategy would appear to have partially anticipated the higher volatility. The fund's two-period market exposure would reflect some average of the before- and after-event positions.

Over two periods, the portfolio would appear to partially "time" the volatility-increasing event because of the move into cash. A returns-based measure over the two periods can confuse this behavior with superior performance. Goetzmann, Ingersoll, and Ivkovic (2000) address interim trading bias by simulating the multiperiod returns generated by the option to trade between return observation dates. Ferson, Henry, and Kisgen (2006) use continuous-time models to address the problem, and this approach is outlined in chapter 27.

A weight-based measure can avoid interim trading bias by examining the conditional covariance between the manager's weights at the beginning of the first period and the subsequent two-period returns. The ability of the manager to trade at the intervening period does not enter into the measure, and thus interim trading creates no bias. Of course, managers may engage in interim trading based on superior information to enhance performance, and a weight-based measure will also not record these interim trading effects. Thus, the cost of using a weight-based measure to avoid bias is a potential loss of power.

25.2.5 Accounting for Costs

As just argued, it is important to recognize the role of costs in any comparison of a managed portfolio return with that of a performance benchmark. There are fund-level costs and investor-level costs. The fund's costs of investing include direct transactions costs, fees paid for portfolio management services and to meet regulatory requirements, sales and marketing expenses. Funds may earn income from securities lending and other interactions with the brokerage community. Investor-level costs may include advisory service, income taxes, other information or time costs, or the costs of behavioral biases. Some funds may charge additional sales charges (load fees), or redemption fees on purchase or sale of the funds' shares. Funds purchased through a broker may involve additional brokerage charges.

Mutual fund returns, in contrast, are measured net of all the expenses summarized in the funds' expense ratio and also the trading costs incurred by the fund but not reported in the expense ratio. The measured returns for other types of funds, such as pension funds and hedge funds, may reflect or exclude these and other costs. Measuring the managed portfolio's returns and the performance benchmark returns on a cost-equivalent basis can get complicated.

The most common situation in academic empirical studies using returns-based measures has the OE benchmark paying no costs, while the measured return of the mutual fund being evaluated is net of the funds' expenses and trading costs. Sometimes the expense ratio is added back to capture part of the costs, and studies have also tried to approximate fund trading costs. Investors' costs is another complex issue.

With weight-based measures, hypothetical returns are constructed by applying the portfolio weights to measured returns on the securities held by the funds. The result is a measure of hypothetical before-cost fund returns, which may be more comparable with an OE benchmark that pays no costs. Thus, weight-based measures seem attractive for measuring investment ability, if not for capturing value added for investors.

It would seem that a logical next step for research on fund performance measures is to more carefully take account of the full range of costs and taxes associated with investing in funds. There are asset pricing models that make predictions about the return-relevant measures of transactions costs and taxes. However, the problem is that the incidence of many of these costs is likely to be different for different investors. For example, a pension plan pays no income taxes on the dividends or capital gains generated by a portfolio, so the manager and the plan client may care little about the form in which the gains are earned. A college endowment fund (until the most recent tax law change) typically pays no tax, but if constrained to spend out of the income component of return, the beneficiaries of the endowment may have a preference for the form of the return. An individual investor may be taxed more favorably on capital gains than on dividends, and the relative tax cost may depend on the investor's total income profile. Thus, for a given fund, the OE portfolio for one investor may not be the same as that for another investor, and different investors may view the performance of the same fund in different ways. Performance measures that reflect costs could be constructed for a range of hypothetical investors with different OE portfolios.

25.3   The Evidence for Professionally Managed Portfolios
Using Classical Measures

This section reviews the evidence on fund performance based on the classical measures, focusing mainly on the evidence prior to the advent of the Center for Research in Security Prices (CRSP) mutual fund database in the mid-1990s. The modern evidence and the evidence from conditional performance evaluation is discussed later.

25.3.1   Data Issues

Historical returns of mutual funds are made available through the NSAR forms that are filed every semiannual period in compliance with the Investment Company Act of 1940. Form NSAR requires each registered investment company to report the month-end net asset values of fund shares for that semiannual period. Net asset values reflect the value of a fund share to investors after expenses resulting from management fee and trading costs, but before the deduction of any load fees or the payment of personal taxes. Net returns therefore reflect the change in net asset value plus any dividend or capital gain distributions over the performance period.

Early studies of managed portfolio performance faced a variety of data-related challenges. Many studies used mutual fund return data that were hand-collected from Wiesenberger's *Investment Companies* series, Moody's Investor Services manuals, and other print media. The methods for computing reported returns were not standardized. Early studies of managed portfolio performance were therefore restricted to small samples over short sample periods. Often, data on funds that had ceased to operate were not available, raising the issue of survivor selection bias in the available returns data. When funds entered the databases, they often brought a track record of historical returns with them. If these are included in a study, a potential back-fill bias is created.

Two important data advances occurred in the mid-1980s. First, historical mutual fund returns data became available in an electronic format provided by Morningstar, and later through the CRSP Survivorship-Bias-Free Mutual Fund Database—currently the two most widely used mutual fund databases. Elton, Gruber, and Blake (2001) compare the CRSP data with Morningstar data and find differences in alpha on the order of −4.5 basis points (bp) and −8.7 bp per year for large and small stocks, respectively, over the period 1979–1998. However, they find that this difference has become smaller over time. Second, the semiannual portfolio holdings reported in Form NSAR became electronically available through a private data vendor, CDA Investment Technologies, Inc. Since 1985, mutual funds have been required to report their stockholdings to the SEC on Form NSAR. In addition to the semiannual reports, most funds voluntarily disclose their holdings more frequently, on a quarterly basis, to the CDA database. Recently, many funds voluntarily report holdings monthly to Morningstar. An explosion in the number of mutual funds studies using return-based measures and studies using weight-based measures occurred as researchers combined the machine-readable data with more efficient computing power.

25.3.2 Survivorship

Survivorship creates potential problems affecting both the average levels of performance and the apparent persistence in performance. One obvious reason for a manager to leave a database is poor performance. To the extent that funds die because of poor performance, the measured performance of the surviving funds is biased upward. If the database includes only those funds that survive for a length of time, the returns of the surviving funds are likely to be higher than those of the population of funds. Elton, Gruber, and Blake (1996) estimate an average survival bias of 0.7–0.9% per year in older mutual fund return data (see also Brown and Goetzmann 1995; Malkiel 1995; Gruber 1996; Carhart 1997). This suggests that the inferences about the average performance of mutual funds in earlier work may have been too optimistic.

Brown et al. (1997), Brown and Goetzmann (1995), and Hendricks, Patel, and Zeck-hauser (1993) consider the effects of survivorship on performance persistence under the simplifying assumptions that the expected returns of all managers are the same, but there are differences in variances, and that managers leave the database when their realized returns are relatively low. Under these assumptions, survivorship is likely to induce a spurious J-shaped relation between future and past relative returns. In particular, past poor performers that survived are likely to have reversed their performance later. Carhart et al. (2002) show that the effects of survival selection on persistence in performance depends on the birth and death process and can be complex.

Linnainmaa (2013) describes a "reverse" survivorship bias in databases such as the CRSP mutual fund data, which includes those funds that did not survive until the end of the sample. The bias is reverse, because it leads to estimates of average returns that are too low compared with the true population mean. Suppose that unlucky funds with realized returns that hit some lower bound leave the database, but there is no upper bound that triggers a fund's death when realized returns are high. The luckier funds live longer than the unlucky funds. Then the returns of the dead funds in the sample are downward biased, because they are selected based on low realized returns. To illustrate, suppose that all funds' returns are a constant plus random mean-zero noise: $r_t = \mu + \varepsilon_t$, and that funds die at some random time T that depends on their realized average returns. The measured average returns are then equal to $(1/T)\sum r_t = \mu + (1/T)\sum \varepsilon_t$, with expected value $\mu + \mathrm{E}\{(1/T)\sum \varepsilon_t\} = \mu + \mathrm{Cov}\{(1/T), \sum \varepsilon_t\} < \mu$. Linnainmaa's simulations suggest that the downward bias in average returns from a purportedly "survivorship bias-free" database could be economically significant.

25.3.3   Early Evidence

Sharpe (1966) studies the performance of 34 equity style mutual funds, using annual returns for 1954–1963, computing both the Sharpe and Treynor ratios to measure performance. The two performance measures have a rank correlation in his sample of 97.4%. Measured net of expense ratios, the funds perform below the Dow Jones 30 stock index. The average Sharpe ratio of the funds is 0.633, while that of the Dow Jones is 0.677, and only 11 of the 34 funds have Sharpe ratios above the index. Adding back the expense ratio, however, 19 of the 34 funds beat the Dow Jones. Sharpe concludes that while some funds may be choosing portfolios with a better risk-return profile than the index, investors are not realizing better returns after costs. He finds that expense ratios in 1953 have some predictive power for subsequent performance; their rank correlation with the future Sharpe ratios is −50%.

Jensen (1968) studies the performance of 115 open end mutual funds over the period 1945–1964. Data are obtained from Wiesenberger's *Investment Companies*. He finds an average alpha of −1.1% per year relative to the S&P 500 index. The distribution of the alphas across funds is skewed to the left, with 66% of funds having a negative alpha. Using data on expense ratios, Jensen (1968) calculates negative average gross performance of −0.4% per year. Jensen concludes that, in aggregate, the investment ability of mutual fund managers is not great enough to recover these costs.

Carlson (1970) studies the performance of common stock, balanced, and income style mutual fund portfolios over the period 1948–1967. He finds lower Sharpe ratios for all mutual fund portfolios than for the S&P 500 and the NYSE Composite indexes. However, he also finds that the results vary, depending on mutual fund style and choice of market index. Specifically, the net Sharpe ratios of common stock and balanced fund portfolios exceed that of the Dow Jones Industrial Average, although income mutual fund portfolios underperform all market indexes. Carlson also calculates Jensen's alpha for individual funds in his sample. In contrast to Jensen (1968), Carlson (1970) finds that the average net alpha is a positive 0.6% per year, and that the distribution of alpha across funds is positively skewed. He concludes that care must be exercised when generalizing from the performance results of a specific mutual fund group and a specific market index.

Cornell (1979) develops a weight-based performance measure where the expected return on the OE benchmark is the portfolio weighted-average of expected returns on the underlying securities. A security's expected return is calculated as the mean realized return over an estimation window. Copeland and Mayers (1982) apply this methodology to evaluate the Value Line Investment Survey for 1965–1978. They construct five stock portfolios based on the survey's ranks of individual stocks and find negative abnormal returns for the bottom quintile portfolio over the 26 weeks following the publication of the survey. This finding is consistent with the idea that the stocks assigned the lowest ranking correspond to negative private information by the Value Line analysts. However, their estimates are only significant when expected returns on the portfolio securities are calculated based on the market model.

25.3.3.1 Selectivity and Timing

Kon (1983) studies the timing performance of 37 mutual funds over the period 1960–1976. He estimates a two- and three-stage regime switching model to calculate the dynamics of a fund's beta relative to the value-weighted CRSP market index. Kon's measure of timing performance is equal to the sample covariance between the fund's beta and the market return, a measure advocated by Fama (1972). Kon finds no evidence of significant timing performance. In fact, the majority (23) of the 37 sample funds display negative timing performance. He concludes that the allocation of effort and expense into market timing activities should be reevaluated.

Henriksson (1984) studies the timing performance of 116 open-end mutual funds over the period 1968–1980. He employs the parametric test developed by Merton and Henriksson (1981) discussed earlier in this chapter. Henriksson finds no evidence of market timing ability. Only three of the sample funds display significant positive timing ability, whereas 62% of the funds have negative timing coefficients. Henriksson also finds evidence of a "disturbing" negative relation between estimates of timing ability and selection ability. For example, 49 of the 59 funds with positive alpha also had a negative timing coefficient. He concludes that the funds that earn superior returns from stock selection appear to have negative market timing ability.

Chang and Lewellen (1984) estimate the Merton and Henriksson (1981) market-timing model for a sample of 67 mutual funds over the period 1971–1979. They find little evidence of timing ability and, if anything, funds display larger portfolio betas in down markets than in up markets. They also find evidence of a negative relation between stock selection and timing skills. They conclude that the evidence supports the general conclusion that mutual funds do not display market timing ability.

25.3.3.2 Persistence

Some studies find that the performance of mutual funds may be persistent. Persistence means that funds that perform relatively well or poorly in the past may be expected to do so again in the future. Obviously, persistent abnormal performance, if it exists, should be of practical interest to fund investors.

Sharpe (1966) conducts one of the first examinations of the question of persistence in fund performance. He finds that the standard deviations of fund returns have some persistence. The rank correlation between the 1944–1953 and the 1954–1963 periods is 53%. However, the Sharpe ratios have less persistence. Their rank correlation is only 36%, with a *t*-ratio of 1.88. Studies by Jensen (1969), Carlson (1970), Ippolito (1992), Grinblatt and Titman (1992), Hendricks, Patel, and Zeckhauser (1993), Goetzmann and Ibbotson (1994), Shukla and Trzcinka (1994), Brown and Goetzmann (1995), Malkiel (1995), Gruber (1996), Elton, Gruber, and Blake (1996), and Carhart (1997) all find some evidence of persistence in mutual fund performance.

One form of persistence is short-term continuation, or "momentum," in funds' relative returns. The presence of such momentum would suggest that investors could obtain better returns by purchasing those funds that have recently performed well, and by avoiding those that have recently performed poorly. Some of the above-cited studies find evidence of continuation over several months after portfolios of high-relative-return funds are formed. However, much of this continuation seems to be explained by funds' holdings of momentum stocks (e.g., Grinblatt, Titman, and Wermers 1995; Carhart 1997).

Much of the empirical evidence on performance persistence for mutual funds suggests a positive relation between future and past performance, concentrated in the poorly performing funds. Poor performance may be persistent. This is not as would be expected under the hypothesis that persistence is a spurious result, at least in a simple model of survivorship bias. Brown and Goetzmann (1995), Elton, Gruber, and Blake (1996), and Carhart (1997) find similar patterns of persistence in samples of mutual funds designed to avoid survivorship bias. To the extent that fund performance persists, it seems to be mainly the poor performers.

The evidence on persistence for pension fund managers is relatively sparse. Christopherson and Turner (1991) study pension managers and conclude that alpha at one time is not predictable from alpha at a previous time. Lakonishok, Shleifer, and Vishny (1992) find some persistence of the relative returns of pension funds for 2–3-year investment horizons. Christopherson, Ferson, and Glassman (1998) study persistence with conditional models, as described below. Similar to the studies of mutual funds, they find some evidence of persistence, but it is concentrated in the poorly performing funds.

25.3.3.3 Fund Flows and Performance

Ippolito (1989) first observed that funds with high past performance tended to attract relatively more new investment money over the next year. Funds whose past returns were positive had a stronger reaction in flows to their past CAPM alphas than did those with negative returns. Further evidence that investor flows chase recent high fund returns, and that the relation of flows to performance was convex, was found by Chevalier and Ellison (1997), Sirri and Tufano (1998), and others. These and subsequent studies examine additional factors that moderate the flow-performance relation. Uncertainty about the fund performance, for example, leads to a muted flow response (Sirri and Tufano 1998; Starks and Sun 2016). Kim (2017) finds that the shape of the flow performance relation varies, depending on market volatility and the dispersion in fund returns.

Del Guercio and Tkac (2002) find that mutual fund investors pay more attention to simple measures of relative return than to more complex measures like alpha, when directing their new money flows. Berk and van Binsbergen (2015) argue that we can learn which asset pricing model is really used by investors if we find the model whose alphas elicit the largest flow response. They look at several models and conclude that the traditional CAPM is as good as any of them. Barber, Huang, and Odean (2016) study more fully the factors that mutual fund investor flows respond to, and again the CAPM dominates the various models.

If past performance does not actually predict future performance, then it would seem that mutual fund investors as a group are behaving strangely in chasing good past performance with fund flows. Gruber (1996) forms portfolios of mutual funds that are weighted according to their recent new money flows. He finds that the new money earns higher average returns than the old money invested in equity style funds. This "smart money" effect is subsequently confirmed by Zheng (1999). However, Sapp and Tiwari (2004) find that the smart money effect is mainly explained by momentum, in the sense that the effect disappears if the factor model controls for a momentum factor. Frazzini and Lamont (2008) find that flows follow stocks that subsequently underperform on a risk-adjusted basis, attributing the effect to investor sentiment. Solomon, Soltes, and Sosyura (2014) find that more new money flows in response to past performance when there has been more media attention.

Huang, Wei, and Yan (2007) present a finely articulated model of the relation between mutual fund flows and performance. They consider that investors bear an information cost to investing in funds, which impedes the flow response to lower levels of performance and leads to a greater response of flows to higher levels of performance. Jiao, Massa, and Zhang (2016) argue that short-sales restrictions can contribute to the nonlinear flow performance relation. Ferson and Lin (2014) argue that heterogeneity among a fund's investors should moderate the flow response to the fund's measured performance.

25.3.3.4 Tournaments and Risk Shifting

Studies of the relation between flows and performance find an interesting relation between past performance and subsequent new money flows. Funds with the highest returns on average realize the largest subsequent inflows of new money, while funds with performance below the median do not experience withdrawals of a similar magnitude. This nonlinearity creates an incentive for funds akin to that of a call option, even if the manager's compensation is a fixed fraction of the assets under management. Brown, Harlow, and Starks (1996) argue that managers may respond to this incentive with a risk-shifting strategy, on the assumption that performance evaluation occurs at annual periods. They find that those funds that are performing relatively poorly near the middle of the year seem to increase their risk in the last 5 months of the year, as if to maximize the value of the optionlike payoff caused by fund flows. Funds whose annual performance to date is relatively high near the middle of the year seem to lower their risk, as if to "lock in" their position in the performance tournament.

Koski and Pontiff (1999) examine the use of derivatives by mutual funds, motivated in part by the idea that derivatives may be a cost-effective tool for risk management. They find evidence of risk shifting similar to what Brown, Harlow, and Starks (1996) report, but little evidence that the risk shifting is related to the use of derivatives. Busse (2001) reexamines the evidence for risk shifting using daily data and argues that the evidence for this behavior in the earlier studies is exaggerated by a bias, related to return autocorrelations, in estimating the standard errors of monthly returns. Using daily data, Busse finds no evidence for risk-shifting behavior. Goriaev, Nijman, and Werker (2005) also find that the evidence for risk shifting is not robust. There are many more recent studies, some of which find evidence of risk shifting. For example, Ferson and Mo (2016) find more risk shifting when funds face greater convexity in their incentives. Kim (2017) finds that risk-shifting behavior varies with some of the same variables that are correlated with the shape of the flow performance relation.

25.3.4 Evidence from Holdings-Based Measures

Grinblatt and Titman (1989b) are the first to apply weight-based performance measures, examining the quarterly holdings of mutual funds. Grinblatt and Titman (1993) take the fund's OE portfolio in a given quarter to be the fund's portfolio, as defined by its weights in the previous quarter. Thus, the underlying model assumes that a manager with no information holds fixed portfolio weights. Fund performance is measured as the average difference in raw returns, over the subsequent quarter, between the fund and this OE portfolio. They estimate abnormal before-cost performance to be a statistically significant positive 2% per year in aggregate. Aggressive growth funds exhibit the strongest performance at 3.4%. Since neither the funds' hypothetical returns nor the OE portfolio returns pay any costs or management fees, the weight-based measure speaks directly to investment ability before costs. Thus, this evidence suggests that fund managers in aggregate do have investment ability.

Daniel et al. (1997) study the holdings of most equity mutual funds that existed during any quarter in the period 1975–1994. For each fund in a given quarter, they define the OE portfolio based on the characteristics of the securities held. Specifically, each security in the fund's portfolio is assigned to one of 125 characteristic groups, depending upon its size, book-to-market ratio, and lagged return, measurable with respect to the beginning of the quarter. They construct passive (value-weighted) portfolios across all NYSE, AMEX, and NASDAQ stocks for each characteristic group. The return on the OE portfolio in a given quarter is the summation, across all securities in the fund's portfolio, of fund portfolio weights times the return on the corresponding matched portfolio for each security. They find that the average fund in their sample has a significantly positive performance of 77 basis points per year. Daniel et al. (1997) conclude that the observed performance is about the same size as the typical management fee. Therefore, the average fund is not expected to deliver value added to investors after costs.

26 Conditional Performance Evaluation

Traditional measures of risk-adjusted performance for mutual funds compare the average return of a fund with an otherwise equivalent benchmark designed to control for the fund's average risk. For example, Jensen's (1968) alpha is the difference between the return of a fund and a portfolio constructed from a market index and cash, typically using weights that are fixed over time. The portfolio has the same average market exposure, or unconditional beta risk, as the fund. This classical approach is unconditional with respect to changes in the state of the economy, but there is pretty strong evidence that the state of the economy does change over time. Conditional performance evaluation (CPE) is designed to allow time-varying risks and performance that depend on public information.

26.1 Overview

In the CPE approach, the state of the economy is measured using predetermined public information variables. For example, provided that the estimation period covers both bull and bear markets, we can estimate the conditional risk and performance in each type of market. This way, knowing that we are now in a bull state of the market, for example, we can estimate the fund's expected performance given a bull state.

Problems associated with variation over time in mutual fund risks and market returns have long been recognized (e.g., Jensen 1972; Grant 1977), but CPE draws an important distinction between variation that can be tracked with public information and variation due to private information on the state of the economy. CPE takes the view that a managed portfolio strategy that can be replicated using readily available public information should not be judged as having superior performance. For example, in a conditional approach, a mechanical market timing rule using lagged interest rate data is not a strategy that requires investment ability. Only managers who correctly use more information than is generally publicly available are considered to have potentially superior investment ability. CPE is therefore consistent with versions of semistrong-form market efficiency as described by Fama (1970). By choosing the lagged variables, it is possible to set the hurdle for superior ability at any desired level of information.

In addition to the lagged state variables, CPE, like any performance evaluation, requires a choice of benchmark portfolios. The first measures used a broad equity index, motivated by the CAPM. Ferson and Schadt (1996) used a market index and a multifactor benchmark for CPE. Current practice is more likely to use a benchmark representing the fund manager's investment style.

26.2   Motivating Example

The appeal of CPE can be illustrated with the following highly stylized numerical example. Assume that there are two equally likely states of the market as reflected in investors' expectations; say, a "Bull" state and a "Bear" state. In a Bull market, assume that the expected return of the S&P 500 is 20%, and in a Bear market, it is 10%. (This differs from the conventional definition of a bear market, which some consider to be a 20% decline off a previous high.) The risk-free return to cash is 5%. Assume that all investors share these views—the current state of expected market returns is common knowledge. In this case, assuming an efficient market, an investment strategy using the current state as its only information will not yield abnormal returns.

Of course, this example is far too stylized to be realistic. It really needs a background risk factor or state variables to avoid being trivial. There should be uncertainty in both the Bull and Bear market states. But I hope you can suspend disbelief for the illustration. Imagine a mutual fund that holds the S&P 500 in a Bull market and holds cash in a Bear market, based on public information. Consider the performance of this fund based on CPE and the CAPM. Conditional on a Bull market, the beta of the fund is 1.0, the fund's expected return is 20%, equal to the S&P 500, and the fund's conditional alpha is zero. Conditional on a Bear market, the fund's beta is 0.0, the expected return of the fund is the risk-free return, 5%, and the conditional alpha is, again, zero. CPE correctly reports an alpha of zero in each state.

By contrast, an unconditional approach to performance evaluation incorrectly reports a nonzero alpha for our hypothetical mutual fund. Without conditioning on the state, the returns of this fund would seem to be highly sensitive to the market return, and the unconditional beta of the fund is 1.5. The calculation is as follows, where M is the market return and F is the fund return. The unconditional beta is $\mathrm{Cov(F, M)/Var(M)}$. The numerator is $\mathrm{Cov(F, M)} = \mathrm{E}\{(F - E(F))(M - E(M))|\mathrm{Bull}\} \times \mathrm{Prob(Bull)} + \mathrm{E}\{(F - E(F))(M - E(M))|\mathrm{Bear}\} \times \mathrm{Prob(Bear)} = \{(.20 - .125)(.20 - .15)\} \times .5 + \{(.05 - .125)(.10 - .15)\} \times .5 = 0.00375$. The denominator is $\mathrm{Var(M)} = \mathrm{E}\{(M - E(M))^2|\mathrm{Bull}\} \times \mathrm{Prob(Bull)} + \mathrm{E}\{(M - E(M))^2|\mathrm{Bear}\} \times \mathrm{Prob(Bear)} = \{(.20 - .15)^2\} \times .5 + \{(.10 - .15)^2\} \times .5 = 0.0025$. The beta is therefore $.00375/.0025 = 1.5$. Note that the unconditional beta is not the same as the average conditional beta, because the latter is 0.5 in this example. The unconditional expected return of the fund is $0.5(.20) + 0.5(.05) = 0.125$. The unconditional expected return of the S&P 500 is

$.5(.20) + .5(.10) = 0.15$, and the unconditional alpha of the fund is therefore $(.125 - .05) - 1.5(.15 - .05) = -7.5\%$. The unconditional approach leads to the mistaken conclusion that the manager has negative abnormal performance.

26.3 Conditional Alphas and Betas

Conditional alphas are developed as a natural generalization of the traditional (i.e., unconditional) alphas. In the CPE approach, the risk adjustment for a Bull market state may be different from that for a Bear market state, if the fund's strategy implies different risk exposures in the different states. Let $r_{m,t+1}$ be the excess return on a market or benchmark index. This could be a vector of excess returns if a multifactor model is used. The model proposed by Ferson and Schadt (1996) is

$$r_{p,t+1} = \alpha_p + \beta_0 r_{m,t+1} + \beta'[r_{m,t+1} \otimes Z_t] + u_{pt+1}, \tag{26.1}$$

where $r_{p,t+1}$ is the return of the fund in excess of a short-term cash instrument, and Z_t is the vector of lagged conditioning variables, in de-meaned form. The symbol $\otimes$ denotes the Kronecker product, or element-by-element multiplication when $r_{m,t+1}$ is a single market index. A special case of equation (26.1) is the classical market model regression, where the terms involving Z_t are omitted. In this case, α_p is Jensen's (1968) alpha.

To see how the model in equation (26.1) arises, consider a conditional market model regression allowing for a time-varying fund beta, $\beta(Z_t)$, that may depend on the public information Z_t:

$$r_{p,t+1} = \alpha_p + \beta(Z_t) r_{m,t+1} + u_{pt+1}, \tag{26.2}$$

with $E(u_{p,t+1}|Z_t) = E\{u_{p,t+1} r_{mt+1}|Z_t\} = 0$. Now assume that the time-varying beta can be modeled as a linear function: $\beta(Z_t) = \beta_0 + \beta' Z_t$. The coefficient β_0 is the average conditional beta of the fund (as Z is normalized to mean zero), and the term $\beta' Z_t$ captures the time-varying conditional beta. The assumption that the conditional beta is a linear function of the lagged instrument can be motivated by a Taylor series approximation or by a model such as that of Admati et al. (1986), in which an optimizing agent would endogenously generate a linear conditional beta by trading assets with constant betas, using a linear portfolio weight function.

Substituting the expression for the conditional beta into equation (26.2), the result is equation (26.1). Note that since $E(Z_t) = 0$, it follows that

$$E\{\beta'[r_{m,t+1} \otimes Z_t]\} = Cov\{\beta(Z_t), r_{m,t+1}\} = Cov\{\beta(Z_t), E(r_{m,t+1}|Z_t)\}, \tag{26.3}$$

where the second equality follows from representing $r_{m,t+1} = E(r_{m,t+1}|Z_t) + u_{m,t+1}$, and using $Cov\{u_{m,t+1}, \beta(Z_t)\} = 0$. Equation (26.3) states that the interaction terms $\beta'[r_{m,t+1} \otimes Z_t]$ in the regression (26.1) control for common movements in the fund's conditional beta and the conditional expected benchmark return. This was the cause of the "bias" in the unconditional alpha in the example we used to motivate CPE. The conditional alpha, α_p in equation (26.1), is thus measured net of these.

In this setting, CPE compares a fund with a "naive" dynamic strategy, formed using the public information Z_t, that has the same time-varying conditional beta as the fund. This strategy has a weight at time t on the market index equal to $\beta_0 + \beta'Z_t$, and a weight of $\{1 - \beta_0 - \beta'Z_t\}$ in the risk-free asset. Equation (26.1) implies that α_p in the Ferson and Schadt model is the difference between the unconditional expected return of the fund and that of the benchmark strategy.

Christopherson, Ferson, and Glassman (1998) propose a refinement to allow for a time-varying conditional alpha:

$$r_{p,t+1} = \alpha_{p0} + \alpha'_p Z_t + \beta_0 r_{m,t+1} + \beta'[r_{m,t+1} \otimes Z_t] + u_{pt+1}. \tag{26.4}$$

In this model, $\alpha_{p0} + \alpha'_p Z_t$ measures the time-varying conditional alpha, and the OE portfolio is the same as in the previous case, but the time-varying alpha is now the difference between the conditional expected return of the fund, given Z_t, and the conditional expected return of the OE portfolio strategy. This refinement of the model may have more power to detect abnormal performance if performance varies with the state of the economy.

The regression (26.4) also has statistical advantages in the presence of lagged instruments that may be highly persistent regressors, with high autocorrelation, as is often the case in practice. Ferson, Sarkissian, and Simin (2008) show that by including the $\alpha'_p Z_t$ term, the regression delivers smaller finite-sample biases in the beta coefficients. They also warn, however, that the t-statistics for the time-varying alphas are likely to be biased in finite samples.

26.4   Conditional Market Timing

The classical market-timing model of Treynor and Mazuy (1966) is presented in chapter 25. Treynor and Mazuy argue that a market-timing manager will generate a return that bears a convex relation to the market. However, a convex relation may arise for other reasons. One of these is common time-variation in the fund's beta risk and the expected market risk premium, related to public information on the state of the economy. Ferson and Schadt (1996) propose a CPE version of the Treynor-Mazuy model to handle this situation:

$$r_{pt+1} = a_p + b_p r_{mt+1} + C'_p(Z_t r_{m,t+1}) + \Lambda_p r^2_{mt+1} + w_{t+1}. \tag{26.5}$$

In equation (26.5), the term $C'_p(Z_t r_{m,t+1})$ controls for common time variation in the market risk premium and the fund's beta, just like it did in regression (26.1). A manager who uses Z_t linearly to time the market has no conditional timing ability, and thus $\Lambda_p = 0$. The coefficient Λ_p measures the market timing ability based on information beyond that contained in a linear function of Z_t. Merton and Henriksson (1981) and Henriksson (1984) describe a model of market timing in which the quadratic term is replaced by an option payoff, $\text{Max}(0, r_{m,t+1})$, as described earlier. Ferson and Schadt (1996) develop a conditional version of this model as well.

Interpreting the intercept in market timing regressions like equation (26.5) raises issues similar to those in the classical Treynor-Mazuy regression. As described above, a_p is not the difference between the fund's return and that of an OE portfolio. The model may be generalized, using a conditional maximum correlation portfolio to r_m^2, as described by Aragon and Ferson (2006) and implemented by Chen, Ferson, and Peters (2010).

In theoretical market-timing models, the timing coefficient is shown to depend on both the precision of the manager's market-timing signal and the manager's risk tolerance. For a given signal precision, a more risk tolerant manager will implement a more aggressive timing strategy, thus generating more convexity. Similarly, for a given risk tolerance, a manager with a more precise timing signal will be more aggressive. Precision probably varies over time, as fund managers are likely to receive information of varying uncertainty about economic conditions at different times. Effective risk aversion may also vary over time, according to arguments describing mutual fund "tournaments" for new money flows (e.g., Brown, Harlow, and Starks 1996), which may induce managers to take more risks when their performance is lagging and to be more conservative when they want to lock in favorable recent performance. Therefore, it seems likely that the timing coefficient that measures the convexity of a fund's conditional relation to the market is likely to vary over time. Ferson and Qian (2004) allow for such effects by letting the timing coefficient vary over time as a function of the state of the economy. Replacing the fixed timing coefficient above with $\Lambda_p = \Lambda_{0p} + \Lambda'_{1p} Z_t$, we arrive at a conditional timing model with time-varying performance:

$$r_{pt+1} = a_p + b_p r_{mt+1} + C'_p(Z_t r_{m,t+1}) + \Lambda_{0p} r_{mt+1}^2 + \Lambda'_{1p}(Z_t r_{m,t+1}^2) + w_{t+1}. \tag{26.6}$$

In this model, the coefficient Λ_{1p} on the interaction term $(Z_t r_{m,t+1}^2)$ captures the variability in the manager's timing ability, if any, over the states of the economy. By examining the significance of the coefficients in Λ_{1p}, it is easy to test the null hypothesis that the timing ability is fixed against the alternative hypothesis that timing ability varies with the economic state. Ferson and Qian (2004) find evidence that market timing ability varies with the economic state.

Becker et al. (1999) further develop conditional market-timing models. In addition to incorporating public information, their model features explicit performance benchmarks for measuring the relative performance of fund managers. In practice, performance benchmarks represent an important component of some fund managers' incentives, especially for hedge funds that incorporate explicit incentive fees. Even for mutual funds, Schultz (1996) reports that Vanguard included incentive-based provisions in 24 of 38 compensation contracts with external fund managers at that time. Blake, Elton, and Gruber (2003) find that about 10% of the managers in a sample of US mutual funds are compensated according to incentive contracts. These contracts determine a manager's compensation by comparing fund performance to that of a benchmark portfolio. The incentive contracts induce a preference for portfolio returns in excess of the benchmark. The model of Becker et al. refines the conditional market-timing models of Ferson and Schadt (1996) to allow for explicit, exogenous performance benchmarks and the separate estimation of parameters for risk aversion and the quality of the market timing signal.

26.5   Conditional Weight-Based Measures

The weight-based performance measures discussed in chapter 25 are unconditional, meaning they do not attempt to control for dynamic changes in expected returns and volatility. Like the classical returns-based performance measures, unconditional weight-based measures have problems handling return dynamics. Grinblatt and Titman (1993) warn that unconditional weight-based measures can show performance when the manager targets stocks whose expected return and risk have risen temporarily (e.g., stocks subject to takeover or bankruptcy); when a manager exploits serial correlation in stock returns or return seasonalities; and when a manager gradually changes the risk of the portfolio over time, as in style drift. These problems may be addressed using a conditional approach.

Ferson and Khang (2002) develop the *conditional weight-based measure* (CWM) of performance and show that it has some advantages. Like other CPE approaches, the measure controls for changes in expected returns and volatility, as captured by a set of lagged economic variables or instruments. However, the CWM uses the information in both the lagged economic variables and the fund's portfolio weights.

The CWM is the average of the conditional covariances between future returns and portfolio weight changes, summed across the securities held:

$$CWM = E\{\sum_j w_j(Z, \Omega)[r_j - E(r_j|Z)]\}, \tag{26.7}$$

where $w_j(Z, \Omega)$ denotes the portfolio weight at the beginning of the period. The weights may depend on the public information, denoted by Z. The weights of a manager with superior information, denoted by Ω, may also depend on the superior information. By definition, superior information is any information that can be used to predict patterns in future returns that cannot be discerned from public information alone.

In equation (26.7), the term $r_j - E(r_j|Z)$ denotes the unexpected, or abnormal, future returns of the securities, indexed by j. Here the abnormal return is defined as the component of return not expected by an investor who only sees the public information Z at the beginning of the period. The sum of the covariances between the weights, measured at the end of December, and the subsequent abnormal returns for the securities in the first quarter is positive for a manager with superior information Ω. If the manager has no superior information then the covariance is zero.

26.6   Interpreting Conditional Performance Evidence

Ferson and Schadt (1996) find evidence that funds' risk exposures change significantly in response to variables like interest rates and dividend yields. Using conditional models, Ferson and Schadt (1996) and Kryzanowski, Lalancette, and To (1997) find that the distribution of mutual fund alphas is shifted to the right, compared to the unconditional alphas, and is centered near zero, while the unconditional model alphas center at negative numbers. Thus, conditional models tend to paint a more optimistic picture of mutual fund performance than do unconditional models. This general pattern is confirmed in studies by Zheng (1999), Ferson and Qian (2004), and others. A zero alpha suggests that funds have enough investment ability to cover their fees and trading costs, but that they do not add value for investors. This is consistent with the model of Berk and Green (2004), in which managers are able to extract the full rents accruing to their investment ability.

Ferson and Warther (1996) attribute differences between unconditional and conditional alphas to predictable flows of public money into funds. Inflows are correlated with reduced market exposure at times when the public expects high returns, which leads to larger cash holdings at such times. Larger cash holdings when returns are predictably higher leads to lower unconditional performance. In pension funds, which are not subject to high-frequency flows of public money, no overall shift in the distribution of fund alphas is found when moving to conditional models (Christopherson, Ferson, and Glassman 1998).

26.7   Conditional Market Timing Evidence

While studies of mutual funds' market timing ability using the classical models found evidence of perverse, negative timing coefficients, the literature on conditional market timing finds different results. Ferson and Schadt (1996) confirm earlier evidence that the classical measures can produce negative timing coefficients for naive dynamic strategies, but the conditional timing measures avoid this bias and suggest neutral timing ability in their sample of equity style mutual funds for 1968–1990.

Becker et al. (1999) simultaneously estimate parameters that describe the public information environment, the risk aversion of the fund manager, and the precision of the fund's market-timing signal. Using a sample of more than 400 US mutual funds from 1976 through 1994, they find that both benchmark investing and conditioning information are important in the model. The point estimates suggest that mutual funds behave as highly risk-averse benchmark investors. Once the public information variables are controlled for, there is little evidence that mutual funds have conditional market timing ability. Only in a subsample of asset-allocation style funds is there a hint of timing ability.

Ferson and Qian (2004) allow for conditional timing coefficients to vary over time. They find that successful market timing is more likely when the stock market and short-term corporate debt markets are highly liquid. Market timing trades may be made at lower cost in highly liquid markets. They find that market timing funds can deliver significant conditional timing performance when the term structure slope is steep. In contrast, the timing funds seem unable to deliver reliable market timing services when the slope of the term structure is flat. This may reflect a larger dispersion in the cross section of asset returns, and therefore more room for asset allocation strategies, when the term structure is steeply sloped. When Ferson and Qian sort the timing funds into groups according to fund characteristics, the best conditional timers have the longest track records, the largest total net assets, and the lowest expense ratios, or the smallest capital gains. Kacperczyk, Van Nieuwerburgh, and Veldkamp (2014) argue that funds with ability can switch their focus between market timing and stock selection depending on market conditions.

26.8   Conditional Weight-Based Evidence

Ferson and Khang (2002) examine the conditional weight-based approach to measuring performance. They first conduct experiments to assess the extent of interim trading bias in returns-based measures. Even though their hypothetical portfolios trade only two times between each quarterly observation date, they find that the interim trading bias can have a huge impact on returns-based performance measures.

Ferson and Khang apply weight-based measures to a sample of pension fund managers for 1985–1994. They find that under the unconditional weight measures, growth managers show small positive performance. Using CWMs to avoid interim trading bias and to control for public information effects, the estimates of the pension funds' performance are close to zero.

26.9   Volatility Timing Evidence

Busse (1999) asks whether fund returns contain information about market volatility. He finds evidence using daily data that funds may shift their market exposures in response to changes in second moments, but little evidence that funds can predict future shocks to market volatility. He finds some evidence that the funds that reduce market exposure in response to higher market volatility deliver larger alphas for investors. Ferson and Mo (2016) also find some evidence for this behavior in mutual funds.

27 Term Structure and Bond Fund Performance

This chapter reviews the methods and evidence for evaluating the performance of bond funds. I argue that bond fund performance has received much less attention from researchers than it deserves, and the area presents a research opportunity for the future.

27.1 Background

The amount of research on the performance of mutual funds, hedge funds, and other fund types has exploded, but most of this research focuses on equity-style funds and hedge funds. The relatively small amount of research on fixed income funds seems curious, given the importance of these funds and assets in the economy. Recently, the total net assets of US bond mutual funds has been about $4.6 trillion. This is up from $1.5 trillion a decade earlier, according to the Investment Company Institute, and compares with about $10 trillion in US equity mutual funds. As of 2017, Vanguard alone had about a trillion dollars of assets in their fixed income funds, and the three largest active bond funds held more than 300 billion dollars of assets. Large amounts of additional fixed income assets are held in professionally managed portfolios outside of mutual funds, for example, in pension funds, trusts, and insurance company accounts. The economic importance of methods to evaluate fixed income funds is huge.

Recent years have seen a trend in investor flows favoring passive investment styles and exchange traded funds over active, open-end funds, but a vast amount of fixed income assets remain actively managed. Recent estimates are that about 2/3 of mutual fund assets in the United States remain actively managed. The turnover of a typical bond mutual fund far exceeds that of a typical equity fund (e.g., see Moneta 2015), suggesting that active portfolio management is common in bond funds. So it is likely that much of interest is going on in fixed income funds, and it is important to understand the performance of bond fund managers.

Early studies that focus on US fixed income funds include Blake, Elton, and Gruber (1993), and Elton, Gruber, and Blake (1995). Ferson, Henry, and Kisgen (2006) cite about 20 studies. In recent years, the number of studies has expanded, but it remains far smaller than the number of studies of hedge funds and equity style mutual funds. We know much less about bond funds than we do about hedge funds and equity mutual funds, and so bond funds present a great research opportunity.

The relatively small amount of research on fixed income funds likely reflects various factors. One is that for bond funds, unlike equity funds, no common set of risk factors has emerged. For equity style funds, the Fama-French three-factor or Carhart (1997) four-factor models have become the recent empirical standard. Researchers who do not wish to address the issue of benchmark specification in their studies are drawn to these conventions. Work on term structure models has identified a standard set of empirical factors, such as the level, slope, and curvature or volatility of the yield curve (e.g., Litterman and Scheinkman 1991). These factors are relevant or even sufficient for US government bond returns, but corporate bond, high-yield, and mortgage security funds seem to require other factors. Elton, Gruber, and Blake (1995) identify and examine some factors for bond funds, and Chen, Ferson, and Peters (2010) discuss additional factors. Still, no standard set of factors has emerged.

Another challenge is the fact that bond-pricing models are usually cast in continuous time, following Vasicek (1977) and Cox, Ingersoll, and Ross (1985), and the SDFs in those models typically are nonlinear functions of the state variables. However, the continuous-time models present a huge advantage. They can be aggregated over time to derive the SDF for, say, monthly returns. This addresses the interim trading problem.

The interim trading problem is important because managed bond portfolio returns are likely to have different dynamics from those of the underlying bonds, since fund managers change their portfolio weights during the month. Managers may also hold interest rate derivatives, whose returns may be replicated by dynamic trading within the month. Addressing interim trading implicitly accommodates the use of fixed income derivatives in the funds' portfolios. The rest of this chapter fleshes out the approach, drawing from joint work with Darren Kisgen and Tyler Henry (Ferson, Kisgen, and Henry 2006), and concludes with a summary of the broader empirical evidence on bond fund performance. This material presumes some familiarity with continuous-time diffusion processes, but you might be able to follow along without that background.

27.2   Term Structure Model Stochastic Discount Factors

Popular term structure models specify a continuous-time stochastic process, such as a diffusion process, for the underlying state variable(s). Let X be the state variable following a diffusion process:

$$dX = \mu(X_t)dt + \sigma(X_t)dw, \qquad\qquad (27.1)$$

where dw is the local change in a standard Weiner process. The state variable(s) may be the level of an interest rate, the slope of the term structure, or the like. The model specifies the form of a market price of risk, q(X), associated with the state variable. The market price of risk is the expected return in excess of the instantaneous interest rate, per unit of state variable risk.

We specialize to time-homogeneous diffusions, where the functions $\mu(\cdot)$ and $\sigma(\cdot)$ in equation (27.1) depend on time only through the level of the state variable at a point in time. Interest rate models, such as Hull and White (1990), allow time variation in the functions, choosing them to fit the term structure of spot or forward rates observed at each time t. Such models are attractive for the practice of pricing interest-dependent derivative securities, because by fitting the current term structure at each date, the models can avoid derivative prices that allow arbitrage opportunities at the current prices. The goal of fund performance measurement does not necessarily require precisely fitting the structure of derivatives prices at each date. We want good models for the covariances of portfolio returns with the SDFs.

The term structure literature has directed a lot of firepower at accurately modeling continuous-time processes like equation (27.1) when X is an interest rate. For example, studies following Chan et al. (1992) debate whether the power in the diffusion function, $\sigma(x_t)$, for a short-term interest rate is 0.5, 1.0, 1.5, or some other number. Other studies ask whether the drift of the process, $\mu(x_t)$, is linear or nonlinear (see, e.g., Ait-Sahalia 1996; Ait-Sahalia and Duarte 2003). Ait-Sahalia rejects most of the parametric models for the spot rate that have been proposed in the literature by comparing their implied density functions with those observed in interest rate data. Of course, no model is an exact representation of the world. What is not yet known is how important these "failures" of the models are for the application to mutual fund portfolio returns.

Term structure models based on equation (27.1) can be shown (using Girsanov's theorem; see Cox, Ingersoll, and Ross 1985) to imply time-aggregated stochastic discount factors of the following form:

$$_t m_{t+1} = \exp(-A_{t+1} - B_{t+1} - C_{t+1}),$$

where

$$A_{t+1} = \int_t^{t+1} r_s \, ds,$$

$$B_{t+1} = \int_t^{t+1} q(X_s) \, dw_s \tag{27.2}$$

$$C_{t+1} = (1/2) \int_t^{t+1} q(X_s)^2 \, ds,$$

where r_s is the instantaneous interest rate at time s. The notation $_t m_{t+1}$ is chosen to emphasize that the SDF refers to a discrete time interval (say, one month) that begins at time t and ends at time $t+1$. When there are multiple state variables, terms like B_{t+1} and C_{t+1} exist for each state variable. Note that, unlike for beta pricing models, where the SDF is linear in the factors, the SDF in equation (27.2) is nonlinear. Dietz, Fogler, and Rivers (1981) argue that tests of bond portfolio performance should allow for nonlinearity.

To use the term structure models with monthly mutual fund data, consider a first-order Euler approximation scheme for equation (27.2):

$$X(t+\Delta) - X(t) \approx \mu(X_t)\Delta + \sigma(X_t)[w(t+\Delta) - w(t)]. \qquad (27.3)$$

The period between t and $t+1$ is divided into $1/\Delta$ increments of length Δ. For example, the period is one month, to match the mutual fund returns, divided into increments of one day. For a given model, we can often get daily data on $X(t+\Delta)$ and $X(t)$, and the functions $\mu(X_t)$ and $\sigma(X_t)$ are specified by the model. We can therefore infer the approximate daily values of $[w(t+\Delta) - w(t)]$ from equation (27.3). The terms A_{t+1}, B_{t+1}, and C_{t+1} in equation (27.2) are approximated using daily data by

$$
\begin{aligned}
A_{t+1} &\approx \textstyle\sum_{i=1,\ldots,1/\Delta} r(t+(i-1)\Delta)\Delta, \\
B_{t+1} &\approx \textstyle\sum_{i=1,\ldots,1/\Delta} q[X(t+(i-1)\Delta)][w(t+i\Delta) - w(t+(i-1)\Delta)], \\
C_{t+1} &\approx (1/2)\textstyle\sum_{i=1,\ldots,1/\Delta} q[X(t+(i-1)\Delta)]^2\Delta.
\end{aligned}
\qquad (27.4)
$$

Farnsworth (1997) and Stanton (1997) evaluate the accuracy of similar first-order approximation schemes. Stanton concludes that with daily data, these approximations are almost indistinguishable from the true functions over a wide range of values, and the approximation errors should be small when the series being studied is observed monthly. He also evaluates higher-order approximation schemes and finds that with daily data, they offer negligible improvements over the first-order approximations.

27.3 Single-Factor Model Example

A special case illustrating the approach is a model with a single state variable in the affine class, where the short-term interest rate r_t is the state variable at time t:

$$
\begin{aligned}
dr &= K(\theta - r_t)dt + \sigma(r)dw, \\
\sigma(r) &= (\gamma + \delta r)^{1/2}, \\
q(r) &= \lambda(\gamma + \delta r)^{1/2}.
\end{aligned}
\qquad (27.5)
$$

Equation (27.5) includes as special cases the single-factor models of Vasicek (1977), where $\delta = 0$, and of Cox, Ingersoll, and Ross (1985), where $\gamma = 0$. The coefficients in the Euler equations in equation (27.4) reduce to

$$
\begin{aligned}
A_{t+1} &\approx \textstyle\sum_{i=1,\ldots,1/\Delta} r(t+(i-1)\Delta)\Delta, \\
B_{t+1} &\approx \lambda\{r_{t+1} - r_t - K\theta + KA_{t+1}), \\
C_{t+1} &\approx (\lambda^2/2)(\gamma + \delta A_{t+1}).
\end{aligned}
\qquad (27.6)
$$

Thus, the empirical SDF for monthly returns depends on both the discrete change in the interest rate and on the average rate over the month, A_{t+1}. The single-factor affine model's SDF is written as

$$_t m_{t+1} = \exp(a + bA_{t+1} + c[r_{t+1} - r_t]), \qquad (27.7)$$

where A_{t+1} is the average value of the short-term interest rate over the month, and $\{a, b, c\}$ are constants.

Ferson, Henry, and Kisgen (2006) show how this approach can be applied to a two-factor affine model, a three-factor affine model, and a two-factor term structure model from Brennan and Schwartz (1979) that is not affine. The motivation for the three-factor model is provided by studies finding that three factors work well in capturing the term structure of government bond returns.

Because of the effects of time aggregation, the SDF in equation (27.7) features both a discrete change in the state variable, the short rate $[r_{t+1} - r_t]$, and an average of the daily short-rate levels over the month, A_{t+1}. In a two-factor affine model, the SDF depends both on the monthly changes in the long and short rates and on monthly averages of the long-rate and short-rate levels. A three-factor model adds a discrete change and an average convexity factor. The time-averaged factors that appear in the time-aggregated SDF are the key to its ability to handle interim trading.

Even with the additional factors that arise from time aggregation, the number of parameters that can be identified in these reduced-form models is smaller than the number of underlying parameters in the theoretical models. For example, the one-factor affine model has five parameters, including those that run the stochastic process for the short-term interest rate (four, in the special cases of the Vasicek and Cox-Ingersoll-Ross models), while only three parameters can be identified using the m-talk equation (1.1) with the m in equation (27.7). It would be possible to incorporate additional moment conditions, derived from the interest rate process specifications, and thereby identify additional parameters. The continuous time processes are not necessary assumptions of the SDF representations (Gourieroux, Monfort, and Polimenis 2006). If we used the extra moment conditions and the interest rate process is misspecified, then the misspecification would spill over to the estimated performance measures. To identify the covariances of funds' discrete-period returns with the factors motivated by the time-aggregated models, it may be sufficient to work with the smaller number of parameters, but this question has not been fully explored.

27.4 Addressing Interim Trading Bias

Stochastic discount factors like equation (27.7) address the interim trading bias in returns-based performance measures. The problem arises when discrete period (say, monthly) returns are available to the researcher, while the manager trades during the month. If public information about returns is revealed during the month and a manager trades on this information, traditional performance measures can record this trading as reflecting "superior" information or ability.

Equation (27.7) deals with the interim trading problem in the following way. Assume that an SDF prices the returns on the primitive assets held by funds for the arbitrarily short period t to $t+\Delta$. Call this SDF $_t m_{t+\Delta}$. This SDF must price all portfolios of the primitive assets formed at time t and held until $t+\Delta$, with weights that may depend on the information generated by the state variables at time t. Similarly, at time $t+\Delta$, there is an SDF $_{t+\Delta} m_{t+2\Delta}$ that prices all portfolios formed at time $t+\Delta$ and held until $t+2\Delta$, with weights that can be any function of $X_{t+\Delta}$. From the Euler equation (1.1) and the law of iterated expectations, if we assume a model with time-separable preferences, it follows that the SDF $(_t m_{t+\Delta})$ $(_{t+\Delta} m_{t+2\Delta})$ prices all nonanticipating strategy returns measured over the period from t to $t+2\Delta$, allowing for dynamic strategies that trade at t and $t+\Delta$ as functions of the public information at these two dates.

To show this result, let $w(X_t)$ be a portfolio weight vector that sums to 1.0 and consider the two-period strategy with gross return $_t R_{pt+2\Delta} = [w(X_t)'_t R_{t+\Delta}][w(X_{t+\Delta})'_{t+\Delta} R_{t+2\Delta}]$. At time $t+\Delta$, the Euler equation implies

$$1 = E_{t+\Delta}\{_{t+\Delta} m_{t+2\Delta}[w(X_{t+\Delta})'_{t+\Delta} R_{t+2\Delta}]\}.$$

At time t, we have

$$
\begin{aligned}
1 &= E_t\{_t m_{t+\Delta}[w(X_t)'_t R_{t+\Delta}]1\} \\
  &= E_t\{_t m_{t+\Delta}[w(X_t)'_t R_{t+\Delta}]E_{t+\Delta}\{_{t+\Delta} m_{t+2\Delta}[w(X_{t+\Delta})'_{t+\Delta} R_{t+2\Delta}]\}\} \\
  &= E_t\{(_t m_{t+\Delta})(_{t+\Delta} m_{t+2\Delta})[_t R_{pt+2\Delta}]\}.
\end{aligned}
\tag{27.8}
$$

The key result is that the functional form of the SDF in equation (27.7) implies that $(_t m_{t+\Delta})$ $(_{t+\Delta} m_{t+2\Delta}) = _t m_{t+2\Delta}$. By induction and the fact that Δ is arbitrary, it follows that a time-aggregated SDF like the one in (27.7) should correctly price all nonanticipating portfolio strategies that trade the primitive assets using only the public information as captured by the state variables. Thus, there should be no interim trading bias in theory with this approach. Of course, in practice we are limited by the data, and with daily data we implicitly assume that managers trade only at the end of each day. It is an unexplored empirical question as to whether that approximation is adequate.

To interpret the empirical factors that arise from time aggregation, consider the example of a two-factor affine model, in which we obtain four empirical factors. If the affine model were literally true, the market would be complete, and all risks could be locally spanned by two factors. In the continuous time model, the market is complete in the space of all nonanticipating dynamic trading strategies. It takes two more time-averaged factors to represent the state space of the nonanticipating functions in the monthly data.

If we limit attention to a subset of the returns space, there can be more than one SDF that correctly prices returns in the subset. In particular, suppose we limit attention to the monthly returns of pure discount bonds and coupon bonds that may be formed as portfolios of discount bonds, with weights that are fixed at the beginning of the month. This is a common practice in empirical studies of bond returns. In this case, it is easy to show that an SDF that depends only on the levels of the state variables, with no time-aggregation terms, will price this subset of returns. This is consistent with empirical studies that find that a few factors are adequate to explain most of the common variance in yield changes or bond returns (e.g., Gultekin and Rogalski 1984; Litterman and Scheinkman 1991; Jagannathan, Kaplin, and Sun 2003). Thus, it is the desire to price dynamic strategies and avoid interim trading bias that motivates the additional time-aggregated factors in the empirical SDFs.

It is interesting that the models imply a role for time-averaged data in pricing the risks of holding-period returns for dynamic strategies. A common instinct in asset pricing research is to focus on end-of-period data. Indeed, the Center for Research in Security Prices (CRSP) data files and many other financial databases are organized this way. But such an approach relies on the assumption that the primitive trading interval in the model is of the same length as the period over which the data are measured. Many macroeconomic data series come as time averages. Such data are often viewed with suspicion (e.g., Working 1960; Breeden, Gibbons, and Litzenberger 1989; Campbell 1993). When dynamic strategies are important it may be necessary to refine this view. The time averaged data may be just what the doctor called for!

27.5 Empirical Findings

Blake, Elton, and Gruber (1993) and Elton, Gruber, and Blake (1995) study US bond mutual fund performance using various bond factors. They do not incorporate time-aggregated factors. They find that the average performance is slightly negative after costs and is largely driven by funds' expenses, and thus is similar in flavor to the classical evidence on active equity fund performance. This might suggest that investors would be better off selecting low-cost passive funds, and Elton, Gruber, and Blake (1995) draw that conclusion. Comer, Larrymore, and Rodriguez (2007) use daily data on hybrid funds holding both stocks and bonds and find that the performance results depend on the choice of factors in the model.

Ferson, Henry, and Kisgen (2006) evaluate time-aggregated SDF models using benchmarks constructed to represent both passive and dynamic bond portfolio strategies. The returns and volatility of these benchmarks vary, conditional on states of the term structure. They use the time-aggregated factor models to evaluate government bond fund performance, including a conditional evaluation.

During 1986–2000, most government bond funds returned less on average than passive benchmarks that don't pay expenses, as observed by Elton, Gruber, and Blake (1995), but not for all states of the term structure. There seem to be states of the economy where the conditional expected returns of funds exceed the benchmarks. Conditioning on the slope and convexity of the term structure, Ferson, Henry, and Kisgen (2006) find more variation in funds' average returns across these states than across fund groups with differing characteristics. Adjusting for risk using the stochastic discount factor models, they find that significant abnormal performance is rare.

Performance may be decomposed into components, such as timing and selectivity. If investors place value on timing ability (e.g., a fund that can mitigate losses in down markets), they would be willing to pay for this insurance with lower average returns. Chen, Ferson, and Peters (2010) provide a comprehensive study of the ability of US bond funds to time their markets. They argue that timing ability on the part of a fund manager is the ability to use superior information about the future realizations of common factors that affect bond market returns. If common factors explain a significant part of the variance of bond returns, consistent with term structure studies such as Litterman and Scheinkman (1991), then a significant fraction of the potential performance of bond funds might be attributed to timing. However, measuring the timing ability of bond funds is a subtle problem.

Traditional models of market timing ability, such as those reviewed in chapter 25, rely on convexity in the relation between the fund's returns and the common factors. Comer (2006) studies multivariate versions of the Treynor-Mazuy regression for hybrid bond funds that hold both bonds and stocks and finds mixed evidence of timing ability that varies across subperiods. Chen, Ferson, and Peters (2010) emphasize that four broad types of potential biases can arise when using nonlinearities to identify timing ability. There might be nonlinear relations between the economic factors and a fund's benchmark. Interim trading or derivatives can generate nonlinearity. Stale pricing that is correlated with economic factors—systematic stale pricing—can generate nonlinearity. Finally, funds' exposures to risk factors may vary with public information.

Chen, Ferson, and Peters (2010) study monthly returns for more than 1,400 bond funds during 1962–2007 and find that controlling for non–timing-related nonlinearity is important. Funds' returns are typically more concave—with respect to a set of nine bond market factors—than are unmanaged benchmarks. Thus, funds would appear to have poor (negative) market timing ability in naive models. They introduce controls for non–timing-related nonlinearities and find that the overall distribution of the timing coefficients appears to be neutral to weakly positive.

Only a few papers have used holdings-based performance measures to study bond funds. This no doubt reflects the significant data challenges. There are just so many bonds! General Motors alone has hundreds of bonds. And many of them don't trade very often. Boney, Comer, and Kelly (2009) examine the ability of investment grade bond funds to shift between bonds and cash and across maturities, and they find perverse "negative" timing ability. Cici and Gibson (2012) use corporate bond holdings to study the performance of corporate bond funds; they find no performance in security selection. Huang and Wang (2014) look for timing ability in a sample of 146 government bond funds. They find some weak evidence that bond funds' holdings can predict subsequent returns for about one month. Comer (2006) looks at the power of holdings-based measures for bonds and finds that they can be better than returns-based measures. Moneta (2015) studies about 1,000 US bond funds during 1997–2006, using holdings aggregated to asset classes defined as different sectors, credit qualities, and maturity ranges. He finds performance, before costs and fees, of about 1% per year, which is roughly the magnitude of expense ratios plus trading costs. He finds no evidence of timing ability. This evidence accords well with the early evidence on equity funds based on holdings.

28 Investment Performance Evaluation: A Modern Perspective

In this chapter, I collect and summarize some ideas about investment performance evaluation from a stream of recent papers and book chapters (Aragon and Ferson 2006; Ferson 2003, 2013; Ferson and Lin 2014; Ferson and Mo 2016; Ferson and Wang 2017), being careful not to step on anyone's toes regarding copyrights. My goal is to make several observations about the current state of the art in investment performance measurement. First, the traditional alphas used in the literature are not to be trusted as normative indicators for when to buy or sell funds, but SDF alphas are better. Traditional alphas can be equivalent to the correct SDF alphas, but to ensure equivalence, an "Appropriate Benchmark" must be used. Despite a traditional focus on mean variance efficient portfolios, these are almost never Appropriate Benchmarks, unless you insist on quadratic utility. However, seemingly in contrast to conventional wisdom, Sharpe ratios can be justified as performance measures when they are used properly, by comparing the Sharpe ratio of the fund to be evaluated with the Sharpe ratio of an Appropriate Benchmark. Finally, current holdings-based approaches to performance measurement are also flawed, although under certain conditions, they can be equivalent to the correct SDF alpha. These conditions are strong and unlikely to be met in practice, so I offer suggestions for how to improve the implementation of holdings-based performance measures, from Ferson and Mo (2016).

French (2008) estimates that investors pay about 2/3 of a percent per year for active management, as compared to passive management. Gruber (1996) describes a puzzle of active management. Why should investors pay for active management if the after-cost unconditional performance is negative, as much of the previous literature has found? As of this writing, there are more equity mutual funds in the United States than there are common stocks. This chapter describes some ideas about this puzzle.

First, a notational convention. The public information "fully reflected" in market prices and known by the "client" of a performance measure is denoted by Z. The potentially superior information held by a fund manager is denoted by Ω. Having and using information better than the client is how a fund manager delivers abnormal performance, or alpha.

28.1   Market Efficiency and Fund Performance

Let's revisit market informational efficiency in the context of managed funds. When applying informational efficiency, it matters whether the portfolio manager or fund investors are under study. At issue is the efficient use of information. Managers use their information to form the fund's portfolio strategies. Evidence about the before-cost performance of a fund therefore relates to the information used by the manager. For example, a manager may use superior information to deliver abnormal performance, conditional on public information, before trading costs, administrative costs, and management fees. Managers who do that have investment ability. This is in violation of the strong form of market efficiency. The compensation of managers with investment ability—the management fee—is a question about the labor market for fund managers. These fees should depend on the supply and demand for investment ability. Grossman and Stiglitz (1980) emphasize that no one would expend resources to gather information if it didn't pay to trade on it. Gârleanu and Pedersen (2018) develop an equilibrium model where managers have superior information, and investors must pay some costs to search them out. Both managers and investors can get paid for superior information. Magkotsios (2017) presents a model with competition for flows among managers with talent, in which managers and investors can each get paid.

The value added for (fund) investors, if any, is traditionally the central issue for studies of investment performance. Much of the evidence in the literature is described in terms of portfolio strategies that use information to combine mutual funds. If funds' alphas measured after costs persist over time, and investors can use the information in the past returns of the funds to form strategies that deliver abnormal performance, this speaks to weak form efficiency. If those strategies rely on additional public information beyond past prices and returns, such as other fund characteristics, then it speaks to semistrong-form efficiency.

A manager may be able to generate risk-adjusted returns before costs and fees, yet after costs, investors' returns may be inferior. If a fund delivers abnormal positive performance on an after-cost basis, the fund adds value for investors. If a manager has investment ability, but either extracts the rents to this ability in the form of fees and expenses, or dissipates it through trading costs, there is no value added for investors. This is the situation predicted in the equilibrium of Berk and Green (2004), where fund investors are perfectly competitive, earning zero abnormal returns, and managers with skills extract all the rents to portfolio management.

28.2   Troubles with Traditional Alphas

Studies of investment performance routinely use various traditional measures of alpha, referring to CAPM alpha, three-factor alpha or four-factor alpha, or even DGTW alpha, assuming that the reader hardly requires a definition. (By now, you probably know what all these alphas are.) Despite the familiarity with these traditional alphas, the literature has only partially resolved a basic question: Given a fund with a positive (negative) alpha, will an investor want to buy (sell) that fund? The simplest intuition for the attractiveness of a positive alpha is taught with the CAPM, where a combination of a positive-alpha fund, the market portfolio, and cash can "beat the market" in a mean variance sense (higher mean return, given the variance). Dybvig and Ross (1985b) show that if a fund has a positive alpha measured relative to a benchmark, then buying some of the fund at the margin will result in a higher Sharpe ratio than that of the benchmark, if the benchmark excess return is positive.

The literature also offers many examples where a traditional alpha is not a reliable buy-sell indicator. When there is differential information, the portfolio of a better-informed manager expands the opportunity set of the less-informed client, so the client would generally like to use the managed portfolio return. The problem is, the client might wish to short the fund, even if it has a positive alpha (Chen and Knez 1996). As Grant (1977), Dybvig and Ross (1985b), and Grinblatt and Titman (1989a) show, a market timing manager with true ability can generate a negative alpha. We can find positive alphas when performance is neutral but leverage or options are used (e.g., Jagannathan and Korajczyk 1986; Leland 1999), or we can find negative alphas when performance is neutral but you don't account for public information (Ferson and Schadt 1996). Goetzmann et al. (2007) show how option-like strategies can lead to absurd measured performance. Roll (1978) and Green (1986) give examples of arbitrary alphas when the benchmark used to define alpha is not mean variance efficient. Dybvig and Ross (1985a) and Hansen and Richard (1987) show that a portfolio can be mean variance efficient, given the informed manager's knowledge, but can appear to be mean variance inefficient to the uninformed client. Glode (2011) argues that investors may wish to buy a fund even when it has a negative traditional alpha. Thus, the existing literature suggests that traditional alphas are not reliable signals for an investor's decision on whether to buy or sell a fund. So, it may not be such a puzzle to find that investors would fail to take to heart the typical finding of negative after costs unconditional performance, and fail to avoid managed funds.

28.3   SDF Alphas and Managers' Information

An investment manager forms a portfolio of the assets with gross return $R_{pt+1} = x(\Omega_t)'R_{t+1}$, where $x(\Omega_t)$ is the vector of portfolio weights, and Ω_t is the manager's information at the beginning of the period, at time t. If Ω_t is more informative than the market's information set, Z_t, then the portfolio R_{pt+1} may not be priced through m-talk. That is, the manager may record "abnormal performance," or nonzero alpha. (In what follows, I will drop the time subscripts unless they are needed for clarity.) Define the SDF alpha for portfolio p in the usual way: $\alpha_p = E(mR_p|Z) - 1$, where Z is the public information. In general, the SDF alpha is a function of the information Z, known to the investment client at the beginning of the period, and depends on the client's SDF. If R_{Bt+1} is any zero-alpha benchmark, the SDF alpha can also be written as $\alpha_p \equiv E(m[R_p - R_B]|Z)$.

Ferson and Lin (2014) provide conditions under which the SDF alpha is a reliable guide for normative investment choice. They assume the client utility function to be time additive in a multiperiod model. The indirect value function is J(W, s), and the SDF is $m_{t+1} = \beta J_w(W_{t+1}, s_{t+1})/u_c(C_t)$. Subscripts on J($\cdot$) and u($\cdot$) denote derivatives, and s is a vector of state variables. The model presents the client with the opportunity to buy or sell a managed portfolio with return $R_p = x(\Omega)'R$, where Ω is the manager's information. The client adjusts to this new opportunity by changing her consumption and portfolio choices until alpha is zero at the new optimum. Ferson and Lin show that the optimal amount purchased is proportional to the SDF alpha of the fund. Thus, the right alpha (which provides a reliable buy-sell signal) is the SDF alpha, and it is, in general, client-specific. The "manipulation proof" performance measure of Goetzmann et al. (2007) is a special case of this analysis, using a single-period model and a specific utility function.

Ferson and Lin (2014) derive bounds on the extent to which a client can disagree with a traditional alpha because it does not use the right SDF for the client. This, of course, assumes that market is incomplete, so the SDF is not equated across investors. They find that the likely magnitude of this disagreement is comparable to the impact of benchmark choice on traditional alphas and comparable to the impact of imprecise estimation of funds' traditional alphas. Chrétien and Kammoun (2017) point out that disagreement and heterogeneity can potentially resolve the puzzle of active management raised by Gruber (1996). All a fund needs to survive is a clientele that likes what the fund delivers. They develop a "best clientele" performance measure, which is the highest SDF alpha for a fund, among clients who assign zero alphas to passive benchmarks and whose SDF volatility is bounded by a maximum Sharpe ratio (or "good deal" bound) similar to Cochrane and Saa-Requejo (2000). They find that augmenting the monthly Sharpe ratio of the market index by 0.04 is enough to remove negative alphas on average, and they conclude that under reasonable parameters, there should be a clientele that likes most mutual funds.

28.4   The Appropriate Benchmark

In empirical practice, a traditional alpha is almost always measured as the expected return of the fund in excess of a benchmark: $E(R_p - R_B|Z)$. A classical example is Jensen's (1968, 1972) alpha, which follows from the CAPM (Sharpe 1964). Jensen's alpha is the expected excess return of the fund over a benchmark that combines the risk-free asset and the market benchmark index so as to have the same beta as the fund to be evaluated: $\alpha_p = E[R_p - \{\beta_p R_b + (1 - \beta_p)R_f\}]$. Jensen's alpha is also the intercept in a time series regression of $R_p - R_f$ on $R_b - R_f$. Jensen's alpha would be a reliable measure of performance if it coincided with the right, SDF alpha. What are the conditions under which this will occur?

Ferson (2013) defines an *Appropriate Benchmark* as one that has the same covariance with the relevant SDF as the portfolio to be evaluated. The Appropriate Benchmark is a special case of the more general OE portfolio defined by Aragon and Ferson (2006), which considers aspects of the problem other than risk, such as costs and taxes. Thus, an Appropriate Benchmark has the same risk as the fund, from the client's perspective. To see this, take the definition of the SDF alpha and rewrite it as

$$\alpha_p = E[m(R_p - R_B)|Z] = E(m|Z)E(R_p - R_B|Z) + \text{Cov}(m, R_p - R_B|Z). \tag{28.1}$$

The term $E(m|Z)$ is identical for all funds and just translates the measures through time. $E(m|Z)$ is the price of a risk-free, one-period bond and is measured at the beginning of the period, while the returns are measured at the end of the period. We see that the expected return in excess of a benchmark is essentially equivalent to the SDF alpha if and only if $\text{Cov}(m, R_p|Z) = \text{Cov}(m, R_B|Z)$; that is, only if R_B is an Appropriate Benchmark.

When will the market portfolio-based benchmark of the CAPM be an Appropriate Benchmark? When the SDF is linear in the market portfolio return, the SDF approach is equivalent to the CAPM. This situation is also equivalent to assuming that the client has a quadratic utility function defined over the benchmark return. This result foreshadows the next one, which claims that mean variance efficient portfolios are almost never Appropriate Benchmarks.

28.5   Mean Variance Efficient Benchmarks Are Almost Never Appropriate Benchmarks

The literature on traditional alphas is strongly influenced by the CAPM and often focuses on mean variance efficient benchmarks. For example, in the CAPM, the market portfolio-based benchmark is mean variance efficient. Grinblatt and Titman (1989a) discuss mean variance efficient benchmarks. Chen and Knez (1996) and Dahlquist and Soderlind (1999) use mean variance efficient benchmarks and SDF alphas. However, mean variance efficient portfolios are almost never Appropriate Benchmarks.

We learned in our discussion of the three paradigms of empirical asset pricing that a portfolio R_B is minimum variance efficient if and only if it maximizes the correlation to the SDF. Then we can write the SDF as $m = a + bR_B + u$, with $E(uR|Z) = 0$, if and only if R_B is minimum variance efficient in the set of assets R conditional on the information Z (the coefficients a and b can be functions of Z). The fund's portfolio, R_p, has $\text{Cov}(R_p, m|Z) = b\text{Cov}(R_p, R_B|Z) + \text{Cov}(R_p, u|Z)$. If R_B is an Appropriate Benchmark, it implies that $\text{Cov}(R_p, u|Z) = 0$. This occurs in two possible situations: (i) R_B is conditionally minimum variance efficient in the more inclusive set of assets (R, R_p), in which case alpha is always zero and not very useful, or (ii) the SDF is exactly linear in R_B (u = 0). This implies that the utility function behind the SDF is a quadratic function of R_B. The quadratic utility, as is well known, has some unappealing characteristics, such as increasing absolute risk aversion and satiation, and so is not very attractive. Nevertheless, outside this setting, a mean variance efficient portfolio is not an Appropriate Benchmark.

28.6   Justifying the Sharpe Ratio

The Sharpe ratio (Sharpe 1992) for a portfolio p is $SR_p = E(r_p)/\sigma(R_p)$, where $r_p \equiv R_p - R_f$ is the return of the portfolio p, net of the return, R_f, to a safe asset, and $\sigma(R_p)$ is the standard deviation of the portfolio return. The Sharpe ratio is traditionally thought to be an inappropriate performance measure when returns are not normal. Leland (1999) shows that it is important to consider higher moments of the distributions if the performance measure is to accurately capture an investor's utility. Goetzmann et al. (2007) show that by selling put options at fair market prices, one can generate very high Sharpe ratios without investment skill. They also give an example where a manager with forecasting skill can have a low Sharpe ratio. Lo (2002) presents examples where high-frequency trading can generate high Sharpe ratios.

Despite these limitations, the Sharpe ratio is simple and is often used in practice as a measure of portfolio performance. Under certain conditions, the Sharpe ratio can be motivated as a valid performance measure. These conditions include the assumption that the return measurement and trading intervals coincide with the decision interval of the client, which rules out problems associated with interim trading, as discussed in chapter 27. Then, the trick is to compare the Sharpe ratio of a fund with that of an Appropriate Benchmark. If the fund's Sharpe ratio exceeds that of the benchmark, we have

$$E(R_p - R_f)/\sigma(R_p) > E(R_B - R_f)/\sigma(R_B). \tag{28.2}$$

Let ρ be the correlation between the fund's return and the benchmark return R_B, and assume that the correlation is positive. Because the correlation is less than 1.0, equation (28.2) implies that $E(R_p - R_f) > [\rho\sigma(R_p)/\sigma(R_B)]E(R_B - R_f)$. Recognizing $[\rho\sigma(R_p)/\sigma(R_B)]$ as the regression beta of the fund's excess return on that of the benchmark, we see that $\alpha_p = E(R_{pt}) - E[R_{ft} + \beta_p(R_{Bt} - R_{ft})] > 0$; that is, the alpha measuring the fund's expected return, in excess of an Appropriate Benchmark portfolio that combines R_B with cash, is positive. The Sharpe ratio is invariant to the use of leverage at a fixed risk-free rate. Thus, comparing the Sharpe ratio of a managed fund to that of an Appropriate Benchmark, the Sharpe ratio of the fund is larger than that of the benchmark only if the SDF alpha is positive. Thus, the Sharpe ratio is a reliable indicator of the client's demand to buy the fund.

Note that the same algebra works for any benchmark, as pointed out by Dybvig and Ross (1985a). A higher Sharpe ratio than the benchmark corresponds to a positive alpha measured on the benchmark. However, if we don't compare to the Appropriate Benchmark, a positive alpha on a benchmark may not mean anything to the client. To say that the client will prefer a fund with a positive alpha, we need an Appropriate Benchmark.

28.7   Comments on Holdings-Based Performance Measures

Grinblatt and Titman (1989a) derive a holdings-based measure in a single-period model, where the returns and managers' information are jointly normally distributed. These classical measures are discussed in chapter 25, and chapter 26 discusses conditional holdings-based measures. Assuming nonincreasing absolute risk aversion on the part of the manager, Grinblatt and Titman show that if a manager has information, then

$$\text{Cov}\{x(\Omega)'R|Z\} > 0, \tag{28.3}$$

where $x(\Omega)$ is the manager's optimal weight vector. The notation $x'R$ is the inner product, so that the inequality (28.3) states that the sum of the covariances between the weights of a manager with private information, Ω, and the returns for the securities in a portfolio is positive. The idea is that a manager who increases the fund's portfolio weight in securities before they perform well, or who anticipates and avoids losers, has investment ability. It is important to note that under the normality assumption that justifies this result, managers get signals about expected returns, but a manager never receives a signal about volatility or other second moments.

When risk adjustment is made explicit through m-talk, holdings-based measures can be generalized to adjust for risk. Consider how a manager with superior information S generates alpha. Substitute $R_p = x(\Omega)'R$ into the definition of the SDF alpha, and use the definition of covariance to see that

$$
\begin{aligned}
\alpha_p &= E(mR'x(\Omega)|Z) - 1 \\
&= E(mR|Z)'E(x(\Omega)|Z) - 1 + \text{Cov}(mR'x(\Omega)|Z) \\
&= \text{Cov}(mR'x(\Omega)|Z).
\end{aligned}
\tag{28.4}
$$

Moving between the second and third lines, I have used the facts that $E(mR|Z)$ is a vector of ones and that the weights $x(\Omega)$ sum to one.

Equation (28.4) shows that the SDF alpha is the sum of the covariances of the manager's weights with the future risk-adjusted abnormal returns of the assets, mR. This is not what the literature on holdings-based performance has typically done. Instead of $\text{Cov}(mR'x(\Omega)|Z)$, other studies have computed versions of $\text{Cov}(mR'x(\Omega)|Z)$, leaving out the risk adjustment. Thus, the right way to do holdings-based performance measures in future work is to introduce risk adjustment.

There are, of course, conditions under which the common approach to holdings-based performance measurement can be justified, as shown by Ferson (2013). The argument is summarized here. Assuming that R_B is an Appropriate Benchmark with $\text{Cov}(m; [R_p - R_B]|Z) = 0$, we can expand equation (28.4) as

$$\alpha_p = E(m|Z)\{E(x(\Omega) - x_B|Z)'E(R|Z) + \text{Cov}([x(\Omega) - x_B]'R|Z)\}. \tag{28.5}$$

We can now ask: When does equation (28.5) reduce to $\text{Cov}([x(\Omega)-x_B]'R|Z)$? That requires $E(x(\Omega)-x_B|Z)=0$: The benchmark weights are the expected weights for the fund using public information. Common holdings-based measures can be justified, in the sense that they are equivalent to the SDF alpha, under three conditions: (i) The return defined by the benchmark weights, $r_B=x_B'r$, is an Appropriate Benchmark return; (ii) the weights also satisfy $E(x(\Omega)|Z)=x_B$; and (iii) the alpha of the benchmark return $r_B=x_B'r$ is zero. These represent a strong set of conditions that are unlikely to hold in practice. I conclude that it is probably better just to do the risk adjustment, as in $\text{Cov}(mR'x(\Omega)|Z)$, than it is to struggle to justify the existing measures.

Grinblatt and Titman (1993) show that the inequality (28.3) holds only when all the fund's holdings are used in the calculation. For example, a manager with information may overweight some assets and underweight others for hedging purposes, and if one side of the hedge is omitted, the measure can be misleading. It is not really justified to apply holdings-based measures to subsets of stocks, but it is done all the time in the literature. This seems analogous to the problem of measuring the correct market portfolio in the CAPM. The data can never be perfect.

28.8   An Example of Holdings-Based Performance with Risk Adjustment

Ferson and Mo (2016) provide an example of a holdings-based performance measure with risk adjustment. They assume that the SDF is given by linear factor models: $m=a-b'r_B$. This assumption is admittedly very strong, and I present it here as an illustration. Consider a factor model regression for the excess returns of the N underlying securities:

$$r=a+\beta r_B+u, \tag{28.6}$$

where β is the $N\times K$ matrix of regression betas, and $E(ur_B)=0$. Define the vector of "abnormal" (or idiosyncratic) returns as the sum of the intercept plus residuals: $v=a+u$. A fund forms a portfolio using weights x, and the excess returns are $r_p=x'r=(x'\beta)r_B+x'v$. Let $w'=x'\beta$ be the asset allocation weights on the factor portfolios. Substituting into the definition of the SDF alpha and using the assumption that r_B has zero alphas, Ferson and Mo (2016) obtain

$$\alpha_p=a\,\text{Cov}(w'r_B)-b'E\{[r_Br_B'-E(r_Br_B')]w\}+E\{(a-b'r_B)x'v\}. \tag{28.7}$$

The first term of equation (28.7) captures market-level timing through the covariance between the portfolio weights and the subsequent factor returns. (One factor is the market.) This is essentially the classical holdings-based measure applied at the "asset allocation" level. The second term captures "volatility timing," through the relation between the portfolio weights and the second-moment matrix of benchmark return innovations. The third term of (28.7) captures selectivity ability. This analysis shows that the original Grinblatt and Titman measure leaves out the timing behavior, because as noted above, it assumes joint normality with homoskedasticity, so that an informed manager never gets a signal that conditional volatility will change. A GMM translation of the model in (28.7) was reviewed in chapter 20.

28.9   Summary

The traditional alphas used in the literature are not to be trusted as normative indicators for when to buy or sell funds, but SDF alphas are better. Traditional alphas can be equivalent to the correct SDF alphas, but this requires that an Appropriate Benchmark be used. Despite a traditional focus on mean variance efficient portfolios, such portfolios are almost never Appropriate Benchmarks, unless you are willing to assume quadratic utility. However, seemingly in contrast to conventional wisdom, Sharpe ratios can be justified as performance measures under certain (strong) assumptions, when they are used properly by comparing the Sharpe ratio of the fund to be evaluated with the Sharpe ratio of an Appropriate Benchmark. Finally, current holdings-based approaches to performance measurement are flawed, because they do not properly adjust for risk. I offer suggestions for how to improve the implementation of holdings-based performance measures in future work, and supply a recent example.

VI SELECTED TOPICS

29 Production-Based Asset Pricing

In a basic economics course, students are exposed to the Fisher diagram, as illustrated below, and learn that optimizing consumers equate their marginal rates of substitution (MRS) for consumption of two goods, X and Y, to the relative market prices of the goods. In asset pricing problems, the good X can be consumption today and the good Y can be consumption tomorrow. The market price is then the one-period interest rate, the slope of the k-k' line in figure 29.1. The figure also shows a concave production frontier (from T to T'). Firms maximize value by picking a point on this frontier that equates their marginal rates of transformation (MRT) in production to the interest rate, the slope of the g-g' line in the figure. Consumption, savings, and production decisions are adjusted until, in general equilibrium, the two slopes are the same.

In an asset pricing problem with uncertainty and complete markets, we can let good X be consumption today and good Y be consumption tomorrow in state s. There are multiple Y-axes, one for each state that can occur tomorrow. The slopes for each state represent the state prices P_s and can be used to price any asset. We saw in chapter 2 that the state price should equal the MRS for state-s consumption.

29.1 Setup and Goals

In a general equilibrium, the MRT should equal the MRS, so we should be able to price assets off of the MRT. This is production-based asset pricing. It's a laudable goal, but I will argue that the literature has not yet fully developed or implemented the idea. This topic is a research opportunity. It seems plausible that firms operate under relatively intense pressure to maximize profit or value, compared with consumers whose optimizations may be clouded by time or borrowing constraints, psychological costs, and more. Perhaps the MRT can work better than the MRS.

Some simple algebra sets this argument up and introduces the main concepts. Imagine a firm in complete markets, allocating its current resources W into various kinds of capital k_s. The capital allocated to state s produces future output $F(k_s)$ if and only if state s occurs.

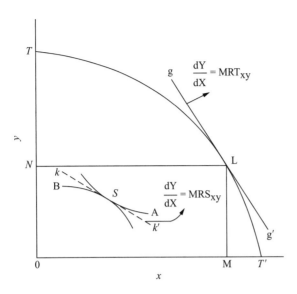

Figure 29.1
Production-based asset pricing.

The state prices are P_s. The firm maximizes the value of its output, allocating all its resources to the various kinds of capital:

$$\text{Max}_{\{ks\}} \sum_s P_s F(k_s) - \lambda(\sum_s k_s - W), \tag{29.1}$$

with first order condition $\qquad P_s F_k(k_s) = \lambda, \tag{29.2}$

where $F_k(k_s)$ is the marginal product of capital in state s, and λ is the shadow value of current resources. In complete markets, consumer optimization implies that the state prices are $P_s = \pi(s)[u'(C_s)/u'(C_0)]$, where $\pi(s)$ is the probability of state s. We have two equivalent ways to compute state prices:

$$P_s = \pi(s)[u'(C_s)/u'(C_0)] = \lambda/F_k(k_s). \tag{29.3}$$

The left-hand equality represents the usual, consumer-based asset pricing. The second equality is production-based asset pricing. In this example, the state price is inversely proportional to the marginal product of capital in the state. In states where capital is very productive, there is a lot of output to eat, and so state prices are low. The consumer problem says that the marginal utility of consumption in such states is low. If we measure the marginal product of capital better than the marginal utility of consumption, then production-based asset pricing could work better than consumption-based asset pricing.

Most studies in this area include costs for adjusting capital in the model. Let's replace the value of the firm's output with the value net of costs, $\sum_s P_s[F(k_s) - (\eta/2)k_s^2]$, assuming a quadratic adjustment cost equal to $(\eta/2)k_s^2$. The costs are increasing with the scale of the capital deployed. With this change, the first order condition for the firm's problem determines the state price: $P_s = \lambda\,/[F_k(k_s) - \eta k_s]$. The marginal benefit of investing is now reduced by the marginal adjustment cost, which is state dependent through the level of capital deployed, k_s. Adjustment costs lead to higher state prices, as the cost of producing future consumption is higher in current consumption units. The higher state prices associated with higher adjustment costs can help with goals like explaining the equity premium puzzle, which can be thought of as stating that the state prices that value the market portfolio in the classical consumer problem are "too low."

A central concept in the production-based asset pricing literature is the *investment return*, denoted by R_{t+1}^I. The investment return is the rate at which a firm can transform time t consumption into time $t+1$ consumption through marginal investment changes. In our stripped-down example, assume that a marginal dollar is allocated to the state capitals Δk_s, with $\sum_s \Delta k_s = 1$. The payout in state s for the marginal dollar of investment capital is $F_k(k_s)\Delta k_s$. The value of this payoff is $\sum_s P_s\{F_k(k_s)\Delta k_s\} = \sum_s (\lambda/F_k(k_s))\{F_k(k_s)\Delta k_s\} = \lambda$. The gross investment return in state s is therefore: $R_{t+1}^I(s) = 1 + F_k(k_s)\Delta k_s/\lambda$. With adjustment costs, the investment return is $1 + [F_k(k_s) - \eta k_s]\,\Delta k_s/\lambda$, with $\lambda = \sum_s P_s\{[F_k(k_s) - \eta k_s]\Delta k_s\}$.

In this simple model, a stock is simply a claim to the net payout of the firm—its liquidating dividend in each state s. The dividend in state s is $F(k_s)$, and the price of the stock is $\sum_s P_s F(k_s)$. The gross equity return in state s is $R_{t+1}^e(s) = 1 + F(k_s)/\sum_s P_s F(k_s)$.

Many studies in this area compare stock returns to investment returns. In particular, studies examine the hypothesis that the stock return is equal to the investment return. We can see from the above that this is typically not the case. However, if we assume constant returns to scale in production, then $F_k(k_s)k_s = F(k_s)$. Then the marginal return to capital, $F_k(k_s)\Delta k_s/\lambda$, is equal to the average return to capital, $F(k_s)/\sum_s[P_s F(k_s)]$, and the investment return is equal to the stock return.

One of the challenges for production-based asset pricing is that the investment and capital accumulation process often do not allow enough flexibility. In particular, the level of capital at time $t+1$ in the model is often deterministic once the investment decision is set at time t. One way to fix that is to introduce a *total factor productivity* (TFP) shock, A_{t+1}. Suppose that the production function depends on capital k_t and labor L_t, and output is given by $y_{t+1} = A_{t+1}F(k_t, L_t)$. Take the logs and think of a regression of $\ln(y_{t+1})$ on $\ln(k_t)$ and $\ln(L_t)$. The TFP is the residual from the regression that tries to predict output with the productive inputs.

These are some of the main concepts, laid out in the simplest possible way that I can imagine. The literature in this area involves some pretty nasty-looking algebraic expressions that articulate these ideas in more realistic settings. It's time to dive into some of that.

29.2   Physical Investment with Capital Adjustment Costs

The first level of realism is physical investment choice. Cochrane (1991) is one of the early papers. In his model, the firm's problem is to maximize at time t the value of its future output net of its investments in capital, subject to a capital accumulation constraint:

$$\text{Max}_{\{ks, It\}} \sum_s P_s [F(k_t, s_t) - I_t] - \sum_t \lambda_t (k_{t+1} - g(k_t, I_t)), \tag{29.4}$$

where $g(\cdot)$ describes the capital accumulation process associated with physical investment I_t. This is deterministic: $k_{t+1} = g(k_t, I_t) = (1 - \delta)(k_t + I_t)$, where δ is the depreciation rate of physical capital. Cochrane includes a labor input and labor choice in the model, but he does not get much out of the labor, so I omit it here for simplicity. The first-order condition for investment I_t at time t is $0 = -(\sum_s P_s) + g_1(t)\lambda_t$, where $g_1(t)$ denotes the partial derivative.

Cochrane (1991) articulates the model for quadratic adjustment costs associated with physical investment. The cost, $\varphi_t = (\alpha/2)(I_t/K_t)^2 I_t$, takes a bite out of your initial investment. The capital accumulation condition becomes $K_{t+1} = g(I_t, K_t) = (1 - \delta)[K_t + (1 - (\alpha/2)(I_t/K_t)^2) I_t]$. The net-of-cost capital placed in service at time t depreciates for one period by $t + 1$. The firm's problem is now

$$\text{Max} \sum_s P_s [F(k_t, s_t) - I_t] - \sum_t \lambda_t (k_{t+1} - g(k_t, I_t)). \tag{29.5}$$

The first-order condition with respect to labor is unchanged. The first-order condition with respect to investment now involves the derivatives of φ_t.

The physical investment restriction is awkward, because firms can only vary investment I_t across time. They can no longer control the state-contingent capital k_s at time $t + 1$, like they could in our stripped-down model with complete markets. Thus, the first-order condition for the firm's problem can no longer identify the state prices P_s. The physical investment process makes it hard to do production-based asset pricing using the MRT.

29.3   Investment Returns

Cochrane (1991), Sharathchandra (1991), and Braun (1991) derive the investment return for production functions with adjustment costs for installing capital. A similar analysis can also be conducted for investment in labor or human capital. The derivation is similar to the perturbation argument used in chapter 3 to find m_{t+1} in consumption-based models. Suppose that the firm has set an optimal investment plan $\{I_t\}$ by maximizing value:

$$\text{Max}_{\{It\}} V_t = E_t \{\sum_{j \geq 0} m_{t+j} [F(t + j) - I_{t+j}]\}, \text{ subject to } k_{t+1} = g(k_t, I_t),$$
$$\equiv E_t \{[F(t) - I_t] + V_{t+1}\}, \tag{29.6}$$

where we normalize $m_t = 1$ to be the numeraire at time t. The perturbation increases investment at time t, relative to the optimal strategy, by a small amount Δ, resulting in $I_t(\Delta) = I_t + \Delta$. This results in $F_k(t + 1)g_1(t)\Delta$ more output at time $t + 1$ and more capital at $t + 2$. Suppose the firm also cleverly reduces investment at time $t + 1$ so as to leave its capital stock and future output unchanged for dates $t + \tau$, for $\tau > 1$. The time $t + 1$ investment that accomplishes this is $I_{t+1}(\Delta) = I_{t+1} - [g_K(t + 1)g_1(t)/g_1(t + 1)]\Delta$. The reduction in investment at $t + 1$ can be paid out and gets added into the investment return at time $t + 1$. The perturbed value function is denoted as $V_t(\Delta)$. The initial investment strategy is optimal if

$$\partial V_t(\Delta)/\partial \Delta|_{\Delta=0} = 0 = -1 + E_t[m_{t+1}\{F_k(t+1)g_I(t) + [g_K(t+1)/g_I(t+1)]g_I(t)\}]. \tag{29.7}$$

This corresponds to the *m*-talk equation (1.2) holding for the investment return

$$R^I(t+1) = [F_k(t+1) + g_K(t+1)/g_I(t+1)]g_I(t). \tag{29.8}$$

So, we invest a dollar through this somewhat tortured perturbation to hold the future output fixed after $t+1$, and we get back the investment return, which should be priced by *m*-talk like any other asset. Of course, this perturbation is not an optimal strategy. Firms facing investment costs are unlikely to optimally make an investment and then reverse it after one period. But that is the point of the perturbation argument. The perturbation is feasible but not optimal. It is a departure from the optimal strategy. In our previous uses of the perturbation argument, we derived *m*. Here we use similar logic to derive another return based on the production function. But the asset pricing here is being done with the MRS, not with the MRT, so this is not quite production-based asset pricing.

29.4   Stock Returns and Investment Returns

We think of the stock price as the value of the future dividends paid by the firm. In the simplest production models, this is the total value of future net output V_t in equation (29.6). The stock return is the current dividend plus the change in the total value of the capital of the firm, which is the return to the average dollar invested in the firm that owns the production process. In contrast, the investment return is a marginal return to investment capital. In general, the two need not be the same, as we saw in our stripped-down model. However, if we assume constant returns to scale, the marginal productivity is constant and equal to the average productivity: $F(k_s) = F_k(k_s)k_s$. If we assume that the marginal value of investing capital in production is the average value of the invested capital, then $\sum_s P_s F(k_s) = \lambda$, and we can characterize a stock return as the return from holding a marginal unit of capital in production at time t and selling it at $t+1$. Under constant returns to scale, this return should equal the change in the total value of the firm's capital.

At time t the marginal cost of one unit of capital at $t+1$, in units of current investment, is $1/g_I(k_t, I_t)$. For this investment, you get output at $t+1$ equal to $F_k(t+1)1$, and you can sell off k_{t+2} in the amount $g_K(k_{t+1}, I_{t+1})$ with a value equal to $g_K(k_{t+1}, I_{t+1})/g_I(k_{t+1}, I_{t+1})$ at time $t+1$. The ratio of the total of these payoffs at $t+1$ to the cost at t is a stylized marginal stock return,

$$R^e(t+1) = [F_k(t+1) + g_K(t+1)/g_I(t+1)]g_I(t), \tag{29.9}$$

and it is equal to the investment return defined in equation (29.8). In a model with multiple types of capital inputs in the production function, Hayashi (1982) shows that the marginal stock return is a weighted average of the investment returns for the various types of capital. Much of the empirical literature in this area explores the relation between stock returns and investment returns, assuming constant returns to scale.

29.5   True Production-Based Asset Pricing

Although it is nice to have a relation between stock returns and investment returns, this is not asset pricing based on the MRT. There are several approaches to "true" production-based pricing in the literature. The first seems natural and is discussed in Jermann (2010). If we assume that there are as many linearly independent production processes as there are states of nature, then the market is complete in production. We could find the state prices, for example, using combinations of the production processes for different firms.

The suggested approach is similar to one for finding state prices when the securities markets are complete—we just substitute here the production payoffs for the security payoffs. Given S states and S linearly independent payoffs per unit of cost, arrayed in the matrix $\underline{R}$, the S-vector of state prices $\underline{P}$ gives us $\underline{1}' = \underline{P}'\underline{R}$, or $\underline{P}' = \underline{1}'\underline{R}^{-1}$, where $\underline{1}$ is an S-vector of ones (the unit costs). This seems simple enough, but I don't think anyone has yet given it a serious empirical try. Jermann (2010) did not go very far with the approach, focusing most of his attention on the stock return in the model and how it is related to the parameters of the production process. Nowadays, we have rich data on firm-level production processes, so it might be possible to use a large panel of firms' outputs and take this idea further. The assumption would be that the output of our large cross section of firms may present as many linearly independent claims as there are independent claims on consumption.

One approach that might deliver a version of true production-based pricing could be adopted from studies that try to estimate the pricing kernel m from stock and options data. In that literature, the assumption is often made that a stock index, taken as a proxy for aggregate wealth, summarizes the relevant states. The level of aggregate output might be a natural state variable here. Data on output and a statistical model can be used to estimate the objective distribution or "p measure," which delivers the state probabilities. Previous literature uses options prices on the market index to infer the q measure or risk-neutral probabilities (Banz and Miller 1978; Breeden and Litzenberger 1978). Combining these with a risk-free rate proxy delivers m as the ratio of the risk-neutral to the objective probabilities (Jackwerth and Rubinstein 1996; Bakshi, Kapadia, and Madan 2003; Chernov 2003; Barone-Adesi, Engle, and Mancini 2006). Possibly, the recovery approach of Ross (2015) can also be adapted. Option prices might not be available for the levels of aggregate output, but there may be some other clever ways to model the risk-neutral probabilities for output.

One approach to production-based asset pricing that has been used in the literature is to close the model, as in "general" equilibrium, making the agents consume the aggregate "dividends" paid by firms. For example, this gives us $m_{t+1} = u'(C_{t+1})/u'(C_t) = u'(F(t+1) - I_{t+1})/u'(F(t) - I_t)$, and we are pricing assets using capital, labor, investment, and other data through the production function. This approach is used in Jermann (1998), Berk, Green, and Naik (1999), Carlson, Fisher, and Giammarino (2004), İmrohoroğlu and Tüzel (2014), and others. Cochrane (1991) and Sharathchandra (1993) assume that consumers have log utility functions and that the aggregate investment return is the return on aggregate wealth, $R_{W,t+1}$, as in the model of Rubinstein (1974). In this model, $m_{t+1} = R_{W,t+1}^{-1}$. These approaches still rely on the form of the representative agent utility function $u'(\cdot)$ and still use the MRS to price things. Ideally, the MRT should be defined independent of investors' utility functions.

In many models, the general equilibrium approach is used in a reduced form. It is common to assume, for example, that $\ln(m_{t+1})$ is an affine function of the shocks in the model, such as the TFP shock $\ln(A_{t+1})$. Another reduced-form approach is taken by Cochrane (1996), who assumes that m is linear in investment returns and prices assets using the corresponding linear factor model. This approach has recently been extended by Fama and French (2016) and Hou, Xue, and Zhang (2015), who introduce investment and profitability factors. Zhang (2017) provides a review of the literature focusing on investment factors.

Perhaps the most satisfying version of production-based asset pricing is where the MRT is directly used to price assets. Belo (2010) pursues a version of this approach. His model allows firms to make state-dependent output choices, harking back to our stripped-down model of complete production markets. At time t, firms can choose their TPF "shocks" to be realized in time $t+1$ output, subject to the restriction $E[(A_{t+1}/\Theta_{t+1})^\alpha]^{1/\alpha} \le 1$, where Θ_{t+1} is an exogenous parameter that determines the ability to choose output shocks. In states where Θ_{t+1} is high, it is easier to choose a large A_{t+1} without hitting the bound. The condition imposes a trade-off: To produce a lot in one state, the firm must produce less in other states. Belo derives the MRT in this model as a function of the parameters Θ_{t+1}. To implement the model, he assumes that the Θ_{t+1} follow a factor structure with a few common factors and finds the MRT in terms of price and output change data.

29.6 Empirical Evidence

The empirical literature on production-based asset pricing has taken several directions. The approach that takes up the most journal pages to date is model calibration. Parameters are chosen for the model, the shocks are fed in exogenously, and the model is used to endogenously generate stock returns, investment, and so forth. The model is considered a success if the artificial data from it can match moments in the actual data.

Calibration studies examine multiple goods, multiple production inputs, different types of productive capital, different scale economies, various adjustment cost specifications, and the like. The models have been used to study simple corporate finance decisions like debt structure (Liu, Whited, and Zhang 2009) and much more. Calibration is something of a black art to me, so I won't say too much about it. Jermann (2010) illustrates the approach. He studies the properties of the stock return implied by the investment process and links the parameters of the model to those of the stock return. For example, for the model to generate a large equity premium, the adjustment cost function must be sufficiently convex. With quadratic adjustment costs, the marginal cost of investment is linear, and the investment-to-capital ratio is proportional to the market-to-book value ratio, a version of Q. Calibration seems to generate a lot of insights about how the structure of the models work.

Empirical studies examine the properties of output and investment data, the estimated investment returns, TFPs, and their relation to asset market returns both over time and in a cross section of firms. Early examples include Cochrane (1991) and Braun (1991), who find that investment returns are highly correlated with the growth rates of aggregate investment, for all the adjustment cost and production functions they examine. İmrohoroğlu and Tüzel (2014) study firm-level TFP for a large sample of US publicly traded firms. They find that firms with relatively low TFP in a given year have lower stock returns that year, with higher future average returns and implied costs of capital going forward. Firms' TFPs are related in the cross section to various firm characteristics, including the market capitalization (size), book/market ratio, investment/capital ratio, asset growth, inventories, a measure of organizational capital, research and development expenditures, the ratio of real estate to total physical capital, leverage, firm age, and other characteristics. This suggests to me that there may be enough richness in the cross-sectional structure to implement production-based asset pricing on cross sections of firms, as suggested previously.

Studies examine the hypothesis that the stock return is equal to the investment return, as implied by constant returns to scale. Cochrane (1991) finds that conditional expectations of future stock returns, estimated by regressing stock returns on lagged variables, behave similarly to expected investment returns. He also finds that expected future gross national product growth, conditioned on lagged investment returns, is similar to that conditioned on lagged stock market returns. Liu, Whited, and Zhang (2009) test the hypothesis that the expected difference between the stock return and the investment return is zero, and that the variance of two returns are equal; they find some evidence in support of these predictions.

Another line of empirical inquiry is to test the pricing of the investment return, as in $E_t\{m_{t+1}R_{1,t+1}\} = 1$, for specified choices of the MRS, m_{t+1}. This test essentially asks whether physical investment returns are priced like financial asset returns. Braun (1991) rejects this hypothesis when m_{t+1} is the inverse of the return on aggregate wealth, using various proxies for aggregate wealth. Sharathchandra (1993) rejects the hypothesis for a simpler production function in which the investment return is a fixed linear function of the output-to-capital ratio.

Cochrane (1996) assumes that the MRS m_{t+1} can be described using a factor model in which investment returns or investment growth are the factors. Such models can be empirically tested just like the other factor models we studied earlier in the book. Cochrane finds that investment returns are useful factors in a multiple-beta asset pricing model, in the sense that they work about as well as the factors proposed by Chen, Roll, and Ross (1986) and better than a simple consumption model. Fama and French (2016) extend this approach, arguing that investment and profitability factors should augment their eponymous three-factor model; they find that it helps price stocks in several international markets when the five factors are used. The new factors include high-low type portfolios of stocks based on profitability (Robust minus Weak) and investment (Conservative minus Aggressive). Hou, Xue, and Zhang (2015) argue that the five-factor model of Fama and French may be improved, and that a four-factor model using the market, size, investment, and profitability factors works better. Zhang (2017) provides a recent review. Many devils lurk in the details of measuring profitability (e.g., Novy-Marx 2013; Ball et al. 2015), but we are now getting a bit far afield.

30 The Campbell-Shiller Approximation and Vector Autoregressions

This chapter reviews two tools together: the Campbell and Shiller (1988a) approximation and vector autoregressions (VARs). Each tool is important separately, but since they are used together so often, it make sense to review them together. Our discussion of VARs includes brief reviews of Granger (1969) causality and impulse-response functions. VARs are used with the Campbell-Shiller approximation to deliver the expected future values that appear in most of the applications. The Campbell-Shiller approximation has been a phenomenal success, and there must be more than 100 papers that have used it. Its key benefit is tractability. It allows a researcher to express the log of the price or the price/dividend ratio of an asset as a linear combination of the expected future returns and dividend growth rates of the asset. The linearity is the magic. It can be conjured up for dividend-to-price, and also for book-to-market, consumption-to-wealth, and more. I review some prominent examples and research that indicates some "gotchas" in the method.

30.1 The Campbell-Shiller Approximation

The easiest way to see the logic of this approximation is to start with a version of the well-known Gordon (1962) growth model. We all learned present-value calculations, where, for a given discount rate r, the price of a stock today, P_t, is equal to its discounted expected future dividends:

$$P_t = \sum_{j=1,\ldots,\infty} E_t(D_{t+j})/(1+r)^j. \tag{30.1}$$

If $E_t(D_{t+j}) = D_t(1+g)^j$, where g is the fixed dividend growth rate, then

$$P_t/D_t = (1+g)/(r-g). \tag{30.2}$$

If equation (30.2) could be applied in a world with time-varying dividend growth rates and discount rates, then when variation over time or across firms in the price/dividend ratio is observed, it would signal that either expected future growth rates or discount rates (or maybe both) had changed. More specifically, if the price/dividend ratio goes up, then the value attached to a given dividend level is higher, meaning that either expected future growth rates have gone up or discount rates for the future have gone down.

The Campbell-Shiller approximation basically extends this idea to a setting where discount rates and expected dividend growth rates vary over time. Here is how it works. Start with the definition of the continuously compounded return:

$$r_t = \ln(R_t) = \ln((P_t + D_t)/P_{t-1})) = \ln((P_t + D_t)/D_t)(D_t/D_{t-1})(D_{t-1}/P_{t-1})). \tag{30.3}$$

Write the first of the three product terms in (30.3) as $\ln((P_t + D_t)/D_t) = \ln(1 + e^{-dpt}) = f(dp_t)$, where $dp_t = \ln(D_t/P_t)$. Note that we have switched from price/dividend to dividend/price ratios. Expand the function $f(\cdot)$ in a first-order Taylor series around $\delta = E(dp_t)$ to obtain

$$\ln(1 + e^{-dpt}) \approx f(\delta) - [e^{-\delta}/(1 + e^{-\delta})](dp_t - \delta). \tag{30.4}$$

Campbell and Shiller (1988a) define $[e^{-\delta}/(1 + e^{-\delta})] \equiv \rho$, and based on historical average US stock market data, estimate it to be slightly less than 1.0. Let $\Delta d_t = \ln(D_t/D_{t-1})$. Now expand the log of the three products in equation (30.3) into the sum of the logs, and use (30.4) for the first term, obtaining:

$$r_t \approx \Delta d_t + dp_{t-1} + k - \rho dp_t, \tag{30.5}$$
where $k = \ln(1 + e^{-\delta}) + \rho\delta$.

Equation (30.5) contains the magic in the Campbell-Shiller approximation: We have an approximate expression for continuously compounded returns that is linear in log dividend growth rates and dividend/price ratios. Solving for dp_{t-1} and iterating forward:

$$\begin{aligned}
dp_{t-1} &\approx r_t - \Delta d_t - k + \rho dp_t, \\
&= r_t - \Delta d_t - k + \rho(r_{t+1} - \Delta d_{t+1} - k + \rho dp_{t+1}) \\
&= (r_t + \rho r_{t+1}) - (\Delta d_t + \rho\Delta d_{t+1}) - k(1 + \rho) \\
&\cdots \\
&= (\textstyle\sum_{j=0,\dots,\infty} \rho^j r_{t+j}) - (\textstyle\sum_{j=0,\dots,\infty} \rho^j \Delta d_{t+j}) - k/(1 - \rho).
\end{aligned} \tag{30.6}$$

We can take the expectation at time $t - 1$ on both sides of the bottom row of equation (30.6) to obtain an interpretation similar to the Gordon growth model. The variance of log dividend/price ratios should reflect variance in discounted expected future returns or in expected future dividend growth rates. If both these expectations don't vary much over time, dividend price ratios also shouldn't vary much over time. But the log dividend price ratio has an annual standard deviation of 0.29, based on NYSE index data for 1926–1986, compared with 0.13 for dividend growth and 0.21 for ex-dividend stock market returns (Campbell and Shiller 1988a, table 2). If dividend/price ratios rise at a point in time, it means that valuations are lower and expected future discount rates have increased or expected future dividend growth rates have decreased, maybe both.

Campbell (1991) presents another version of the approximation that allows us to express return variation in terms of cash flow news and discount rate news. Take the expectation at time $t-1$ of equation (30.5) and subtract it from equation (30.5) to see that the unexpected stock return is $r_t - E_{t-1}(r_t) \approx (\Delta d_t - E_{t-1}(\Delta d_t)) - \rho(dp_t - E_{t-1}(dp_t))$. Using the bottom row of equation (30.6) to evaluate $(dp_t - E_{t-1}(dp_t))$, we find that

$$r_t - E_{t-1}(r_t) \approx N_{CFt} - N_{DRt}, \tag{30.7}$$

where $N_{DRt} \equiv (E_t \sum_{j=1,...,\infty} \rho^j r_{t+j}) - (E_{t-1} \sum_{j=1,...,\infty} \rho^j r_{t+j}) \equiv (E_t - E_{t-1})(\sum_{j=1,...,\infty} \rho^j r_{t+j})$, cleverly denoting the change in expected future discounted returns, or news about discount rate. N_{CFt} is defined similarly in terms of changes in expected future discounted dividend growths, except the j subscript in the summations starts at $j=0$. Equation (30.7) states that unexpected stock returns are approximately equal to news about cash flows less news about discount rates.

Campbell and Vuolteenaho (2004) use equation (30.7) to decompose stock market betas. A beta on, say, the market r_m is $\beta = Cov(r_t - E_{t-1}(r_t), r_{mt})/Var(r_m)$, so

$$\beta \approx Cov(N_{CFt}, r_{mt})/Var(r_m) + Cov(N_{DRt}, r_{mt})/Var(r_m) = \beta_{CF} - \beta_{DR}. \tag{30.8}$$

In a version of the Merton (1973) model, the cash flow beta has a positive risk premium, and the discount rate beta has a negative risk premium. They cleverly denote the cash flow beta as the "good" beta and the discount rate beta as the "bad" beta, drawing an analogy with "good" and "bad" forms of cholesterol in your blood. Small market capitalization and value stocks (those with high ratios of book/market value) are found to have higher cash flow betas than large and growth stocks, which helps explain the size and book/market anomalies—the high returns on small and value stocks that translate into positive alphas for small and value stocks in the traditional CAPM.

30.2   Return and Dividend Growth Predictability

Chapter 32 discusses several kinds of predictability and reviews attempts in the literature to predict the levels of stock market returns in particular. Predicting the market return typically involves regressions of future stock returns over time on lagged variables, and the lagged dividend/price ratio has been a popular predictor variable. Table 30.1 is an example from Cochrane (2008).

Table 30.1 shows that returns appear to be predictable in this sample using lagged dividend/price ratios, but cash-flow growth rates, proxied by Δd_{t+1}, do not appear to be predictable using dividend/price ratios. (I think that r_{t+1} in this table refers to the log of the return, but it doesn't say. This just goes to show that even the best among us can't write with complete clarity all the time.)

The decomposition in equation (30.6) suggests an interesting interpretation of the return and dividend growth regressions. If we could actually observe the discounted future returns and dividend growth rates, regressions of these two on the lagged dividend/price ratio would be "complementary regressions": Their slopes, intercepts, and residuals are

Table 30.1
Forecasting regressions.

Regression	b	t	R^2(%)	σ(bx)(%)
$R_{t+1} = a + b(D_t/P_t) + \varepsilon_{t+1}$	3.39	2.28	5.8	4.9
$R_{t+1} - R_{Ft} = a + b(D_t/P_t) + \varepsilon_{t+1}$	3.83	2.61	7.4	5.6
$D_{t+1}/D_t = a + b(D_t/P_t) + \varepsilon_{t+1}$	0.07	0.06	0.0001	0.001
$r_{t+1} = a_r + b_r(d_t - p_t) + \varepsilon_{rt+1}$	0.097	1.92	4.0	4.0
$\Delta d_{t+1} = a_d + b_d(d_t - p_t) + \varepsilon_{dpt+1}$	0.008	0.18	0.00	0.003

Source: Cochrane (2008), table 1. Copyright Society for Financial Studies.
Notes: R_{t+1} is the real return, deflated by the consumer price index, D_{t+1}/D_t is real dividend growth, and D_t/P_t is the dividend-price ratio of the Center for Research in Security Prices (CRSP) value-weighted portfolio. R_{Ft} is the real return on 3-month Treasury bills. Small letters are logs of corresponding capital letters. Annual data, 1926–2004. σ(bx) gives the standard deviation of the fitted value of the regression.

related in such a way that you only have to run one of the regressions to know what the other one would say. To illustrate simply, assume $y = x + z$. The two regressions are $x = a + by + u$, and $z = c + dy + v$. The identity linking y with x and z implies that $d = 1 - b$, $c = -a$, and $v = -u$. Cochrane (2008) uses similar logic to interpret the relative lack of evidence for predictability of future dividend growth rates using dividend/price ratios, as implying stronger evidence for predictability of future returns using dividend/price ratios.

30.3   Vector Autoregressions

VARs have long been popular in economics, as exemplified by the influential work of Sims (1980), in modeling the dynamics of variables in economic systems. VARs are also quite popular in macrofinance these days and pretty good at forecasting. (Not to be confused with VaR, which stands for "value at risk.") Let y_t be an $N \times 1$ vector of variables and write the pth-order vector autoregression as

$$y_t = c + \sum_{j=1,\ldots,p} \varphi_j y_{t-j} + \epsilon_t, \tag{30.9}$$

where c is an $N \times 1$ vector of intercepts, the φ_j are $N \times N$ coefficient matrices, and the regression errors ϵ_t are assumed to be mean zero and uncorrelated with the lagged y values.

Of course, estimating the VAR is a special case of a GMM problem, like most of the empirical methods in finance, and we can proceed to estimate and test hypotheses about the VAR using the GMM. Of course, ML assuming normality is a popular special case.

A useful example of testing hypotheses about VAR coefficients is *Granger causality* (Granger 1969), which addresses the question of how one set of variables forecasts another. Partition the vector of variables as $y_t = (y_{1t}', y_{2t}')'$. We say that y_1 Granger-causes y_2 when lagged values of y_1 predict future values of y_2, conditioning on the lagged values of y_2. We can test the null hypothesis that y_2 does not Granger-cause y_1 by testing for zeros in the coefficient matrices. In particular, each φ matrix should be upper triangular if the null hypothesis is true. Granger causality isn't really about causality, as discussed in chapter 23, it's about forecasting.

A useful trick with VARs is impulse-response analysis. This is easy to illustrate by specializing equation (30.9) to a first-order VAR, or VAR(1):

$$y_t = c + \varphi y_{t-1} + \epsilon_t. \tag{30.10}$$

Iterating this regression forward, we see that

$$
\begin{aligned}
y_t &= c + \varphi(c + \varphi y_{t-2} + \epsilon_{t-1}) + \epsilon_t \\
&= (c + \varphi c) + \varphi^2 y_{t-2} + (\epsilon_t + \varphi \epsilon_{t-1}) \\
&\qquad \cdots \\
&= c/(1 - \varphi) + \sum_{k=0,\dots,\infty} \varphi^k \epsilon_{t-k}.
\end{aligned} \tag{30.11}
$$

The last line expresses the model in moving average form, as a function of the iid error shocks only. The expressions for the moving-average coefficients are especially simple for the AR(1), but the idea works for VAR(p) and vector autoregressive, moving average models, too. Now, suppose that we are interested in the effect of a shock to the vector of errors at time t on the future values of the variables at $t + s$. The last line of equation (30.11) reveals that $\partial y_{t+s}/\partial \epsilon_t = \varphi^s$, so the *impulse response function* is $= \varphi^s \Delta \epsilon_t$. We can study the impact $\Delta y_{t+s} \approx (\partial y_{t+s}/\partial \epsilon_t)\Delta \epsilon_t = \varphi^s \Delta \epsilon_t$ when the shocks are independent over time. In the VAR(1) model, the impact of the shocks damps out exponentially into the future, but richer models can be easily studied.

30.4   VARs Meet Campbell-Shiller Approximations

Define the vector $y_t = (r_t, \Delta d_t, dp_t)'$, and use the VAR(1) model. This could be generalized to include more variables; in particular, to higher order VARs by putting other lags of the variables into y_t. From equation (30.10), we see that $E_{t-1}(y_{t+j}) = \varphi^{j+1} y_{t-1}$, and $E_t(y_{t+j}) = \varphi^j y_t$. Therefore, the change in expectation is $(E_t - E_{t-1})y_{t+j} = \varphi^j(y_t - \varphi y_{t-1}) = \varphi^j \epsilon_t$. Let e1 be the vector $(1, 0, 0)'$. Then, $(E_t - E_{t-1})r_{t+j} = e1'\varphi^j \epsilon_t$. Taking the weighted infinite sum, we have

$$
\begin{aligned}
N_{\mathrm{DR}t} &= (E_t - E_{t-1})(\sum_{j=1,\dots,\infty} \rho^j r_{t+j}) = e1'(\sum_{j=1,\dots,\infty} \rho^j \varphi^j)\epsilon_t \\
&= e1'\rho\varphi(I - \rho\varphi)^{-1}\epsilon_t.
\end{aligned} \tag{30.12}
$$

Now, $N_{\mathrm{CF}t}$ may be computed as a "residual:"

$$
\begin{aligned}
N_{\mathrm{CF}t} &= (r_t - E_{t-1}(r_t)) + N_{\mathrm{DR}t} \\
&= e1'(y_t - E_{t-1}(y_t)) + N_{\mathrm{DR}t} \\
&= e1'\epsilon_t + N_{\mathrm{DR}t}.
\end{aligned} \tag{30.13}
$$

Campbell (1991) uses this VAR to decompose the variance of unexpected stock returns in the United States into the variance of news about cash flows, the variance of news about discount rates, and the covariance between the two types of news. He finds that during 1927–1988, about a third of the variance of a stock market index is attributed to the variance of cash flow news, about a third to the variance of discount rate news, and about a third to the covariance. Vuolteenaho (2002) applies the decomposition to portfolios of stocks sorted by market capitalization and argues that the cash flow news is the dominant component for individual stocks.

30.5   Examples

With about a hundred papers using Campbell-Shiller decompositions, there are plenty of examples. I want to mention just a few that extend the basic ideas. First, just like a stock price is the discounted present value of its dividends, a consumer's wealth can be thought of as the present value of her future consumptions. Lettau and Ludvigson (2001a) use this insight to derive an equation analogous to (30.6):

$$c_t - w_t \approx (\textstyle\sum_{j=0,\ldots,\infty} \rho^j r_{t+j}) + (\textstyle\sum_{j=0,\ldots,\infty} \rho^j \Delta c_{t+j}), \tag{30.14}$$

where $c_t - w_t = \ln(C_t/W_t)$ is the log of the consumption/wealth ratio, and Δc_{t+j} is the continuously compounded growth rate of consumption. They develop a proxy for wealth using income and dividend data in cointegrating regressions, and they call the resulting consumption/wealth proxy *Cay*. They implement the model using a VAR and quarterly data. They find that *Cay* can predict stock market returns for 1 to 6 years and is also useful for predicting dividend growth rates out to about 5 years during 1948–2001.

Vuolteenaho (2002) derives a similar approximation for the log of the book/market ratio. He uses the assumption of "clean surplus" accounting, which is to say, there are no accruals. If B_t is the book value of a stock, D_t the dividends, and X_t the earnings at time t, the assumption is

$$B_t = B_{t-1} - D_t + X_t. \tag{30.15}$$

Define the accounting return on equity as $r_{OEt} = \ln((B_t + D_t)/B_{t-1})$, and an approximation similar to Campbell and Shiller (1988a) leads to

$$\ln(B_{t-1}/M_{t-1}) \approx (\textstyle\sum_{j=0,\ldots,\infty} \rho^j r_{t+j}) - (\textstyle\sum_{j=0,\ldots,\infty} \rho^j r_{OEt+j}), \tag{30.16}$$

where M_t is the market value, equal to price times the number of shares. Thus, earnings over book equity is used as the cash flow proxy in place of dividend growth rates. Vuolteenaho concludes that while cash flow news is larger, relative to discount rate news, at the firm level, the cash flow news diversifies quickly in a portfolio, which explains why it is not the dominant component for the market index.

While some studies refer to the Campbell-Shiller approximation as an "identity" or an "accounting identity," it is not. It is a first-order Taylor series approximation. Campbell and Shiller (1988a) examine the accuracy of the approximation using simulations. In a simulation calibrated to annual data, where the exact dp_t has a mean of -2.87 and a standard deviation of 0.18, the approximation error has a mean of 0.16 and a standard deviation of 0.003.

Chen and Zhao (2009) argue that because the cash flow news component in the Campbell-Shiller approximation is computed as a residual, it can't be estimated precisely enough to infer its relative importance for stock return variances. For example, if there is a missing factor in the model, it can appear in the cash flow news measure. To illustrate, they apply the approach to US Treasury bond returns, which should have zero cash flow news, and find that the estimated variance of cash flow news is larger than the variance of discount rate news, more so for the longer maturity bonds. They modify the approach to directly calculate the cash flow news and find that the results can be substantially different. In particular, they challenge the previous findings that good and bad betas can explain the return differences between value and growth stocks. They also document that the variance decomposition can be highly sensitive to the choice of lagged predictor variables.

31 Long-Run Risk Models

The long-run risk (LRR) model, first introduced by Bansal and Yaron (2004), has been a phenomenal success. It is characterized by a representative agent with Epstein-Zin preferences (reviewed in chapter 5) and a stochastic process for consumption growth featuring expected growth or volatility (or both) that is highly persistent. The persistent LRR variable, combined with preference parameters that imply a preference for the early resolution of uncertainty, make the agent in the model care a lot about shocks to the LRR process. The shocks get large risk premiums in the model. A rapidly expanding literature finds the models useful for fitting the equity premium puzzle, size and book-to-market effects, momentum, long-term return reversals in stock prices, risk premiums in bond markets, real exchange rate movements, and more (see the review by Bansal 2007).

The LRR variable, usually including the conditional expected consumption growth, is quite important for pricing assets in the model. However, it explains only a small part of the variance of consumption growth and is hard to see in the data. This aspect of the model made it controversial, and it took the authors a long time to get the original paper published. I remember when it was first presented at National Bureau of Economic Research meetings as a working paper. The feeling in the room was that nobody could find anything wrong with the model, but it was hard to believe.

Well, Bansal and Yaron have been vindicated, and the LRR framework has now become very popular. As of early 2018, the Bansal and Yaron (2004) paper had more than 2,700 citations, according to Google Scholar. This is "rock star" status in terms of the citation impact. As the LRR literature has expanded, papers have offered different versions of this model. This chapter reviews the basic structure of the LRR model and its interpretation, working through a stationary and a cointegrated version of the model. It then briefly reviews some criticisms and extensions of the LRR approach.

The LRR model assumes a representative agent with recursive preferences, introduced by Kreps and Porteus (1978) and further developed by Epstein and Zin (1989, 1991). In chapter 5, our discussion of these preferences emphasizes that they allow separate parameters to control risk aversion and the intertemporal elasticity of substitution. Here are the preferences represented in recursive form using the notation of Bansal and Yaron (2004):

$$U_t = \{(1-\beta)\, C_t^{(1-\gamma)/\theta} + \beta[E_t(U_{t+1}(1-\gamma))]^{(1-\gamma)/\theta}\}^{\theta/(1-\gamma)}, \tag{31.1}$$

where C_t is consumption, γ is the coefficient of relative risk aversion, $\theta = (1-\gamma)/(1-1/\psi)$, and ψ is the elasticity of intertemporal substitution (EIS). We saw in chapter 5 that when risk aversion is large relative to the EIS, the agent is averse to variance in the future expected utility in the sense that m is larger when the variance of the future expectation is larger. In this case, the agent is relatively unwilling to substitute consumption across time and prefers an early resolution of uncertainty.

Epstein and Zin (1991) derive the SDF for this model:

$$m_{t+1} = \beta^\theta (C_{t+1}/C_t)^{-\theta/\psi} R_{wt+1}^{(\theta-1)}, \tag{31.2}$$

where R_{wt+1} is the gross return of a claim to aggregate consumption. In their version of the model, this is equivalent to the aggregate wealth portfolio.

31.1   A Stationary Model

Bansal and Yaron add to the model a specification of the stochastic processes for consumption growth, $\Delta c_t = \ln(C_{t+1}/C_t)$; expected consumption growth; and its volatility:

$$\begin{aligned}
\Delta c_t &= x_{t-1} + \sigma_{t-1}\varepsilon_{ct}, \\
x_t &= \mu + \rho_x x_{t-1} + \varphi\sigma_{t-1}\varepsilon_{xt}, \\
\sigma_t^2 &= \underline{\sigma} + \rho_\sigma(\sigma_{t-1}^2 - \underline{\sigma}) + \varepsilon_{\sigma t},
\end{aligned} \tag{31.3}$$

where x_{t-1} is the conditional mean of consumption growth. The conditional mean is the latent LRR variable. It is assumed to be a stationary but persistent stochastic process, with ρ_x less than but close to 1.0. For example, in Bansal and Yaron (2004), $\rho_x = 0.98$. Consumption growth is conditionally heteroskedastic, with conditional volatility σ_{t-1}, given the information at time $t-1$. The shocks $\{\varepsilon_{ct}, \varepsilon_{xt}, \varepsilon_{\sigma t}\}$ are assumed to be homoskedastic and independent over time, although they may be correlated. A GMM formulation of this model was reviewed in chapter 20.

31.2   Interpreting the Model

To interpret the model, Bansal and Yaron use the log-linear approximation of Campbell and Shiller (1988a), applied to approximate the return on the aggregate wealth portfolio as a function of consumption growth and the log price/dividend ratio:

$$\ln(R_{wt+1}) \approx k_0 + k_1 \ln(P_{t+1}/D_{t+1}) + \Delta c_t - \ln(P_t/D_t). \tag{31.4}$$

As shown by Constantinides and Ghosh (2011), the log price/dividend ratio is an affine function of the conditional variance and expected consumption growth:

The coefficients in equation (31.5) can we found

$$\ln(P_t/D_t) = A_0 + A_1 x_t + A_2 \sigma_t^2. \tag{31.5}$$

The coefficients in equation (31.5) can be found in the model by substituting this expression into the approximation for $\ln(R_{wt+1})$ and writing the Euler equation as $1 = E_t\{\exp(\ln(m_{t+1}) + \ln(R_{wt+1}))\}$. Now, assuming joint normality of $\ln(m_{t+1})$ and $\ln(R_{wt+1})$, the normal moment generator implies that

$$E_t\{\ln(m_{t+1}) + \ln(R_{wt+1})\} + (1/2)\mathrm{Var}_t\{\ln(m_{t+1}) + \ln(R_{wt+1})\} = 0. \tag{31.6}$$

If we plug $\ln(m_{t+1}) = \ln(\beta^\theta(C_{t+1}/C_t)^{-\theta/\psi} R_{wt+1}^{(\theta-1)})$ and the expression (31.4) for $\ln(R_{wt+1})$ into equation (31.6), we see that the coefficients on the constant, x_t, and σ_t^2 must be zero if the expression is to equal zero for all t. This delivers three equations in three unknowns that can be solved for the coefficients A_0, A_1, and A_2. In particular:

$$\begin{aligned} A_1 &= (1 - 1/\psi)/(1 - k_1\rho_x), \\ A_2 &= (\theta/2)[(1 - 1/\psi)^2 + (k_1 A_1 \varphi)^2]/(1 - k_1\rho_\sigma). \end{aligned} \tag{31.7}$$

The coefficient A_1 measures the sensitivity of the price/dividend ratio to changes in the expected consumption growth x_t. It changes sign depending on whether the EIS ψ is greater or less than 1.0. If the EIS is less than 1.0, the sign of A_1 is negative and stock price/dividend ratios are "countercyclical," in the sense that they are negatively related to expected consumption growth. If the EIS is greater, price/dividend ratios are "procyclical." The coefficient A_2 measures the sensitivity of the price/dividend ratio to changes in the conditional variance of consumption growth, σ_t^2. When $\theta < 0$ ($\gamma > 1/\psi$), risk aversion is larger than the inverse of the EIS, and higher consumption risk is bad news for stock prices, $A_2 < 0$.

The implications of the LRR model for the equity premium puzzle can be seen through the expected excess return on the aggregate wealth portfolio. Using the most general model for risk premiums from chapter 4, the expected risk premium can be found approximately as:

$$\begin{aligned} \mathrm{Cov}_t(-m_{t+1}, \ln(R_{wt+1})) &= \gamma\sigma_t^2 + (1-\theta)(k_1 A_1 \varphi)^2 \sigma_t^2 \\ &\quad + (1-\theta)(k_1 A_2)^2 \mathrm{Var}_t(\varepsilon_{\sigma t+1}). \end{aligned} \tag{31.8}$$

When $\theta = 1$, we have the classical power utility model, and the risk premium is the first term only in equation (31.8). The premiums for the LRRs do not appear in the standard power-utility of consumption model when $\theta = 1$ ($\gamma = 1/x$). When $\theta < 1$, the risk premium is higher than in the power utility case and is increasing in both the variance of the expected consumption growth and the conditional volatility of consumption growth.

Bansal and Yaron (2004) show that in this model, the innovations in the log of the stochastic discount factor are, to good approximation, linear in the three-vector of shocks $u_t = [\sigma_{t-1}\varepsilon_{ct}, \varphi\sigma_{t-1}\varepsilon_{xt}, \varepsilon_{\sigma t}]$. The linear function has constant coefficients. Because the coefficients in the stochastic discount factor are constant, unconditional expected returns are approximately linear functions of the unconditional covariances of returns with the shocks.

That is, a linear beta pricing model with three factors approximately describes unconditional expected returns. Expressions for the risk premiums for the three shocks may be found as functions of the deeper structural parameters of the model.

Constantinides and Ghosh (2011) use unconditional means, variances, autocovariances, and higher moments of consumption and dividend growth to identify the structural parameters. They find that the model is rejected when this additional structure is imposed. Ferson, Nallareddy, and Xie (2013) examine the stationary LRR model for step-ahead prediction of asset returns. They find that the model does not work much better than the classical CAPM, except for momentum returns, where the model has an uncanny ability to fit and forecast well. Imposing the deeper structure on the risk premium parameters results in models with smaller variances but larger forecast errors.

31.3 A Cointegrated Model

Bansal and Yaron (2004) include an equation that specifies the process for aggregate dividend growth as an affine function of the expected consumption growth. They show that this model allows the aggregate price/dividend ratio to be even more sensitive to the expected consumption growth. However, it makes sense that the aggregate consumption and the aggregate dividends in this model should be linked. Indeed, in the model of Mehra and Prescott (1985), the model was closed by assuming that the aggregate consumption must equal the aggregate dividend.

Bansal, Dittmar, and Lundblad (2005), Bansal, Gallant, and Tauchen (2007), and Bansal, Dittmar, and Kiku (2007) examine models in which it is assumed that the natural logarithms of aggregate consumption and dividend levels are *cointegrated*. Two random variables x and y are cointegrated if both are nonstationary variables (like random walks), but there is a linear combination of the two that is a stationary process. The implication is that x and y can't in some sense get "too far" apart. The individual consumption and dividend series are nonstationary, so they can wander off arbitrarily far from a starting point, but the difference between consumption and dividends is a stationary process that cannot be explosive. The model is:

$$
\begin{aligned}
d_t &= \delta_0 + \delta_1 c_t + \sigma_{t-1}\varepsilon_{dt}, \\
\Delta c_t &= a + \gamma's_{t-1} + \varphi_c\sigma_{t-1}\varepsilon_{ct} \equiv x_{t-1} + \varphi_c\sigma_{t-1}\varepsilon_{ct}, \\
\sigma_t^2 &= \underline{\sigma} + \rho_\sigma(\sigma_{t-1}^2 - \underline{\sigma}) + \varepsilon_{\sigma t},
\end{aligned}
\tag{31.9}
$$

where d_t is the natural logarithm of the aggregate dividend level, and s_t is a vector of state variables at time t. Bansal, Dittmar, and Kiku (2007) allow for time trends in the levels of dividends and consumption, but find that they don't have much effect.

Constantinides and Ghosh (2011) show that in this version of the model, the innovations in the log stochastic discount factor are approximately linear in the heteroskedastic shocks to consumption growth, the state variables s_t, and the cointegrating residual $\sigma_{t-1}\varepsilon_{dt}$. The coefficients in this linear relation are constant over time. Therefore, unconditional expected excess returns are approximately linear in the covariances of return with these shocks, and we have a beta pricing model with two more factors than the dimension of the state vector s_t.

The cointegrated model can be formulated for GMM estimation as follows:

$$u_{1t} = d_t - \delta_0 - \delta_1 c_t,$$

$$u_{2t} = \Delta c_t - [a_0 + a_1 rf_{t-1} + a_2 \ln(P/D)_{t-1} + a_3 u_{1t-1}] \equiv \Delta c_t - x_{t-1},$$

$$u_{3t} = rf_t - [g_0 + g_1 rf_{t-1} + g_2 \ln(P/D)_{t-1} + g_3 u_{1t-1}],$$

$$u_{4t} = \ln(P/D)_t - [h_0 + h_1 rf_{t-1} + h_2 \ln(P/D)_{t-1} + h_3 u_{1t-1}], \tag{31.10}$$

$$u_{5t} = r_t - \mu - \beta(u_{1t}, u_{2t}, u_{3t}, u_{4t})',$$

$$u_{6t} = \lambda - [\beta'V(r)^{-1}\beta]^{-1}\beta'V(r)^{-1}r_t.$$

Identification and estimation are similar to those for the stationary model, discussed in chapter 20. The GMM estimator of the cointegrating parameter δ_1 is superconsistent and has a nonstandard limiting distribution, as shown by Stock (1987). This implies that the u_{1t-1} shock estimates may be more precise than standard regression estimates.

Constantinides and Ghosh (2011) and Ferson, Nallareddy, and Xie (2013) estimate cointegrated versions of the model. The latter study finds that the cointegrated model fits the data and forecasts better than the stationary model.

31.4 Extensions

The LRR model has come to have quite an impact. While the model has had phenomenal success fitting various aspects of financial market data, such as the equity premium puzzle, size and book-to-market effects, momentum, long-term return reversals in stock prices, risk premiums in bond markets, real exchange rate movements and more, it was initially controversial, as described above. Much of the work in this area involves calibration, which I admit appears to me to be a black art. Myriad instances have been noted of the model not being able to fit certain asset return and macroeconomic data moments very well. By now, many of those shortcomings have been addressed with extensions of the basic models. Here are a few examples of the criticisms and extensions.

There has been some debate about where to set the utility function parameters. The model requires that the elasticity of intertemporal substitution is small relative to the effect of risk aversion, so the agent is relatively unwilling to substitute future for present consumption and is averse to LRRs. The coefficient of relative risk aversion needs to be larger than the inverse of the EIS. Also, the EIS has to be greater than 1.0. Previous studies, such as Hall (1978) and Campbell (2003), regressed consumption on short-term interest rates and interpreted the coefficients to imply that the EIS was less than 1.0. In response to questions about the magnitude of the EIS, Bansal, Kiku, and Yaron (2009) argue strongly through simulations and calibration that the EIS should be greater than 1.0. Bansal, Kiku, and Yaron (2016) point out that earlier regression estimates could find values less than 1.0 because of measurement error, even if the true EIS was greater than 1.0.

Epstein, Farhi, and Strzalecki (2014) argue that the literature has not asked the introspective questions about the EIS as it has for risk aversion. They propose a thought experiment for the EIS, asking: What fraction of your lifetime consumption would you give up to resolve all uncertainty about the future values? They call this percentage the "timing premium." This approach is similar to the experiment Lucas (2003) used to conclude that the uncertainty associated with cyclical fluctuations might not be economically very large. They find in model calibrations that the timing premium is 20–27% if the EIS is set to 1.0 but is much larger in the LRR model. With EIS = 1.5, an LRR agent might be willing to give up as much as 50% of her lifetime consumption.

Beeler and Campbell (2012) provide a detailed diagnostic on the LRR model performance, comparing two previous calibrations with the data. They point out that the conditional variance of stock returns in the model is too predictable, using a lagged dividend yield, relative to what we should find in the data. The model seems inconsistent with the weak relation between consumption growth and short-term interest rates. The model also generates a negatively sloped term structure of real interest rates, whereas we see positive slopes most of the time in the data, at least for nominal interest rates.

Bansal, Kiku, and Yaron (2009) respond to some of these criticisms. They compare the LRR with the external habit model of Campbell and Cochrane (1999), extend the data to pre–World War II, and use annual frequencies. The predictability patterns in the data line up better in this longer horizon data. For example, consumption growth predictability is higher, with regression R-squares as large as 15% in 5-year consumption growth rates. Other predictability patterns also look more favorable for the model. Hansen, Heaton, and Li (2008) characterize the LRR model in terms of permanent and temporary components to m. Using VARs, they find significant predictability in the expected consumption growth variable.

Early calibrations questioned the relative importance of the LRR aspects of expected consumption growth versus consumption volatility. Bansal, Kiku, and Yaron (2009) leaned toward the volatility component as the more important one, and they built that into their calibration in Bansal, Kiku, and Yaron (2016). Dittmar, Palomino, and Yang (2016) seem to agree. Boguth and Kuehn (2013) find empirically that the conditional volatility of consumption significantly varies over time and is a negatively priced risk factor. The lagged volatility can predict future returns on stock portfolios. Zhou and Zhu (2014) extend the model to have both short- and long-run components of volatility. They find that this helps the model to fit a negative volatility risk premium and to better match the low predictability of consumption growth using dividend yields.

Production sectors are brought into the LRR framework by Croce (2014) and Ai (2010). Here the expected production growth has a persistent LRR component. Ai finds this inclusion improves a calibrated model's ability to fit the consumption/wealth ratio and the return on aggregate wealth. Croce finds that stock prices are highly sensitive to shocks to the expected consumption growth.

Leisure choice is brought into the LRR problem by Dittmar, Palomino, and Yang (2016), using a two-good model in consumption and leisure, with substitution between the two. They find that this model does not produce negatively sloped term structures. Inflation risks have been introduced in the LRR setup by Koijen et al. (2010) and Bansal and Shaliastovich (2013). Koijen et al. find that this helps the model to better explain the consumption/wealth ratio. Bansal and Shaliastovich find that with an LRR component for expected inflation, the model no longer implies downward-sloping term structures of interest rates. In their model, the consumption and inflation process are dependent, so there are real effects of inflation.

Information quality is studied in the LRR set up by Ai (2010) and Ai et al. (2018). Ai et al. find that when the representative agent can't distinguish between short- and long-run risk shocks, the model can generate a downward-sloping term structure of equity yields, as observed in the data. Ai finds that the effect helps to better fit the consumption/wealth ratio.

Drechsler (2013) incorporates rare but potentially large jumps in the persistent components of consumption growth, dividend growth, and volatility. LRR models with these features can generate a negative volatility risk premium and a VRP whose lagged values can predict stock returns. (Recall that the VRP is the difference between an option-implied variance and the expected realized variance of a stock market index.)

Barro and Jin (2016) bring rare disasters into the LRR model. The rare disaster model of Barro (2006) uses standard power utility functions and assumes that the levels of consumption can get hit, with low probability, by a shock that takes a permanent bite of size b out of the level of consumption. Gabaix (2012) develops a variable rare disaster model, where the size of the bite taken by a disaster can vary over time. Barro and Jin set calibrations suggesting that investors with a preference for early resolution of uncertainty would be willing to give up 20% of GDP to eliminate the rare disaster risk. As this brief review suggests, this is an active literature with much more room to run.

32 Predictability: An Overview

When pondering predictability in financial markets, three simple questions come to mind. Predict what? With what? So what? This chapter is organized accordingly.

In terms of what to predict, we first discuss attempts to predict the levels of asset returns. We then discuss predicting return variances or volatility, including autoregressive conditional heteroskedasticity (ARCH) and Generalized ARCH (GARCH)–style models, followed by attempts to predict second co-moments, betas, and higher moments of returns. The chapter concludes with a discussion of cross-sectional, or relative return, predictability. Here the attempt is to predict the difference between two returns, as in a spread portfolio. The literature in this area has also spent some energy trying to predict the future growth rates of various cash flow or real economic growth rate proxies, like consumption growth rates, dividends, and earnings per share. Some of this is discussed in chapter 30 and some is briefly mentioned in this one, but here I focus mainly on predicting asset returns and their moments.

In terms of what is used to make predictions, I organize the discussion along the lines suggested in Fama (1970). Fama describes increasingly fine information sets that can be used for prediction. Weak-form predictability uses the information in past market prices. Semistrong-form predictability uses variables that are obviously publicly available, and strong form uses anything else. While there is a literature characterizing strong-form predictability (e.g., analyzing the profitability of corporate insiders' trades), this chapter concentrates on the first two categories of information.

That covers the questions, Predict what? And with what? Now, we have the "So what?" Interest in predicting stock prices or returns is probably as old as the markets themselves, and the literature on the subject is enormous. Investors are obviously interested in predicting asset returns so they can use the forecasts to form better-performing portfolios. We learned a lot in chapter 9 about how to use forecasts, interpreted as conditional expected returns and covariances, in mean variance efficient portfolios. The issue of predictability in stock returns has broader economic implications. For example, it relates to the efficiency of capital markets in allocating resources to their highest valued uses. We discussed the m-talk version of interpreting predictability and market informational efficiency in chapter 6.

The interpretation of predictability, and even the evidence for its existence, have been controversial. However, for the field of financial economics and asset pricing in particular, allowing for predictability through time-variation in expected returns, risk measures, and volatility has been one of the most significant developments of the past three decades for model formulation. Conditional asset pricing models have provided a rich setting for the study of the dynamic behavior of asset markets in almost every area. To pick one setting dear to my heart, CPE, reviewed in chapter 26, is the application of models with predictability to the problem of evaluating the performance of portfolio managers. Models that allow for time-varying conditional moments produce different inferences about fund performance than do traditional measures that do not allow for predictability, and conditional models have influenced both academic views and professional investment practice. Research on predictability has stimulated numerous advances in the statistical and econometric methods of financial economics. Research on predictability in asset markets is likely to continue and will likely remain both useful and controversial for some time. This is a long chapter, but it only scratches the surface.

32.1   Predicting the Levels of Returns

I can think of three views about predicting the levels of returns. The first view argues that it should not be possible to predict returns in an efficient market. This is the most traditional view. A second view argues that predictability is a natural outcome of an efficient capital market. A third view ascribes predictability to imperfectly rational investors and other market frictions.

Early studies, reviewed by Fama (1970), concluded that a martingale or random walk was a good model for stock values or their logarithms. Thus, the best forecast of the future value was the current value, and the return of a stock could not otherwise be predicted. We can understand this view in terms of the joint hypothesis as described in chapter 6, assuming that m is a constant over time, or that the conditional covariances of returns with m divided by $E_t(m_{t+1})$ are fixed over time. Although these assumptions are pretty strong, the current financial economics literature still reflects this traditionalist view to a certain, but I think diminishing, extent. With this view, any predictability in the levels of returns represents exploitable inefficiencies or other problems in the way capital markets function. The traditionalist view argues that traders will bid up the prices of stocks with predictably high returns, thus lowering their return and removing any predictability at the new price (e.g., Friedman 1953; Samuelson 1965). Active portfolio managers find this view attractive, as it means they can make money by making the markets more efficient.

The efficient markets view of predictability was described by Fama (1970). According to this view, returns might be predictable if required or expected returns vary over time in association with changing interest rates, risk, or the aggregate risk aversion in the market. If expected returns vary over time, there may be no abnormal trading profits and thus, no incentive to exploit the predictability. Predictability can be expected in an efficient capital market. Write the return as $R = E(R|Z) + u$, where Z is the publicly available information at the beginning of the period, and u is the unexpected return. Since $E(u|Z) = 0$, the unexpected return cannot be predicted ahead of time, but $E(R|Z)$ can be tracked using information in Z, and thus R can be predicted to some extent. Thus, predictability, in the efficient markets view, rests on systematic variation through time in the expected return.

The third view is that imperfections are assumed to impede price-correcting, or arbitrage trading. Predictable patterns can thus emerge when there are important imperfections. Behavioral studies focus on human imperfections, but similar arguments apply to imperfections like trading costs, taxes, or important costs for processing or responding to information. The human and market frictions might interact. Predictable patterns are thought to be exploitable, in the sense that an investor who could avoid the friction or cognitive imperfection could profit from the predictability at the expense of other traders. Some active money managers like this view, too. In this view, things might be improved by education to control behavioral biases, which is an attractive idea for some professors.

32.1.1 Weak-Form Return Prediction

Early studies examine the ability to predict stock market returns using lagged returns and other weak-form information, as in so-called technical analysis, for example. Fama (1970) surveys the earlier evidence and finds it lacking. Later studies find evidence that bond returns can be predicted using lagged bond yields (e.g., Fama 1984b; Stambaugh 1988; Fama and Bliss 1987; Campbell and Shiller 1991; Dai and Singleton 2002; Cochrane and Piazzesi 2005). Predictability in exchange rate changes is found by Hansen and Hodrick (1983), Backus, Foresi, and Telmer (2001), and Bansal and Shaliastovich (2013), among others.

Predicting bond returns is one area where a lot of work has been done on weak-form return predictability. Recall from chapter 6 that pure discount bonds of any two maturities define a forward rate. The current continuously compounded forward rate f_t is the difference between the future spot interest rate, r_{st+1}, and the excess return on the long-term bond, R_{Lt+1}, over that of the short-term bond with rate r_{st}. Taking expectations at time t, $f_t = E_t(r_{st+1}) + E_t(R_{Lt+1} - r_{st})$. Subtracting the current spot rate from both sides, Fama (1984b) arrives at a relation between the current slope of the term structure and two expectations: $f_t - r_{st} = E_t(r_{st+1} - r_{st}) + E_t(R_{Lt+1} - r_{st})$. Any variation in the slope must correspond to variation in expected spot rate changes or to variation in expected risk premiums. Regressing the future excess returns on the current slope for Treasury bills up to 1 year to maturity, Fama finds that most of the variation in the term slope is associated with changing expected risk premiums. Fama and Bliss (1987) extend the idea to a monthly sample of 1- to 5-year spot rates that they approximate from Treasury bond prices, and they find that the term slope can predict future spot rate changes out to 4 years, with regression R-squares increasing with the horizon.

Much of the literature on weak-form predictability of the levels of stock returns can be characterized through autoregressions. Let r_t be the continuously compounded rate of

return over the shortest measurement interval ending at time t. Let $r(t, t+H) = \sum_{j=1,\ldots,H} r_{t+j}$
Then

$$r(t, t+H) = a_H + \rho_H r(t-H, t) + \varepsilon(t, t+H) \tag{32.1}$$

is the autoregression, and H is the return horizon. The autoregression is related to the variance ratio statistic,

$$\mathrm{Var}\{r(t, t+H)\}/H\mathrm{Var}(r_t), \tag{32.2}$$

proposed by Working (1949) and studied for stock returns by Lo and MacKinlay (1988, 1990) and others. If the returns have no autocorrelation, the variance ratio should equal 1.0, because the variance of a sum is the sum of the variances. Cochrane (1988) studies the variance ratio as a function of the autocorrelation in returns. Kaul (1996) provides an analysis of various statistics that have been used to evaluate weak-form return predictability, showing how they can be viewed as combinations of autocorrelations at different lags, with different weights assigned to the lags.

Conrad and Kaul (1988, 1989) study serial dependence in weekly stock returns. They point out that if expected returns follow an autoregressive process, the actual returns would be described by the sum of an autoregressive process and white noise, and thus follow an ARMA process. The autoregressive and moving average coefficients would be expected to have the opposite signs: If current expected returns increase, it may signal that future expected returns are higher, but stock prices fall in the short run, because the future cash flows are discounted at a new, higher rate. The two effects can offset each other, and returns could produce small autocorrelations. Estimating ARMA models, Conrad and Kaul find that the autoregressive coefficient for weekly returns on stock portfolios are positive, near 0.5, and can explain up to 25% of the variation in the returns on a portfolio of small-firm stocks.

Fama and French (1988) use autoregressions like equation (32.1) to study predictability in portfolio returns measured over 1-month to multiyear horizons. They report autocorrelations for various return horizons for a variety of industry, size, and index portfolios. They find autocorrelations as large as −50% at the 4-year horizon. Fama and French show that if stock prices contain both a permanent (i.e., random walk) and a stationary shock component, returns will follow an autoregressive process, and the regression slope is expected to fall as the return horizon lengthens, to about −0.5, and then approach zero from below as the horizon goes to infinity. Fama and French (1988) find such U-shaped patterns in the autocorrelations as a function of the horizon, with seemingly strong negative serial dependence, or mean reversion, at 4- to 5-year horizons.

Mean reversion can be consistent with either time-varying expected returns or mispricing. If mispricing, mean reversion would be expected if stock values depart temporarily from the fundamental, or correct prices, but are drawn back to that level. Cecchetti, Lam, and Mark (1994) simulate a standard power utility consumption model with a regime-shifting process and find that commonly used measures of mean reversion usually lie within a 60% confidence interval of the median value produced by the simulations. So, mean reversion matching what is observed in the data can be consistent with a simple rational asset pricing model.

The autocorrelations in portfolio returns are driven by the underlying stocks in the portfolio. When the number of stocks is large enough, it is not the autocorrelations of the individual stocks but their lead-lag cross correlations that drive the portfolio autocorrelation. This is similar to the observation that the variance of a portfolio is driven by the average covariance of the stocks and not by the variance of the stocks, as the number of stocks gets large. Lo and MacKinlay (1990) examine weekly portfolio autocorrelations and find that large firms "lead" small firm stocks, in the sense that future small firm stock returns are correlated with past large firm stock returns, but the reverse is not observed.

32.1.2   Semistrong-Form Return Prediction

Semistrong-form predictability studies document predictability using lagged variables that are publicly available information. Early examples include Keim and Stambaugh (1986), Poterba and Summers (1988), and Fama and French (1989). These studies describe the time variation in expected returns that the lagged predictors reveal. They focus on predicting the levels of stock market returns over various return horizons. Evidence for predictability using lagged variables soon emerged for bond returns (e.g., Fama and French 1989) and in other asset classes. Several studies explore restrictions on the amount of predictability that can arise in asset pricing models. Prominent examples include Kirby (1998), Ross (2005), and Huang and Zhou (2017). These studies derive restrictions on the slope coefficients and R-squares in predictive regressions for stock returns based on asset pricing models.

Studies of semistrong-form predictability can be described using the regression

$$r(t, t + H) = \alpha_H + \beta'_H Z_t + v(t, t + H), \tag{32.3}$$

where Z_t is a vector of variables that are publicly available by time t. Many predictor variables have been analyzed in published studies, and it is useful to group them into categories. Chang and Ferson (2018) divide them into valuation ratios, interest rates and spreads, anomaly formation spreads, market activity measures, macroeconomic variables, and miscellaneous and technical indicators.

Valuation ratios are measures of cash flows or fundamental value divided by the stock price. Keim and Stambaugh (1986) use a constant as the numerator of the ratio and de-trend the price. Rozeff (1984), Campbell and Shiller (1988a), and Fama and French (1989) use dividend/price ratios; Pontiff and Schall (1998) and Kothari and Shanken (1997) use the book value of equity divided by price. Boudoukh et al. (2006) and Lei (2006) add share repurchases and other noncash payouts, respectively, to the dividend measure and find that it predicts better when defined this way. Lettau and Ludvigson (2001b) propose a macroeconomic version of a valuation ratio: aggregate consumption divided by a measure of aggregate wealth (discussed in chapter 30) that they call *Cay*. All these studies find the regression coefficients β_H to be significant.

Malkiel (2004) reviews a valuation ratio approach that he calls the "Federal Reserve Model." Here, the market price/earnings ratio is empirically modeled as a function of producer prices, Treasury yields, and other variables, and the difference between the model's output and the ratios observed in the market are used to predict the market's direction. (Malkiel finds that the model does not outperform a buy-and-hold strategy.)

Rozeff (1984) and Berk (1995) argue that valuation ratios should generally predict stock returns. Consider the simplest model of a stock price P as the discounted value of a fixed flow of expected future cash flows or dividends: $P = c/R$, where c is the expected cash flow, and R is the expected rate of return. Then, $R = c/P$, and the dividend/price ratio is the expected return of the stock. If predictability is attributed to the expected return, as in the efficient markets view, then a valuation ratio should be a good predictor variable. Chapter 30 shows how the Campbell-Shiller approximation generalizes this idea.

The second category of semistrong-form return predictors is bond yields and yield spreads. Fama and Schwert (1977) were among the first to observe that the level of short-term Treasury yields predicts returns in equation (32.3) with a negative coefficient. They interpreted the short-term yield as a measure of expected inflation. Ferson (1989) argues that these regressions imply that the covariances of stock returns with *m* must vary over time with changes in interest rates. Keim and Stambaugh (1986) study the yield spreads of low-quality over high-quality bonds and find predictive ability for stock returns, and Campbell (1987) studies several different yield spreads in shorter-term Treasury securities. Fama and French (1989) assemble a list of variables from studies in the 1980s and describe their relations with US business cycles.

Another category of semistrong-form predictor variables for stock returns includes calendar and seasonal effects. These are among the first predictors identified in the literature, such as the January effect in small stock returns (Keim 1983). The list of effects that have been related to stock returns and the list of studies is too long to cite here. (For reviews, see Haugen and Lakonishok 1988; Schwert 2003.) Some examples include the season (winter versus summer), the month of the year (especially high returns in January), the time of the month (first versus last half), holidays, the day of the week (low returns on Mondays), the time of the day, the amount of sunlight (as in seasonal affective disorder), and even the frequency of geomagnetic storms.

Of course, many other semistrong-form predictor variables have been proposed for the levels of returns, and more will doubtless be proposed in the future. Additional variables include the fraction of equity issues in new issues of corporate securities (Baker and Wurgler 2000), firms' investment plans (Lamont 2000), the average "idiosyncratic," or firm-specific component of past return volatility (Polk, Thompson, and Vuolteenaho 2004), the level of corporate cash holdings (Greenwood 2005), the aggregate rate of dividend initiation (Baker and Wurgler 2000), share issuance (Pontiff and Woodgate 2008), and the political party currently in office (Santa-Clara and Valkanov 2003). Chang and Ferson (2018) study a sample of more than 40 predictors from the literature for monthly stock market returns.

Studies of semistrong-form predictability in stock index returns typically report regressions with small R-squares over annual or shorter horizons, as the fraction of the variance in returns that can be predicted with the lagged variables is small. The R-squares are larger for longer-horizon returns. For example, Goyal and Welch (2003) run regressions for stock market index returns with a combined set of common predictors and find in-sample R-squares that vary from just over 1% in monthly data to 8% in quarterly, 19% in annual, and 61% for 5-year returns. This is interpreted as the result of expected returns that are more persistent than returns themselves, as would be expected if expected returns vary relatively slowly over time. The variance of the sum of the expected returns accumulates with longer horizons faster than the variance of the sum of the returns, and the R-squares increase with the horizon (e.g., Fama and French 1989).

Small changes in expected returns can produce economically significant changes in asset values, because stocks are long-duration assets. Studies such as Kandel and Stambaugh (1996), Campbell and Viceira (2002), and Fleming, Kirby, and Ostdiek (2001) show that optimal portfolio decisions can be affected to an economically significant degree by return predictability, even when the amount of predictability, as measured by R-squared, is small. Campbell and Thompson (2005) show how the R-squared of a predictive regression is related to the Sharpe ratio that a mean-variance agent using the information to form a conditionally mean variance efficient portfolio can obtain. As an example, they suggest that an investor can use a predictive R-square of 8% to increase the Sharpe ratio by as much as 2/3.

The evidence of semistrong-form predictability has survived some "out-of-sample" tests, working in other countries and over different time periods, and it has failed in others. Welch and Goyal (2007) find limited step-ahead predictive ability in a sample of standard lagged predictors. Campbell and Thompson (2005), however, using step-ahead tests, find that semistrong-form predictability holds up in recent data when forecasts are constrained to be positive.

Some studies find that semistrong-form stock market predictability, measured using lagged variables, has weakened in recent samples. It may be that the predictability was never really there, or that it was real when first publicized but diminished as traders attempted to exploit it. Ferson, Heuson, and Su (2005) examine semistrong-form predictability by regressing individual stocks on firm-specific predictors and then measure the average of the covariances of the fitted values. It can be shown that the variance of the expected return on a large portfolio is approximately the average covariance of the individual expected returns. As the firm-specific predictors had not been examined extensively in the literature at that time, these predictors may be less subject to naive data mining biases. These authors find no evidence that predictability, measured in this way, is weaker in recent subperiods.

A recently growing literature focuses on predicting the levels of returns by combining the forecasts from a large set of lagged variables. For example, Rapach, Strauss, and Zhou (2010) combine the forecasts from the individual predictors used in Welch and Goyal (2007) and find that the average forecast has a smaller measurement error and improved out-of-sample performance. Bali, Nichols, and Weinbaum (2017) use the cross section of dividends, earnings, and book values. Light, Maslov, and Rytchkov (2107) use 26 firm characteristics. Yang (2018) uses eight accounting variables (chapter 12 describes the partial least squares method that he uses to combine the predictors). Timmerman (2006) reviews combined forecasts in other studies.

Asset pricing theory associates predictability in the levels of returns to changes in the second moments, that is, the covariances with m. Studies find that predictability using lagged variables is largely explained by asset pricing models with multiple risk factors, if these models allow the premiums associated with those risks to vary over time (e.g., Ferson and Harvey 1991; Ferson and Korajczyk 1995; Avramov and Chordia 2006).

Some studies consider the innovations in standard lagged predictors as risk factors in beta pricing models, following the insight of Campbell (1996). Han and Lee (2006) use innovations in a term spread and default spread as risk factors in a three-beta model. Koijen, Lustig, and Van Nieuwerburgh (2017) use innovations in a term spread and the Cochrane and Piazzesi (2005) factor, which is a combination of Treasury forward rates. Fama and French (1993) consider innovations in a term and default spread as additional factors in their three-factor model. Petkova (2006) finds that innovations in dividend yields, a default-related yield spread, a term spread and short-term interest rates provide factor proxies in a multiple-beta model that performs about as well as the Fama and French (1993) factor model. Chang and Ferson (2018) evaluate several predictors from this perspective.

Another stream in the literature uses lagged variables that were initially proposed to predict returns to also predict fundamentals, such as dividend growth rates, earnings growth, and future consumption growth. This idea is motivated by a central intuition of asset pricing theory. If a variable predicts stock returns R, then it should also predict the stochastic discount factor (SDF) m of the model, so that the Euler equation error, $mR - 1$, is unpredictable, as the model requires. The stochastic discount factor depends on consumption or cash flow growth, so the predictability of stock returns should be related to the predictability of consumption or cash flow growth. For example, the Campbell and Shiller (1988a) decomposition explicitly links the variance of dividend yields to either changes in expected future returns or dividend growth rates, as discussed in chapter 30.

32.1.3 Data and Methodological Issues in Predictive Regressions

Even though the regressions (32.1) and (32.3) seem pretty straightforward, interpreting the predictability evidence for stock returns based on these regressions is not. It can be argued that one of the greatest contributions of the literature on semistrong-form stock return predictability is the methodological lessons it has taught researchers in the field. The methodological issues include the various ways to run the regression, issues of data snooping, spurious regression, standard error estimation, multiple comparisons across return horizons, efficient estimation, finite-sample bias, microstructure effects, regime shifts, and more. This section discusses these issues.

32.1.3.1 Three Ways to Go

The first issue is how to run the predictive regressions. The literature has employed three basic approaches. The first is to run regression (32.3) with overlapping data on $r(t, t+H)$, observed each period t, which is usually 1 month (e.g., Fama and French 1989). This approach uses all the data, compared with a sampling scheme that measures nonoverlapping returns every H months, and should therefore be more efficient than the sampling scheme. In a 70-year sample, you only get 14 nonoverlapping 5-year returns. However, the overlap induces autocorrelation in the error terms in the form of an $(H-1)$-order moving average process. This was discussed in the general GMM case in chapter 17. To conduct inference about β_H, it is necessary to estimate the coefficients and the standard errors without bias in the presence of the moving average error terms. Such estimations are complicated by the evidence that stock return data are conditionally heteroskedastic. The standard errors from Newey and West (1987a) that allow for autocorrelation and heteroskedasticity, discussed in chapter 17, are popular here.

A second approach recognizes that what we really want to estimate is the numerator of the regression coefficient, $\mathrm{Cov}\{r(t,t+H),Z_t\} = \mathrm{Cov}\{\sum_{j=1,\ldots,H} R_{t+j}, Z_t\}$. Hodrick (1992) realized that if the variables are covariance stationary, then $\mathrm{Cov}\{\sum_{j=1,\ldots,H} R_{t+j}, Z_t\} = \mathrm{Cov}\{R_t, \sum_{j=1,\ldots,H} Z_{t-j}\}$. This suggests using the return horizon $H=1$ on the left-hand side of the regression (32.3) and replacing Z_t on the right-hand side with $\sum_{j=1,\ldots,H} Z_{t-j}$. This approach is also taken by Cochrane (1988). In this approach, the error terms in the regression are not overlapping, and there is no induced moving average structure to account for when obtaining consistent standard errors. There may be efficiency gains compared with the first approach, but these depend on the stochastic process that drives the variables. The moving averaged Z predictors are likely to be highly persistent, which raises its own concerns.

A third approach to estimating the predictive regression coefficient is to model the single-period data $\{R_t, Z_{t-1}\}$ using a VAR and then infer the value of the long-horizon coefficient β_H from the autoregression parameters. This approach is taken by Kandel and Stambaugh (1990), Campbell (1993), and others. It can be efficient if the VAR is correctly specified but is subject to error if the autoregression is not correctly specified. Boudoukh, Richardson, and Whitelaw (1994) present a statistical analysis comparing the three main approaches to estimating long-horizon return regressions.

32.1.3.2 Data Snooping

"Data mining" refers to sifting through the data in search of predictive or associative patterns. There are two kinds of data mining. Sophisticated data mining accounts for the number of searches undertaken when evaluating the statistical significance of the finding (e.g., White 2000). Accounting for the number of searches is important, because if 100 independent variables are examined, we expect to find five that are "significant" at the 5% level, even if there is no predictive relation. Naive data mining, or "data snooping," does not account for the number of searches.

Data snooping may be the most significant methodological concern in the current asset pricing literature. This problem was discussed in chapter 24, along with methods to account for multiple comparisons. Concerns about data snooping for lagged variables to predict the levels of returns are expressed by Lo and MacKinlay (1990), Foster, Smith, and Whaley (1997), Ferson, Sarkissian, and Simin (2003), Novy-Marx (2014), and others.

Given the strong interest in predicting stock returns among academics and practitioners, and the many studies using the same data, the big problem is that it is difficult to account for the number of searches. By now, there probably have been at least as many regressions run using the Center for Research in Security Prices (CRSP) database as there are numbers in the database. Compounding this problem are various selection biases. Perhaps most difficult is the fact that only "significant" results get circulated and published in academic papers. No one knows how many insignificant regressions were run before those results were found.

A reasonable response to these concerns is to see whether the predictive relations hold out-of-sample. This kind of evidence is mixed, and the exercises are usually not really out-of-sample. Instead, a step-ahead approach is typically applied, where only data available at the forecast date are used to estimate the model and form the forecast. Practitioners call this a "back-test." The problem here is that often the data have already been looked at, and back-test results can be data snooped just like in-sample regressions. Some studies find support for predictability in step-ahead exercises (e.g., Fama and French 1989; Pesaran and Timmerman 1995). Semistrong-form variables show some ability to predict returns beyond the US data where they were originally studied (e.g., Harvey 1991; Solnik 1993; Ferson and Harvey 1993, 1999). Other studies conclude that predictability using many of the semistrong-form variables does not hold outside of the original samples (e.g., Goyal and Welch 2003, and Welch and Goyal 2007). Even this evidence is difficult to interpret, because a variable could have real predictive power yet still fail to outperform a naive benchmark when predicting step-ahead (Campbell and Thompson 2005; Hjalmarsson 2006).

32.1.3.3 Spurious Regression

A potential issue with predictive regressions is spurious regression bias, where the usual *t*-ratio appears significant but there is no predictive relation. Spurious regression has been studied by Yule (1926) and Granger and Newbold (1974), who warn that empirical relations may be found between the levels of trending time series that are actually independent. For example, given two independent random walks, it is likely that a regression of one on the other will produce a "significant" slope coefficient, evaluated by the usual *t*-statistics.

In equation (32.3), the dependent variables are stock returns, which are not highly persistent. However, recall that the returns may be considered to be the sum of the expected returns plus unpredictable noise. If the expected returns are persistent, even if stationary time series, there is a risk of spurious regression in finite samples. Because the unpredictable noise represents a substantial portion of the variance of stock returns, spurious regression effects will differ for stock returns from the classical setting of Granger and Newbold. The classical setting, with unit root or near–unit root processes, has received a lot of attention in the economics and econometrics literature. The spurious regression problem for stock returns, accounting for the white noise implied by the unexpected returns, has received some attention in recent econometric studies (e.g., Deng 2013), but more attention is probably warranted.

Spurious regression results in t-ratios for predictive regressions that are too large in finite samples when the null hypothesis of no predictability is true. This problem might be caused by the numerator or the denominator of the t-ratio. Ferson, Sarkissian, and Simin (2003) conduct simulations that demonstrate that the problem is not the numerator of the t-ratio. The point estimates of the slope coefficient are well specified in finite samples. The problem is the denominator. The estimates of the standard errors are too small in finite samples under spurious regression.

Ferson, Sarkissian, and Simin study the interaction between data mining and spurious regression. They find that data mining for predictor variables interacts with spurious regression bias in equation (32.3). The two effects reinforce each other, because more highly persistent series are more likely to be found significant when snooping for predictor variables. Simulations suggest that many of the regressions in the literature, based on individual predictor variables, could be the result of a spurious mining process. Ferson, Sarkissian, and Simin compute the critical number of searches, just large enough to render a lagged variable insignificant in the predictive regression, and they find that for a sample of 13 common predictors, all it would take is between two and five searches.

Powell et al. (2006) extend the analysis of the interaction between data mining and spurious regression to conclude that recent studies of presidential regimes (Santa-Clara and Valkanov 2003), the "Halloween Indicator" (Bouman and Jacobsen 2002), and business cycle effects in momentum (Chordia and Shivakumar 2002) appear insignificant in view of their combined effects. Chun (2009) derives a bias correction accounting for both spurious regression and lagged stochastic regressor bias when there is data mining. He applies his corrections to international equity market return regressions and finds, with data up to 2012, that 8 out of the 18 national equity market returns have at least one predictor that remains significant. Novy-Marx (2014) criticizes some lagged predictors using simulation—in particular, the sentiment index of Baker and Wurgler (2006)—and finds that such predictors may be subject to a spurious mining bias.

Ferson, Sarkissian, and Simin (2003) address the problem of what to do about spurious regression with data mining in practice. They examine several standard error estimators and several potential solutions by using simulations. They find that the key is to get the autocorrelation of the predictor variables down to about 0.95 or less in monthly data, which effectively removes the spurious regression problem. Then the problem of data mining can be addressed using standard multiple comparisons statistics. One of the best ways to do this when the original predictor is too persistent is to transform it by subtracting a lagged moving average, as advocated by Campbell (1991).

32.1.3.4   Multiple Comparisons Across Return Horizons

From the perspective of testing the null hypothesis of no predictability, we are interested in whether β_H is zero. We studied this as a GMM problem, and there are several ways to proceed. The most common is to form a t-ratio, dividing the slope coefficient estimator by a standard error. The discussion in chapter 17 of consistent standard error estimation for GMM problems applies here. Various GMM tests developed in chapter 18 can be applied. Several studies examine the finite-sample properties of predictive regressions, their standard errors, and tests when multiple-horizon regressions are run.

Kim, Nelson, and Startz (1991) examine predictive regressions using overlapping data with simulations and find that finite-sample biases lead to understated standard errors, which makes the t-ratios too high. The evidence for predictability is diminished after adjusting for these problems, but the regressions remain significant for the longer-horizon returns. Goetzmann and Jorion (1993) bootstrap returns for multiple-horizon regressions on dividend yields with overlapping returns data. Their simulations incorporate the fact that the dividend/price predictor variable is formed using the lagged price that appears in the left-hand-side return. They find that the predictive regressions remain significant for all except the 36-month horizon. These studies generally support the existence of semistrong-form predictability at longer horizons. However, Boudoukh, Richardson, and Whitelaw (2005) argue that high correlation of test statistics across the return horizons renders suspect much of the evidence for semistrong-form long-horizon predictability.

When $H > 1$ and several horizons are examined together, the issue of multiple comparisons arises. If 20 independent horizons are examined, one expects to find one "significant" t-statistic at the 5% level when the null hypothesis of no predictability is true. However, the slope coefficients for the different horizons are correlated, which complicates the inference. Richardson (1993) shows that this implies that the U-shaped patterns in the autocorrelations across return horizons, observed by Fama and French (1988), are likely to be observed by chance. He notes that when multiple horizons are examined, the largest autocorrelation observed could easily be as large as the values found by Fama and French, just by chance. Furthermore, the autocorrelations for adjacent horizons are highly correlated. He provides a nice derivation of the asymptotic standard errors for the regression coefficients at different horizons, and he estimates the correlation between the 48- and 60-month horizon to be 0.92. When a large value of the autocorrelation coefficient is observed, a hump-shaped or U-shaped pattern across the horizons is to be expected. Richardson also conducts joint tests across the horizons to determine whether all the autoregression slopes are zero. He finds little evidence to reject the hypothesis that all autocorrelations are zero in the Fama and French (1988) design.

32.1.3.5   Stambaugh Bias

Even when H = 1 and there is no overlap, autocorrelation in the error terms can lead to finite-sample biases in estimates of the slope in a predictive regression. Stambaugh (1999) studies a bias that arises because the regressor is stochastic and its future values are correlated with the error term in the regression. This problem differs from that of spurious regression, which is a finite-sample bias in the standard errors. The Stambaugh bias impacts the slopes. For example, if Z_t is a dividend/price ratio, then shocks to the dividend price ratio at time $t+1$ are related to stock returns at time $t+1$ through the stock price. Stambaugh provides corrections for this bias, which he shows is related to the bias in a sample autocorrelation coefficient, as derived by Kendal (1954). Amihud and Horvitz (2004) explore solutions for this bias in a multiple regression setting. We addressed this problem in the context of panel regressions in chapter 23.

32.1.3.6   Microstructure Effects

Many studies measure small but statistically significant serial dependence in daily or intra-daily stock return data. Serial dependence in individual daily stock returns can arise from end-of-day price quotes that fluctuate between bid and ask (Roll 1984); autocorrelation in an index can arise from nonsynchronous trading of the stocks in the index (e.g., Fisher 1966; Scholes and Williams 1977). These effects do not represent predictability that can be exploited with any feasible trading strategy. Spurious predictability due to such data problems should probably not be attributed to time variation in the expected discount rate for stocks. Kaul and Nimalendran (1990) find that bid-ask bounce is a major source of the measured negative serial dependence in individual daily stock returns on the NASDAQ.

Even with weekly returns, measured autocorrelation could reflect nontrading or nonsynchronous trading. Mech (1993) tries to eliminate nontrading effects from size-sorted portfolios of stocks, and he finds that they have only a small impact on the weekly portfolio autocorrelations. He points to other delays in price adjustment induced by transactions costs. In particular, he describes market-maker strategies that could induce autocorrelation in portfolios of NASDAQ stocks, because ask quotes lead bid quotes in some periods and follow bid quotes in others. Boudoukh, Richardson, and Whitelaw (1994) illustrate that heterogeneity across stocks in the nontrading frequencies can exacerbate the autocorrelations in weekly portfolio returns.

Lo and MacKinlay (1990) and Muthaswamy (1988) use statistical models that attempt to separate out the various effects in measured portfolio return autocorrelations. Mech (1993) and Boudoukh, Richardson, and Whitelaw (1994) use stock index futures contracts, which are not subject to nonsynchronous trading, and find little evidence for weak-form predictability in their returns at a weekly frequency.

32.1.3.7   Structural Breaks

Several studies question the stability over time of the coefficients in regressions like equation (32.3). In particular, the underlying relations may be subject to regime shifts. The issue of structural breaks in predictive regressions is raised by Kim, Nelson, and Startz

(1991). Lettau and Ludvigson (2009) find that the identification of structural breaks can explain how a variable can predict the levels of returns in-sample but fail to predict in step-ahead exercises. Smith and Timmerman (2017) explore a panel regression approach with multiple portfolios that have structural breaks. They find that they get more power to detect breaks by using the information in the cross section.

32.2   Predicting Variances: Volatility Modeling

A statement falsely attributed to Mark Twain is that "if risk be constant, life be simple." We all know that modern life is not simple; most scholars would agree that the evidence for changes and predictable variation in risk is stronger than for predictability in the levels of returns. Early studies of stock market data like Mandelbrot (1963) and Fama (1965) note that stock market risk seems to change over time. Officer (1973) notes that the variances of stocks, money supply, and industrial output all look larger during the Great Depression (1929–1934) and that high volatility periods tend to cluster together in time. Black (1976) describes a "leverage effect" in stock market variances, where the variances seem to go up after the market return goes down. (When equity values go down, a firm typically finds itself with higher leverage.) Time-varying betas are studied by Blume (1968), Gonedes (1973), Fisher (1966), and others in the early empirical literature on the CAPM.

32.2.1   Early Volatility Modeling

Attempts to model and measure time-varying market variance or volatility appear next. Parkinson (1980) proposed a "high-low" estimator of market variance: $[(1/T)\sum_t \ln(H_t/L_t)]^2$, where H_t and L_t are respectively the high and low prices observed over a period, say, 1 day. This measure was studied by Garman and Klass (1980) and Marsh and Rosenfeld (1983), who found that it is subject to microstructure-related biases. Officer (1973) used a rolling sample standard deviation of stock returns to model volatility, and Fama (1976a) used a moving average of absolute spot rate changes as an estimator of volatility, foreshadowing the modern literature on "realized" volatility.

32.2.2   Estimating Conditional Variances

32.2.2.1   Regression-Based Approaches

A good way to get started on this huge topic is with some simple regressions. Before we model the conditional variance, we want to see whether it varies significantly over time. A classical way to test for this in a predictive regression like (32.3) is White's (1980) test for conditional heteroskedasticity. A predictive regression is used to model the conditional mean. With excess returns, the regression is $r_t = \delta' Z_{t-1} + \varepsilon_t$, $t = 1, \ldots, T$. The conditional variance given Z_{t-1} is $E(\varepsilon_t^2 | Z_{t-1})$. The test for conditional heteroskedasticity regresses the fitted squared residuals ε_t^2 on Z_{t-1}. White shows that the regression R-squared is asymptotically distributed as a chi-squared (dim(Z)) under the null hypothesis that the variance is not predictable using Z. Engle (1982) used this approach with eight lags of inflation rates, rejecting a constant conditional variance, in his famous paper motivating ARCH models.

The literature on time-varying volatility is rife with acronyms. To get in the swing of things, let's introduce the first one. This is the simplest version of Engle's (1982) ARCH model, the ARCH(1). The acronym stands for "autoregressive conditional heteroskedasticity." The "(1)" indicates that the model is of first order; thus, it only accounts for a single lag. The model is

$$\epsilon_t^2 = \alpha_0 + \alpha_1 \epsilon_{t-1}^2 + v_t. \tag{32.4}$$

The parameters of the model can be estimated by replacing the residuals with their fitted values from the predictive regression for the returns. The fitted conditional variance is $E(\epsilon_t^2 | Z_{t-1}) = \alpha_0 + \alpha_1 \epsilon_{t-1}^2$. Engle estimates the model by starting with OLS on the mean equation, using the residuals to estimate the parameters of the variance equation, and iterating between the two until things converge. Note that this is almost a model of weak-form predictability in the variance, since it uses only past returns in the variance model. If the mean equation specified a constant mean over time, this would be a weak-form model. In high-frequency applications, the mean equation often is a constant mean model, because misspecification of the mean is of second order at high frequency. This is discussed more fully in section 32.2.3.

Let's take this approach one step further and describe a semistrong-form variance model. I call this the "PARCH(Z, 1) model." It stands for "Poor man's autoregressive conditional heteroskedasticity (model) with a lagged Z" (I am being a little tongue in cheek here):

$$\epsilon_t^2 = \alpha_0 + \alpha_1 \epsilon_{t-1}^2 + b\delta' Z_{t-1} + v_t. \tag{32.5}$$

Davidian and Carroll (1987) take a more sophisticated regression approach to volatility modeling. They regress the absolute residuals from the mean equation, $|\epsilon_t|$, on Z_{t-1} and use the fitted values to model the conditional standard deviation. They multiply the fitted absolute residuals by $(\pi/2)^{1/2}$, motivated by a normality approximation, because under normality, the standard deviation is approximately $(\pi/2)^{1/2}$ times the expected absolute value. They motivate the absolute residuals based on the idea that the absolute value is more robust to extreme values than is the square. Ding, Granger, and Engle (1993) find that low-frequency components of volatility may be better captured by absolute returns than by squared returns. Ghysels, Santa-Clara, and Valkanov (2004) study high-frequency foreign exchange rate data and find that absolute residuals work better than squared returns.

Glosten, Jagannathan, and Runkle (1993) adopt a regression approach to model the conditional variance of the CRSP stock market index, where Z_{t-1} is taken to be the lagged Treasury yield and these are seasonal dummy variables. The market return is regressed on Z_{t-1} in the mean equation. The squared residuals are regressed on Z_{t-1} in the variance equation, and the lagged Treasury yield is found to be a significant predictor of the conditional variance.

Schwert (1990) combines a version of high-frequency estimation with variance regressions. He first takes daily stock return data (the high frequency) and computes a sample variance for the daily returns in a given month, producing a monthly time series of variances. These are not quite the conditional variances of monthly returns, because they are of the wrong scale and would not be known conditional on information at the beginning of the month. He regresses the monthly time series on 12 lagged values and dummy variables for the month of the year to capture seasonal effects. He scales the fitted values by 30 (days per month) to obtain his conditional variance estimates. He uses these conditional variance estimates to study volatility around the market crash of 1987.

Schwert (1990) uses a similar approach with the absolute residuals, following Davidian and Carroll (1987), and uses the market variance estimates to study its relation to some macro factors. He finds that the stock variance series predicts some of the macro variance series and is related to bond market volatility, lagged bond yield spreads, and stock market price/earnings ratios, but the overall explanatory power of the macro variables for stock volatility is weak. In particular, a measure of aggregate corporate leverage has little explanatory power, casting some doubt on the idea that the "leverage effect" of Fisher Black is actually caused by leverage. French, Schwert, and Stambaugh (1987), Schwert and Seguin (1990), and Campbell et al. (2001) also mix daily and monthly frequency data to study volatility.

Ghysels, Santa-Clara, and Valkanov (2004) introduce a mixed data sampling regression (Midas) for volatility modeling. In this model, the lagged Z values in the variance equation can be measured at different frequencies. For example, the variance measure or squared residual on the left-hand side might be for a monthly frequency. Many macroeconomic time series are only available quarterly. But there may be variables with information about the conditional variance observed at daily or higher frequencies, such as stock returns, option-implied volatilities, and exchange rates. The variables on the right-hand side of the variance equation can be observed at a mix of frequencies, and there can be a long coefficient vector in the variance regression. Ghysels, Santa-Clara, and Valkanov constrain the coefficient vector, representing it as a function of some smaller parameter vector. For example, if imposing positive coefficients, then when the lagged predictors are positive, the fitted variance will be positive at each date. Ghysels, Santa-Clara, and Valkanov (2004) use a beta function that depends on only two parameters. They study the asymptotic properties of Midas regression models and compare them to distributed lag models.

The literature provides more than a fair bit of evidence comparing the performance of variance models. It would be nearly impossible to provide a comprehensive review, since the number of models to be compared is mind boggling, as you will see shortly. If we had the true residuals from the mean model using all public information Ω_{t-1}, then because $\text{Var}(r_t|\Omega_{t-1}) = E(\varepsilon_t^2|\Omega_{t-1})$, we could regress the squared residuals on various forecasts and public information available at time $t-1$. The model with the highest R-squared wins. Pagan and Schwert (1990) take this approach and a long view of the data. They use monthly US stock market return data back to 1835, focusing on the pre-CRSP period ending in 1925. Seven models are compared, including a two-step model that regresses the squared monthly residuals on eight monthly lags, a GARCH(1, 2), an EGARCH(1, 2), a Markov regime-switching model, and three nonparametric kernel models. In sample, the more complex models offer a better fit, but in a step-ahead analysis, the GARCH and EGARCH models perform better. Andersen and Bollerslev (1998) argue that ARCH models provide accurate forecasts over daily or longer forecast horizons. Bollerslev, Chou, and Kroner (1992), Andersen et al. (2006), and Ghysels, Harvey, and Renault (1996) provide reviews of the empirical evidence.

32.2.2.2 ARCH and GARCH

It's time to explain more of these volatility model acronyms. This is Nobel Prize–winning stuff. Rob Engle won the prize in 2003 for his ARCH model and related work. The field exploded, and these kinds of models are now widely used in academics and in practical risk modeling. Engle's (1982) ARCH(q) model can be written as a mean equation like (32.3) combined with

$$h_{t-1} \equiv E(\varepsilon_t^2 | \Omega_{t-1}) = \alpha_0 + \alpha_1 \varepsilon_{t-1} + \cdots + \alpha_q \varepsilon_{t-q}, \tag{32.6}$$

where q denotes the order of the process, controlling the length of its memory. While it is not always standard convention in this literature, I subscript things at the date when they would be, theoretically, measurable. This regression captures the idea that large shocks to volatility tend to persist in future values of volatility, as the earlier studies observe and more recent work confirms. The model is analogous to an MA(q) time series model applied to the squared residuals. If all the $\alpha_i > 0$, it implies a positive conditional variance. The model has covariance stationarity when $\alpha_0 > 0$, $\sum_{j>0} \alpha_j < 1$. Engle found that he needed lots of lags to capture inflation rate data.

Tim Bollerslev (1986) was Engle's student and, as doctoral students often do, carried the torch farther, proposing and analyzing the GARCH(p, q) model. This is "generalized autoregressive conditional heteroskedasticity." We have the mean equation plus the following:

$$h_{t-1} \equiv E(\varepsilon_t^2 | \Omega_{t-1}) = \alpha_0 + \sum_{j=1,\ldots,q} \alpha_j \varepsilon_{t-j}^2 + \sum_{j=1,\ldots,P} \beta_j h_{t-j-1}. \tag{32.7}$$

This model is analogous to an ARMA(p, q) in time series, applied to the squared residuals and lagged conditional variances. The autoregressive part captures infinite but declining memory, unlike the finite memory of the ARCH model. If all coefficients are positive, the model produces positive conditional variances at each date, but this is not necessary. For example, a GARCH(1, 2) model can have $\alpha_2 < 0$ and still deliver positive variances for some parameter values. Bollerslev shows that the model is covariance stationary when $\alpha_0 > 0$, $(\sum_{j>0} \alpha_j + \sum_{j>0} \beta_j) < 1$. By recursive substitution into equation (32.7), we can express the expected value of a future variance, $E(h_{t+j} | \Omega_{t-1})$, as a function of only h_{t-1} and the model parameters. This allows for forecasts of variance for dates in the future, and a "term structure" of the conditional variance.

32.2.2.3   Lots of GARCH-type Models!

Now we are starting to get the hang of this volatility model acronym business. A famous extension is Nelson's (1991) EGARCH model: "exponential generalized autoregressive moving average." We have a mean equation as before, and the variance equation is

$$\ln(h_{t-1}) = \alpha_0 + \sum_{j=1,\ldots,P} \beta_j \ln(h_{t-j-1}) + \sum_{j=1,\ldots,q} \alpha_j [\varphi u_{t-j} + (|u_{t-j}| - \sqrt{(2/\pi)})], \qquad (32.8)$$

where $u_t = \varepsilon_t / \sqrt{(h_{t-1})}$ is the Studentized residual from the mean equation. Note that here we model the natural logarithm of the variance, so the fitted variance will be positive without any restrictions on the coefficients, as e^x is positive for any x. This model captures a "leverage effect" in volatility through the parameter φ, which allows for an asymmetric response of the future volatility to the past shocks, depending on whether they are positive or negative. The last term on the right-hand side of equation (32.8) is called the "news response function." It records the impact of a shock in the returns on volatility. Lamoureux and Lastrapes (1990) show that the news response is related to the volume of market trading. Glosten, Jagannathan, and Runkle (1993) propose to model $h_{t-1} = \alpha_0 + \alpha_1 \varepsilon_{t-12} + b[\varepsilon_{t-12} \text{Max}(\varepsilon_{t-1}, 0)]$, also incorporating asymmetric response, and find that it works well for monthly returns where seasonality is at issue.

The creation of ARCH-type models has exploded since Engle kicked this thing off, and they all seem to have acronyms. Here is an incomplete list. AARCH (Duan 1997): augmented ARCH. APARCH (Ding, Granger, and Engle 1993): asymmetric power ARCH. ARCH-M (Engle, Lilien, and Robins 1987): ARCH-in-mean. FARCH (Engle, Ng, and Rothschild 1990): factor ARCH. FIGARCH (Bollerslev, Baillie, and Mikkelsen 1996): fractionally integrated GARCH. FIEGARCH (Baillie, Bollerslev, and Mikkelsen 1996): fractionally integrated EGARCH. IGARCH (Bollerslev and Engle 1993): integrated GARCH. GARCH-NIG (Bollerslev and Forsberg 2002): GARCH with normal inverse Gaussian errors. MARCH (Milhoj 1987): multiplicative ARCH. MC-GARCH (Engle and Sokalska 2012): multiplicative component GARCH. MSG-ARCH (Engle and Kroner 1995): multivariance simultaneous generalized ARCH (not monosodium glutamate–ARCH). NARCH (Higgins and Bera 1992): nonlinear ARCH. NAGARCH (Engle and Ng 1993): nonlinear asymmetric GARCH. PNP-ARCH (Engle and Ng 1993): partially nonparametric ARCH. SP-ARCH (Engle and Gonzalez-Rivera 1991): semiparametric ARCH. QARCH (Sentana 1995): quadratic ARCH. QTARCH (Gourieroux and Montfort 1992): qualitative threshold ARCH. STARV (Andersen 1994): stochastic autoregressive volatility. TARCH (Zakoian 1994): threshold ARCH. This proliferation has led some to exclaim "NAARCH" (not another ARCH!). Now you can see why I can't possibly review all of this.

32.2.2.4   Some Aspects of GARCH Estimation

The first question is: What is the right order of the GARCH(p, q) model? Since the studentized residual, $u_t = \varepsilon_t / \sqrt{(h_{t-1})}$, should have mean zero and variance equal to 1.0, it can be used to form diagnostic tests with simple asymptotic t-ratios. These can be accumulated over several lags to form chi-square tests. These kinds of diagnostics might be of low power. Engle, Ng, and Rothschild (1990) jack up the power by regressing the fitted u_t on its lags multiplied by dummy variables for positive or negative lagged residuals, thus using asymmetry as an alternative hypothesis to the model specification. As discussed in chapter 18, conditional moment tests can be formed from orthogonality conditions like $E\{h_{t-1}(\epsilon_t^2 - h_{t-1})\} = 0$, reflecting the idea that the forecast error should be orthogonal to the forecast. Fortunately, the literature typically finds that for asset return data, the second-order GARCH model is usually enough, and a GARCH(1, 1) often works pretty well for most applications. The fitted variances from different models tend to plot closely together.

Of course, estimation of GARCH type models, like everything else, is a GMM problem. Consider the ARCH(p) model and the following system of moments:

$$\epsilon_t = r_t - \delta'Z_{t-1},$$
$$v_t = \epsilon_t^2 - \alpha_0 - \sum_{j=1,\ldots,p} \alpha_j \epsilon_{t-j}^2. \tag{32.9}$$

Let $g_{1t} = \epsilon_t Z_{t-1}'$ and $g_{2t} = v_t Z_{t-1}'$, then $g_t = (g_{1t}' g_{2t}')'$, and $g = (1/T)\sum_t g_t$ is the sample mean moment condition. The parameters are $\varphi = (\delta', \alpha').'$ Provided $\dim(Z) > (p+1)$ the model is overidentified.

Note that the gradient matrix of this GMM problem has upper right block $\partial g_1/\partial \alpha = 0$. The parameters of the mean equation are not affected by the parameters of the variance equation. We can use OLS separately on the mean equation to obtain consistent, asymptotically normal and efficient (in the GMM sense) estimates of δ. The residuals from the mean equation will be consistent, resulting in consistent estimates of the variance equation in the second step.

The system can be simplified by going to an ML setup. Assuming normality, $\epsilon_t \sim N(0, h_{t-1})$, and the log likelihood function for time t is, apart from a constant,

$$\ln L_t(\varphi) = -\ln(h_{t-1}(\varphi)) - \{\epsilon_t^2(\varphi)/h_{t-1}(\varphi)\}, \tag{32.10}$$

where the notation emphasizes the dependence on the parameter vector, $\varphi = (\delta,'\alpha,'\beta'),'$ when using the variance equation (32.7), and the mean equation in (32.9). The sample log likelihood function is $(1/T)\sum_t \ln L_t$. The first-order condition for the ML estimator is $(1/T)\sum_t \partial \ln L_t(\varphi)/\partial \varphi = 0$. This is an exactly identified GMM problem, as described in chapter 20. The asymptotic covariance matrix is therefore given by

$$ACov[\varphi] = [(\partial g/\partial \varphi)'W(\partial g/\partial \varphi)]^{-1}$$
$$= [E(\partial^2 \ln L_t(\varphi)/\partial \varphi \partial \varphi')'\{E(\partial \ln L(\varphi)/\partial \varphi)(\partial \ln L(\varphi)/\partial \varphi)'\}^{-1} \tag{32.11}$$
$$(E\partial^2 \ln L_t(\varphi)/\partial \varphi \partial \varphi')]^{-1}.$$

Under normality, $E\{(\partial \ln L(\varphi)/\partial \varphi)(\partial \ln L(\varphi)/\partial \varphi)'\} = E(-\partial^2 \ln L_t(\varphi)/\partial \varphi \partial \varphi')$, and two of the three terms cancel out. If you don't like the normality assumption, keep all three terms and you have quasi-ML (Bollerslev and Wooldridge 1992).

Inspecting the gradients for the GARCH(p, q) problem reveals that the information matrix is block diagonal, just like the GMM problem of system (32.9). This property has three useful implications. First, the ML estimate of δ, evaluated at consistent estimates for (α, β), and the ML estimates of (α, β), evaluated at consistent estimates of δ, are asymptotically ML estimators and efficient, as if the full system were done simultaneously. Second, because the asymptotic covariance matrix of (δ, α, β) is block diagonal, the asymptotic variances for δ and (α, β) can be computed separately. Third, and for the same reason, numerical search procedures like the popular Berndt et al. (1974) algorithm simplify into separate systems for each parameter block.

One complication arises in the GARCH model that does not occur in the simpler ARCH model. Because of the lagged conditional variances in the GARCH model, $\partial h_{t-1}(\varphi)/\partial\varphi$ is a function of lagged $\partial h_{t-j}(\varphi)/\partial\varphi$. As a result, you have to solve recursively from the beginning to the end of your times series sample, updating each period, each time you try a new parameter value. The usual approach is to "precondition" on the first q lags. Be prepared for some numerical instability in practice.

32.2.3 Realized Volatility

Take the interval between times t and $t+1$, and break it into ($1/\Delta$) pieces of length Δ. Let r_j be the continuously compounded return over the jth interval. Then the continuously compounded return over the whole period is $\ln(R_{t+1}) = \sum_j r_j$, where the summation runs over the intervals, and $\sum_j \Delta = 1$. Of course, the mean of the sum is the sum of the means. To get the scale right, suppose that the $\{r_j\}_j$ are independent; then the variance of the sum is the sum of the variances. Both the mean and the variance grow with the length of the return horizon. We can capture this by

$$r_j = \mu_{j-1}\Delta + \sigma_{j-1}\sqrt{\Delta}\,\epsilon_j, \tag{32.12}$$

where μ_{j-1} is the conditional mean return given information at the instant $j-1$, and the ϵ_j are iid(0, 1) shocks. With this representation, the mean is proportional to the length of the interval, Δ, and the conditional variance at the instant $j-1$ is $\mathrm{Var}(\sigma_{j-1}\sqrt{\Delta}\,\epsilon_j) = \sigma_{j-1}^2\Delta$, also proportional to the length of the interval. (I am dancing around a continuous-time diffusion representation, where if you like, you can replace r with dP/P; $\mu_{j-1}\Delta$ with $\mu_{j-1}dt$; and $\sqrt{\Delta}\epsilon_j$ with dW_j, the change in a Wiener process.)

With the representation (32.12) in hand, the first concept is the *quadratic variation*. This is the variance around the local conditional mean, depicted as the fuzz around the solid curve that represents the time-varying conditional mean in figure 32.1:

$$\text{Quadratic variation} \equiv \mathrm{Var}\{\sum_j (r_j - \mu_{j-1}\Delta)\} = \mathrm{Var}(\sum_j \sigma_{j-1}\sqrt{\Delta}\epsilon_j) = \sum_j \sigma_{j-1}^2\Delta. \tag{32.13}$$

The movement of the solid curve in the figure depicts the variation through time in the local, conditional mean. The quadratic variation is the sum of the local conditional variances,

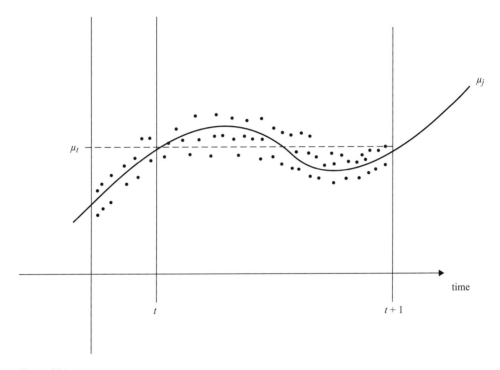

Figure 32.1
Quadratic variation versus conditional variance.

σ^2_{j-1}, over the interval between times t and $t+1$. This turns out to be the variance that is implicit in options prices, because when we replicate options by trading dynamically at each instant, the replicating strategy can condition on the information at each instant j.

The quadratic variation can be estimated using versions of the *realized variance* over the period. For example, studies including Schwert (1990) and French, Schwert, and Stambaugh (1987) mentioned earlier, use the sample variances of daily returns in a month. Nelson (1992) and Foster and Nelson (1996) describe a "continuous-record asymptotics," where the length of the interval, Δ, goes to zero, and the number of intervals, $1/\Delta$, goes to infinity. These authors show that the sample realized variances consistently estimate the quadratic variation.

Often, high-frequency studies simply ignore the mean, setting it to zero when computing the high-frequency sample variance. In fact, the most common realized variance estimator is simply $\sum_j r_j^2$. This is justified in the limit as Δ goes to zero, because the contribution of the conditional mean to the variance is of order Δ^2, so it goes to zero very fast as Δ goes to zero. Consider the expected value of the realized variance:

$$\begin{aligned}
E(\textstyle\sum_j r_j^2) &= E\{\textstyle\sum_j (\mu_{j-1}\Delta + \sigma_{j-1}\sqrt{\Delta}\epsilon_j)^2\} \\
&= \textstyle\sum_j E(\mu_{j-1}^2 \Delta^2) + \textstyle\sum_j \sigma_{j-1}^2 E((\sqrt{\Delta}\epsilon_j)^2) \\
&= \Delta^2 \textstyle\sum_j E(\mu_{j-1}^2) + \Delta E(\textstyle\sum_j \sigma_{j-1}^2),
\end{aligned} \tag{32.14}$$

where the second line uses the fact that ϵ_j is mean zero and is uncorrelated with μ_{j-1}. In the limit, we can ignore the Δ^2 term as second order, so the expected mean of the squared high-frequency returns is a consistent estimator for the quadratic variation. Nelson (1992) and Andersen et al. (2003) show that these arguments work in more general settings, allowing asymmetric GARCH news response functions, jumps, and other features.

In a finite sample, the continuous-record approximation is close when the average squared mean in (32.14), $\Delta^2 \sum_j E(\mu_{j-1}^2)$, is a small fraction of $E(\sum_j r_j^2)$. Consider daily data in which the standard deviation of daily returns is about 1% per day, or 18.9% per year. Suppose the average expected return is large: 30% per year. The squared daily expected return corresponding to $\Delta^2 \sum_j E(\mu_{j-1}^2)$ is about $(0.30/360)^2$, which is about 0.69% of the daily realized variance. Thus, the approximation should be a good one in most daily data. Simulations in Foster and Nelson (1996) suggest that the continuous-record asymptotics is a good approximation for daily stock market return data. When the mean does not much impact the realized variance, it justifies the use of a constant mean model on daily data for the purposes of modeling the quadratic variation.

Of course, at a high-enough frequency, we run into microstructure problems in the data, such as nontrading, bid-ask bounce, and other forms of illiquidity impacts on measured returns. The literature has tried several strategies for dealing with these issues. One is to use ARMA-type models to first filter the high-frequency data of autocorrelation (on the assumption that any autocorrelation is a microstructure bias) and remove it from the data before moving on to volatility modeling. Microstructure events might cause jumps. The literature shows that realized volatility can be consistent in the presence of jumps, and studies have been able to separate the jump component (e.g., Barndorff-Nielsen and Shephard 2004).

In an asset pricing study, we are typically interested in the conditional moments of the holding period return, like $\mathrm{Var}_t\{\sum_j r_j\}$. This quantity is not the same thing as the quadratic variation and is not accurately measured by the realized variance. The conditional moments for the return over the period t to $t+1$, conditioned on the information at the beginning of the period t, include the quadratic variation and the variance of the conditional mean. To see this, write the conditional variance as $E_t\{[\sum_j r_j - E_t(\mu_{j-1})\Delta]^2\}$. By iterated expectations, $E_t(\mu_{j-1}) = \mu_t$, so this becomes $E_t\{(\sum_j r_j - \mu_t)^2\}$. Adding and subtracting the unconditional means μ, we have a decomposition:

$$\mathrm{Var}_t\{\textstyle\sum_j r_j\} = E_t\{[(\textstyle\sum r_j - \mu) + (\mu - \mu_t)]^2\}, \tag{32.15}$$

Taking the unconditional expectation, we have

$$E(\mathrm{Var}_t\{\textstyle\sum_j r_j\}) = E(\textstyle\sum r_j - \mu)^2 - E(\mu_t - \mu)^2. \tag{32.16}$$

The first term is a version of realized variance, a consistent estimate of the quadratic variation for small μ. The second term is the expected variance of the local conditional mean. This term is depicted in figure 32.1 as the movement of the solid line around a horizontal bar at μ_t. Chang (2016) shows that the variance of the conditional mean can be larger than the quadratic variation in instances with annual stock portfolio return data.

32.3   Predicting Covariances, Correlations, and Higher Moments

That covariances and higher moments vary over time is not very controversial, as the evidence has mounted over time. In asset pricing models, time-varying second moments should translate into time-varying expected returns. It seems pretty natural that the mean equation in a GARCH-type model should include a second moment or variance term: $r_{mt} = \delta' Z_{t-1} + \beta h_{mt-1} + \varepsilon_{mt}$, when, for example, applied to the market portfolio return. This results in a GARCH-in-mean model. A classical example is the Merton (1980) model.

32.3.1   GARCH-in-Mean and Merton's Model

Merton (1980) shows that a simple version of his Intertemporal Asset Pricing model implies that expected returns on the aggregate stock market should be positively related to the conditional variance of the market returns; that is, there should be a positive risk-return trade-off for the market as a whole. The coefficient on the market variance is a risk aversion term shown as $\{-WJ_{WW}/J_W\}$ in equation (13.14) of chapter 13. Sharpe's (1964) CAPM, interpreted as a conditional model, also makes a similar prediction. Thus, the mean equation for the market return should have a market variance in it. In particular, Merton's model implies that $\delta = 0$ and $\beta > 0$ in the mean equation above, which includes the market variance.

Studies tried to predict market returns with predetermined market volatility or conditional variance measures and found mixed results—weak or even negative coefficients (e.g., French, Schwert, and Stambaugh 1987; Glosten, Jagannathan, and Runkle 1993). Scruggs (1998) showed that if an additional bond risk factor was included, as suggested by Merton's (1973) model, the partial coefficients became positive, as predicted by the theory. Ghysels, Santa-Clara, and Valkanov (2004), mentioned earlier, find that when the variance is estimated using Midas models, the coefficient on the market variance is positive.

32.3.2   Regression-Based Approaches to Higher Moments

We have discussed regression-based approaches for conditional variances. For conditional variance and beta estimation, rolling regressions have long been popular. Here we use only data available up to the forecast date to estimate the model. Fama and MacBeth (1973) use rolling regression betas. Ferson and Harvey (1991) interpret the rolling betas as conditional betas. Braun, Nelson, and Sunier (1995) justify interpreting rolling regression betas as conditional betas. Regressions can also be used to estimate conditional covariances and other moments. For example, imagine regressing the products of two residuals from the mean equation on Z_{t-1}. The fitted value is an estimator of the conditional covariance. In chapter 20, we looked at some GMM examples, going back to Harvey (1989) and others, where the products of the residuals are essentially regressed on the lagged Z values to generate time-varying conditional covariances. Davidian and Carroll's (1987) method has been used, regressing the products of two absolute residuals on Z_{t-1} and putting the sign of the product of the residuals on the fitted value.

Realized volatility can also be extended for covariances, simply estimating $E\sum_j(r_{1j}, r_{2j})$ the realized covariance, for returns 1 and 2. ARMA-type models are often applied to the time series of realized volatility as if it had no estimation error. Typically, the persistence over time is very high, and ARIMA-type models seem to fit realized volatility better (e.g., Andersen et al. 2001). These are ARMA models applied to the first differences to take out a unit root. Standard time series methods can then be used for volatility forecasting.

32.3.3 Multivariate GARCH

Many studies work with multivariate GARCH-style models to capture covariances, where an entire covariance matrix $\sum_t$ has elements that follow GARCH processes. Since an $N\times N$ covariance matrix only has $(N^2+N)/2$ unique elements, it is convenient to put the unique elements in the vector, vech($\sum_t$). Even so, there are a lot of moments to model, and multivariate GARCH-type models suffer the "curse of dimensionality." Consider the multivariate GARCH(1, 1) model:

$$\text{Vech}(\sum_{t-1}) = C + A'\text{vech}(\epsilon_{t-1}\epsilon'_{t-1}) + B'\text{vech}(\sum_{t-2}), \tag{32.17}$$

where ϵ_{t-1} is the N-vector of return shocks, C is a parameter vector of length $(N^2+N)/2$, and A and B are matrices of dimension $(N^2+N)/2\times(N^2+N)/2$. For $N=5$, the number of parameters is 465, which seems like a lot, but when $N=100$, there are more than 51 million parameters! Studies are usually limited to a small number of assets, and simplifying assumptions are made to reduce the number of parameters. Another challenge for multivariate GARCH is to make sure that the parameter values imply positive definite fitted covariance matrices. Bollerslev, Engle, and Wooldridge (1988) describe the parameter conditions that deliver positive definite matrices.

To reduce the number of parameters in multivariate GARCH models, studies have assumed, for example, that the A and B matrices are diagonal. Bollerslev (1990) proposes a constant-correlation model. Here the $N\times N$ conditional variance matrix is written as

$$\sum_{t-1} = \text{Diag}(\sum_{t-1}^{1/2})'\underline{\underline{\rho}}\text{Diag}(\sum_{t-1}^{1/2}), \tag{32.18}$$

where $\text{Diag}(\sum_{t-1}^{1/2})$ is an $N\times N$ diagonal matrix with the conditional standard deviations on the diagonal, and $\underline{\underline{\rho}}$ is the conditional correlation matrix, which is assumed to be a constant matrix. This drastically reduces the number of parameters at the expense of assuming that all the dynamics of conditional covariances come from the product of the two conditional standard deviations. Bollerslev (1990) describes a simple three-step method for estimating the constant correlation model. First, individual GARCH models are fitted for the conditional variances in the vector $\underline{h}_{t-1}$ and the vector of shocks in the mean equations, $\underline{\epsilon}_t$. The fitted shocks are Studentized to obtain $\underline{\tilde{e}}_t = \underline{\epsilon}_t./\sqrt{\underline{h}_{t-1}}$, where $./$ indicates element-by-element division. The Studentized shocks are used to estimate $\underline{\underline{\rho}}$ with $(1/T)\sum_t(\underline{\tilde{e}}_t\underline{\tilde{e}}'_t)$.

Engle (2002) extends the model to allow for dynamics in the correlations. This dynamic conditional correlation (DCC) model is

$$\underline{\underline{\rho}}_{t-1} = \underline{\underline{\rho}}(1-a-b) + a\,(\underline{\tilde{e}}_{t-1}\underline{\tilde{e}}'_{t-1}) + b\underline{\underline{\rho}}_{t-2}, \tag{32.19}$$

which is like a GARCH(1, 1) for the correlations. Engle finds that the estimates of the conditional correlations need not be between zero and one, so he recommends normalizing them to obtain that property. Kroner and Ng (1998), Ghysels, Harvey and Renault (1996), and Bauwens, Laurent, and Rombouts (2006) provide surveys of multivariate GARCH models.

32.3.4   Linear Beta Models

A very popular way to model conditional betas has been to parameterize them as linear functions of the conditioning information: $\beta_{t-1} = b_0 + b_1'Z_{t-1}$. This form can be motivated as a first-order Taylor series or derived in some normal exponential models of optimal portfolio choice (e.g., Admati and Ross 1985). Plugging into the factor model regression,

$$r_t = a + \beta_{t-1}r_{mt} + u_t \tag{32.20}$$

we obtain

$$r_t = a + b_0 r_{mt} + b_1'(r_{mt}Z_{t-1}) + u_t. \tag{32.21}$$

The time-varying beta can be estimated from a simple OLS time series regression for an asset r_t on the market return and its product with the lagged Z_s.

The linear beta approach is developed by Shanken (1990) and used by Cochrane (1996), Ferson and Schadt (1996), Ferson and Harvey (1997, 1999), and Lettau and Ludvigson (2001a), among many others. Christopherson, Ferson, and Glassman (1998) include a linear term in Z_{t-1}, interpreted as a time-varying alpha. Ferson, Sarkissian, and Simin (2008) evaluate the sampling properties of regressions like (32.21) with interaction terms when the lagged Z values are persistent variables. Lettau and Ludvigson (2001a) and Ferson and Harvey (1999) combine the linear beta approach with Fama and MacBeth (1973)-style regressions and find that the coefficients on the interaction are strong predictors in cross sections of portfolio returns. Jagannathan and Wang (1996) present an example using labor income and provide some analysis of the asymptotic properties of the estimators. Ghysels (1998) and Simin (2008) find that time-varying beta models don't improve the step-ahead forecasting performance, compared with fixed beta models or even no-beta models with constant returns. Brooks, Faff, and McKenzie (1998) compare some time-varying beta models for forecasting, including a regression approach, multivariate GARCH, and the Kalman filter.

32.3.5 Predicting Higher Moments

Much of the machinery for predicting second moments can be used or adapted to higher moments and co-moments. Harvey and Siddique (2001) present autoregressive conditional skewness and co-skewness, and Harvey and Siddique (2000) find that conditional co-skewness with the market helps explain cross sections of portfolio returns and momentum returns. Conrad, Dittmar, and Ghysels (2013) use option prices to model higher moments and find that conditional skewness and kurtosis are related to future returns. Bera and Premaratne (2001) find that simultaneously varying mean, variance, skewness, and kurtosis provides a better fit to stock returns than holding any of them constant. Chabi-Yo (2012) studies fourth- and fifth-order moments. Ergun and Jun (2010) let the whole distribution of returns and thus all its higher moments vary over time. Wachter (2013) finds that time-varying jump risk, modeling rare economic disasters, can help in modeling market volatility. This remains an active area of research, and I am sorry if I have left out your favorite paper in the area in this brief discussion.

32.4 Predicting Cross-Sectional Differences in Returns

The third type of predictability is cross-sectional or relative predictability. We try to predict the differences between portfolio returns. The portfolios are typically formed with weights that depend on some characteristic of a firm or its stock. Often, stocks are sorted on the basis of some characteristic into deciles or quintiles, and the difference between the high- and low-characteristic portfolio returns is what is being predicted. When the characteristics used in the sorts are predetermined relative to the returns, the characteristics predict the future cross-sectional or high-low return differences. Let $w_H(Z_{t-1})$ be the portfolio weight for the high-characteristic portfolio, where Z_{t-1} is the information in the vector of firm characteristics, and let $w_L(Z_{t-1})$ be the portfolio weight for the low-characteristic portfolio. The high minus low (HML) excess return is

$$w_H(Z_{t-1})'R_t - w_L(Z_{t-1})'R_t = (w_H(Z_{t-1}) - w_L(Z_{t-1}))'R_t. \tag{32.22}$$

The weights may be determined by sorting stocks but also can be found using Fama and MacBeth (1973)-style regressions of returns on the characteristics. Recall from chapter 22 that the slope coefficient in a cross-sectional regression of the returns in R_t on a vector of characteristics measured at time $t-1$ is a portfolio of the returns whose weights depend on the predetermined vector of characteristics at time $t-1$. The weights sum to zero, so the coefficient is an excess return. The weights must be long in the high-characteristic stocks and short in the low-characteristic stocks. Thus, Fama-MacBeth regression slopes are versions of $(w_H(Z_{t-1}) - w_L(Z_{t-1}))'R_t$.

The literature presents an embarrassment of riches in studies that have proposed predictable relative returns. Early examples include the momentum and reversal effects, which use lagged stock returns as the characteristic. Subsequent studies include a wide array of firm characteristics. Many of these relative returns have been considered anomalies. A relative return is an anomaly when its predictability cannot be explained by an accepted asset pricing model. We know from m-talk and the joint hypothesis of market informational efficiency that

$$E\{m_t(w_H(Z_{t-1}) - w_L(Z_{t-1}))'R_t | Z_{t-1}\} = 0, \qquad (32.23)$$

for any Z_{t-1} that is publicly known at time $t-1$, provided that we have an m that works in the m-talk equation for the underlying assets in R_t. Thus, the cross-sectional predictability should be removed after we multiply the excess return by the right m. The early anomalies, like the size effect, were anomalous relative to the CAPM, where m_t is a linear function of the market portfolio return. Subsequently, many of the models for m reviewed in chapter 5 have been applied to rationalize cross-sectional anomalies. Hou, Xue, and Zhang (2015) use a model including an investment factor to rationalize about 80 anomalous spread portfolio factors.

Relative return prediction roughly differences out predictive ability for the levels of returns. To the extent that all stocks move up and down predictably with the market over business cycles, for example, such predictability in the market return can be differenced out of the relative returns of two stock portfolios, provided their betas are of a similar magnitude. In portfolio formation, both return level prediction and relative return prediction determine the portfolio performance. We can think of predicting the levels as market timing and predicting cross-sectional differences in returns as security selection. These ideas are reviewed in part V of this book.

32.4.1 Weak-Form Cross-Sectional Return Prediction

In a weak-form version of relative predictability, the Z_{t-1} in the portfolio weights in equation (32.22) is based on the lagged stock returns. Jegadeesh (1990) uses the cross-sectional regression approach, running stock returns on their lags, and finds some evidence for weak-form relative predictability.

The best-known examples of weak-form relative predictability are "momentum" (Grinblatt and Titman 1993) and "reversals" (DeBondt and Thaler 1985). These are both based on sorting stocks on past returns. If past winner stocks have predictably higher returns, we have momentum. If past winner stocks can be predicted to have lower future returns, we have reversals. Lehmann (1990a) finds some evidence for reversals in the weekly returns of US stocks. Jegadeesh (1990) finds some evidence for reversals at a monthly return frequency. DeBondt and Thaler (1985) concentrate on longer-horizon reversals, over several years, and find reversals.

Jegadeesh and Titman (1993) find that relatively high-past-return stocks tend to repeat their performance over 3- to 12-month horizons. They study US data for 1927–1989 but focus on the 1965–1989 period. The magnitude of the effect is striking. The top 20% winner stocks over the last 6 months can outperform the loser stocks by about 1% per month for the next 6 months. Jegadeesh and Titman (2001) find that the momentum effect holds out of sample, using fresh data.

The momentum effect has spawned a huge literature that is largely supportive of the momentum effect but has not reached a consensus about its causes. Someone could write an entire book about it. Here I mention a few of the various ideas about momentum that are being explored.

One area of interest is the broad search for m factors that can rationalize momentum profits. The efficient markets view of predictability suggests that momentum trading strategies should be subject to greater risk, which justifies their enhanced returns. Liu and Zhang (2008) find that momentum winners have temporarily higher loadings on industrial production. Ang, Chen, and Xing (2001) find that momentum winners have a greater exposure to downside risks. However, most efforts at explaining the effect by risk adjustments have failed. Indeed, after adjustment for exposure to the common factors in many models, the momentum strategy alpha typically gets larger, as the loser stocks often have higher betas than the winner stocks have.

The role of trading costs and liquidity in momentum has also been examined. Lesmond, Schill, and Zhou (2004) and Korajczyk and Sadka (2004) measure the trading costs of momentum strategies and conclude that the apparent excess returns to the strategies are consumed by trading costs. Korajczyk and Sadka estimate that 5 billion dollars could be invested in momentum before price impact costs dwarfed the momentum profits. Avramov, Cheng, and Hameed (2016) find that momentum profits are higher in more liquid market states. McLean (2010) examines several proxies for arbitrage costs, including short-sales constraints, transactions costs, and idiosyncratic risks. He finds that these costs have little relation to momentum profits.

Menzly and Ozbas (2010) find that momentum is associated with the position of a firm in the supply chain; in particular, suppliers and customer industry stock returns tend to predict each other. This relation is stronger when there is less analyst coverage. Studies have sliced and diced common stocks into portfolios based on many characteristics of the data, at which point the momentum effect often retreats into subsets of stocks, subperiods of time, phases of the business cycle, or other parts of the data. Chordia and Shivakumar (2002) find that momentum based on earnings surprises dominates momentum based on past stock returns. Studies find momentum to be concentrated according to the season (December and January), industries, the size of the firm (more momentum in large stocks), the price of the share (more when the price is above $5 per share), and so forth. The effect does not appear prior to 1940, appears stronger after 1968, and appears stronger during economic expansions than during contractions. There has been a lot of work on the curious and enigmatic momentum effect, and more work no doubt remains to be done.

DeBondt and Thaler (1985) find that past high-return stocks perform poorly over the next 5 years. They interpret reversals in long-horizon relative returns as evidence that the market overreacts to news about stock values and then eventually corrects the mistake. The reversal effect was shown to occur mainly in the month of January by Zarowin (1990) and Grinblatt and Moskowitz (2003), interpreted as related to "tax loss selling." In this story, investors sell loser stocks at the end of the year for tax reasons, thus depressing their prices, and buy them back in the new year, subsequently raising their prices. McLean (2010) finds that reversals are concentrated in stocks with high idiosyncratic risks, which is thought to present a deterrent to arbitrage traders, who might otherwise correct tempo-rary errors in the market prices.

Reversals have been found to be concentrated in small stocks, low-priced stocks, the month of January, high idiosyncratic risk stocks, and are found to be more pronounced in earlier samples than in more recent data. The effect does not appear to be as robust as momentum.

32.4.2 Semistrong-Form Cross-Sectional Return Prediction

In this case, the lagged characteristic in the portfolio weights, $w_H(Z_{t-1})$ and $w_L(Z_{t-1})$ in equation (32.22), can be any publicly available data. Studies typically are careful to ensure that the weights are predetermined relative to the returns. For example, in several studies, Fama and French wait to use accounting data until it is at least 6 months old. Banz (1981) describes the small firm effect, where small market capitalization stocks earn higher returns on average than do large stocks. Fama and French (1993) immortalized the size effect with the SMB, the excess return of small over large stocks. They also added HML, the excess returns of stocks with high book/market ratios over those growth stocks with low ratios, to obtain their famous three-factor model. More recently, Fama and French (2016) added two more factors: CMA, conservative minus aggressive investment policy, and RMW, robust minus weak profitability. These factors are challenged by Hou, Xue, and Zhang (2015), who argue that a model with a market factor, a small stock, and a single investment factor works at least as well empirically and is better motivated in theory.

The literature began to judge anomalies relative to previously known anomalies. A new anomaly had to resist "explanation" by the Fama and French (1996) three-factor model, with the SMB and HML excess returns as factors. Sometimes a momentum factor is included. From the triality of the three paradigms of empirical asset pricing, these models correspond to a specification for m that is linear in the three Fama-French factors. Andy Lo is said to have claimed that yesterday's alpha is tomorrow's beta. This makes sense in asset management, because as small stock and value stock strategies became popular among investors, they became asset classes, and those who evaluate asset manager performance want to look at portfolio performance, adjusting for the fund's exposure to these well-known strategies.

Empirical work on cross-sectional predictors developed early in the investment management community. Such studies as Rosenberg and Marathe (1975) and Rosenberg and Guy (1976) examine cross-sectional return predictors, which they viewed as predictors for betas. This is reasonable, because in a beta pricing model, the only thing that varies in the cross section that should predict returns is the betas. The investment firm Barra provides a model with more than 40 such predictors, which they call "factors." S&P Capital IQ advertises a data library with more than 400 "factors." In academic studies, Haugen and Baker (1996) examine a large set of predictors in cross-sectional regressions, and Lewellen (2015) studies 15 firm characteristics as cross-sectional predictors in cross-sectional regressions. The number of cross-sectional return predictors in the academic literature has exploded in recent years. Harvey, Liu, and Zhu (2016) examine more than 300 of them from previous studies. Bali, Engle, and Murray (2016) devote much of their book to the study of portfolios of stocks sorted on various characteristics. Stambaugh and Yuan (2016) find a factor structure in a large set of anomalies, suggesting that common factors explain their returns.

32.4.3 Methodological Issues

Data snooping for cross-sectional predictability is a serious concern. There may be even more available choices to search over for cross-sectional predictors than there are potential variables for market return predictive regressions, given the huge amount of available data on firms' characteristics. We can't know how many searches have been undertaken. All the issues discussed in chapter 24 are relevant here, such as the incentives created by publication bias, the temptations to *p*-hack, and the use of canned software packages. Some scholars have argued for changes in the conventional standards for statistical significance. Harvey, Liu, and Zhu (2016) famously suggest that we should require *t*-ratios larger than three for new factors in asset pricing, but Harvey (2017) points out that this fix won't really solve the problem of data snooping, arguing instead for a Bayesian approach.

Cooper, Gutierrez, and Marcum (2005) conduct real-time simulations of the ability of an investor to identify the cross-sectional predictability of size, book/market, and momentum if they had to pick them out of a universe of 170 investment rules and had no prior beliefs about them. If average monthly returns or end-of-period wealth are used to evaluate the rules over a rolling, 10-year sample of monthly data, the real-time portfolios do not beat the market. They conclude that the evidence for cross-sectional predictability is likely overstated by the hindsight employed in studies. McLean and Pontiff (2016) present an interesting analysis of anomaly returns after a paper's publication and out of sample.

Portfolios sorted on anomalies have a potentially pernicious problem. The HML spread portfolios can soak up the anomaly and appear to be a priced risk factor. This possibility is examined by Ferson, Sarkissian, and Simin (1999), and the argument is briefly reviewed here. The argument uses our previous observation that HML spread portfolios are a lot like Fama and MacBeth (1973)-style cross-sectional regression coefficients. Consider a cross-sectional regression of returns on some anomalous cross-sectional predictor, call it α_i:

$$R_{it} = \lambda_{0t} + \lambda_{1t}\alpha_i + u_{it}, \ I = 1, \ldots, N. \tag{32.24}$$

The cross-sectional regression coefficient delivers a "risk premium" estimate, or a high-low spread portfolio, as we saw in chapter 22. Let $A_i \equiv \alpha_i - (1/N)\sum_i\alpha_i$. Then the OLS estimator constructs the portfolio: $R_{pt} = \lambda_{1t} = \sum_i w_i R_{it}$, $w_i = A_i/\sum_i A_i^2$. Suppose we used R_{pt} as a "factor" in an asset pricing model. Would it appear to "price"? We have that $\text{Cov}(\underline{R}_t, R_{pt}) = \text{Cov}(R)w = \text{Cov}(R)\underline{A}/\sum_i A_i^2$, where $\underline{A}$ is the vector of de-meaned anomaly characteristics. Now, if $\text{Cov}(R)\underline{A}$ is roughly proportional to $\underline{A}$, then the vector of covariances with R_{pt} will appear to "explain" the cross section of expected asset returns. For example, suppose that $\text{Cov}(R) = I$. Then returns are independent, and there is no systematic risk. But $\text{Cov}(R)\underline{A} = \underline{A}$, so the "factor" R_{pt}, formed by the Fama-MacBeth approach, will appear to work perfectly, in the sense that covariances with the factor return will exactly explain the cross section of expected returns! Ferson, Sarkissian, and Simin (1999) show more generally that it is precisely when the characteristic used in the cross-sectional regression is an anomaly, unrelated to risk, that it will appear to serve as a risk factor when loadings on the portfolio are used as if they were risk-factor betas.

In a practical setting, characteristics might be correlated with risks and they might be partly anomalous. Ferson and Harvey (1997) present an approach to disentangle the two kinds of information from characteristics. Similar approaches are used in Brennan, Chordia, and Subrahmanyam (1998) and Chordia, Goyal, and Shanken (2015).

32.5   Behavioral Perspective

A strong behavioral perspective pervades much of the literature on return prediction, both for levels and for cross-sectional prediction. The momentum effect has inspired various behavioral models suggesting that momentum may occur because markets underreact to news. One argument (Daniel, Hirshleifer, and Subrahmanyam 1998) is that traders have "biased self-attribution," meaning that they think their private information is better than it really is, at least for some stocks. As a result, they do not react fully to public news about the value of those stocks, so good news takes time to be incorporated in market prices, resulting in momentum. Some papers assume that traders form expectations by extrapolation (e.g., Greenwood and Shleifer 2014; Hirshleifer and Yu 2013), leading to underreaction to news. In another argument, traders suffer a "disposition effect," implying that they tend to hold on to their losing stocks longer than they should, which can lead to momentum (Grinblatt and Han 2003). Hirshleifer, Lim, and Teoh (2011) present a model where limited attention leads to underreaction to different earnings components, which can lead to earnings momentum and accrual anomalies. These arguments suggest that traders who can avoid these cognitive biases may profit from momentum trading strategies.

Like momentum, behavioral models also attempt to explain reversals as the result of cognitive biases. The models of Barberis, Shleifer, and Vishny (1998), Daniel, Hirshleifer, and Subrahmanyam (1998), and Hong and Stein (1999) argue that both short-run momentum and long-term reversals can reflect biases in under- and overreacting to news about stock values.

Another behavioral argument is salience. Investors may underreact to information that is not salient. Hirshleifer, Hsu, and Li (2013), for example, argue that patent activity is likely to be ignored or underweighted by investors, so that the stock prices of those firms with higher levels of patent activity will be too low, and subsequent returns to those stocks will be abnormally high relative to firms with less patent activity. They find empirical evidence to support this view. Yang (2018) argues that investors may underreact to changes in accounting numbers when they are paying less attention, and uses Google search volumes to measure attention. The behavioral literature on return predictability is large and growing. Fama (1998) on one hand, and DeBondt (1998) and Subrahmanyam (2008) on the other, provide reviews from contrasting perspectives.

32.6   Concluding Remarks

Does the current body of evidence lead to the conclusion that there actually is predictability in stock returns? It depends on whether you are talking about predicting the levels of returns, the second moment of returns, or other higher moments and co-moments. All of these are important for asset pricing. The direct evidence for predictability in the levels of returns may be the weakest, especially for weak-form predictability. However, I think most researchers would agree that there is predictability in second moments like the conditional market variance and conditional covariances or betas. Since asset pricing theory links expected returns with these second moments, it might lead one to put more weight on the evidence that the first moments are predictable. The evidence for predictable variation in skewness, kurtosis, and higher co-moments is more sparse, but it leans toward the conclusion that there is predictability. Overall, I am in the camp that thinks there is predictability. It is much harder to think of worlds where expected returns and risks would be constant over time than it is to think of worlds where they vary over time.

Cross-sectional predictability likely exists but is troubling. While the evidence for some predictability goes back a long way in investment practice, I think that a lot of data snooping has been conducted. Even the most famous cross-sectional predictors would have been hard to find ex ante (Cooper, Gutierrez, and Marcum 2005), and the relative predictability at least partly disappears after papers are published on it (McLean and Pontiff 2016). Many scholars are working on this problem right now, and we all still have a lot to learn.

33 Characteristics versus Covariances

This short chapter reviews the "characteristics versus covariances" debate, which addresses the question as to whether risk factor loadings or stock characteristics are the more important determinant of their expected returns. Historically, this issue has been associated with a behavioral versus rational risk-pricing debate, although these approaches do not address the same questions, and the debate likely can't be resolved.

33.1 The Issue

The characteristics of firms, such as their market capitalization (size), book-to-market ratios, past return performance, and a host of accounting measures, are related to the cross section of their stock returns. Studies going back at least to Rosenberg and Marathe (1975) document this fact. Famously, Fama and French (1992) document the cross-sectional explanatory power of firms' market capitalizations and book-to-market ratios.

In a cross-sectional regression of future stock returns on firm characteristics, the regression slope measures the excess return premium associated with a characteristic. As discussed in chapter 22, Fama and MacBeth (1973) point out that the slope coefficient is the excess return of a portfolio, long those stocks with large values of the characteristic, and short stocks with small values of the characteristic. The null hypothesis of a zero slope says that the excess return associated with a characteristic is zero. In the case of the size and book-to-market characteristics, hundreds of studies find that the excess return is not zero, and the well-known Fama and French (1993, 1996) factors capture the excess returns that codify the cross-sectional explanatory power. A big question in this literature has been: Is the explanatory power related to risk? Or, alternatively, is it related to some systematic mispricing that savvy investors can exploit to generate abnormal returns?

Fama and French (1993, 1996) sort stocks on their size and book/market ratios and propose their famous three-factor model as an empirical representation for the patterns in the cross section of the returns of the sorted portfolios. As is well known, this model uses a market index and two spread or high-minus-low (HML) portfolios constructed with groups of stocks sorted by characteristics. Fama and French argue that the factors are proxies for risk, as in the classical Merton (1973) model.

Daniel and Titman (1997) and Davis, Fama, and French (2000) debate whether firms' size and book-to-market characteristics or their loadings (betas) on the associated HML factors provide a better explanation for the cross section of stock returns sorted into portfolios. Daniel and Titman (1997, 2012) sort stocks first on size and book/market, in a 3×3 design, and then sort each of the nine portfolios into five sub-portfolios on the basis of past beta estimates, with the goal of introducing independent variation in the beta estimates. They find that the beta estimates have little explanatory power, given the characteristics, and suggest that the explanatory power is related to systematic mispricing and not to risk.

33.2 A Quick Survey of the Evidence

Many studies provide evidence that the Fama-French factors are correlated with plausible risks. Fama and French (1995), Liew and Vassalou (2000), Ludvigson and Ng (2007), and Yao (2008) examine correlations of the Fama-French factors with earnings and other economic variables. Other studies find a relation between the Fama-French factors and the covariances of stock returns. Moskowitz (2003) uses weekly returns for 1963–1997 in GARCH models, finding that size and book-to-market factors are correlated with time-varying second moments. Mei (1993) models the common factor structure in returns, assuming constant betas and using lagged returns, and finds that the model can capture the size effect but not the book-to-market effect. Bai and Ng (2006) develop tests that determine whether observables like the Fama-French factors are the common factors extracted by factor analysis. Their findings vary with the data frequency and the level of aggregation of the stocks. Kuntara and Roll (2013) propose a multistep protocol for factor identification that allows characteristics to be related to loadings. Cochrane (2005) argues that the ability of factors to spread returns in portfolios is prima facie evidence that the factors are related to risks.

Other studies argue that the Fama-French factors are repackaged anomalies. Lo and MacKinlay (1990) show that if portfolios are sorted on an anomaly related to average returns, standard tests are highly likely to find a "significant" factor. Berk (1997) points out that an anomaly that impacts price levels will appear in the cross section of the returns of portfolios sorted with respect to price (like size and book-to-market). MacKinlay (1995) argues that the Fama-French factors are unlikely to be explained by risk, given their high marginal Sharpe ratios. Ferson (1996) and Ferson, Sarkissian, and Simin (1999) show that factors formed by sorting stocks on an anomaly and forming an HML spread portfolio, similar to the cross-sectional regression coefficient, will generate patterns of betas on that factor that can appear to provide a "risk-based" explanation for portfolio returns sorted on the characteristic, even if there is no relation of the characteristic to risk.

33.3 Interpretation

As Fama (1973, 1991) emphasizes, the interpretation of this evidence must confront a joint hypothesis. A model for the risk factors that m depends on must be specified. If the model misses relevant risk factors, we might falsely conclude that characteristics are related to returns because of mispricing. For example, if idiosyncratic risk is priced, and if the characteristics are correlated with idiosyncratic risk in the cross section, then the explanatory ability that we attribute to risk would be understated. In the presence of arbitrage costs and other trading costs, idiosyncratic risk is likely to be related to the cross section of stock returns (Levy 1978; Merton 1987). In the cross section, idiosyncratic risk is correlated with total risk, size, and book-to-market characteristics and risk loadings.

As argued by Ferson and Harvey (1997), conditioning is important to control for the information in characteristics about risk factor loadings, as loadings are subject to more estimation error than characteristics are. A huge body of empirical evidence indicates that loadings are time varying, conditional on lagged public information. Ferson and Harvey (1997) argue that to test whether characteristics enter as a determinant of expected returns, given betas, one has to first extract from the characteristics the information that they contain about betas. A beta pricing model only allows the characteristics to be associated with the cross section of returns to the extent that they are related to betas. Berk, Green, and Naik (1999), Gomez, Kogan, and Zhang (2003), and Lin and Zhang (2012) present models with production where characteristics like size and book/market are endogenously related to the true betas. These authors emphasize that when empirical estimates of the betas are noisy, the characteristics can have explanatory power.

Ferson and Harvey propose using instrumental variables, where the loadings are conditioned on the characteristics. Using country-level return data, they find that characteristics are highly significant determinants of loadings. Lewellen (1999) instruments betas on characteristics for US stocks and finds that after controlling for risk using the conditional betas, he cannot reject the null hypothesis that the book-to-market factor has no incremental explanatory power for the cross section of returns. Chordia, Goyal, and Shanken (2015) use a similar instrumental variables approach on individual US stocks and find that conditional betas are significantly related to the size and book-to-market characteristics. They find that conditioning allows the betas to better explain the cross section of average stock returns, but that the characteristics retain some explanatory power in the presence of the loadings.

Many scholars have associated the characteristics versus covariances debate with questions of rational beta pricing versus behavioral explanations for returns. These are obviously not the same issues. Characteristics could be related to returns, even in the absence of behavioral bias, because they are good instruments for the rationally priced risks. Covariances with the Fama-French factors could be related to returns, even in the absence of risk pricing, if the return spreads associated with size and book-to-market factors are pure behavioral biases with no relation to risk (Ferson, Sarkissian, and Simin 1999).

Much of the confusion in the debate about characteristics versus covariances seems to be related to portfolio sorting. Berk (1997) points out that since the total sum of squares must equal the within group plus the between group, then by sorting first on characteristics and then on beta estimates in the characteristics groups, the amount of independent variation in beta is smaller than that of the characteristics. This results in low power to find a relation between returns and beta. Lewellen, Nagel, and Shanken (2010) point out that the Fama and French (1993) 25 portfolios have a strong factor structure, such that almost any set of factors correlated with the FF3 factors will be able to capture the cross section of average returns on the sorted portfolios. Zhang (2009a) finds that factors extracted from 100 portfolios sorted on size and book-to-market also have a strong three-factor structure, related to the FF3 factors. Obviously, the FF25 portfolios concentrate the factor structure to emphasize size and book-to-market factors.

Brennan, Chordia, and Subrahmanyam (1998) argue that using individual stock returns instead of portfolios is an attractive way to address the question of whether characteristics explain the cross section of returns, given loadings. Brennan, Chordia, and Subrahmanyam use the asymptotic principal components as risk factors. Regressing the factor risk-adjusted returns of individual stocks in the cross section on the characteristics for 1966–1995, these authors find that the characteristics have marginal explanatory power for the risk-adjusted returns, but that neither set of factors—the asymptotic principal components or the FF3—can "price" the other set.

Chan, Karceski, and Lakonishok (1998) examine a large number of characteristics—both accounting-related variables and loadings on macroeconomic factors. They include asymptotic principal components among their factors and use the unconditional loadings as alternative stock attributes. They sort stocks on each lagged attribute, form the HML portfolio, and examine its future return. They argue that the variance of the HML portfolio, when compared to that of a randomly sorted HML portfolio, is a good measure of how much systematic return comovement is captured by the HML factor. If the attribute-sorted portfolios have loadings on the true risk factors in returns that are uncorrelated with the attribute, then the HML loading on the true risk factors should be close to zero, as for a randomly sorted portfolio. If sorting on the attribute creates a spread in the true risk loadings, the difference between the variance of the HML portfolio return and a randomly sorted HML portfolio should be roughly the squared beta difference times the true common factor variance. These authors find that common factor variance is related to the market portfolio, size, book-to-market, and other factors. Chan, Karceski, and Lakonishok use unconditional loadings as the attributes, and if conditional loadings are important, as the evidence suggests, the inferences might be different with conditional loadings.

In recent work, Kelly, Pruit, and Su (2017) develop an instrumented principal components analysis, extending their partial least-squares approach (described in chapter 12) and allowing the loadings on the common factors to depend on firm characteristics. If the conditional loadings do not correspond to common risk factors, the characteristic is assigned to the intercept or alpha. These authors examine individual stocks and find that a conditional three-factor model works well on individual stock returns, compared to other factor models like the Fama-French models. A model with seven characteristics, which enter only as drivers of conditional betas, produces alphas for the individual stocks that are close to zero. Freyberger, Neuhierl, and Weber (2016) develop a least-absolute-shrinkage and selection operator to evaluate the incremental explanatory power of a large cross section of characteristics, allowing for nonlinearity. Running the model on 24 characteristics over 1963–2015, only eight of the characteristics offer marginal explanatory power.

In summary, the received literature has examined many aspects of the cross-sectional explanatory power of the size, book-to-market, and other stock characteristics and seems to find that while they are related to risks, it is hard to find a set of risks that completely explains the explanatory power of the characteristics themselves. The distinction between rational risk pricing and behavioral explanations for the explanatory power of characteristics is not resolved and likely won't be any time soon.

34 Volatility and the Cross Section of Stock Returns

The basic prediction of asset pricing models through m-talk is that expected returns are proportional to their covariances with m. In particular, only the systematic part of the variance of stock return—the part that covaries with m—should get a risk premium. However, for many models of m, this is not true empirically, and studies find a relation between average stock returns and the idiosyncratic volatilities. Some studies find a positive and some find a negative relation, but recently, attention has focused on the negative relation between idiosyncratic volatility and subsequent returns as observed by Ang et al. (2006). This chapter presents a brief review of a large literature that attempts to understand how stock returns, stock volatility, and idiosyncratic volatility are related in the cross section of stock returns. This is a fast-moving literature with many seemingly contradictory findings, and I can't presume to divine the truth of it all. I just hope to organize the main ideas.

34.1 Overview

Broadly speaking, the relation between risk and return is the core of empirical asset pricing. Beta pricing models and m-talk say that only a portion of the variance of an asset return should be priced: the part that covaries with the stochastic discount factor. The idiosyncratic part of the risk should not be priced in well-functioning markets. This is a strong intuition, but it relies on frictionless trading in markets. Sadly, markets do have frictions, and several kinds of frictions could result in a relation between idiosyncratic volatility and expected returns.

Early studies that bring frictions to the table tend to suggest that stock returns should be positively related to their idiosyncratic volatilities. Levy (1978) shows that idiosyncratic risk has a positive premium, given the market beta, if constraints limit portfolio holdings to a fixed number of stocks. If investors can't fully diversify, they can't eliminate idiosyncratic risk. And if they have quadratic utility, as Levy assumed, they are averse to the remaining idiosyncratic variance in their portfolios and must be paid a positive premium to hold stocks that make the portfolio idiosyncratic variance larger. Hirshleifer (1988) argues that trading costs that limit some groups of traders in commodities markets can lead to a positive risk premium for idiosyncratic volatility.

Merton (1987) presents a model in which investors must pay an information cost to invest in each stock, which leads the investor to choose a subset of the available assets, resulting in less-than-perfect diversification. Malkiel and Xu (2002) generalize Merton's model, assuming exogenously that a group of investors can't hold the market. They emphasize the implication that the unconstrained investors will also not be able to hold the market portfolio in equilibrium, and these authors derive a positive risk premium for the covariance between the idiosyncratic volatility of a stock and the average idiosyncratic volatility.

Mehra, Wahal, and Xie (2016) present a model with information acquisition costs that make diversification costly. The representative investor's choice of diversification is endogenous, which affects the premium for covariance with idiosyncratic volatility. The premium is increasing and convex as a function of the level of average idiosyncratic volatility.

Mitton and Vorkink (2007) argue that skewness preference can lead to underdiversification in equilibrium, so that idiosyncratic volatility is priced. Herskovic et al. (2016) point to nontraded human capital. Ewens, Jones, and Rhodes-Kropf (2012) argue for a positive price for idiosyncratic risks in venture capital, based on the three-way agency problem involving investors, entrepreneurs, and venture fund managers.

Malkiel and Xu (2002) argue that the Prospect theory of Kahneman and Tversky (1979) can kick in when fund managers must explain their best and worst stocks to the board or to fund company directors, reducing the supply of investors for highly idiosyncratic volatility stocks and causing a positive premium for bearing idiosyncratic volatility. Hsu, Kudoh, and Yamada (2013) and Bootra and Hur (2015) further develop this behavioral tack. De Long et al. (1990) argue that noise trader risk can be a priced factor when there are limits to arbitrage, and noise trader risk could be associated with idiosyncratic volatility.

In models like those of Levy and Merton, the idiosyncratic risk of an individual stock commands a return premium, because it directly contributes to the idiosyncratic risk of an investor's portfolio—something the investor does not like. There should be a direct relation between a stock's idiosyncratic risk and its expected return. Relating this to the previous chapter, it's the characteristic that matters.

In models like Malkiel and Xu (2002), Mehra, Wahal, and Xie (2016), and others, the covariance of idiosyncratic risk with the average idiosyncratic volatility gets a risk premium. The average level of idiosyncratic volatility is a common factor that varies over time, and higher average idiosyncratic volatility is associated with bad and scary times. Jiang, Xu, and Yao (2009) find that idiosyncratic volatility is related to negative earnings surprises, which supports the intuition, and several studies described below find that the average idiosyncratic volatility tracks the market volatility through time. A stock whose return is positively correlated with the average idiosyncratic volatility provides a hedge against bad times, so its idiosyncratic volatility beta earns a negative risk premium. A stock's own idiosyncratic volatility may be related to its idiosyncratic volatility beta and thus serve as a proxy for the underlying hedging feature. This situation creates a "characteristics versus covariances" problem for idiosyncratic volatility, similar to that discussed in chapter 33.

Other studies argue that idiosyncratic volatility is a proxy for some other underlying causal factor. If we could measure the underlying factor in such models, the explanatory power of idiosyncratic volatility would disappear once we control for it.

34.2   Time Series Evidence

We reviewed some of this evidence in chapter 32, in a discussion of predictability in the second moments of asset returns. Merton (1980) studied his model, which predicts a positive time series relation between the future market return (as a proxy for the current expected market return) and the current conditional variance of the market portfolio. He uses a two-sided rolling sample variance, centered on the current period, as his proxy for the conditional variance. He finds a negative coefficient on the market variance. Glosten, Jagannathan, and Runkle (1993) find a negative relation between conditional variance and market returns when the conditional variance is conditioned on a short-term interest rate. Goyal and Santa-Clara (2003) find that average idiosyncratic volatility predicts aggregate market excess returns with a positive coefficient. Bali et al. (2005) show that idiosyncratic volatility fails to predict market excess return after controlling for a liquidity premium. Guo and Savickas (2010) find that after controlling for market risk, an average idiosyncratic volatility predicts market excess returns with a negative coefficient. Controlling for the serial correlation in idiosyncratic volatility, Jiang, Xu, and Xao (2009) find idiosyncratic volatility predicts market excess returns with a positive coefficient.

This is a bit of a mess. And it might be a more important mess in recent data. Campbell et al. (2001) examine the trends in measures of aggregate market, industry, and average idiosyncratic volatility during 1962–1997. They use simple monthly measures of sample standard deviations from daily data. They find no trends in the levels of the market, the VIX, or industry volatility, but they do find a significant increase in the average idiosyncratic volatility over their sample period. Although they don't examine the cross section of returns, their evidence suggests that if idiosyncratic volatility is important for the cross section of returns, it might be more important in more recent data. However, Brandt et al. (2009) find that the higher idiosyncratic volatility observed by Campbell et al. (2001) seems to fall back to pre-1990s levels by 2003. The reversal is found to be concentrated in firms with high retail ownership.

French, Schwert, and Stambaugh (1987) break the time series of monthly sample market variances from daily data into a conditionally expected part that would be measurable at time t and an unexpected part. They run time series regressions of the monthly variances on lagged values, taking the fitted values and the residuals as separate measures. They find a positive time series relation between the filtered conditional or expected market variances and future market returns, and a strong negative relation between market returns and the unanticipated component of market variance. The two relations are consistent. If there is a positive relation to the conditional variance, then contemporaneous shocks to variance should be associated with a negative price reaction through a discount rate effect, as the market is repriced to a higher discount rate because of a higher future variance. This suggests that studies whose measure of market volatility contains both ex ante and unexpected parts would produce mixed results. Studies may have "overconditioned" their realized volatility measures, allowing them to depend on information that would not be available at the current date.

I am not sure that the literature has nailed down this connection yet, but it seems that a similar pattern emerges in the cross section as in the time series, where the expected or common movements in idiosyncratic risk are associated with a premium of the opposite sign as the unexpected or idiosyncratic movements.

34.3   Cross-Sectional Evidence

Fama and MacBeth (1973) include a measure of the average residual standard deviation in a portfolio as an alternative hypothesis to market beta pricing in their famous cross-sectional regression analysis. They find some negative and some positive coefficients on the idiosyncratic variance for different subperiods, but the absolute *t*-ratios are less than 2.0. Lehmann (1990b), Bali et al. (2005), and Huang et al. (2009a) examine the empirical relation between idiosyncratic volatility and average returns in cross sections. They find that the results are sensitive to the sample period, return frequency, and measurement techniques. Spiegel and Wang (2006) find a positive relation at the stock level. Goyal and Santa-Clara (2003) also find a positive premium. Boheme et al. (2009) find a positive conditional relation given low volatility and low short-selling activity. Malkiel and Xu (2002) repeat Fama and MacBeth's analysis with more recent data and find that idiosyncratic volatility is a strong cross-sectional predictor of returns that is not subsumed by size, book/market, or liquidity measures. They also find that it shows up in Japanese as well as US stock return data.

There is corroborating evidence on nonzero idiosyncratic risk premiums in other settings. Green and Rydqvist (1997) find that Swedish lottery bonds seem to reflect risk aversion for idiosyncratic risk. Begin, Dorion, and Gauthier (2017) develop an option pricing model with systematic and idiosyncratic jump risk and find that the idiosyncratic jump risk premium can be as large as 39% of the risk premium on the average individual stock.

Ang et al. (2006) is a blockbuster paper on this topic, with more than 2,900 Google Scholar citations as of July 2018. (French, Schwert, and Stambaugh (1987) show more than 4,300 citations). Ang et al. examine the pricing of aggregate volatility risk in the cross section of stock portfolio returns. They find that value-weighted portfolios with high average idiosyncratic volatility relative to the Fama and French (1993) model in 1 month have low average returns in the next month. Size, book-to-market, momentum, liquidity effects, volume, turnover, and even momentum do not account for either the low average returns earned by stocks with high exposure to systematic volatility risk or for the low average returns of stocks with high idiosyncratic volatility. They find a negative premium for both the "characteristic," idiosyncratic volatility, and the "covariance" with the average idiosyncratic volatility. The effect of the characteristic is not explained by the covariance when both are used in the cross-sectional regressions.

Ang et al. (2009) extend the evidence of a negative relation between return and idiosyncratic volatility to international stock markets. This negative relation of returns to idiosyncratic volatility is puzzling in view of the theories that predict a positive premium due to underdiversification.

34.4   Explaining a Return-Idiosyncratic Volatility Relation That Is Negative

The negative relation between stock returns and idiosyncratic volatility in the cross section has received much scrutiny since Ang et al. (2006) came on the scene. Some studies explore various measurement issues. Others ask whether idiosyncratic volatility proxies for other factors. Stambaugh, Yu, and Yaun (2015) offer an explanation that further develops the idea of Miller (1977), where short sales costs interact with disagreement. They provide a strong empirical case for the story.

One of the measurement issues appears in both the time series (French, Schwert, and Stambaugh 1987) and cross-sectional evidence. The proxy for idiosyncratic volatility may be a biased measure for the ex ante conditional volatility. The measure may be contaminated with unexpected volatility, which is negatively related to returns, and thus a negative relation can appear in the data when using the biased measure. Using the exponential GARCH models to estimate ex ante expected idiosyncratic volatilities, Fu (2009) recovers a significantly positive relation between the fitted idiosyncratic volatilities and expected returns. Peterson and Smedema (2010) find similar results using GARCH models. Duarte et al. (2012) also find a similar result, regressing realized volatility on macro variables. In a related and somewhat behavioral argument, Rachwalski and Wen (2016) argue that if the market underreacts to changes in an average volatility risk factor that carries a positive premium, the measured idiosyncratic volatility may be biased, so that a negative premium is measured. The effect should be stronger in less liquid firms.

Numerous studies explore the possibility that idiosyncratic volatility is a proxy for other factors. Ang et al. (2009) suggest a missing factor explanation. One seemingly natural choice is changes in the market volatility itself as mentioned earlier. Chen and Petkova (2012) argue that idiosyncratic variance is associated with changes in the average conditional variance. Barinov (2012, 2015b) and Barinov and Wu (2014) find that betas on average idiosyncratic volatility are associated with lower returns on stocks with higher short interest, new issues of equity, and distressed firms. Idiosyncratic volatility could be a proxy for dispersion or heterogeneity in the real economy. Chapter 2 describes how a central planner or representative agent utility dislikes a shock, which increases heterogeneity in consumption allocations, because with concave utility functions, the agents with the negative shock are hurt more than the agents with a positive shock benefit. An example is the model of Constantinides and Duffie (1996), where a higher cross-sectional variance of consumption shocks causes a higher m and thus commands a risk premium.

Avramov and Cederburgh (2014) argue that idiosyncratic volatility is a proxy for firms' dividend size and expected dividend growth rates. They assume that firms with smaller dividend size face higher uncertainty in their dividend growth rates, and that the equity discount rate structure is upward sloping. While consistent with many models, van Binsbergen, Brandt, and Koijen (2012) find negatively sloped term structures of equity discount rates. Financial frictions should be related to idiosyncratic volatility, as argued by Atkeson, Eisfeldt, and Weill (2017). Babenko, Boguth, and Tserlukevich (2016) argue that a sequence of idiosyncratic cash flow shocks to different types of firm assets changes the relative values of the assets and the systematic risk of the firm. Idiosyncratic volatility could be a proxy for changes in the systematic risk of the firm. In particular, in high volatility times, the value of growth options may be higher, causing changes in the systematic risks of growth stocks.

Studies have examined many correlates, and the idiosyncratic volatility effect has been found to be correlated with firm size, the level of institutional ownership, the level of firm volatility, and more. Bekaert, Hodrick, and Zhang (2012) find average idiosyncratic volatility to be associated with cash flow variables, business cycle variables, and marketwide volatility. Fu (2009) suggests that Ang et al.'s (2006) findings are largely explained by the return reversal of a subset of small stocks with high idiosyncratic volatilities. Spiegel and Wang (2006) find a positive relation between idiosyncratic volatility and illiquidity measures. Huang et al. (2009b) argue that leaving out the previous month's stock return can bias the conditional idiosyncratic volatility measure and produce a negative relation, driven by the short-term reversal of the extreme winner and loser stocks that populate the high idiosyncratic risk portfolios. Baker and Wurgler (2006) find the relation between returns and idiosyncratic volatility to be related to their investor sentiment index. Barinov (2013, 2015a) finds that stocks with higher dispersion of analysis forecasts and more growth options have larger betas on average idiosyncratic volatility.

Jiang, Xu, and Yao (2009) examine several alternative variables that do not explain the idiosyncratic volatility-return relation, including underreaction to accounting earnings, accruals, growth, or investment. The idiosyncratic volatility effect is stronger among stocks with more retail ownership. Hou and Loh (2016) examine a wide range of variables proposed to explain the stock return-idiosyncratic volatility relation during 1963–2012. They find little evidence that co-skewness, dispersion of analysts' forecasts, fraction of zero returns (a measure of transaction costs from Lesmond, Ogden, and Trzcinka 1999), the Amihud (2002) liquidity measure, or the beta on average variance can explain much of the effect. However, retail trading, return reversals, bid-ask spreads, and past earnings surprises have some explanatory power. Collectively, the covariates capture less than 50% of the variance in the puzzle. The authors' preferred explanation based on investor lottery preferences explains a larger fraction.

Miller (1977) argues for a positive relation between investor disagreement and over-pricing for a stock, and thus a negative relation to the future return. When there is a dispersion of opinion in the presence of short sale constraints, the most optimistic market participants set a stock's price. The pessimists' views get downweighted, because it costs more to short a stock than it does to buy one. The implication of Miller's (1977) theory is that a negative correlation exists between risk-adjusted returns and dispersion of beliefs, which, if associated with idiosyncratic volatility, would predict a negative relation between idiosyncratic volatility and subsequent stock returns. But the idiosyncratic volatility is not the cause, just a proxy, and its effect would disappear if we could measure the underlying dispersion in beliefs. This story also predicts an interaction effect: Mispricing should be higher for stocks that are more costly to short sell.

Stambaugh, Yu, and Yuan (2015) build on this idea. Pontiff (2006) argues that modern portfolio manager–type arbitrageurs (we discussed them in chapter 8 and in other chapters) will take smaller positions in stocks with greater idiosyncratic volatility. If they hedge out the market factors, this is the risk they face. Stambaugh, Yu, and Yuan build a model in which it is more costly to short a position than to hold it long and the trader is averse to residual risk. Noise traders in the model generate mispricing, and when it requires short selling to correct, the arbitrageurs do not react as strongly as they do when it requires a long position. Thus, mispricing is concentrated on the short side and stocks are overpriced. The model predicts a negative premium for idiosyncratic volatility for overpriced stocks and a positive premium for underpriced stocks. The negative premium is bigger and dominates the unconditional relation. A key aspect of their study is the development of an index of mispricing for a stock, based on its average ranks according to a set of eleven anomalies. The intensity of the idiosyncratic variance-to-stock return relation should be greater when the extent of mispricing is greater. Stambaugh, Yu, and Yuan provide a battery of tests supporting the implications of this idea.

34.5   Common Factors in Idiosyncratic Volatility

Are you ready for this? By definition, idiosyncratic volatility is the volatility of return that could be diversified away in a large portfolio, which limits its correlation across stocks, as we saw in chapter 12. Suppose that idiosyncratic volatility moves over time. There is nothing that says idiosyncratic volatility can't move together with market volatility or have other common components, and this seems to be the case. So we can have systematic and idiosyncratic components in idiosyncratic volatility. Which is more important: systematic idiosyncratic volatility or idiosyncratic idiosyncratic volatility? I don't think we know for sure yet.

That there may be common components in idiosyncratic volatility is suggested in the evidence of Campbell et al. (2001), who find that their time series for market variance and the average idiosyncratic variance move together. Duarte et al. (2012) examine the principal components of firms' idiosyncratic volatility and find that the first common factor explains 32% of the variance. The first common factor is correlated with business cycle variables like the NBER business cycle indicator, a junk bond yield spread, the market dividend yield, and the level of market volatility. Connor, Korajczyk, and Linton (2006) model idiosyncratic volatility as a function of macro and firm-specific latent factors. Gourier (2016) uses an affine jump model with idiosyncratic and common components, and options prices together with stock market data. She finds a strong factor structure in idiosyncratic volatility, with the first principal component explaining 80% of the variance. Boloorforoosh (2014) uses equity, index, and options markets data to find a strong factor structure in idiosyncratic volatilities that persists even after removing the VIX as a common factor.

Boloorforoosh (2014) finds that the common component in idiosyncratic volatility is highly correlated through time with the average idiosyncratic volatility. Bartram, Brown, and Stulz (2017) find a strong relation over time between the average idiosyncratic volatility and the regular old market volatility, and they provide a detailed analysis of the economic and empirical explanations for this relation. Herskovic et al. (2016) find that the common factor in idiosyncratic volatility is strongly related to market volatility, is not explained by the correlations of stock returns, and is related to common factors in sales growth volatility. The first common factor in stock return volatility and sales growth volatility each explain about a third of the variance in the average firm's idiosyncratic volatility or sales growth volatility. Like stock returns, the common factors in sales growth volatility move closely with the aggregate sales growth volatility. Herskovic et al. present a model in which common factors in firms' idiosyncratic risks lead households to face common factors in the idiosyncratic volatilities of their consumption growth. These authors examine household income risk data for 1978–2011 and measure idiosyncratic income risk as the cross-sectional standard deviation of incomes. This factor is correlated with the common factor in stock return idiosyncratic volatility, as are measures of cross-sectional dispersions of housing price growth and wages per capita.

Studies examine the pricing of the common component or average idiosyncratic volatility to see whether it can shed light on the puzzling findings of Ang et al. (2006). The evidence to date seems mixed. Positive premiums for the average idiosyncratic volatility are found by Chen and Petkova (2012), Boloorforoosh (2014), and Gourier (2016). Herskovic et al. (2016) find a negative premium for betas on the common component in idiosyncratic volatility, consistent with the argument that high average idiosyncratic volatility is associated with bad times. Duarte et al. (2012) find that when sorting on the residual, or the idiosyncratic part of idiosyncratic volatility, the negative relation of Ang et al. (2006) is no longer found. It seems there is more work to do on this question.

Appendix

A.1 Eigenvectors (x) and Eigenvalues (λ) for Square Symmetric $n \times n$ Matrices A

Definition

Ordered eigenvalues: $\lambda_{max}(A) = \lambda_1 > \lambda_2 > \ldots > \lambda_{min}(A)$.
Orthonormal eigenvectors: $x_i' x_j = 0, i \neq j, x_i' x_i = 1$,

$\lambda_1 = Max(x'Ax)$ s.t. $x'x = 1$ (first order condition: $Ax = \lambda x$),
$\lambda_2 = Max(x'Ax)$ s.t. $x'x = 1$, $x'x_1 = 0$,
$\lambda_3 = Max(x'Ax)$ s.t. $x'x = 1$, $x'x_1 = 0$, $x'x_2 = 0$,

and so on.

Some Properties

These properties follow from the above definitions, where I is the identity matrix:

(a) For c scalar, $\lambda_i(cA) = c\lambda_i(A)$.
(b) $\lambda_i(A^{-1}) = 1/\lambda_{n-i}(A)$. In particular, $\lambda_{max}(A^{-1}) = 1/\lambda_{min}(A)$, and $\lambda_{min}(A^{-1}) = 1/\lambda_{max}(A)$.
(c) $\lambda_i(A + cI) = c + \lambda_i(A)$.
(d) $x_i(A + cI) = x_i(A)$.
(e) For B $n \times K$, with B'B rank K, then $\lambda_i(BB') = \lambda_i(B'B)$, $i = 1, \ldots, K$, and $\lambda_i(BB') = 0$, for $i = K+1, \ldots, n$.

(f) **Representation Theorem:**

Let $H = \begin{pmatrix} -X_1- \\ \vdots \\ -X_n- \end{pmatrix}$

be the matrix of eigenvectors for matrix A. Then $HH' = I(H' = H^{-1})$, $H'H = H^{-1}H = I$, and

$$A = H' \begin{pmatrix} \lambda_1 & 0 \\ & \ddots & \\ 0 & \lambda_n \end{pmatrix} H$$

In other words, $A = \sum_i \lambda_i(x_i x_i')$ is the eigenvector-eigenvalue decomposition.

A.2   Direct Sums

Consider $((I_n))$, the set of all linear combinations of the columns of the n-dimensional identity matrix. This spans R^n, which is the space of all n-dimensional vectors of real numbers. Let $((\beta))$ denote the span (the set of all linear combinations of the columns) of β. If β is $K \times n$, with $K < n$, this defines a subspace of R^n.

Define $((\beta))^\perp$, where $\perp$ denotes the orthogonal complement (i.e., the part of R^n spanned by vectors orthogonal to the columns of β). We can therefore write R^n as the *direct sum*:

$$R^n = ((In)) = ((\beta)) + ((\beta))^\perp.$$

This is used in the simple algebra for the arbitrage pricing theory.

A.3   Gauss Code for the Generalized Method of Moments

```
new,64000;
output file=c:\kill\temp.out reset;
" —— output of gmmoid.prg——— ";

@****************************************************
Gauss Code for GMM Estimation of
Overidentified Linear or Nonlinear Models
*********************************************** @
load path=c:\kill;
iflin=0;

@
iflin=1;
"—— iflin = 1 if linear model, iflin= ";; iflin;
@

@ ———————— load input data here—— @

load dshocks[865,6]=dshocks.dat; @ example load statement @
r=derets[.,2:10];
z=shocks[.,2:5];

" ——example of summary stats ";
" r mean min max std auto1 ";
i=1;
format 8,4;
do while i<=cols(r);
gosub autocorr(r[.,i],1); pop rho;
(i~meanc(r[.,i])~minc(r[.,i])~maxc(r[.,i])~stdc(r[.,i])~rho);
i=i+1;
endo;
" ";
```

```
@ ———regress returns on lagged z——— @
i=1;
do while i<=cols(r);
" ——— return ";; i;
gosub ols1(r[.,i],z,0); pop resid;
i=i+1;
endo;

@ ——— begin GMM code——————— @

coeff=zeros(K,1); @ put initial coefficient values here @

K=rows(coeff); @ set up for typical asset pricing problem @
n=cols(r); @ number of test assets @
L=cols(z); @ dim(z) including constant @
T=rows(r);
north=n*L;

tol=(0.0000001)*T;
maxit=30;
w=eye(north);
g=zeros(north,1);
gd=zeros(north,k);
mterms=0;
mtype= 2; @ Hansen matrix=2, Newey-West = 1 @
fullout=1; @ set to zero for reduced output, 1 for full @
view=1;
noconv=30;
maxweigh=10;
oldsbj=5000;
gosub formg;
obj=g'w*g;
stage=1;
iter=1;
begtime=hsec;

/*
@ ———linear model gmm block——— @
if iflin==1;

ilin=1;
do while ilin<maxit and maxc(abs(b-oldb)) > tol;

gosub optw;
oldb=b;
b=-inv(BigB'W*bigB)*BigB'W*BigA;
c=b;
```

```
"  ilin  =  ";;  ilin;;"  b=  ";;b';

  if  maxc(b-oldb)  <  tol;
        vc  =  inv(gd'w*gd);
        stderr=sqrt(diag(vc));
        tstat=coeff./stderr;
      obj  =  g'w*g;
        dfchi=north-k;
        if  dfchi;  probchi=cdfchic(obj,dfchi);  endif;
        "Chi-square=";;  obj;
        "right-tail  p-value=";;  probchi;
        "df=";;dfchi;
        goto  fingmm;
    endif;

ilin=ilin+1;
endo;
"  ***********  linear  model  conv  at  iter=  ";;  (ilin-1);
goto  fingmm;
endif;
@  ————  end  of  linear  model  block————  @
*/

/*
@—————————this  block  for  nonlinear  models—————————@
ITERATE:  @  use  for  nonlinear  models  @

oldobj=10000;
delob=1.0;
dc=1.0;
iter=1;

do  while  (abs(dc)  >  tol)  and  (abs(delob)  >  tol)  and  (iter  <  noconv);
    oldobj=obj;
    c=coeff;
    gosub  formg;
    gosub  formgd;
    gdwgd=gd'w*gd;
    gosub  formg;
    gdwg=gd'w*g;
    obj=g'w*g;
  direc=inv(gd'w*gd)*gd'w*g;
  stp=0;
   gosub  optstp;
  coeff=coeff  -  direc*stp;
  iter=iter+1;
  c=coeff;
```

```
   gosub formg;
   obj= g'w*g;
 delob= obj - oldobj;
 dc=g'w*gd*inv(gd'w*gd)*gd'w*g;

/*
 if view==1;
   "end of iter=";; iter;; "stage=";;stage;;"obj=";; obj;
   format 10,6;
   "delta obj= ";;delob;;"step= ";;stp;
   "_____";
 endif;
*/
endo; @ iterations for this stage loop @

gosub optw;
vc = inv(gd'w*gd);

stage=stage+1;
 stderr=sqrt(diag(vc));
 tstat=coeff./stderr;

if view ==1;
 format 3,2;
 " *********************************";
 " ** GMM Results, Stage ";; stage-1;; " **";
 " *********************************";
 "coeff=alfo";
 " ";
 "Coeff Value Std Err T-stat";
 j=1;
 do while j <= k;
   format 1,0; "coeff";; j;;
   format 14,6; (c[j,1]);; (stderr[j,1]);; (tstat[j,1]);
   j=j+1;
 endo;

 obj = g'w*g; @ quad form with new weighing matrix @

 dfchi=north-k;
 if dfchi; probchi=cdfchic(obj,dfchi); endif;
   "J-stat=";; obj;
   "right-tail p-value=";; probchi;
   "df=";;dfchi;
endif;
```

```
if stage > maxweigh; goto fingmm; endif;
if abs(oldsbj - obj) <= tol; goto fingmm; endif;
oldsbj=obj;
goto ITERATE;
@ ———————- end of nonlinear models code block——— @
*/

FINGMM:

" ————final output here ——— ";
" ————model ——— ";; i;
" ";
" jstat (pvalue ";; obj;; probchi;
" ";
" c, std errors on second line";
" ";
(c');
stderr';
" ";
" ————average pricing errors ——— ";
" ";
gosub formg;
(g./T)';
" ";
"—average average ——— ";; meanc(g./T);
"—MAD of mean pricing errors ";; meanc(abs(g./T));
" ";
end;

@ ********** subroutines ****************************@

@ ******* ols1.inc **************************************
include this code after program end statement

CALL this by GOSUB OLS1(Y,X,P); pop residu;

where y is T x 1 vector of dependent variable
   x is (T,k) matrix of T obs on K indep variables
   (A CONSTANT SHOULD BE INCLUDED IN x IF DESIRED)
   p is 1 x 1 with type of standard errors wanted:
      0 = Hansen Matrix with 0 moving average terms
      n = Newey West matrix with n moving average terms
OLS t-ratios are provided automatically
************************************************** @
ols1:
pop subp; @ p will be in subp @
```

```
pop subx; @ x will be in subx @
pop suby; @ y will be in suby @
subT = rows(subx);
xtxinv=invpd(subx'subx);
subbeta=xtxinv * subx'suby;
subu=suby - subx * subbeta;
subk=cols(subx);
subvcov=zeros(subK,subK);
@ ————— ols std errors here————— @

subsigma=stdc(subu); @ uses T-1 weighing @
subvcov= (subsigma^2)*xtxinv;
slserr=sqrt(diag(subvcov));
rsq = 1 - (subsigma^2 /(stdc(suby)^2));

@ ————— white or nw std errors here —————@
subz=subu .* subx;
subj= (-1)*subp;
s=zeros(subk,subk);
submcov=zeros(subk,subk);
subwt = 1;
do while subj < subp+1;
   subwt = 1 - (abs(subj) / (subp+1));
   if subj < 0; subi=1; subm= subT+subj+1; endif;
   if subj >= 0; subi=subj+1; subm= subT+1; endif;
   do while subi < subm;
     s = s + subz[subi,.]'subz[subi-subj,.];
     subi=subi+1;
   endo;
submcov = s .* subwt + submcov;
s=zeros(Subk,Subk);
subj=subj+1;
endo;

subvcovc =(xtxinv * submcov * xtxinv);
stderr=sqrt(diag(subvcovc));
tratc = subbeta ./ stderr;

" ";
@ ——————————output—————————— @

" OLS WITH HET-MA-CONSISTENT STD ERRORS, NUMBER OF MA TERMS=";;subp;

"————————————————————————————————————————————————————————————";
```

```
"Coeff Value NW Std Err T-stat OLS Std Err";
j=1;
do while j <= subk;
  format 1,0; "beta";; j;;
  format 14,6; subbeta[j,1];; stderr[j,1];; tratc[j,1];; slserr[j,1];
  j=j+1;
endo;
" ";
"Rsquared is";; rsq;
"ADJ RSQUARED IS";; (1- (1-rsq)*(subt-1)/(subt-subk));
"_____";
clear subp,subx,suby,xtxinv,subbeta,subcov,stderr,tratc,slserr;
clear subk,subT,subj,subm,subwt,subvcovc,submcov,s,subz,subm,subi;
clear rsq,subsigma;
return(subu);

@ ************************* autocorr.inc ***********
INCLUDE THIS CODE AFTER PROGRAM END STATEMENT
CALL this by gosub autocorr(x,p); pop rho;

where x is (T,k) matrix of T obs on K variables
   p is 1 x 1 with order of autocorrelation wanted
   rho (use any name here) comes back as column of
     k p-th order autocorrelations
**************************************************** @
autocorr:
pop subp; @ p will be in subp @
pop subx; @ x will be in subx @
subT = rows(subx);
subxp = subx[subp+1:subT,.];
submu = meanc(subxp);

subrho = diag((subxp - submu')'((subx[1:subt-subp,.]) - submu'))
     ./ ((stdc(subx)^2) .* (subt-subp));

clear subp,subx,subt,subxp,submu;
return(subrho);

@ ********** scov.inc ********************
call this subroutine by gosub scov(x); pop covmat;
where x is (T,K) data matrix
estimates use T-1 weighing
*****************************************@
scov:
pop ssubx;
```

```
ssubT=rows(ssubx);
ssubm=meanc(ssubx);
covmat=(ssubx'ssubx)./(ssubT-1) - ssubm*ssubm';
clear ssubm;
return(covmat);

@ ****************************************************
CORR.INC
INCLUDE THIS CODE AFTER PROGRAM END STATEMENT
CALL this by gosub corr(x); pop corr;

where x is (T,k) matrix of T obs on K variables
   corr (use any name here) comes back as K by K
   correlation matrix
**************************************************** @
corr:
pop subx; @ x will be in subx @
subT = rows(subx);
subcorr = ((subx-meanc(subx)')'(subx-meanc(subx)')./(subT-1) ./
(stdc(subx)*stdc(subx)'));
clear subx,subt;
return(subcorr);

@ *************************** @
formg:

@ m=m(c) - update model at new parameter values, c, here @
u=m.*r - 1;
u=u*~z;
g=meanc(u)*T;
return;

@ ──── formgd numerically. Sometimes, analytical derivatives are better ─@
formgd:

dh=1e-8; @set precision@
e=eye(k);
fn bp(c0,aa,ii)=c0+aa*e[.,ii]*dh;
gd=zeros(north,k);

ii=1;
c0=c;
do until ii>k;
   c=bp(c0,1,ii); gosub formg; gplus = g;
   c=bp(c0,-1,ii); gosub formg; gminus = g;
   gd[.,ii]=(gplus-gminus)./(2*dh);
   ii=ii+1;
endo;
```

```
c=c0;
clear dh,e,c0,gplus,gminus;
return;

@ ── procedure finds the optimal weighting matrix ──@
optw:

u=m.*r - 1;
gg=u*~z;
mug=meanc(gg);

if mtype==2 and mterms == 0;
w=inv(gg'gg - mug'mug); @ this is Hansen MA(0) matrix @
endif;

if mterms>0;
wt=1;
iw=mterms+1;
jw=-mterms;

do while iw <= rows(bigg)-mterms;
do while jw <= mterms;

if mtype==1; wt = 1-abs(jw/(mterms+1)); endif; @ nw weights @

w = w + wt*(gg[iw,.]'-mug)*(gg[iw-jw,.]-mug');

jw=jw+1;
endo;
iw=iw+1;
endo;
endif;

w=invpd(w);
clear gg,iw,jw,wt,mug;
return;

@ This subroutine computes the optimal step size by forming a @
@ bracket and using quadratic interpolation. @
@ this is one of two alternative blocks of code with different step size
algorithms
@

/*
optstp:

 fcn0=g'g;
 coeff0=coeff;
 try=1;
 stp0=0;
```

```
  c=coeff0 - try*direc;
  gosub formg;
  fcn1=g'g;
  if fcn1 >= fcn0;
    fcn2=fcn1;
    stp2=1.0;
    goto back;
    endif;
  stp1=try;
  forward: @ go forward to find third step @
  try=try*2.0;
  c=coeff0 - try*direc;
  gosub formg;
  fcn2=g'g;
  stp2=try;
  if fcn2 > fcn1; goto compute; else; goto forward; endif;
back: @ go backward to find third step @
  try=try*0.5;
  c=coeff0 - try*direc;
  gosub formg;
  fcn1=g'g;
  stp1=try;
compute: @ compute optimal step @
 stp0sq=stp0*stp0;
 stp1sq=stp1*stp1;
 stp2sq=stp2*stp2;
 a=(stp1-stp2)*fcn0 + (stp2-stp0)*fcn1 + (stp0-stp1)*fcn2;
 b=(stp2sq-stp1sq)*fcn0 + (stp0sq-stp2sq)*fcn1 + (stp1sq-stp0sq)*fcn2;
 stp=-b/(2*a);
 return);
*/

@ This subroutine computes the optimal stepsize with linesearch @

optstp:

stp=.01;
coeff0=c;
 fcn0=g'w*g;
 try = stp;
 c=coeff - try*direc;
 gosub formg;
 fcn1=g'w*g;
 if fcn1 >= fcn0; goto backs; endif;
 chang=fcn0-fcn1;
```

```
do while chang >= 0;
   fcn0=fcn1;
   stp = try;
   try = try*2.0;
   c=coeff - try*direc;
   gosub formg;
   fcn1=g'w*g;
   chang=fcn0 - fcn1;
 endo;
 c=coeff0;
clear coeff0,fcn0,try,fcn1,chang;
 return;

 backs:
   iters=0;
   do while fcn1 > fcn0 and iters < 10;
      iters = iters + 1;
      try = try*0.5;
      c=coeff - try*direc;
      gosub formg;
      fcn1=g'w*g;
   endo;
if fcn1 >= fcn0; stp = 0.0; else; stp = try; endif;
c= coeff0;
clear iters,try,fcn0,fcn1;
return;
```

A.4   Matlab Code for the Generalized Method of Moments (Contributed by Davidson Heath)

This note and the accompanying files illustrate one way to do GMM estimation in Matlab. I certainly don't claim that it will always work or that it's anywhere close to optimal. We will do the simplest model there is—OLS—but the code supplied is a template that should work with most problems and introduces a few useful commands in Matlab. We want to estimate $y_t = a + bx_t + u_t$. Make sure your working directory is in the Matlab path.

Make a blank m-file, and save it as work_file.m. I usually keep it open, and then you can run the whole file by typing "work_file" in the command window, or else run individual commands and pieces of the file by highlighting in the editor and hitting F9.

```
Initializing,

clear all; close all; clc;
Import your data. In this case we'll just simulate some.
T = 100;
x = randn(T,1);
```

```
eps_raw = trnd(10,T-1,1);
y=2*x + [eps_raw;0]+exp(x).*[0;eps_raw];
figure; plot(x,y, '.'); title('scatterplot Y vs X')
```

OLS_results = regstats(y,x); _=0, _=2, and our errors are fat tailed, heteroskedastic and autocorrelated. Now we need to do the estimate. We will need another m-file, a function that takes in 1) parameters 2) data and 3) a weighting matrix and gives us a J-stat.

Open another blank m-file and save it as OLS_J.m .

```
function [Jstat, g_t, g_T] = OLS_J(param_vec, data, W)
```

OLS_J takes in the three arguments and gives us the J-stat, and *gt* and *gT* optionally.

param_vec is the vectorized parameters _ = [_ _]0. Data is [*y x*].

```
alpha = param_vec(1); beta = param_vec(2);   1
y=data(:,1); x = data(:,2);
T=size(y,1);
```

Given those, we want to compute the moments

$ut = yt - \_ - \_ xt$

$gt = ut \; [1 \; xt]$

```
u_t=nan(T,1);
g_t=nan(T,2);
for t=1:T
u_t(t)=y(t) - alpha - beta*x(t);
g_t(t,:)=kron(u_t(t),[1 x(t)]);
end
g_T=mean(g_t);
Jstat=g_T*W*g_T';
```

That's a bit clunky but just to illustrate the general way with the kron() function.

Save OLS_J.m and let's make sure it works. Go back to work_file.m:

```
testparams=[0;1];
Jstat_test=OLS_J(testparams, [y,x], eye(2));
```

Eye(2) is the 2x2 identity matrix. Evaluating gives us a Jstat for _ = [0 1]0. Okay, now

we need to search for the best fit.

To do this we need to specify a function call that takes *only* a parameter vector, and

gives the J stat, for the specified data and weighting matrix:

```
OLS_J1=@(param_vec) OLS_J(param_vec, [y,x], eye(2));
Jstat_test2 = OLS_J1([0;1]);
```

Okay, now to search. I like to do a grid search or shotgun search to find a good starting

point.

```
bestgridJ = 1e10; bestgridtheta = [NaN; NaN];
for gridalpha = -10:1:10
for gridbeta = -10:1:10
```

```
Jstat = OLS_J1([gridalpha;gridbeta]);
if Jstat < bestgridJ
bestgridJ = Jstat;
bestgridtheta = [gridalpha;gridbeta];
2
end
end %i
end %j
```

That steps over the parameter space from [-10,-10] to [+10,+10] in increments of one and picks the best point. We use that as our starting point for the real search. We will not use the built in fmin functions but rather a user-defined minimizer. Do NOT use the custom minimizer for any other estimation; it is here only as an illustration.

```
theta_hat = fmin_custom(OLS_J1, bestgridtheta);
disp('OLS coeffs 1st stage coeffs')
disp([OLS_results.beta, theta_hat]);
```

We get exactly the same estimate as OLS.

Okay, now on to the second stage. First we have to evaluate the model at the first stage

to get the $gt(\_)$ for our optimal weighting matrix.

```
[Jstat, g_t, g_T] = OLS_J1(theta_hat);
figure; plot(g_t); title('Time series of the moments')
figure; plot(g_t(1:end-1,1),g_t(2:end,1),'+'); title('g1_t vs g1_{t-1}')
figure; plot(g_t(1:end-1,2),g_t(2:end,2),'o'); title('g2_t vs g2_{t-1}')
```

Always a good idea to look at the pricing errors / moments. They potentially look

autocorrelated from the last two plots (plus we set it up that way), so let's do Newey West:

```
Acovg = g_t.'*g_t/T;
num_lags = 1;
for n = 1:num_lags
NWweight = 1 - n/(num_lags+1);
lag_cov = g_t(1+n:end,:).'*g_t(1:end-n,:)/T;
Acovg = Acovg + NWweight*(lag_cov+lag_cov');
end
W2 = inv(Acovg);
```

Now let's define our function call that evaluates the J stat using the optimal weighting matrix:

```
OLS_J2 = @(param_vec) OLS_J(param_vec, [y,x], W2);
```

The first stage ^_ should be a good guess for a starting point. (Of course since we're exactly identified, it's a perfect guess.)

3

```
theta_hat2 = fmin_custom(OLS_J2, theta_hat);
disp('OLS coeffs 2nd stage coeffs')
disp([OLS_results.beta, theta_hat2]);
```

And finally, standard errors. With the optimal weighting matrix the formula is simple.

We just need the gradient of the moment estimates. Again, we could easily get them analytically but as an illustration of the more general approach, let's do finite difference:

```
stepsize = 1e-10;
[ans, ans, g_T] = OLS_J2(theta_hat2);
for i = 1:2
theta_hat2_fd = theta_hat2;
theta_hat2_fd(i) = theta_hat2(i)+stepsize;
[ans, ans, g_T_fd] = OLS_J2(theta_hat2_fd);
dgT(:,i) = (g_T_fd - g_T)'/stepsize;
end
```

gives the gradient matrix via finite difference, and

```
thetahat2_SE = sqrt(diag(inv(dgT'*W2*dgT)))/T;
```

gives the standard errors. In the end,

• We exactly recapitulate the OLS estimate (because we are exactly identified and can always set the Jstat to zero)
• The standard error for _ is about the same size in OLS vs GMM
• The standard error for _ is larger in GMM reflecting the autocorrelation in the errors

Some specific code:

```
clear all; close all; clc

%simulate data and run regular OLS
T=100;
x=randn(T,1);
eps_raw = trnd(10,T-1,1);
y=2*x+[eps_raw;0]+exp(x).*[0;eps_raw];
figure;plot(x,y, '.'); title('scatterplot Y vs X')
OLS_results = regstats(y,x);

%just evaluate our J function at some parameters,
%make sure it's working
testparams = [0;1];
Jstat_test = OLS_J(testparams, [y,x], eye(2))

%define function call that only takes params
%using the identity weighting matrix
OLS_J1 = @(param_vec) OLS_J(param_vec, [y,x], eye(2));
Jstat_test2 = OLS_J1([0;1])

%okay, time to search for best params
%initial grid search
bestgridJ = 1e10; bestgridtheta = [NaN; NaN];
```

```
for gridalpha = -10:1:10
   for gridbeta = -10:1:10
      Jstat = OLS _ J1([gridalpha;gridbeta]);
      if Jstat < bestgridJ
         bestgridJ = Jstat;
         bestgridtheta = [gridalpha;gridbeta];
      end
      end %i
end %j

%do a proper search starting from best grid point
theta _ hat = fmin _ custom(OLS _ J1, bestgridtheta);
disp('OLS coeffs 1st stage coeffs')
disp([OLS _ results.beta, theta _ hat]);

%%%%%%%%%%%%%
%second stage
%%%%%%%%%%%%%
%Evaluate model at first stage thetahat
[Jstat, g _ t, g _ T] = OLS _ J1(theta _ hat);

%look at the errors
figure; plot(g _ t); title('Time series of the moments g _ t')
figure; plot(g _ t(1:end-1,1),g _ t(2:end,1),'+'); title('g1 _ t vs g1 _ {t-1}')
figure; plot(g _ t(1:end-1,2),g _ t(2:end,2),'o'); title('g2 _ t vs g2 _ {t-1}')

%estimate Acovg
Acovg = g _ t.'*g _ t/T;
num _ lags = 1;
for n = 1:num _ lags
   NWweight = 1 - n/(num _ lags+1);
   lag _ cov = g _ t(1+n:end,:).'*g _ t(1:end-n,:)/T;
   Acovg = Acovg + NWweight*(lag _ cov+lag _ cov');
end

%optimal weight matrix, and function call that uses that
W2 = inv(Acovg);
OLS _ J2 = @(param _ vec) OLS _ J(param _ vec, [y,x], W2);

%2nd stage estimate, efficient under heteroskedasticity and autocorrelation
theta _ hat2 = fmin _ custom(OLS _ J2, theta _ hat);

disp('OLS coeffs 2nd stage coeffs')
disp([OLS _ results.beta, theta _ hat2]);

%standard errors, robust to heteroskedasticity and autocorrelation
%for this we need gprime the change in g _ T w.r.t. theta
%delta method, or "standard errors for anything"…
```

```
stepsize = 1e-10;
[ans, ans, g_T] = OLS_J2(theta_hat2);
for i = 1:2
   theta_hat2_fd = theta_hat2;
   theta_hat2_fd(i) = theta_hat2(i)+stepsize;
   [ans, ans, g_T_fd] = OLS_J2(theta_hat2_fd);
   dgT(:,i) = (g_T_fd - g_T)'/stepsize;
end

thetahat2_SE = sqrt(diag(inv(dgT'*W2*dgT))/T);

%compare GMM vs OLS standard errors
disp('OLS SEs GMM SEs')
disp([OLS_results.tstat.se thetahat2_SE]);

function best_param = fmin_custom(func, start_point)

% takes in a function func, and a vector that is
% a valid starting point for func.

% just steps around until the home point
% is the best in the neighbourhood.

size_small = 1e-5;
size_med = 1e-3;
size_big = 1e-1;

n_params = size(start_point,1);

best_param = start_point;
best_val = func(start_point);
not_done = 1;
while not_done == 1;

   not_done = 0;

   for i = 1:n_params
        step_up = best_param;
        step_up(i) = step_up(i)+size_big;
        if func(step_up) < best_val
           best_param = step_up;
           best_val = func(step_up);
           not_done = 1;
        end
        step_down = best_param;
        step_down(i) = step_down(i)-size_big;
        if func(step_down) < best_val
            best_param = step_down;
```

```
      best _ val = func(step _ down);
      not _ done = 1;
   end
   step _ up = best _ param;
   step _ up(i) = step _ up(i)+size _ med;
   if func(step _ up) < best _ val
      best _ param = step _ up;
      best _ val = func(step _ up);
      not _ done = 1;
   end
   step _ down = best _ param;
   step _ down(i) = step _ down(i)-size _ med;
   if func(step _ down) < best _ val
      best _ param = step _ down;
      best _ val = func(step _ down);
      not _ done = 1;
   end
   step _ up = best _ param;
   step _ up(i) = step _ up(i)+size _ small;
   if func(step _ up) < best _ val
      best _ param = step _ up;
      best _ val = func(step _ up);
      not _ done = 1;
   end
   step _ down = best _ param;
   step _ down(i) = step _ down(i)-size _ small;
   if func(step _ down) < best _ val
      best _ param = step _ down;
      best _ val = func(step _ down);
      not _ done = 1;
   end
   end %i

end %while not-done
```

References

Abel, A. B. 1990. "Asset prices under habit formation and catching up with the Joneses." NBER Working Paper w3279, National Bureau of Economic Research, Cambridge, MA.

Abhay, Abhyankar, Devraj Basu, and Alexander Stremme. 2002. "Portfolio efficiency and stochastic discount factor bounds with conditioning information: A unified approach." Working paper, University of Warwick, Coventry, England.

Admati, A. R., S. Bhattacharya, P. Pfleiderer, and S. A. Ross. 1986. "On timing and selectivity." *Journal of Finance* 41: 715–730.

Admati, A. R., and S. A. Ross. 1985. "Measuring investment performance in a rational expectations equilibrium model." *Journal of Business* 58(1): 1–26.

Ahn, S. C., and A. R. Horenstein. 2013. "Eigenvalue ratio test for the number of factors." *Econometrica* 81(3): 1203–1227.

Ai, H. 2010. "Information quality and long-run risk: Asset pricing implications." *Journal of Finance* 65: 1333–1367.

Ai, H., M. M. Croce, A. M. Diercks, and K. Li. 2018. "News shocks and the production-based term structure of equity returns." *Review of Financial Studies* 31(7): 2423–2467.

Ait-Sahalia, Y. 1996. "Testing continuous-time models of the spot interest rate." *Review of Financial Studies* 9(2): 385–426.

Ait-Sahalia, Y., and J. Duarte. 2003. "Nonparametric option pricing under shape restrictions." *Journal of Econometrics* 116(1–2): 9–47.

Ait-Sahalia, Y., and A. W. Lo. 2000. "Nonparametric risk management and implied risk aversion." *Journal of Econometrics* 94(1): 9–51.

Akbas, F., W. J. Armstrong, S. Sorescu, and A. Subrahmanyam. 2016. "Capital market efficiency and arbitrage efficacy." *Journal of Financial and Quantitative Analysis* 51: 387–413.

Allen, E. R. 1991. "Evaluating consumption-based models of asset pricing." PhD dissertation, University of Chicago.

Almeida, C., and R. Garcia. 2016. "Economic implications of nonlinear pricing kernels." *Management Science* 63(10): 3361–3380.

Alvarez, Fernando, and Urban J. Jermann. 2005. "Using asset prices to measure the persistence of the marginal utility of wealth." *Econometrica* 73: 1977–2016.

Amihud, Y. 2002. "Illiquidity and stock returns: cross-section and time-series effects." *Journal of Financial Markets* 5(1): 31–56.

Amihud, Y., and C. Hurvich. 2004. "Predictive regressions: A reduced-bias estimation method." *Journal of Financial and Quantitative Analysis* 39: 813–842.

Amihud, Y., C. M. Hurvich, and Y. Wang. 2008. "Multiple-predictor regressions: Hypothesis testing." *Review of Financial Studies* 22(1): 413–434.

Amihud, Y., C. M. Hurvich, and Y. Wang. 2010. "Predictive regression with order-p autoregressive predictors." *Journal of Empirical Finance* 17(3): 513–525.

Andersen, T. G. 1994. "Stochastic autoregressive volatility: A framework for volatility modeling." *Mathematical Finance* 4: 75–102.

Andersen, T. G., and T. Bollerslev. 1998. "Answering the skeptics: Yes, standard volatility models do provide accurate forecasts." *International Economic Review*, November 1, 885–905.

Andersen, T. G., T. Bollerslev, P. F. Christoffersen, and F. X. Diebold. 2006. "Volatility and correlation forecasting." In *Handbook of Economic Forecasting* 1, 777–878. Amsterdam: Elsevier.

Andersen, T. G., T. Bollerslev, F. X. Diebold, and H. Ebens. 2001. "The distribution of realized stock return volatility." *Journal of Financial Economics* 61(1): 43–76.

Andersen, T. G., T. Bollerslev, F. X. Diebold, and P. Labys. 2003. "Modeling and forecasting realized volatility." *Econometrica* 71: 579–625.

Anderson, T. G., and B. E. Sorensen. 1996. "GMM estimation of a stochastic volatility model: A Monte Carlo study." *Journal of Business and Economic Statistics* 14(3): 328–352.

Anderson, T. W., and Cheng Hsiao, 1981. "Estimation of dynamic models with error components." *Journal of the American Statistical Association* 76(375): 598–606.

Anderson, T. W., and Cheng Hsiao. 1982. "Formulation and estimation of dynamic models using panel data." *Journal of Econometrics* 18: 47–82.

Andrews, D. K. 1997. "A stopping rule for the computation of generalized method of moments estimators." *Econometrica* 65: 913–931.

Andrews, D. K., and J. H. Stock. 2007. "Inference with weak instruments." In *Advances in Economics and Econometrics: Theory and Applications: Ninth World Congress of the Econometric Society 3*, ed. R. Blundell, W. Newey, and T. Persson, Ch. 6. Cambridge: Cambridge University Press.

Andrews, D. W. 2005. "Higher-order improvements of the parametric bootstrap for Markov Processes." In *Identification and Interference for Econometric Models: Essays in Honor of Thomas Rothenberg*, 171–215. Cambridge: Cambridge University Press.

Andrews, D. W., and M. Buchinsky. 2001. "Evaluation of a three-step method for choosing the number of bootstrap repetitions." *Journal of Econometrics* 103: 345–386.

Andrews, D. W., and J. C. Monahan. 1992. "An improved heteroskedasticity and autocorrelation consistent covariance matrix estimator." *Econometrica* 60(4): 953–966.

Ang, A., J. Chen, and Y. Xing. 2001. "Downside correlation and expected stock returns." Working paper, New York University, New York.

Ang, A., R. J. Hodrick, Y. Xing, and X. Zhang. 2006. "The cross-section of volatility and expected returns." *Journal of Finance* 61: 259–299.

Ang, A., R. J. Hodrick, Y. Xing, and X. Zhang. 2009. "High idiosyncratic volatility and low returns: International and further US evidence." *Journal of Financial Economics* 91(1): 1–23.

Ang, A., J. Liu, and K. Schwartz. 2017. "Using individual stocks or portfolios in tests of factor models." Working paper, Columbia University, New York.

Aragon, G. 2005. "Three essays on investments." PhD dissertation, Boston College.

Aragon, G. O., and W. E. Ferson. 2006. "Portfolio performance evaluation." *Foundations and Trends in Finance* 2(2): 83–190.

Arellano, Manuel, and Stephen Bond. 1991. "Some tests of specification for Panel Data: Monte Carlo evidence and an application to unemployment." *Review of Economic Studies* 58: 277–297.

Ashenfelter, O., and D. Card. 1986. "Why have unemployment rates in Canada and the United States diverged?" *Economica* 53(210): S171–S195.

Athreya, K. B. 1987. "Bootstrap of the mean in the infinite variance case." *Annals of Statistics* 15(2): 724–731.

Atkeson, A. G., A. L. Eisfeldt, and P. O. Weill. 2017. "Measuring the financial soundness of US firms, 1926–2012." *Research in Economics* 71: 613–635.

Avramov, D., and Scott Cederburg. 2014. "The idiosyncratic volatility-expected return relation: Reconciling the conflicting evidence." Working paper, University of Arizona, Tucson.

Avramov, D., S. Cheng, and A. Hameed. 2016. "Time-varying liquidity and momentum profits." *Journal of Financial and Quantitative Analysis* 51: 1897–1923.

Avramov, D., and T. Chordia. 2006. "Asset pricing models and financial market anomalies." *Review of Financial Studies* 19: 1001–1040.

Avramov, D., and R. Wermers. 2006. "Investing in mutual funds when returns are predictable." *Journal of Financial Economics* 81: 339–377.

Babenko, Ilona, Oliver Boguth, and Yuri Tserlukevich. 2016. "Idiosyncratic cash flows and systematic risk." *Journal of Finance* 71: 425–455.

Backus, D., N. Boyarchenko, and M. Chernov. 2016. "Term structures of asset prices and returns." NBER Working Paper w22162, National Bureau of Economic Research, Cambridge, MA.

Backus, D., M. Chernov, and I. Martin. 2011. "Disasters implied by equity index options." *Journal of Finance* 66: 1969–2012.

Backus, David, Silverio Foresi, and Chris Telmer. 2001. "Affine term structure models and the forward premium anomaly." *Journal of Finance* 56: 279–304.

Bai, Jushan, and Serena Ng. 2006. "Evaluating latent and observed factors in macroeconomics and finance." *Journal of Econometrics* 131: 507–537.

Bai, Jushan, and Guofu Zhou. 2015. "Fama MacBeth two-pass regressions: Improving risk premium estimates." *Finance Research Letters* 15: 31–40.

Baillie, R. T., T. Bollerslev, and H. O. Mikkelsen. 1996. "Fractionally integrated generalized autoregressive conditional heteroskedasticity." *Journal of Econometrics* 74: 3–30.

Baker, M., and J. Wurgler. 2000. "The equity share in new issues and aggregate stock returns." *Journal of Finance* 55: 2219–2257.

Baker, M., and J. Wurgler. 2006. "Investor sentiment and the cross-section of stock returns." *Journal of Finance* 61: 1645–1680.

Bakke, T. E., C. E. Jens, and T. M. Whited. 2012. "The real effects of delisting: Evidence from a regression discontinuity design." *Finance Research Letters* 9(4): 183–193.

Baks, K. P., A. Metrick, and J. Wachter. 2001. "Should investors avoid all actively managed mutual funds? A study in Bayesian performance evaluation." *Journal of Finance* 56: 45–85.

Bakshi, G., and F. Chabi-Yo. 2012. "Variance bounds on the permanent and transitory components of stochastic discount factors." *Journal of Financial Economics* 105: 191–208.

Bakshi, G., N. Kapadia, and D. Madan. 2003. "Stock return characteristics, skew laws, and the differential pricing of individual equity options." *Review of Financial Studies* 16: 101–143.

Bakshi, G., and D. Madan. 2006. "A theory of volatility spread." *Management Science* 52: 1945–1956.

Bakshi, G., and D. Madan. 2008. "Investor heterogeneity and the non-monotonicity of the aggregate marginal rate of substitution in the market index." Working paper, University of Maryland, College Park.

Bakshi, G., D. Madan, and G. Panayotov. 2010. "Returns of claims on the upside and the viability of U-shaped pricing kernels." *Journal of Financial Economics* 97: 130–154.

Balduzzi, Pierluigi, and Heidi Kallal. 1997. "Risk premia and variance bounds." *Journal of Finance* 52: 1913–1949.

Bali, T. G., N. Cakici, X. S. Yan, and Z. Zhang. 2005. "Does idiosyncratic risk really matter?" *Journal of Finance* 60(2): 905–929.

Bali, T. G., and R. F. Engle. 2010. "Resurrecting the conditional CAPM with dynamic conditional correlations." Working paper, New York University, New York.

Bali, T. G., R. F. Engle, and S. Murray. 2016. *Empirical Asset Pricing: The Cross Section of Stock Returns.* Hoboken, NJ: John Wiley & Sons.

Bali, T. G., D. C. Nichols, and D. Weinbaum. 2017. "Inferring aggregate market expectations from the cross-section of stock prices." Working paper, Georgetown University, Washington, DC.

Bali, T. G., and H. Zhou. 2016. "Risk, uncertainty and expected returns." *Journal of Financial and Quantitative Analysis* 51: 707–735.

Ball, R., J. Gerakos, J. T. Linnainmaa, and V. V. Nikolaev. 2015. "Deflating profitability." *Journal of Financial Economics* 117: 225–248.

Ball, Ray, and Phillip Brown. 1968. "An empirical evaluation of accounting numbers." *Journal of Accounting Research* 6: 159–178.

Bansal, R. 2007. "Long-run risks in financial markets." *Federal Reserve of St. Louis Review* 89: 1–17.

Bansal, R., R. Dittmar, and D. Kiku. 2007. "Cointegration and consumption risks in asset returns." *Review of Financial Studies* 22(3): 1343–1375.

Bansal, R., R. F. Dittmar, and C. T. Lundblad. 2005. "Consumption, dividends, and the cross section of equity returns." *Journal of Finance* 60: 1639–1672.

Bansal, R., A. R. Gallant, and G. Tauchen. 2007. "Rational pessimism, rational exuberance, and asset pricing models." *Review of Economic Studies* 74: 1005–1033.

Bansal, R., D. Kiku, I. Shaliastovich, and A. Yaron. 2011. "Volatility, the macroeconomy, and asset prices." Working paper, Duke University, Durham, NC.

Bansal, R., D. Kiku, and A. Yaron. 2009. "An empirical evaluation of the long-run risks model for asset prices." NBER Working Paper w15504, National Bureau of Economic Research, Cambridge, MA.

Bansal, R., D. Kiku, and A. Yaron. 2010. "Long run risks, the macroeconomy, and asset prices." *American Economic Review* 100: 542–546.

Bansal, R., D. Kiku, and A. Yaron. 2016. "Risks for the long run: Estimation with time aggregation." *Journal of Monetary Economics* 82: 52–69.

Bansal, R., and B. N. Lehmann. 1997. "Growth-optimal restrictions on asset pricing models." *Macroeconomic Dynamics* 1: 1–22.

Bansal, R. and I. Shaliastovich. 2013. "A long-run risks explanation of predictability puzzles in bond and currency markets." *Review of Financial Studies* 26: 1–33.

Bansal, R. and A. Yaron. 2004. "Risks for the long run: A potential resolution of asset pricing puzzles." *Journal of Finance* 59: 1481–1509.

Banz, R. W. 1981. "The relationship between return and market value of common stocks." *Journal of Financial Economics* 9: 3–18.

Banz, R. W., and M. H. Miller. 1978. "Prices for state-contingent claims: Some estimates and applications." *Journal of Business* 51(4): 653–672.

Barber, B. M., X. Huang, and T. Odean. 2016. "Which factors matter to investors? Evidence from mutual fund flows." *Review of Financial Studies* 29: 2600–2642.

Barberis, N., A. Shleifer, and R. Vishny. 1998. "A model of investor sentiment." *Journal of Financial Economics* 49: 307–343.

Barinov, Alexander. 2012. "Aggregate volatility risk: Explaining the small growth anomaly and the new issues puzzle." Working paper, University of California at Riverside.

Barinov, Alexander. 2013. "Idiosyncratic volatility, growth options and the cross-section of stock returns." Working paper, University of California at Riverside.

Barinov, Alexander. 2015a. "Analyst disagreement and aggregate volatility risk." Working paper, University of California at Riverside.

Barinov, Alexander. 2015b. "The bright side of distress risk." Working paper, University of California at Riverside.

Barinov, Alexander, and Juan Wu. 2014. "High short interest effect and aggregate volatility risk." Working paper, University of California at Riverside.

Barndorff-Nielsen, O. E., and N. Shephard. 2004. "Econometric analysis of realized covariation: High frequency based covariance, regression, and correlation in financial economics." *Econometrica* 72: 885–925.

Barone-Adesi, G., R. Engle, and L. Mancini. 2006. "GARCH options in incomplete markets." Working paper No. 2005-12, Center for Economic Institutions, Tokyo.

Barone-Adesi, G., R. Engle, and L. Mancini. 2008. "A GARCH option pricing model in incomplete markets." *Review of Financial Studies* 21: 1233–1258.

Barovicka, Jaroslav, Lars Peter Hansen, and Jose Scheinkman. 2016. "Misspecified recovery." *Journal of Finance* 71: 2493–2544.

Barras, L., O. Scaillet, and R. Wermers. 2010. "False discoveries in mutual fund performance: Measuring luck in estimated alphas." *Journal of Finance* 65: 179–216.

Barro, R. J. 2006. "Rare disasters and asset markets in the twentieth century." *Quarterly Journal of Economics* 121: 823–866.

Barro, R. J., and T. Jin. 2016. "Rare events and long-run risks." NBER Working Paper w21871, National Bureau of Economic Research, Cambridge, MA.

Bartram, Sohnke, Gregory Brown, and Rene Stulz. 2017. "Why does idiosyncratic risk increase with market risk?" Working paper, Ohio State University, Columbus.

Basak, S., and G. Chabakauri. 2010. "Dynamic mean-variance asset allocation." *Review of Financial Studies* 23: 2970–3016.

Bauwens, L., S. Laurent, and J. V. Rombouts. 2006. "Multivariate GARCH models: a survey." *Journal of Applied Econometrics* 21: 79–109.

Becker, C., W. Ferson, D. Myers, and M. Schill. 1999. "Conditional market timing with benchmark investors." *Journal of Financial Economics* 52: 119–148.

Beeler, Jason, and John Y. Campbell. 2012. "The long run risks model and aggregate asset prices: An empirical experiment." *Critical Finance Review* 1: 141–182.

Begin, Jean-Francois, Christian Dorion, and Genevieve Gauthier. 2017. "Idiosyncratic jump risk matters: Evidence from equity returns and options." Working paper, Simon Fraser University, Vancouver, BC.

Beja, Avraham. 1971. "The structure of the cost of capital under uncertainty." *Review of Economic Studies* 38: 359–368.

Bekaert, G., and Jun Liu. 2004. "Conditioning information and variance bounds on pricing kernels." *Review of Financial Studies* 17: 339–378.

Bekaert, G., R. J. Hodrick, and X. Zhang. 2012. Aggregate idiosyncratic volatility. Journal of Financial and Quantitative Analysis, 47(6), pp.1155–1185.

Bekaert, G., and M. S. Urias. 1996. "Diversification, integration and emerging market closed-end funds." *Journal of Finance* 51: 835–869.

Belo, Fredericko. 2010. "Production-based measures of risk for asset pricing." *Journal of Monetary Economics* 57: 146–163.

Benjamin, D. J., J. O. Berger, M. Johannesson, B. A. Nosek, E. J. Wagenmakers, R. Berk, K. A. Bollen, et al. 2018. "Redefine statistical significance." *Nature Human Behaviour* 2: 6.

Benjamini, Y., and Y. Hochberg. 1995. "Controlling the false discovery rate: A practical and powerful approach to multiple testing." *Journal of the Royal Statistical Society* B 57(1): 289–300.

Bera, Anil, and Gamini Premaratne. 2001. "Modeling asymmetry and excess kurtosis in stock returns data." Working paper, University of Illinois, Champaign.

Berk, J. B. 1995. "A critique of size-related anomalies." *Review of Financial Studies* 8(2): 275–286.

Berk, J. B. 1997. "Necessary conditions for the CAPM." *Journal of Economic Theory* 73: 245–257.

Berk, J. B., and R. C. Green. 2004. "Mutual fund flows and performance in rational markets." *Journal of Political Economy* 112: 1269–1295.

Berk, J. B., R. C. Green, and V. Naik. 1999. "Optimal investment, growth options, and security returns." *Journal of Finance* 54: 1553–1607.

Berk, J. B., and J. H. van Binsbergen. 2015. "Measuring skill in the mutual fund industry." *Journal of Financial Economics* 118: 1–20.

Berndt, E. R., B. H. Hall, R. E. Hall, and J. A. Hausman. 1974. "Estimation and inference in nonlinear structural models." *Annals of Economic and Social Measurement* 3(4): 653–665.

Bhattacharya, S. 1981. "Notes on multiperiod valuation and the pricing of options." *Journal of Finance* 36: 163–180.

Binder, John J. 1985. "On the use of the multivariate regression model in event studies." *Journal of Accounting Research* 23: 370–383.

Black, Fischer. 1972. "Capital market equilibrium with restricted borrowing." *Journal of Business* 45: 444–455.

Black, Fischer. 1976. "Studies of stock price volatility changes." Proceedings of the 1976 Meetings of the American Statistical Association, Business and Economic Statistics Section. 177–181.

Black, Fischer, Michael Jensen, and Myron Scholes. 1972. "The capital asset pricing model: Some empirical tests." In *Studies in the Theory of Capital Markets*, ed. M. C. Jensen, 79–121. New York: Praeger.

Black, Fischer, and M. Scholes. 1974. "The effects of dividend yield and dividend policy on common stock prices and returns." *Journal of Financial Economics* 1(1): 1–22.

Blake, C. R., E. J. Elton, and M. J. Gruber. 1993. "The performance of bond mutual funds." *Journal of Business* 66(3): 371–403.

Blake, C., E. Elton, and M. Gruber. 2003. "Incentive fees and mutual funds." *Journal of Finance* 58: 779–804.

Blume, M. E. 1968. "The assessment of portfolio performance: An application of portfolio theory." PhD dissertation, University of Chicago.

Boguth, O., and L. Kuehn. 2013. "Consumption volatility risk." *Journal of Finance* 68: 2589–2615.

Boheme, R. D., B. R. Danielsen, P. Kumar, and S. M. Sorescu. 2009. "Idiosyncratic risk and the cross-section of stock returns: Merton (1987) meets Miller (1977)." *Journal of Financial Markets* 12: 438–468.

Bollerslev, T. 1986. "Generalized autoregressive conditional heteroskedasticity." *Journal of Econometrics* 31: 307–327.

Bollerslev, T. 1990. "Modelling the coherence in short-run nominal exchange rates: A multivariate generalized ARCH model." *Review of Economics and Statistics* 72(3): 498–505.

Bollerslev, T., Richard T. Baillie, and Hans O. Mikkelsen. 1996. "Fractionally integrated generalized autoregressive conditional heteroskedasticity." *Journal of Econometrics* 74: 3–30.

Bollerslev, T., Ray Chou, and Kenneth Kroner. 1992. "ARCH modeling in finance: A review of the theory and empirical evidence." *Journal of Econometrics* 52: 5–59.

Bollerslev, T., and R. F. Engle. 1993. "Common persistence in conditional variances." *Econometrica* 61(1): 167–186.

Bollerslev, T., R. F. Engle, and J. M. Wooldridge. 1988. "A capital asset pricing model with time-varying covariances." *Journal of Political Economy* 96: 116–131.

Bollerslev, T., and Lars E. Forsberg. 2002. "Bridging the gap between the distribution of realized (ECU) volatility and ARCH modeling (of the euro): The GARCH-NIG model." *Journal of Applied Econometrics* 17: 535–548.

Bollerslev, T., M. Gibson, and H. Zhou. 2011. "Dynamic estimation of volatility risk premia and investor risk aversion from option-implied and realized volatilities." *Journal of Econometrics* 160: 235–245.

Bollerslev, T., J. Marrone, L. Xu, and H. Zhou. 2014. "Stock return predictability and variance risk premia: Statistical inference and international evidence." *Journal of Financial and Quantitative Analysis* 49: 633–661.

Bollerslev, T., and J. M. Wooldridge. 1992. "Quasi-maximum likelihood estimation and inference in dynamic models with time-varying covariances." *Econometric Reviews* 11: 143–172.

Booloorfoosh, Ali. 2014. "Is idiosyncratic volatility risk priced: Evidence from the physical and risk-neutral distribution." Working paper, Concordia University, Montreal.

Boney, V., G. Comer, and L. Kelly. 2009. "Timing the investment grade securities market: Evidence from high quality bond funds." *Journal of Empirical Finance* 16: 55–69.

Bootra, Ajay, and Jungshik Hur. 2015. "High idiosyncratic volatility and low returns: A prospect theory explanation." *Financial Management* 44: 295–322.

Boskin, Michael J. 1996. "Toward a more accurate measure of the cost of living: Final report to the Senate Finance Committee from the Advisory Commission to Study the Consumer Price Index." Washington, DC: Advisory Commission to Study the Consumer Price Index.

Boudoukh, J., R. Michaely, M. Richardson, and M. Roberts. 2006. "On the importance of measuring payout yield: Implications for empirical asset pricing." NBER Working Paper 10651, National Bureau of Economic Research, Cambridge, MA.

Boudoukh, J., M. Richardson, and T. Smith. 1993. "Is the ex-ante risk premium always positive?" *Journal of Financial Economics* 34: 387–408.

Boudoukh, J., M. Richardson, and R. Whitelaw. 1994. "A tale of three schools: Insights on the autocorrelations of short-horizon returns." *Review of Financial Studies* 7: 539–573.

Boudoukh, J., M. Richardson, and R. Whitelaw. 2005. "The myth of long-horizon predictability." NBER working paper 11841, National Bureau of Economic Research, Cambridge, MA.

Bouman, S., and B. Jacobsen. 2002. "The Halloween indicator, 'Sell in May and go away': Another puzzle." *American Economic Review* 92: 1618–1635.

Brandt, M. W. 1999. "Estimating portfolio and consumption choice: A conditional Euler equations approach." *Journal of Finance* 54: 1609–1645.

Brandt, M. W., A. Brav, J. R. Graham, and A. Kumar. 2009. "The idiosyncratic volatility puzzle: Time trend or speculative episodes?" *Review of Financial Studies* 23: 863–899.

Brandt, M. W., J. H. Cochrane, and P. Santa-Clara. 2006. "International risk sharing is better than you think, or exchange rates are too smooth." *Journal of Monetary Economics* 53: 671–698.

Braun, P. A., G. M. Constantinides, and W. E. Ferson. 1993. "Time nonseparability in aggregate consumption: International evidence." *European Economic Review* 37: 897–920.

Braun, P. A., D. B. Nelson, and A. M. Sunier. 1995. "Good news, bad news, volatility, and betas." *Journal of Finance* 50: 1575–1603.

Braun, Philip. 1991. "Asset pricing and capital investment: Theory and evidence." PhD dissertation, University of Chicago.

Brav, Alon, John B. Heaton, and Si Li. 2010. "The limits of the limits of arbitrage." *Review of Finance* 14: 157–187.

Breeden, D. 1979. "An intertemporal asset pricing model with stochastic consumption and investment opportunities." *Journal of Financial Economics* 7: 265–296.

Breeden, Douglas, and Robert H. Litzenberger. 1978. "The prices of state-contingent claims implicit in options prices." *Journal of Business* 51: 621–651.

Breeden, Douglas T., Michael R. Gibbons, and Robert H. Litzenberger. 1989. "Empirical tests of the consumption-oriented CAPM." *Journal of Finance* 44: 231–262.

Breen, William, Lawrence R. Glosten, and Ravi Jagannathan. 1989. "Economic significance of predictable variations in stock index returns." *Journal of Finance* 44: 1177–1190.

Brennan, M. J., T. Chordia, and A. Subrahmanyam. 1998. "Alternative factor specifications, security characteristics, and the cross-section of expected stock returns." *Journal of Financial Economics* 49: 345–373.

Brennan, M. J., and A. Kraus. 1978. "Necessary conditions for aggregation in securities markets." *Journal of Financial and Quantitative Analysis* 13: 407–418.

Brennan, M. J., and E. Schwartz. 1979. "A continuous time approach to the pricing of bonds." Working paper, University of California, Los Angeles.

Brennan, M. J., and B. Solnik. 1989. "International risk sharing and capital mobility." *Journal of International Money and Finance* 8: 359–373.

Brooks, Robert, Robert Faff, and Michael McKenzie. 1998. "Time-varying beta risk of Australian Industry portfolios: A comparison of modelling techniques." *Australian Journal of Management* 5: 45–66.

Brown, D. 1988. "The implications of nonmarketable income for consumption-based models of asset pricing." *Journal of Finance* 43: 867–880.

Brown, K. C., V. W. Harlow, and L. T. Starks. 1996. "Of tournaments and temptations: An analysis of managerial incentives in the mutual fund industry." *Journal of Finance* 51: 85–110.

Brown, S. J. 1976. "Optimal portfolio choice under uncertainty: A Bayesian approach." PhD dissertation, Graduate School of Business, University of Chicago.

Brown, S. J., and W. N. Goetzmann. 1995. "Performance persistence." *Journal of Finance* 50: 679–698.

Brown, S. J., W. N. Goetzmann, R. G. Ibbotson, and S. A. Ross. 1997. "Rejoinder: The J-shape of performance persistence given survivorship bias." *Review of Economics and Statistics* 79: 167–170.

Brown, Stephen J., and Jerold B. Warner. 1980. "Using daily stock returns: The case of event studies." *Journal of Financial Economics* 14: 3–31.

Brunnermeier, Markus, Stefan Nagel, and Lasse Pedersen. 2008. "Carry trades and currency crashes." *NBER Macroeconomics Annual* 23: 313–347.

Burnside, Craig. 1994. "Hansen-Jagannathan Bounds as classical tests of asset-pricing." *Journal of Business and Economic Statistics* 12(1): 57–79.

Burnside, Craig, Martin Eichenbaum, and Sergio Rebelo. 2011. "Carry trade and momentum in currency markets." *Annual Review of Financial Economics* 3: 511–535.

Buse, A. 1982. "The likelihood ratio, Wald, and Lagrange multiplier tests: An expository note." *American Statistician* 36(3a): 153–157.

Busse, J. 1999. "Volatility timing in mutual funds: Evidence from daily returns." *Review of Financial Studies* 12: 1009–1041.

Busse, J. 2001. "Another look at mutual fund tournaments." *Journal of Financial and Quantitative Analysis* 36: 53–73.

Busse, J. A., and P. J. Irvine. 2006. "Bayesian alphas and mutual fund persistence." *Journal of Finance* 61: 2251–2288.

Campbell, John Y. 1987. "Stock returns and the term structure." *Journal of Financial Economics* 18: 373–399.

Campbell, John Y. 1991. "A variance decomposition for stock returns." *Economic Journal* 101: 157–179.

Campbell, John Y. 1993. "Intertemporal asset pricing without consumption data." *American Economic Review* 83: 487–512.

Campbell, John Y. 1996. "Understanding risk and return." *Journal of Political Economy* 104: 298–345.

Campbell, John Y. 2003. "Consumption-based asset pricing." In *Handbook of the Economics of Finance* 1, edited by George Constantinides, Milton Harris, and Rene Stulz, 803–887. Amsterdam, North Holland.

Campbell, John Y., and J. H. Cochrane. 1999. "By force of habit: A consumption-based explanation of aggregate stock market behavior." *Journal of Political Economy* 107: 205–251.

Campbell, John Y., Martin Lettau, Burton G. Malkiel, and Yexiao Xu. 2001. "Have individual stocks become more volatile? An empirical exploration of idiosyncratic risk." *Journal of Finance* 56: 1–43.

Campbell, John Y., A. Lo, and A. C. MacKinlay. 1997. *The Econometrics of Financial Markets*. Princeton, NJ: Princeton University Press.

Campbell, John Y., and Robert J. Shiller. 1988a. "The dividend price ratio and expectations of future dividends and discount factors." *Review of Financial Studies* 1: 195–228.

Campbell, John Y., and Robert J. Shiller. 1991. "Yield spreads and interest rate movements: A bird's eye view." *Review of Economic Studies* 58: 495–514.

Campbell, John Y., and S. Thompson. 2005. "Predicting the equity premium out of sample: Can anything beat the historical average?" NBER Working Paper 11468, National Bureau of Economic Research, Cambridge, MA.

Campbell, John Y., and L. Viceira. 1999. "Consumption and portfolio decisions when expected returns are time-varying." *Quarterly Journal of Economics* 114: 433–495.

Campbell, John Y., and L. Viceira. 2002. *Strategic Asset Allocation: Portfolio Choice for Long-Term Investors*. New York: Oxford University Press.

Campbell, John Y., and T. Vuolteenaho. 2004. "Bad beta, good beta." *American Economic Review* 94: 1249–1275.

Carhart, M. M., J. N. Carpenter, A. W. Lynch, and D. K. Musto. 2002. "Mutual fund survivorship." *Review of Financial Studies* 15: 1439–1463.

Carhart, Mark. 1997. "On persistence in mutual fund performance." *Journal of Finance* 52: 57–82.

Carlson, M., A. Fisher, and R. Giammarino. 2004. "Corporate investment and asset price dynamics: Implications for the cross-section of returns." *Journal of Finance* 59: 2577–2603.

Carlson, R. S. 1970. "Aggregate performance of mutual funds, 1948–1967." *Journal of Financial and Quantitative Analysis* 5: 1–32.

Cass, D. and J. E. Stiglitz. 1970. "The structure of investor preferences and asset returns, and separability in portfolio allocation: A contribution to the pure theory of mutual funds." *Journal of Economic Theory* 2(2): 122–160.

Cattell, Raymond B. 1966. "The Scree test for the number of factors." *Multivariate Behavioral Research* 1(2): 245–276.

Cecchetti, S. G., P. S. Lam, and N. C. Mark. 1994. "Testing volatility restrictions on intertemporal marginal rates of substitution implied by Euler equations and asset returns." *Journal of Finance* 49(1): 123–152.

Chabi-Yo, Fousseni. 2007. "Conditioning information and variance bounds on pricing kernels with higher-order moments: Theory and evidence." *Review of Financial Studies* 21(1): 181–231.

Chabi-Yo, Fousseni. 2011. "Explaining the idiosyncratic volatility puzzle using stochastic discount factors." *Journal of Banking and Finance* 35: 1971–1983.

Chabi-Yo, Fousseni. 2012. "Pricing kernels with stochastic skewness and volatility risk." *Management Science* 58(3): 624–640.

Chabi-Yo, Fousseni, Dietmar Leisen, and Eric Renault. 2014. "Aggregation of preferences for skewed asset returns." *Journal of Economic Theory* 154: 453–489.

Chamberlain, Gary, and M. Rothschild. 1983. "Arbitrage, factor structure and mean variance analysis on large asset markets." *Econometrica* 51: 1281–1304.

Chan, K. C., and N. F. Chen. 1988. "An unconditional asset-pricing test and the role of firm size as an instrumental variable for risk." *Journal of Finance* 43: 309–325.

Chan, K. C., G. A. Karolyi, F. A. Longstaff, and A. B. Sanders. 1992. "An empirical comparison of alternative models of the short-term interest rate." *Journal of Finance* 47: 1209–1227.

Chan, L. K., J. Karceski, and J. Lakonishok. 1998. "The risk and return from factors." *Journal of Financial and Quantitative Analysis* 33: 159–188.

Chang, E. C., and W. G. Lewellen. 1984. "Market timing and mutual fund investment performance." *Journal of Business* 57: 57–72.

Chang, Hui-shyong, and C. F. Lee. 1977. "Using pooled time-series and cross-section data to test the firm and time effects in financial analysis." *Journal of Financial and Quantitative Analysis* 12: 457–471.

Chang, Wayne. 2016. "The term structure of CAPM alphas and betas." Working paper, University of Southern California, Los Angeles.

Chang, Wayne, and Wayne Ferson. 2018. "Predictor variables versus the ICAPM." Working paper, University of Southern California, Los Angeles.

Chang, Yen-hen, Harrison Hong, and Inessa Liskovich. 2014. "Regression discontinuity and the price effects of stock market indexing." *Review of Financial Studies* 28(1): 212–246.

Chen, Long, and Xinlei Zhao. 2009. "Return decomposition." *Review of Financial Studies* 22: 5213–5249.

Chen, N., R. Roll, and S. Ross. 1986. "Economic forces and the stock market." *Journal of Business* 59: 383–403.

Chen, Nai-fu, T. Copeland, and D. Mayers. 1987. "A comparison of single and multifactor performance methodologies." *Journal of Financial and Quantitative Analysis* 22: 401–417.

Chen, X., and S. C. Ludvigson. 2009. "Land of addicts? An empirical investigation of habit-based asset pricing models." *Journal of Applied Econometrics* 24: 1057–1093.

Chen, Yi-li, Hanno Lustig, and Kanda Naknoi. 2016. "Why are exchange rates so smooth? A segmented asset markets explanation." Working paper, University of California, Los Angeles.

Chen, Yong, Michael Cliff, and Haibei Zhao. 2017. "Hedge funds: The good, the bad and the lucky." *Journal of Financial and Quantitative Analysis* 52: 1081–1109.

Chen, Yong, and Wayne E. Ferson. 2017. "How many good and bad funds are there, really?" Working paper, Texas A&M University, College Station.

Chen, Yong, Wayne Ferson, and H. Peters. 2010. "Measuring the timing ability of fixed income funds." *Journal of Financial Economics* 98: 72–89.

Chen, Z., and P. J. Knez. 1996. "Portfolio performance measurement: Theory and applications." *Review of Financial Studies* 9: 511–556.

Chen, Z., and R. Petkova. 2012. "Does idiosyncratic volatility proxy for risk exposure?" *Review of Financial Studies* 25: 2745–2787.

Chernov, M. 2003. "Empirical reverse engineering of the pricing kernel." *Journal of Econometrics* 116(1–2). 329–364.

Chevalier, J., and L. Ellison, 1997. "Risk taking by mutual funds as a response to incentives." *Journal of Political Economy* 105: 1167–1200.

Chiang, I-Hsuan Ethan. 2015. "Modern portfolio management with information." *Journal of Empirical Finance* 33: 114–134.

Chiang, I-Hsuan Ethan. 2016. "Skewness and cowskewness in bond returns." *Journal of Financial Research* 34: 145–178.

Chordia, T., A. Goyal, and J. Shanken. 2015. "Cross-sectional asset pricing with individual stocks: Betas versus characteristics." Working paper, Emory University, Atlanta, GA.

Chordia, T., and L. Shivakumar. 2002. "Momentum, business cycle and time-varying expected returns." *Journal of Finance* 57: 985–1019.

Chrétien, S. 2012. "Bounds on the autocorrelation of admissible stochastic discount factors." *Journal of Banking and Finance* 36: 1943–1962.

Chrétien, S., and M. Kammoun. 2017. "Mutual fund performance evaluation and best clienteles." *Journal of Financial and Quantitative Analysis* 52: 1577–1604.

Christoffersen, Peter, Steven Heston, and Krisk Jacobs. 2013. "Capturing option anomalies with a variance-dependent pricing kernel." *Review of Financial Studies* 26(8): 1963–2006.

Christopherson, Jon A., Wayne Ferson, and Debra A. Glassman. 1998. "Conditioning manager alpha on economic information: Another look at the persistence of performance." *Review of Financial Studies* 11: 111–142.

Christopherson, Jon A., and A. L. Turner. 1991. "Volatility and predictability of manager alpha." *Journal of Portfolio Management* 18: 5–12.

Chun, S. 2009. "Are international equity market returns predictable?" Working paper, Boston University, Boston, MA.

Cici, G., and S. Gibson. 2012. "The performance of corporate bond mutual funds: Evidence based on security-level holdings." *Journal of Financial and Quantitative Analysis* 47: 159–178.

Cochrane, John H. 1988. "How big is the random walk in GNP?" *Journal of Political Economy* 96: 893–920.

Cochrane, John H. 1991. "Production-based asset pricing and the link between stock returns and economic fluctuations." *Journal of Finance* 46: 209–237.

Cochrane, John H. 1996. "A cross-sectional test of an investment-based asset pricing model." *Journal of Political Economy* 104: 572–621.

Cochrane, John H. 2005. *Asset Pricing*, second ed. Princeton, NJ: Princeton University Press.

Cochrane, John H. 2008. "The dog that did not bark: A defense of return predictability." *Review of Financial Studies* 21(4): 1533–1575.

Cochrane, John H., and Lars Hansen. 1992. "Asset pricing lessons for macroeconomics." In *The Macroeconomics Annual*, ed. O. Blanchard and S. Fisher, 115–165. Cambridge, MA: MIT Press.

Cochrane, John H., and Monica Piazzesi. 2005. "Bond risk premia." *American Economic Review* 95: 138–160.

Cochrane, John H., and Jesus Saa-Requejo. 2000. "Beyond arbitrage: Good deal asset pricing bounds in incomplete markets." *Journal of Political Economy* 108: 79–119.

Coggin, T. D., F. J. Fabozzi, and S. Rahman. 1993. "The investment performance of US equity pension fund managers: An empirical investigation." *Journal of Finance* 48: 1039–1055.

Colacito, Riccardo, and Mariano Croce. 2013. "International asset pricing with recursive preferences." *Journal of Finance* 58: 2651–2686.

Comer, G. 2006. "Hybrid mutual funds and market timing performance." *Journal of Business* 79: 771–797.

Comer, G., N. Larrymore, and J. Rodriguez. 2007. "Controlling for fixed-income exposure in portfolio evaluation: Evidence from hybrid mutual funds." *Review of Financial Studies* 22: 481–507.

Connor, G., and R. A. Korajczyk. 1986. "Performance measurement within the Arbitrage Pricing Theory." *Journal of Financial Economics* 15: 373–394.

Connor, G., and R. A. Korajczyk. 1988. "Risk and return in an equilibrium APT: Applications of a new test methodology." *Journal of Financial Economics* 21: 255–289.

Connor, G., R. A. Korajczyk, and O. Linton. 2006. "The common and specific components of dynamic volatility." *Journal of Econometrics* 132: 231–255.

Connor, Gregory. 1984. "A unified beta pricing theory." *Journal of Economic Theory* 34: 13–31.

Conrad, Jennifer, Robert Dittmar, and Eric Ghysels. 2013. "Ex ante skewness and expected stock returns." *Journal of Finance* 68: 85–124.

Conrad, Jennifer, and Gautam Kaul. 1988. "Time-variation in expected returns." *Journal of Business* 61: 409–425.

Conrad, Jennifer, and Gautam Kaul. 1989. "Mean reversion in short-horizon expected returns." *Review of Financial Studies* 2: 225–240.

Constantinides, George M. 1982. "Intertemporal asset pricing with heterogeneous consumers and without demand aggregation." *Journal of Business* 55: 253–267.

Constantinides, George M. 1990. "Habit formation: A resolution of the equity premium puzzle." *Journal of Political Economy* 98: 519–543.

Constantinides, George M., and D. Duffie. 1996. "Asset pricing with heterogeneous consumers." *Journal of Political Economy* 104: 219–240.

Constantinides, George M., and A. Ghosh. 2011. "Asset pricing tests with long-run risks in consumption growth." *Review of Asset Pricing Studies* 1: 96–136.

Cooper, M., R. C. Gutierrez, Jr., and B. Marcum. 2005. "On the predictability of stock returns in real time." *Journal of Business* 78(2): 469–500.

Copeland, T. E., and D. Mayers. 1982. "The value line enigma (1965–1978): A case study of performance evaluation issues." *Journal of Financial Economics* 10(3): 289–321.

Cornell, B., 1979. "Asymmetric information and portfolio performance measurement." *Journal of Financial Economics* 7: 381–390.

Cox, John C., Jonathan E. Ingersoll, and Stephen A. Ross. 1985. "A theory of the term structure of interest rates." *Econometrica* 53: 385–408.

Cragg, J. G., and B. G. Malkiel. 2009. *Expectations and the Structure of Share Prices*. Chicago: University of Chicago Press.

Crane, A. D., S. Michenaud, and J. P. Weston. 2016. "The effect of institutional ownership on payout policy: Evidence from index thresholds." *Review of Financial Studies* 29: 1377–1408.

Cremers, M., and A. Petajisto. 2009. "How active is your fund manager? A new measure that predicts performance." *Review of Financial Studies* 22: 3329–3365.

Croce, M. M. 2014. "Long-run productivity risk: A new hope for production-based asset pricing?" *Journal of Monetary Economics* 66: 13–31.

Culbertson, J. M. 1957. "The theory of the term structure of interest rates." *Quarterly Journal of Economics* 71: 485–577.

Cuñat, V., M. Gine, and M. Guadalupe. 2012. "The vote is cast: The effect of corporate governance on shareholder value." *Journal of Finance* 67: 1943–1977.

Cvitanić, J., and F. Zapatero. 2004. *Introduction to the Economics and Mathematics of Financial Markets*. Cambridge, MA: MIT Press.

Dahlquist, M., and P. Soderlind. 1999. "Evaluating portfolio performance with stochastic discount factors." *Journal of Business* 72: 347–384.

Dai, Q., and K. J. Singleton. 2002. "Expectation puzzles, time-varying risk premia, and affine models of the term structure." *Journal of Financial Economics* 63: 415–441.

Dai, Q., K. J. Singleton, and W. Yang. 2004. "Predictability of bond risk premia and affine term structure models." Manuscript, Stanford University, Stanford, CA.

Daniel, K., M. Grinblatt, S. Titman, and R. Wermers. 1997. "Measuring mutual fund performance with characteristic-based benchmarks." *Journal of Finance* 52: 1035–1058.

Daniel, K., D. Hirshleifer, and A. Subrahmanyam. 1998. "Investor psychology and security market under- and over-reaction." *Journal of Finance* 53: 1839–1885.

Daniel, Kent, and Sheridan Titman. 1997. "Evidence on the characteristics of cross-sectional variation in stock returns." *Journal of Finance* 52: 1–33.

Daniel, Kent, and Sheridan Titman. 2012. "Testing factor model explanations of market anomalies." *Critical Finance Review* 1: 103–139.

Davidian, M., and R. J. Carroll. 1987. "Variance function estimation." *Journal of the American Statistical Association* 82: 1079–1091.

Davis, J. L., E. F. Fama, and K. R. French. 2000. "Characteristics, covariances, and average returns: 1929 to 1997." *Journal of Finance* 55: 389–406.

DeBondt, W. F. 1998. "A portrait of the individual investor." *European Economic Review* 42: 831–844.

DeBondt, Werner, and R. Thaler. 1985. "Does the stock market overreact?" *Journal of Finance* 40: 793–805.

Del Guercio, D., and P. A. Tkac. 2002. "The determinants of the flow of funds of managed portfolios: Mutual funds vs. pension funds." *Journal of Financial and Quantitative Analysis* 37: 523–557.

Delikouras, Stefanos. 2017. "Where's the kink? Disappointment events in consumption growth and equilibrium asset prices." *Review of Asset Pricing Studies* 1: 2851–2888.

De Long, J. B., A. Shleifer, L. H. Summers, and R. J. Waldmann. 1990. "Noise trader risk in financial markets." *Journal of Political Economy* 98: 703–738.

DeMiguel, V., L. Garlappi, F. J. Nogales, and R. Uppal. 2009. "A generalized approach to portfolio optimization: Improving performance by constraining portfolio norms." *Management Science* 55: 798–812.

Deng, A. 2013. "Understanding spurious regression in financial economics." *Journal of Financial Econometrics* 12: 122–150.

De Roon, Frans, and Theo E. Nijman. 2001. "Tests for mean variance spanning: A survey." *Journal of Empirical Finance* 8: 111–155.

De Roon, Frans, Theo E. Nijman, and Bas J. M. Werker. 2001. "Testing for mean variance spanning with short sales constraints and transactions costs: The case of emerging markets." *Journal of Finance* 56: 721–742.

De Santis, G. 1993. "Volatility bounds for stochastic discount factors: Tests and implications from international stock returns." Working paper, University of Southern California, Los Angeles.

Dietz, P. O., H. R. Fogler, and A. U. Rivers. 1981. "Duration, nonlinearity, and bond portfolio performance." *Journal of Portfolio Management* 7(3): 37–41.

Ding, Z., C. W. Granger, and E. F. Engle. 1993. "A long memory property of stock market returns and a new model." *Journal of Empirical Finance* 1: 83–106.

Dittmar, R. F. 2002. "Nonlinear pricing kernels, kurtosis preference, and evidence from the cross section of equity returns." *Journal of Finance* 57: 369–403.

Dittmar, Robert F., Fransisco Palomino, and Wei Yang. 2016. "Leisure preferences, long run risks and human capital returns." *Review of Asset Pricing Studies* 6: 88–134.

Dolley, James C. 1933. "Characteristics and procedure of common stock split ups." *Harvard Business Review* 11: 316–326.

Doshi, Hitesh, Redouane Elkamhi, and Mikhail Simutin. 2015. "Managerial activeness and mutual fund performance." *Review of Asset Pricing Studies* 2: 156–184.

Drechsler, I. 2013. "Uncertainty, time-varying fear, and asset prices." *Journal of Finance* 68: 1843–1889.

Driscoll, J. C., and A. C. Kraay. 1998. "Consistent covariance matrix estimation with spatially dependent panel data." *Review of Economics and Statistics* 80: 549–560.

Duan, J. C. 1997. "Augmented GARCH (p, q) process and its diffusion limit." *Journal of Econometrics* 79: 97–127.

Duan, Y., G. Hu, and David MacLean. 2005. "Idiosyncratic risk and short-sellers: A test of the costly arbitrage hypothesis." Working paper, Boston College.

Duarte, Jefferson, Avraham Kamara, Stephan Siegel, and Celine Sun. 2012. "The common component of idiosyncratic volatility." Working paper, University of Washington, Seattle.

Duffee, Gregory R. 2011. "Information in (and not in) the term structure." *Review of Financial Studies* 24: 2895–2934.

Duffie, Darrel, and Kenneth J. Singleton. 1993. "Simulated moments estimation of Markov models of asset prices." *Econometrica* 61: 929–952.

Dunn, Kenneth B., and Kenneth J. Singleton. 1986. "Modeling the term structure of interest rates under non-separable utility and durability of goods." *Journal of Financial Economics* 17: 27–55.

Dybvig, P. H. 1983. "An explicit bound on individual assets' deviations from APT pricing in a finite economy." *Journal of Financial Economics* 12: 483–496.

Dybvig, Philip H., and J. Ingersoll. 1982. "Mean variance theory in complete markets." *Journal of Business* 55: 233–252.

Dybvig, Philip H., and Stephen A. Ross. 1985a. "Performance measurement using differential information and a security market line." *Journal of Finance* 40: 383–399.

Dybvig, Philip H., and Stephen A. Ross. 1985b. "The analytics of performance measurement using a security market line." *Journal of Finance* 40: 401–416.

Dybvig, Philip H., and Stephen A. Ross. 1985c. "Yes, the APT is testable." *Journal of Finance* 40: 1173–1188.

Eeckhoudt, L., and H. Schlesinger. 2006. "Putting risk in its proper place." *American Economic Review* 96: 280–289.

Efron, B. 1987. "Better bootstrap confidence intervals." *Journal of the American Statistical Association* 82: 171–185.

Efron, B., and R. J. Tibshirani. 1994. *An Introduction to the Bootstrap*. Boca Raton, FL: CRC Press.

Eichenbaum, M. S., L. P. Hansen, and K. J. Singleton. 1988. "A time series analysis of representative agent models of consumption and leisure choice under uncertainty." *Quarterly Journal of Economics* 103(1): 51–78.

Elton, Edwin J., Martin J. Gruber, and Christopher R. Blake. 1995. "Fundamental economic variables, expected returns and bond fund performance." *Journal of Finance* 50: 1229–1256.

Elton, Edwin J., Martin J. Gruber, and Christopher R. Blake. 1996. "Survivor bias and mutual fund performance." *Review of Financial Studies* 9: 1097–1120.

Elton, Edwin J., Martin J. Gruber, and Christopher R. Blake. 2001. "A first look at the accuracy of the CRSP mutual fund database and a comparison of the CRSP and Morningstar mutual fund databases." *Journal of Finance* 56: 2415–2430.

Elton, Edwin J., Martin J. Gruber, and Christopher R. Blake. 2011. "Holdings data, security returns and the selection of superior mutual funds." *Journal of Financial and Quantitative Analysis* 46: 341–367.

Engle, R. 2002. "Dynamic conditional correlation: A simple class of multivariate generalized autoregressive conditional heteroskedasticity models." *Journal of Business & Economic Statistics* 20: 339–350.

Engle, R. F. 1982. "Autoregressive conditional heteroscedasticity with estimates of the variance of United Kingdom inflation." *Econometrica* 50(4): 987–1007.

Engle, R. F., and G. Gonzalez-Rivera. 1991. "Semiparametric ARCH models." *Journal of Business & Economic Statistics* 9: 345–359.

Engle, R. F., and K. F. Kroner. 1995. "Multivariate simultaneous generalized ARCH." *Econometric Theory* 11: 122–150.

Engle, R. F., D. M. Lilien, and R. P. Robins. 1987. "Estimating time varying risk premia in the term structure: The ARCH-M model." *Econometrica* 55(2): 391–407.

Engle, R. F., and V. K. Ng. 1993. "Measuring and testing the impact of news on volatility." *Journal of Finance* 48: 1749–1778.

Engle, R. F., V. K. Ng, and M. Rothschild. 1990. "Asset pricing with a factor-ARCH covariance structure: Empirical estimates for treasury bills." *Journal of Econometrics* 45: 213–237.

Engle, R. F., and M. E. Sokalska. 2012. "Forecasting intraday volatility in the US equity market. Multiplicative component GARCH." *Journal of Financial Econometrics* 10: 54–83.

Epstein, Larry G., Emmanuel Farhi, and Tomasz Strzalecki. 2014. "How much would you pay to resolve long run risk?" *American Economic Review* 104: 2680–2697.

Epstein, Larry G., and Martin Schneider. 2010. "Ambiguity in asset markets." *Annual Review of Financial Economics* 2: 315–346.

Epstein, Larry G., and Stanley E. Zin. 1989. "Substitution, risk aversion and the temporal behavior of asset returns: A theoretical framework." *Econometrica* 57: 937–969.

Epstein, Larry G., and Stanley E. Zin. 1991. "Substitution, risk aversion and the temporal behavior of asset returns: An empirical analysis." *Journal of Political Economy* 99: 263–286.

Ergun, Tolga, and Jongbyund Jun. 2010. "Time-varying higher order conditional moments and forecasting intraday VaR and expected shortfall." *Quarterly Review of Economics and Finance* 50: 264–272.

Ewens, Michael, Charles Jones, and Matthew Rhodes-Kropf. 2012. "The price of diversifable risk in venture capital and private equity." Working paper, California Institute of Technology, Pasadena, CA.

Fama, E. F. 1965. "The behavior of stock-market prices." *Journal of Business* 38: 34–105.

Fama, E. F. 1970. "Efficient capital markets: A review of theory and empirical work." *Journal of Finance* 25: 383–417.

Fama, E. F. 1972. "Components of investment performance." *Journal of Finance* 27: 551–567.

Fama, E. F. 1973. "A note on the market model and the two-parameter model." *Journal of Finance* 28: 1181–1185.

Fama, E. F. 1976a. "Forward rates as predictors of future spot rates." *Journal of Financial Economics* 3: 361–377.

Fama, E. F. 1976b. *Foundations of Finance: Portfolio Decisions and Securities Prices.* New York: Basic Books.

Fama, E. F. 1981. "Stock returns, real activity, inflation, and money." *American Economic Review* 71: 545–565.

Fama, E. F. 1984a. "Forward and spot exchange rates." *Journal of Monetary Economics* 14: 319–338.

Fama, E. F. 1984b. "The information in the term structure." *Journal of Financial Economics* 13: 509–528.

Fama, E. F. 1991. "Efficient capital markets II." *Journal of Finance* 46: 1575–1617.

Fama, E. F. 1996. "Multifactor portfolio efficiency and multifactor asset pricing." *Journal of Financial and Quantitative Analysis* 31(4): 441–464.

Fama, E. F. 1998. "Market efficiency, long-term returns, and behavioral finance." *Journal of Financial Economics* 49: 283–306.

Fama, E. F., and Robert R. Bliss. 1987. "The information in long-maturity forward rates." *American Economic Review* 77: 680–692.

Fama, E. F., L. Fisher, M. C. Jensen, and R. Roll. 1969. "The adjustment of stock prices to new information." *International Economic Review* 10: 1–21.

Fama, E. F., and K. R. French. 1988. "Permanent and temporary components of stock prices." *Journal of Political Economy* 96: 246–273.

Fama, E. F., and K. R. French. 1989. "Business conditions and expected returns on stocks and bonds." *Journal of Financial Economics* 25: 23–49.

Fama, E. F., and K. R. French. 1992. "The cross-section of expected stock returns." *Journal of Finance* 47: 427–465.

Fama, E. F., and K. R. French. 1993. "Common risk factors in the returns of stocks and bonds." *Journal of Financial Economics* 33: 3–56.

Fama, E. F., and K. R. French. 1995. "Size and book-to-market factors in earnings and returns." *Journal of Finance* 50: 131–155.

Fama, E. F., and K. R. French. 1996. "Multifactor explanations of asset pricing anomalies." *Journal of Finance* 51: 55–87.

Fama, E. F., and K. R. French. 2008. "Dissecting anomalies." *Journal of Finance* 63(4): 1653–1678.

Fama, E. F., and K. R. French. 2010. "Luck versus skill in the cross-section of mutual fund returns." *Journal of Finance* 65: 1915–1947.

Fama, E. F., and K. R. French. 2015. "A five-factor asset pricing model." *Journal of Financial Economics* 116: 1–22.

Fama, E. F., and K. R. French. 2016. "Dissecting anomalies with a five-factor model." *Review of Financial Studies* 29: 69–103.

Fama, E. F., and James D. MacBeth. 1973. "Risk, return and equilibrium: Empirical tests." *Journal of Political Economy* 81: 607–36.

Fama, E. F., and G. W. Schwert. 1977. "Asset returns and inflation." *Journal of Financial Economics* 5: 115–146.

Farnsworth, H. K. 1997. "Evaluating stochastic discount factors from term structure models." PhD dissertation, University of Washington, Seattle.

Farnsworth, H. K., W. Ferson, D. Jackson, and S. Todd. 2002. "Performance evaluation with stochastic discount factors." *Journal of Business* 75: 473–504.

Ferson, Wayne E. 1983. "Expectations of real interest rates and aggregate consumption: empirical tests." *Journal of Financial and Quantitative Analysis* 18: 477–497.

Ferson, Wayne E. 1989. "Changes in expected security returns, risk and the level of interest rates." *Journal of Finance* 44: 1191–1217.

Ferson, Wayne E. 1995. "Theory and empirical testing of asset pricing models." In *Finance, Handbooks in Operations Research and Management Science*, ed. R. A. Jarrow, V. Maksimovic, and W. T. Ziemba, 145–200. New York: Elsevier.

Ferson, Wayne E. 1996. "Warning: Attribute-sorted portfolios can be hazardous to your research." In *Modern Finance Theory and Its Applications*, ed. Susumo Saito, Katsushige Sawaki, and Keiichi Kubota, 21–32. Osaka: Center for Academic Societies.

Ferson, Wayne E. 2003. "Investment performance: A review and synthesis." In *Handbook of the Economics of Finance*, vol. 2, ed. G. M. Constantinides, M. Harris, and R. M. Stulz, 969–1010. New York: Elsevier.

Ferson, Wayne E. 2013. "Ruminations on investment performance measurement." *European Financial Management*, *19*(1): 4–13.

Ferson, Wayne E., and George M. Constantinides. 1991. "Habit persistence and durability in aggregate consumption: Empirical tests." *Journal of Financial Economics* 29: 199–240.

Ferson, Wayne E., and S. R. Foerster. 1994. "Finite sample properties of the generalized method of moments in tests of conditional asset pricing models." *Journal of Financial Economics* 36: 29–55.

Ferson, Wayne E., S. R. Foerster, and D. B. Keim. 1993. "General tests of latent variable models and mean-variance spanning." *Journal of Finance* 48: 131–156.

Ferson, Wayne E., and Campbell R. Harvey. 1991. "The variation of economic risk premiums." *Journal of Political Economy* 99: 385–415.

Ferson, Wayne E., and Campbell R. Harvey. 1992. "Seasonality and consumption-based asset pricing." *Journal of Finance* 47: 511–552.

Ferson, Wayne E., and Campbell R. Harvey. 1993. "The risk and predictability of international equity returns." *Review of Financial Studies* 6: 527–566.

Ferson, Wayne E., and Campbell R. Harvey. 1997. "Fundamental determinants of national equity market returns: A perspective on conditional asset pricing." *Journal of Banking & Finance* 21: 1625–1665.

Ferson, Wayne E., and Campbell R. Harvey. 1999. "Conditioning variables and cross-section of stock returns." *Journal of Finance* 54: 1325–1360.

Ferson, Wayne E., T. R. Henry, and D. J. Kisgen. 2006. "Evaluating government bond fund performance with stochastic discount factors." *Review of Financial Studies* 19: 423–456.

Ferson, Wayne E., A. Heuson, and T. Su. 2005. "Weak and semi-strong form stock return predictability revisited." *Management Science* 51: 1582–1592.

Ferson, Wayne E., and Ravi Jagannathan. 1996. "Econometric evaluation of asset pricing models." In *Handbook of Statistics: Statistical Methods in Finance*, vol. 14, ed. G. S. Maddala and C. R. Rao, 1–30. Amsterdam: North Holland.

Ferson, Wayne E., R. Kan, A. Siegel, and J. Wang. 2018. "Asymptotic distributions for tests of asset pricing models with conditioning information." Working paper, University of Southern California, Los Angeles.

Ferson, Wayne E., and K. Khang. 2002. "Conditional performance measurement using portfolio weights: Evidence for pension funds." *Journal of Financial Economics* 65: 249–282.

Ferson, Wayne E., and M. S. Kim. 2012. "The factor structure of mutual fund flows." *International Journal of Portfolio Analysis and Management* 1: 112–143.

Ferson, Wayne E., and R. A. Korajczyk. 1995. "Do arbitrage pricing models explain the predictability of stock returns?" *Journal of Business* 68(3): 309–349.

Ferson, Wayne E., and J. Lin. 2014. "Alpha and performance measurement: The effects of investor heterogeneity." *Journal of Finance* 69: 1565–1598.

Ferson, Wayne E., and H. Mo. 2016. "Performance measurement with selectivity, market and volatility timing." *Journal of Financial Economics* 121: 93–110.

Ferson, Wayne E., S. Nallareddy, and B. Xie. 2016. "The 'out-of-sample' performance of long run risk models." *Journal of Financial Economics* 107: 537–556.

Ferson, Wayne E., and M. Qian. 2004. *Conditional Performance Evaluation Revisited*. Charlottesville, VA: Research Foundation of CFA Institute.

Ferson, Wayne E., Sergei Sarkissian, and Timothy Simin. 1999. "The Alpha Factor Asset Pricing Model: A Parable." *Journal of Financial Markets* 2: 49–68.

Ferson, Wayne E., Sergei Sarkissian, and Timothy Simin. 2003. "Spurious regressions in Financial Economics?" *Journal of Finance* 4: 1393–1413.

Ferson, Wayne E., Sergei Sarkissian, and Timothy Simin. 2008. "Asset pricing models with conditional alphas and betas: The effects of data snooping and spurious regression." *Journal of Financial and Qualitative Analysis* 43: 331–354.

Ferson, Wayne E., and Rudi W. Schadt. 1996. "Measuring fund strategy and performance in changing economic conditions." *Journal of Finance* 51: 425–462.

Ferson, Wayne E., and Andrew F. Siegel. 2001. "The efficient use of conditioning information in portfolios." *Journal of Finance* 56: 967–982.

Ferson, Wayne E., and A. F. Siegel. 2003. "Stochastic discount factor bounds with conditioning information." *Review of Financial Studies* 16: 567–595.

Ferson, Wayne E., and A. F. Siegel. 2009. "Testing portfolio efficiency with conditioning information." *Review of Financial Studies* 22: 2735–2758.

Ferson, Wayne E., and A. F. Siegel. 2015. "Optimal orthogonal portfolios with conditioning information." In *Handbook of Financial Econometrics and Statistics*, ed. C. F. Lee, 977–1002. New York: Springer.

Ferson, Wayne E., A. F. Siegel, and P. T. Xu. 2006. "Mimicking portfolios with conditioning information." *Journal of Financial and Quantitative Analysis* 41: 607–635.

Ferson, Wayne E., and Junbo Wang. 2018. "Holdings based fund performance measures: Estimation and inference." Working paper, University of Southern California, Los Angeles.

Ferson, Wayne E., and V. Warther. 1996. "Evaluating fund performance in a dynamic market." *Financial Analysts Journal* 52: 20–28.

Fisher, L. 1966. "Some new stock-market indexes." *Journal of Business* 39: 191–225.

Flammer, C. 2015. "Corporate social responsibility and the allocation of procurement contracts: Evidence from a natural experiment." Working paper, Boston University.

Fleming, J., C. Kirby, and B. Ostdiek. 2001. "The economic value of volatility timing." *Journal of Finance* 56: 329–352.

Fletcher, J., and D. Basu. 2016. "An examination of the benefits of dynamic trading strategies in UK closed-end funds." *International Review of Financial Analysis* 47: 109–118.

Foster, D. P., and D. B. Nelson. 1996. "Continuous record asymptotics for rolling sample variance estimators." *Econometrica* 64(1): 139–174.

Foster, F. Douglas, Matthew Richardson, and Tom Smith. 1993. "Small sample properties of simulated method of moments estimators." Working paper, Duke University, Durham, North Carolina.

Foster, F. Douglas, T. Smith, and R. Whaley. 1997. "Assessing goodness-of-fit of asset pricing models: The distribution of the maximal *R*-squared." *Journal of Finance* 52: 591–607.

Frazzini, A., and O. A. Lamont. 2008. "Dumb money: Mutual fund flows and the cross-section of stock returns." *Journal of Financial Economics* 88: 299–322.

Frazzini, Andrea, and Lasse Pedersen. 2014. "Betting against beta." *Journal of Financial Economics* 111: 1–25.

French, K., G. W. Schwert, and R. Stambaugh. 1987. "Expected stock returns and volatility." *Journal of Financial Economics* 19: 3–29.

French, Kenneth R. 2008. "The cost of active investing." *Journal of Finance* 63: 1537–1573.

Freyberger, Joachim, Andreas Neuhierl, and Michael Weber. 2016. "Dissecting characteristics nonparametrically." Working paper, University of Wisconsin, Madison.

Friedman, M. 1953. *Essays in Positive Economics.* Chicago: University of Chicago Press.

Frisch, Ragnar, and Frederick Waugh. 1933. "Partial Time Regressions as Compared with Individual Trends." *Econometrica* 1: 387–401.

Frost, P. A., and J. E. Savarino. 1986. "An empirical Bayes approach to efficient portfolio selection." *Journal of Financial and Quantitative Analysis* 21: 293–305.

Fu, F. 2009. "Idiosyncratic risk and the cross-section of expected stock returns." *Journal of Financial Economics* 91: 24–37.

Gabaix, Xavier. 2012. "Variable rare disasters: An exactly solved framework for ten puzzles in macro finance." *Quarterly Journal of Economics* 127: 645–700.

Gagliardini, Patrick, Elisa Ossola, and Olivier Scaillet. 2016. "Time-varying risk premium in large cross-sectional data sets." *Econometrica* 84: 985–1046.

Gallant, Ron, and George Tauchen. 1996. "Which moments to match?" *Econometric Theory* 12: 657–681.

Gallant, Ronald A., Lars Peter Hansen, and George Tauchen. 1990. "Using conditional moments of asset payoffs to infer the volatility of intertemporal marginal rates of substitution." *Journal of Econometrics* 45: 141–179.

Gârleanu, N., and L. H. Pedersen. 2018. "Efficiently inefficient markets for assets and asset management." *Journal of Finance* 73: 1663–1712.

Garman, M. B., and M. J. Klass. 1980. "On the estimation of security price volatilities from historical data." *Journal of Business* 1: 67–78.

Ghysels, Eric, 1998. "On stable factor structures in the pricing of risk: Do time-varying betas help or hurt?" *Journal of Finance* 53: 549–573.

Ghysels, Eric, Andrew Harvey, and Eric Renault. 1996. "Stochastic volatility." In *Handbook of Statistics* 14, ed. G. S. Maddala, 119–191. Amsterdam: North Holland.

Ghysels, Eric, Pedro Santa-Clara, and Rossen Valkanov. 2004. "The MIDAS touch: Mixed Data sampling regression models." Working paper, University of California, Los Angeles.

Gibbons, Michael R. 1980. "Econometric methods for testing a class of financial models: An application of the nonlinear multivariate regression model." PhD dissertation, University of Chicago.

Gibbons, Michael R. 1982. "Multivariate tests of financial models." *Journal of Financial Economics* 10: 3–27.

Gibbons, Michael R., and W. Ferson. 1985. "Testing asset pricing models with changing expectations and an unobservable market portfolio." *Journal of Financial Economics* 14: 217–236.

Gibbons, Michael R., Stephen A. Ross, and Jay Shanken. 1989. "A test of the efficiency of a given portfolio." *Econometrica* 57: 1121–1152.

Gilbert, Thomas, Christopher Hrdkicka, Jonathan Kalodimos, and Stephan Siegel. 2014. "Daily data is bad for beta: Opacity and frequency dependent betas." *Review of Asset Pricing Studies* 4: 78–117.

Glode, V. 2011. "Why mutual funds 'underperform.'" *Journal of Financial Economics*, 99: 546–559.

Glosten, L., and R. Jagannathan. 1994. "A contingent claims approach to performance evaluation." *Journal of Empirical Finance* 1: 133–166.

Glosten, L., R. Jagannathan, and D. E. Runkle. 1993. "On the relation between the expected value and the volatility of the nominal excess return on stocks." *Journal of Finance* 48: 1779–1801.

Goetzmann, W., J. Ingersoll, M. Spiegel, and I. Welch. 2007. "Portfolio performance manipulation and manipulation-proof performance measures." *Review of Financial Studies* 20: 1503–1546.

Goetzmann, W. N., and R. G. Ibbotson. 1994. "Do winners repeat?" *Journal of Portfolio Management* 20: 9–18.

Goetzmann, W. N., J. Ingersoll, and Z. Ivkovic. 2000. "Monthly measurement of daily timers." *Journal of Financial and Quantitative Analysis* 35: 257–290.

Goetzmann, W. N., and P. Jorion. 1993. "Testing the predictive power of dividend yields." *Journal of Finance* 48(2): 663–679.

Gollier, C., and J. W. Pratt. 1996. "Risk vulnerability and the tempering effect of background risk." *Econometrica* 64(5): 1109–1123.

Gomez, Joao, Leonid Kogan, and Motohiro Yogo. 2009. "Durability of output and expected stock returns." *Journal of Political Economy* 117: 941–986.

Gomez, Joao, Leonid Kogan, and Lu Zhang. 2003. "Equilibrium cross section of returns." *Journal of Political Economy* 111: 693–732.

Gomez, Juan Pedro, Richard Priestly, and Fernando Zapatero. 2016. "Labor income, relative wealth concerns and the cross-section of stock returns." *Journal of Financial and Quantitative Analysis* 51: 1111–1133.

Gonedes, N. J. 1973. "Evidence on the information content of accounting numbers: Accounting-based and market-based estimates of systematic risk." *Journal of Financial and Quantitative Analysis* 8(3): 407–443.

Gonzales-Gaverra, N. 1973. "Inflation and capital market prices: Theory and tests." PhD dissertation, Stanford University, Stanford, CA.

González-Urteaga, A., and G. Rubio. 2016. "The cross-sectional variation of volatility risk premia." *Journal of Financial Economics* 119: 353–370.

Gordon, Myron J. 1962. *The Investment, Financing, and Valuation of the Corporation.* Homewood, IL: R. D. Irwin.

Gordon, S., L. Samson, and B. Carmichael. 1996. "Bayesian estimation of stochastic discount factors." *Journal of Business & Economic Statistics* 14(4): 412–420.

Goriaev, M., T. Nijman, and B. Werker. 2005. "Yet another look at mutual fund tournaments." *Journal of Empirical Finance* 12: 127–137.

Gorman, W. M. 1953. "Community preference fields." *Econometrica* 21(1): 63–80.

Gormley, T. A., and D. A. Matsa. 2013. "Common errors: How to (and not to) control for unobserved heterogeneity." *Review of Financial Studies* 27: 617–661.

Gourier, Elise. 2016. "Pricing of idiosyncratic equity and variance risks." Working paper, Queen Mary University of London.

Gourieroux, C., and A. Monfort. 1992. "Qualitative threshold ARCH models." *Journal of Econometrics* 52: 159–199.

Gourieroux, C., A. Monfort, and V. Polimenis. 2006. "Affine models for credit risk analysis." *Journal of Financial Econometrics* 4: 494–530.

Goyal, Amit, and P. Santa-Clara. 2003. "Idiosyncratic risk matters!" *Journal of Finance* 58: 975–1007.

Goyal, Amit, and Ivo Welch. 2003. "Predicting the equity premium with dividend ratios." *Management Science* 49: 639–654.

Granger, C. W. 1969. "Investigating causal relations by econometric models and cross-spectral methods." *Econometrica* 37(3): 424–438.

Granger, C. W. K., and P. Newbold. 1974. "Spurious regressions in econometrics." *Journal of Econometrics* 2: 111–120.

Grant, D. 1977. "Portfolio performance and the 'cost' of timing decisions." *Journal of Finance* 32: 837–846.

Green, R. C. 1986. "Benchmark portfolio inefficiency and deviations from the security market line." *Journal of Finance* 41: 295–312.

Green, R. C., and B. Hollifield. 1992. "When will mean-variance efficient portfolios be well diversified?" *Journal of Finance* 47: 1785–1809.

Green, R. C. and K. Rydqvist. 1997. "The valuation of nonsystematic risks and the pricing of Swedish lottery bonds." *Review of Financial Studies* 10: 447–480.

Greenwood, R., and A. Shleifer. 2014. "Expectations of returns and expected returns." *Review of Financial Studies* 27: 714–746.

Greenwood, R. M. 2005. "Aggregate corporate liquidity and stock returns." Working paper, Harvard University, Cambridge, MA.

Grinblatt, M., and B. Han. 2003. "The disposition effect and momentum." NBER Working Paper 8734, National Bureau of Economic Research, Cambridge, MA.

Grinblatt, M., and T. Moskowitz. 2003. "Predicting price movements from past returns: The role of consistency in tax-loss selling." *Journal of Financial Economics* 71: 541–579.

Grinblatt, M., and S. Titman. 1983. "Factor pricing in a finite economy." *Journal of Financial Economics* 12: 497–507.

Grinblatt, M., and S. Titman. 1987. "The relation between mean-variance efficiency and arbitrage pricing." *Journal of Business* 60: 97–112.

Grinblatt, M., and S. Titman. 1989a. "Mutual fund performance: An analysis of quarterly portfolio holdings." *Journal of Business* 62: 393–416.

Grinblatt, M., and S. Titman. 1989b. "Portfolio performance evaluation: Old issues and new insights." *Review of Financial Studies* 2: 393–422.

Grinblatt, M., and S. Titman. 1992. "The persistence of mutual fund performance." *Journal of Finance* 47: 1977–1984.

Grinblatt, M., and S. Titman. 1993. "Performance measurement without benchmarks: An examination of mutual fund returns." *Journal of Business* 60: 97–112.

Grinblatt, M., S. Titman, and R. Wermers. 1995. "Momentum strategies, portfolio performance and herding: A study of mutual fund behavior." *American Economic Review* 85: 1088–1105.

Grinold, R. C. 1989. "The fundamental law of active management." *Journal of Portfolio Management* 15(3): 30–37.

Grinold, R. C., and R. N. Kahn. 2000. *Active Portfolio Management.* New York: McGraw Hill.

Grishchenko, O. V. 2010. "Internal vs. external habit formation: The relative importance for asset pricing." *Journal of Economics and Business* 62:176–194.

Gromb, D., and D. Vayanos. 2010. "Limits of arbitrage." *Annual Review of Finance and Economics* 2: 251–275.

Grossman, S. J., and R. J. Shiller. 1982. "Consumption correlatedness and risk measurement in economies with non-traded assets and heterogeneous information." *Journal of Financial Economics* 10: 195–210.

Grossman, S. J., and J. E. Stiglitz. 1980. "On the impossibility of informationally efficient markets." *American Economic Review* 70: 393–408.

Gruber, M. 1996. "Another puzzle: The growth in actively managed mutual funds." *Journal of Finance* 51: 783–810.

Gultekin, N. B., and R. J. Rogalski. 1984. "Alternative duration specifications and the measurement of basis risk: Empirical tests." *Journal of Business* 57: 241–264.

Guo, H., and R. Savickas. 2010. "Relation between time-series and cross-sectional effects of idiosyncratic variance on stock returns." *Journal of Banking & Finance* 34(7): 1637–1649.

Hall, A. R. 2000. "Covariance matrix estimation and the power of the overidentifying restrictions test." *Econometrica* 68: 1517–1527.

Hall, Robert E. 1978. "Stochastic implications of the life cycle permanent income hypothesis: Theory and evidence." *Journal of Political Economy* 86: 971–987.

Hall, Robert E. 1987. "Intertemporal substitution in consumption." *Journal of Political Economy* 96: 339–357.

Hansen, L. P., and T. J. Sargent. 2001a. "Acknowledging misspecification in macroeconomic theory." *Review of Economic Dynamics* 4: 519–535.

Hansen, L., and T. J. Sargent. 2001b. "Robust control and model uncertainty." *American Economic Review* 91: 60–66.

Hansen, L. P., and T. J. Sargent. 2008. *Robustness.* Princeton, NJ: Princeton University Press.

Hansen, L. P., and K. J. Singleton. 1996. "Efficient estimation of linear asset-pricing models with moving average errors." *Journal of Business & Economic Statistics* 14: 53–68.

Hansen, Lars, and Kenneth J. Singleton. 1982. "Generalized instrumental variables estimation of nonlinear rational expectations models." *Econometrica* 50: 1269–1286.

Hansen, Lars, and Kenneth J. Singleton. 1983. "Stochastic consumption, risk aversion and the temporal behavior of asset returns." *Journal of Political Economy* 91: 249–265.

Hansen, Lars P. 1982. "Large sample properties of generalized method of moments estimators." *Econometrica* 50: 1029–1053.

Hansen, Lars P. 1985. "A method for calculating bounds on the asymptotic covariance matrices of generalized method of moments estimators." *Journal of Econometrics* 30: 203–238.

Hansen, Lars P., John Heaton, and Nan Li. 2008. "Consumption strikes back? Measuring long-run risk." *Journal of Political Economy* 116: 260–302.

Hansen, Lars P., J. Heaton, and E. Luttmer. 1995. "Econometric evaluation of asset pricing models." *Review of Financial Studies* 8: 237–274.

Hansen, Lars P., John Heaton, and Amir Yaron. 1996. "Finite sample properties of some alternative GMM estimators." *Journal of Business & Economic Statistics* 14: 262–280.

Hansen, Lars P., and Robert J. Hodrick. 1983. "Risk averse speculation in the forward foreign exchange market: An econometric analysis of linear models." In *Exchange Rates and International Macroeconomics*, ed. J. A. Frenkel, 113–152. Chicago: University of Chicago Press.

Hansen, Lars P., and Ravi Jagannathan. 1991. "Implications of security market data for models of dynamic economies." *Journal of Political Economy* 99: 225–262.

Hansen, Lars P., and Ravi Jagannathan. 1997. "Assessing specification errors in stochastic discount factor models." *Journal of Finance* 52(2): 557–590.

Hansen, Lars P., and Scott F. Richard. 1987. "The role of conditioning information in deducing the testable restrictions implied by dynamic asset pricing models." *Econometrica* 55: 587–613.

Harrison, M., and D. Kreps. 1979. "Martingales and arbitrage in multi-period securities markets." *Journal of Economic Theory* 20: 381–408.

Hartzmark, S. M. 2016. "Economic uncertainty and interest rates." *Review of Asset Pricing Studies* 6: 179–220.

Harvey, C. R. 1989. "Time-varying conditional covariances in tests of asset pricing models." *Journal of Financial Economics* 24: 289–317.

Harvey, C. R. 1991. "The world price of covariance risk." *Journal of Finance* 46(1): 111–157.

Harvey, C. R. 2017. "Presidential address: The scientific outlook in financial economics." *Journal of Finance* 72(4): 1399–1440.

Harvey, C. R., and C. M. Kirby. 1996. "Instrumental variables estimation of conditional beta pricing models." In *Handbook of Statistics*, vol 14, ed. G. S. Maddala and C. R. Rao, 35–60. New York: Elsevier.

Harvey, C. R., and Y. Liu. 2016. "Rethinking performance evaluation." Working paper, Duke University, Durham, NC.

Harvey, C. R., Y. Liu, and H. Zhu. 2016. "… and the cross-section of expected returns." *Review of Financial Studies* 29: 5–68.

Harvey, C. R., and A. Siddique. 2000. "Conditional skewness in asset pricing tests." *Journal of Finance* 55: 1263–1295.

Harvey, C. R., and A. Siddique. 2001. "Autoregressive conditional skewness." *Journal of Financial and Quantitative Analysis* 34: 465–487.

Haugen, R. A., and N. L. Baker. 1991. "The efficient market inefficiency of capitalization-weighted stock portfolios." *Journal of Portfolio Management* 17(3): 35–40.

Haugen, R. A., and N. L. Baker. 1996. "Commonality in the determinants of expected stock returns." *Journal of Financial Economics* 41: 401–439.

Haugen, R., and J. Lakonishok. 1988. *The Incredible January Effect*. Homewood, IL: Dow Jones–Irwin.

Hayashi, Fumio. 1982. "Tobin's marginal q and average q: A neoclassical interpretation." *Econometrica* 50: 213–224.

Heaton, J. 1993. "The interaction between time-nonseparable preferences and time aggregation." *Econometrica* 61(2): 353–385.

Hendricks, D., J. Patel, and R. Zeckhauser. 1993. "Hot hands in mutual funds: Short-run persistence of relative performance, 1974–1988." *Journal of Finance* 48: 93–130.

Henriksson, Roy D. 1981. "Market timing and investment performance II: Statistical procedures for evaluating 1981 forecasting skills." *Journal of Business* 1: 513–533.

Herskovic, Bernard, Bryan Kelly, Hanno Lustig, and Stijn Van Niewerburgh. 2016. "The common factor in idiosyncratic volatility: Quantitative asset pricing implications." *Journal of Financial Economics* 19: 249–283.

Hicks, L. R. 1946. *Value and Capital*, second ed. London: Oxford University Press.

Higgins, M. L., and A. K. Bera. 1992. "A class of nonlinear ARCH models." *International Economic Review* 33(1): 137–158.

Hirshleifer, D. 1988. "Residual risk, trading costs and commodity futures risk premia." *Review of Financial Studies* 1: 173–193.

Hirshleifer, D., P. D. Hsu, and D. Li. 2013. "Innovative efficiency and stock returns." *Journal of Financial Economics* 107: 632–654.

Hirshleifer, D., and J. Yu. 2013. "Asset pricing with extrapolative expectations and production." Working paper, University of California, Irvine, and University of Minnesota, Minneapolis.

Hirshleifer, David, Sonya S. Lim, and Siew Gong Teoh. 2011. "Limited investor attention and stock market misreaction to accounting information." *Review of Asset Pricing Studies* 1: 35–73.

Hjalmarsson, E. 2006. "Predictive regressions with panel data." Working paper, Department of Economics, Yale University, New Haven, CT.

Hjalmarsson, E. 2008. "The Stambaugh bias in panel predictive regressions." *Finance Research Letters* 5: 47–58.

Hjalmarsson, E. 2010. "Predicting global stock returns." *Journal of Financial and Quantitative Analysis* 45: 49–80.

Hodrick, R. 1992. "Dividend yields and expected stock returns: Alternative preocedures for inference and measurement." *Review of Financial Studies* 5: 257–286.

Hong, H., and J. Stein. 1999. "A unified theory of underreaction, momentum trading and overreaction in asset markets." *Journal of Finance* 45: 265–295.

Hou, Kewei, and Roger Loh. 2016. "Have we solved the idiosyncratic volatility puzzle?" Working paper, Singapore Management University.

Hou, Kewei, Chen Xue, and Lu Zhang. 2015. "Digesting anomalies: An investment approach." *Review of Financial Studies* 28: 650–705.

Hsu, J., H. Kudoh, and T. Yamada. 2013. "When sell-side analysts meet high-volatility stocks: An alternative explanation for the low-volatility puzzle." *Journal of Investment Management* 11(2): 28–46.

Huang, Chi-fu, and Robert H. Litzenberger. 1988. *Foundations for Financial Economics*. Amsterdam: North Holland.

Huang, D., and G. Zhou. 2017. "Upper bounds on return predictability." *Journal of Financial and Quantitative Analysis* 52: 401–425.

Huang, J. Z., and Wang, Y. 2014. "Timing ability of government bond fund managers: Evidence from portfolio holdings." *Management Science* 60(8): 2091–2109.

Huang, J., K. D. Wei, and H. Yan. 2007. "Participation costs and the sensitivity of fund flows to past performance." *Journal of Finance* 62: 1273–1311.

Huang, W., Q. Liu, S. G. Rhee, and L. Zhang. 2009a. "Another look at idiosyncratic volatility and expected returns." Working paper, University of Hawaii, Manoa.

Huang, W., Q. Liu, S. G. Rhee, and L. Zhang. 2009b. "Return reversals, idiosyncratic risk, and expected returns." *Review of Financial Studies* 23: 147–168.

Huberman, G., and S. Kandel. 1987. "Mean variance spanning." *Journal of Finance* 42: 383–388.

Huberman, Gur. 1982. "A simple approach to arbitrage pricing theory." *Journal of Economic Theory* 28: 183–191.

Huberman, Gur, S. A. Kandel, and R. F. Stambaugh. 1987. "Mimicking portfolios and exact arbitrage pricing." *Journal of Finance* 42: 1–10.

Hull, J., and A. White. 1990. "Pricing interest-rate-derivative securities." *Review of Financial Studies* 3: 573–592.

İmrohoroğlu, A., and Ş. Tüzel. 2014. "Firm-level productivity, risk, and return." *Management Science* 60: 2073–2090.

Ingersoll, J. E. 1984. "Some results in the theory of arbitrage pricing." *Journal of Finance* 39: 1021–1039.

Ingersoll, Jonathan E. 1987. *Theory of Financial Decision Making*. Totowa, NJ: Roman and Littlefield.

Ippolito, R. A. 1989. "Efficiency with costly information: A study of mutual fund performance, 1965–1984." *Quarterly Journal of Economics* 104: 1–23.

Ippolito, R. A. 1992. "Consumer reaction to measures of poor quality: Evidence from the mutual fund industry." *Journal of Law and Economics* 35(1): 45–70.

Izan, Haji Y. 1978. "An empirical analysis of the economic effects of mandatory government audit requirements." PhD dissertation, University of Chicago.

Jackwerth, J. C. 2000. "Recovering risk aversion from option prices and realized returns." *Review of Financial Studies* 13: 433–451.

Jackwerth, J. C., and M. Rubinstein. 1996. "Recovering probability distributions from option prices." *Journal of Finance* 51: 1611–1631.

Jagannathan, R., A. Kaplin, and S. Sun. 2003. "An evaluation of multi-factor CIR models using LIBOR, swap rates, and cap and swaption prices." *Journal of Econometrics* 116(1–2): 113–146.

Jagannathan, R., and R. Korajczyk. 1986. "Assessing the market timing performance of managed portfolios." *Journal of Business* 59: 217–236.

Jagannathan, R., and T. Ma. 2003. "Risk reduction in large portfolios: Why imposing the wrong constraints helps." *Journal of Finance* 58: 1651–1683.

Jagannathan, R., and Y. Wang. 2007. "Lazy investors, discretionary consumption and the cross-section of stock returns." *Journal of Finance* 62: 1623–1661.

Jagannathan, R., and Z. Wang. 1996. "The conditional CAPM and the cross-section of expected returns." *Journal of Finance* 51: 3–53.

Jegadeesh, N. 1990. "Evidence of predictable behavior of security returns." *Journal of Finance* 45: 881–898.

Jegadeesh, N., J. Noh, K. Pukthuanthong, and R. Roll. 2017. "Empirical tests of asset pricing models with individual assets: Resolving the errors-in-variables bias in risk premium estimation." Working paper, Louisiana State University, Baton Rouge.

Jegadeesh, N., and S. Titman. 1993. "Returns to buying winners and selling losers: Implications for stock market efficiency." *Journal of Finance* 48: 65–91.

Jegadeesh, N., and S. Titman. 2001. "Profitability of momentum portfolios: An evaluation of alternative explanations." *Journal of Finance* 56: 699–720.

Jensen, M. C. 1968. "The performance of mutual funds in the period 1945–1964." *Journal of Finance* 23: 389–416.

Jensen, M. C. 1969. "Risk, the pricing of capital assets, and the evaluation of investment portfolios." *Journal of Business* 42: 167–247.

Jensen, M. C. 1972. "Optimal utilization of market forecasts and the evaluation of investment performance." In *Mathematical Methods in Finance*, ed. G. P. Szego and Karl Shell, 1–34. Amsterdam: North-Holland.

Jermann, Urban J. 1998. "Asset pricing in production economies." *Journal of Monetary Economics* 41: 257–275.

Jermann, Urban J. 2010. "The equity premium implied by production." *Journal of Financial Economics* 98: 279–296.

Jiang, G. J., T. Yao, and T. Yu. 2007. "Do mutual funds time the market? Evidence from portfolio holdings." *Journal of Financial Economics* 86: 724–758.

Jiang, George J., Danielle Xu, and Tong Yao. 2009. "The information content of idiosyncratic volatility." *Journal of Financial and Quantitative Analysis* 44: 1–28.

Jiao, Y., M. Massa, and H. Zhang. 2016. "Short selling meets hedge fund 13F: An anatomy of informed demand." *Journal of Financial Economics* 122(3): 544–567.

Jobson, J. D., and B. Korkie. 1980. "Estimation for Markowitz efficient portfolios." *Journal of the American Statistical Association* 75(371): 544–554.

Jobson, J. D., and B. Korkie. 1982. "Potential performance and tests of portfolio efficiency." *Journal of Financial Economics* 10: 433–466.

Johannes, M., A. Korteweg, and N. Polson. 2014. "Sequential learning, predictability, and optimal portfolio returns." *Journal of Finance* 69: 611–644.

Jones, C. S., and H. Mo. 2016. "Out-of-sample performance of mutual fund predictors." Working paper, University of Southern California, Los Angeles.

Jones, Chris S., and Jay Shanken. 2005. "Mutual fund performance with learning across funds." *Journal of Financial Economics* 78: 507–552.

Jorion, P. 1986. "Bayes-Stein estimation for portfolio analysis." *Journal of Financial and Quantitative Analysis* 21: 279–292.

Jorion, P. 2003. "Portfolio optimization with tracking-error constraints." *Financial Analysts Journal* 59(5): 70–82.

Joslin, Scott, Marcel Priebsch, and Kenneth J. Singleton. 2014. "Risk premiums in dynamic term structure models with unspanned macro risks." *Journal of Finance* 69: 1197–1233.

Jurek, Jakub. 2008. "Crash-neutral carry trades." Working paper, Princeton University, Princeton, NJ.

Kacperczyk, M., S. Van Nieuwerburgh, and L. Veldkamp. 2016. "A rational theory of mutual funds' attention allocation." *Econometrica* 84: 571–626.

Kacperczyk, M. C., C. Sialm, and L. Zheng. 2008. "Unobserved actions of mutual funds." *Review of Financial Studies* 21: 2379–2416.

Kahneman, Daniel, and A. Tversky. 1979. "Prospect theory: An analysis of decisions under risk." *Econometrica* 47: 263–292.

Kan, R., and C. Robotti. 2008. "Model comparison using the Hansen-Jagannathan distance." *Review of Financial Studies* 22: 3449–3490.

Kan, R., and G. Zhou. 2004. "Hansen-Jagannathan Distance: Geometry and Exact Distribution." Working paper, School of Economics and Finance, University of Hong Kong.

Kan, R., and G. Zhou. 2006. "A new variance bound on the stochastic discount factor." *Journal of Business* 79: 941–961.

Kan, R., and G. Zhou. 2007. "Optimal portfolio choice with parameter uncertainty." *Journal of Financial and Quantitative Analysis* 42: 621–656.

Kan, Raymond, C. Robotti, and Jay Shanken. 2013. "Pricing model performance and the two-pass cross-sectional regression methodology." *Journal of Finance* 68: 2617–2688.

Kan, Raymond, Xiaolu Wang, and Guofu Zhou. 2016. "Optimal portfolio selection with and without (a) risk-free asset." Working paper, University of Toronto.

Kandel, S., and R. F. Stambaugh. 1989. "A mean variance framework for tests of asset pricing models." *Review of Financial Studies* 2: 125–156.

Kandel, S., and R. F. Stambaugh. 1990. "Expectations and volatility of consumption and asset returns." *Review of Financial Studies* 3: 207–232.

Kandel, S., and R. F. Stambaugh. 1995. "Portfolio inefficiency and the cross-section of expected returns." *Journal of Finance* 50: 157–184.

Kandel, S., and R. F. Stambaugh. 1996. "On the predictability of stock returns: An asset allocation perspective." *Journal of Finance* 51: 385–424.

Karolyi, Andrew. 1992. "Predicting risk: Some generalizations." *Management Science* 38: 1–57.

Kaul, G. 1996. "Predictable components in stock returns." In *Handbook of Statistics*, vol. 14, ed. G. S. Maddala and C. R. Rao, 269–296. Amsterdam: Elsevier Science.

Kaul, G., and M. Nimalendran. 1990. "Price reversals: Bid-ask errors or market overreaction?" *Journal of Financial Economics* 28: 67–93.

Keim, D., and R. Stambaugh. 1986. "Predicting returns in the stock and bond markets." *Journal of Financial Economics* 17: 357–390.

Keim, Donald. 1983. "Size-related anomalies and stock return seasonalities: Further evidence." *Journal of Financial Economics* 12: 13–32.

Kelly, Bryan, and Seth Pruitt. 2012. "The three-pass regression filter: A new approach to forecasting with many predictors." Working paper, Booth School of Business, University of Chicago.

Kelly, Bryan, and Seth Pruit. 2013. "Market expectations in the cross-section of present values." *Journal of Finance* 68: 1721–1756.

Kelly, Bryan, Seth Pruit, and Yinan Su. 2017. "Some characteristics are risk exposures, and the rest are irrelevant." Working paper, University of Chicago.

Kendal, M. G. 1954. "A note on bias in the estimation of autocorrelation." *Biometrika* 41: 403–404.

Kerr, W. R., J. Lerner, J. and A. Schoar. 2010. "The consequences of entrepreneurial finance: A regression discontinuity analysis." NBER Working Paper w15831, National Bureau of Economic Research, Cambridge, MA.

Kessel, Ruben. 1965. "The cyclical behavior of the term structure of interest rates." NBER Working Paper 91, National Bureau of Economic Research, New York.

Kilic, Mete, and Ivan Shaliatovich. 2018. "Good and bad variance premia and expected returns." *Management Science* (forthcoming).

Kim, Dongcheol. 1995. "The errors-in-the-variables problem in the cross-section of stock returns." *Journal of Finance* 50: 1605–1634.

Kim, Min S. 2017. "Changes in mutual fund flows and managerial incentives." Working paper, University of New South Wales, Kensington, Australia.

Kim, Myung, C. R. Nelson, and R. Startz. 1991. "Mean reversion in stock returns? A reappraisal of the statistical evidence." *Review of Economic Studies* 58: 515–528.

Kim, Soohun, and George Skoulakis. 2014. "Estimating and testing linear factor models using large cross-sections: The regression-calibration approach." Working paper, Georgia Institute of Technology, Atlanta.

Kimball, Miles. 1990. "Precautionary saving in the small and in the large." *Econometrica* 58: 53–73.

Kirby, C. 1998. "The restrictions on predictability implied by rational asset pricing models." *Review of Financial Studies* 11: 343–382.

Kirby, C., and B. Ostdiek. 2012. "It's all in the timing: Simple active portfolio strategies that outperform naive diversification." *Journal of Financial and Quantitative Analysis* 47(2): 437–467.

Knight, Frank. 1921. *Risk, Uncertainty and Profit.* Boston: Houghton Mifflin.

Koijen, R. S., H. Lustig, and S. Van Nieuwerburgh. 2017. "The cross-section and time series of stock and bond returns." *Journal of Monetary Economics* 88: 50–69.

Koijen, R. S., H. Lustig, S. Van Nieuwerburgh, and A. Verdelhan. 2010. "Long run risk, the wealth-consumption ratio, and the temporal pricing of risk." *American Economic Review* 100: 552–556.

Kon, S. J., 1983. "The market timing performance of mutual fund managers." *Journal of Business* 56: 323–347.

Korajczyk, R., and R. Sadka. 2004. "Are momentum profits robust to trading costs?" *Journal of Finance* 59: 1039–1082.

Koski, J. L., and J. Pontiff. 1999. "How are derivatives used? Evidence from the mutual fund industry." *Journal of Finance* 54: 791–816.

Kosowski, R., N. Y. Naik, and M. Teo. 2007. "Do hedge funds deliver alpha? A Bayesian and bootstrap analysis." *Journal of Financial Economics* 84: 229–264.

Kothari, S. P., and Jay Shanken. 1997. "Book-to-market, dividend yield and expected market returns: A time-series analysis." *Journal of Financial Economics* 44: 169–203.

Kowsowski, R., A. Timmerman, R. Wermers, and H. L. White. 2006. "Can mutual fund 'stars' really pick stocks? Evidence from a bootstrap analysis." *Journal of Finance* 61: 2251–2295.

Kraus, A., and R. H. Litzenberger. 1976. "Skewness preference and the valuation of risk assets." *Journal of Finance* 31: 1085–1100.

Kreps, D. M., and E. L. Porteus. 1978. "Temporal resolution of uncertainty and dynamic choice theory." *Econometrica* 46(1): 185–200.

Kroencke, Tim A. 2017. "Asset pricing without garbage." *Journal of Finance* 72: 47–68.

Kroner, K. F., and V. K. Ng. 1998. "Modeling asymmetric comovements of asset returns." *Review of Financial Studies* 11(4): 817–844.

Kryzanowski, L., S. Lalancette, and M. C. To. 1997. "Performance attribution using an APT with prespecified macrofactors and time-varying risk premia and betas." *Journal of Financial and Quantitative Analysis* 32: 205–224.

Kuntara, Pukthauanthong, and Richard Roll. 2013. "A protocol for factor identification." Working paper, California Institute of Technology, Pasadena.

Lakonishok, J., A. Shleifer, and R. W. Vishny. 1992. "The impact of institutional trading on stock prices." *Journal of Financial Economics* 32: 23–43.

Lakonishok, J., A. Shleifer, and R. W. Vishny. 1994. "Contrarian investment, extrapolation, and risk." *Journal of Finance* 49: 1541–1578.

Lamont, Owen A. 2000. "Investment plans and stock returns." *Journal of Finance* 55: 2719–2745.

Lamoureux, C. G., and W. D. Lastrapes. 1990. "Persistence in variance, structural change, and the GARCH model." *Journal of Business & Economic Statistics* 8: 225–234.

Ledoit, O., and M. Wolf. 2003. "Honey, I shrunk the sample covariance matrix." Working paper, Universitat Pompeu Fabra, Barcelona.

Lee, C. F., and H. Y. Chen. 2013. "Alternative errors-in-variables methods and their applications in finance." Working paper, Rutgers University, New Brunswick, NJ.

Lee, C. F., C. Wu, and K. J. Wei. 1990. "The heterogeneous investment horizon and the capital asset pricing model: Theory and implications." *Journal of Financial and Quantitative Analysis* 25: 361–376.

Lehmann, B. N. 1990a. "Fads, martingales and market efficiency." *Quarterly Journal of Economics* 105: 1–28.

Lehmann, B. N. 1990b. "Residual risk revisited." *Journal of Econometrics* 45: 71–97.

Lehmann, B. N., and D. Modest. 1987. "Mutual fund performance evaluation: A comparison of benchmarks and benchmark comparisons." *Journal of Finance* 42: 233–265.

Lehmann, E. L., and Romano, J. P. 2005. "Generalizations of the Familywise Error Rate." *Annals of Statistics* 33: 1138–1154.

Lei, Qin. 2006. "Pricing dynamics of dividend and nondividend payout yields: On the stock return predictability." Working paper, Southern Methodist University, Dallas.

Leland, Hayne. 1999. "Performance beyond mean-variance: Performance measurement in a nonsymmetric world." *Financial Analysts Journal* 55: 27–36.

LeRoy, Stephen F., and Richard D. Porter. 1981. "The present value relation: Tests based on implied variance bounds." *Econometrica* 49: 555–574.

Lesmond, D., M. Schill, and C. Zhou. 2004. "The illusory nature of momentum profits." *Journal of Financial Economics* 71: 349–380.

Lesmond, D. A., J. P. Ogden, and C. A. Trzcinka. 1999. "A new estimate of transaction costs." *Review of Financial Studies* 12(5): 1113–1141.

Lettau, M., and S. Ludvigson. 2001a. "Consumption, aggregate wealth, and expected stock returns." *Journal of Finance* 56: 815–849.

Lettau, M., and S. Ludvigson. 2001b. "Resurrecting the (C)CAPM: A cross-sectional test when risk premia are time-varying." *Journal of Political Economy* 109: 1238–1287.

Lettau, M., and S. Ludvigson. 2009. "Measuring and modeling variation in the risk-return tradeoff." In *Handbook of Financial Econometrics*, ed. Yacine Ait-Sahalia and Lars Peter Hansen, 618–684. Amsterdam: Elsevier Science.

Lettau, M., S. C. Ludvigson, and J. A. Wachter. 2007. "The declining equity premium: What role does macroeconomic risk play?" *Review of Financial Studies* 21: 1653–1687.

Levy, H. 1978. "Equilibrium in an imperfect market: A constraint on the number of securities in the portfolio." *American Economic Review* 68: 643–658.

Lewellen, J. 1999. "The time-series relations among expected return, risk, and book-to-market." *Journal of Financial Economics* 54(1): 5–43.

Lewellen, Jonathan. 2015. "The cross-section of expected stock returns." *Critical Finance Review* 4: 1–44.

Lewellen, Jonathan, and Stefan Nagel. 2006. "The conditional CAPM does not explain asset pricing anomalies." *Journal of Financial Economics* 82: 289–314.

Lewellen, Jonathan, Stefan Nagel, and Jay Shanken. 2010. "A skeptical appraisal of asset pricing tests." *Journal of Financial Economics* 96(2): 175–194.

Lewis, Karen. 1999. "Trying to explain home bias in equities and consumption." *Journal of Economic Literature* 37: 571–608.

Lewis, Karen, and Edith X. Liu. 2015. "Evaluating international consumption risk sharing gains: An asset return view." *Journal of Monetary Economics* 71: 84–98.

Liew, J., and M. Vassalou. 2000. "Can book-to-market, size and momentum be risk factors that predict economic growth?" *Journal of Financial Economics* 57: 221–245.

Light, N., D. Maslov, and O. Rytchkov. 2017. "Aggregation of information about the cross section of stock returns: A latent variable approach." *Review of Financial Studies* 30: 1339–1381.

Lin, Xiaoji, and Lu Zhang. 2012. "The investment manifesto." *Journal of Monetary Economics* 60(3): 351–366.

Linnainmaa, J. T. 2013. "Reverse survivorship bias." *Journal of Finance* 68: 789–813.

Linnainmaa, Juhani, and Michael Roberts. 2018. "The history of the cross-section of stock returns." *Review of Financial Studies* 7: 2606–2649.

Lintner, John. 1965. "Security prices, risk, and maximal gains from diversification." *Journal of Finance* 20: 587–615.

Litterman, R., and J. Scheinkman. 1991. "Common factors affecting bond returns." *Journal of Fixed Income* 1: 54–61.

Litzenberger, R., and K. Ramaswamy. 1979. "The effect of personal taxes and dividends on capital asset prices: Theory and evidence." *Journal of Financial Economics* 7: 163–196.

Litzenberger, R., and K. Ramaswamy. 1982. "The effects of dividends on common stock prices: Tax effects or information effects?" *Journal of Finance* 37: 429–433.

Liu, L. 2008. "It takes a model to beat a model: Volatility bounds." *Journal of Empirical Finance* 15: 80–110.

Liu, L. X., T. M. Whited, and L. Zhang. 2009. "Investment-based expected stock returns." *Journal of Political Economy* 117: 1105–1139.

Liu, L. X., and L. Zhang. 2008. "Momentum profits, factor pricing, and macroeconomic risk." *Review of Financial Studies* 21: 2417–2448.

Lo, A., and A. C. MacKinlay. 1988. "Stock market prices do not follow random walks: Evidence from a simple specification test." *Review of Financial Studies* 1(1): 41–66.

Lo, A., and A. C. MacKinlay. 1990. "An econometric analysis of infrequent trading." *Journal of Econometrics* 45: 181–211.

Lo, A. W. 2002. "The statistics of Sharpe ratios." *Financial Analysts Journal* 58(4): 36–52.

Long, J. B. 1974. "Stock prices, inflation, and the term structure of interest rates." *Journal of Financial Economics* 1: 131–170.

Longstaff, F. A. 1989. "Temporal aggregation and the continuous-time capital asset pricing model." *Journal of Finance* 44: 871–887.

Lucas, Robert E., Jr. 1978. "Asset prices in an exchange economy." *Econometrica* 46: 1429–1445.

Lucas, Robert E., Jr. 2003. "Macroeconomic priorities." *American Economic Review* 93: 1–14.

Ludvigson, S. C., and S. Ng. 2007. "The empirical risk–return relation: A factor analysis approach." *Journal of Financial Economics* 83: 171–222.

Ludvigson, Sydney, and S. Ng. 2009. "Macro factors in bond risk premiums." *Review of Financial Studies* 22: 5027–5067.

Lynch, Anthony. 2000. "Portfolio choice and equity characteristics: Characterizing the hedging demands induced by return predictability." *Journal of Financial Economics* 62(1): 67–130.

Lynch, Anthony, and Jessica Wachter. 2013. "Using samples of unequal length in generalized method of moments estimation." *Journal of Financial and Quantitative Analysis* 48: 277–307.

MacBeth, James D. 1979. "Tests of the two-parameter model of market equilibrium." PhD dissertation, University of Chicago.

MacKinlay, A. Craig. 1995. "Multifactor models do not explain deviations from the CAPM." *Journal of Financial Economics* 38: 3–28.

MacKinlay, A. Craig, and Matthew P. Richardson. 1991. "Using the generalized method of moments to test mean-variance efficiency." *Journal of Finance* 46: 511–528.

Magill, M., and M. Quinzii. 1996. *The Theory of Incomplete Markets.* Cambridge, MA: MIT Press.

Magkotsios, Georgios. 2017. "Asset management when today's alpha is tomorrow's beta." Working paper, University of Southern California, Los Angeles.

Maio, Paulo, and Pedro Santa Clara. 2012. "Multifactor models and their consistency with ICAPM." *Journal of Financial Economics* 106: 586–613.

Malkiel, B. G. 1995. "Returns from investing in equity mutual funds, 1971 to 1991." *Journal of Finance* 50: 549–572.

Malkiel, B. G. 2004. "Models of stock market predictability." *Journal of Financial Research* 27: 449–459.

Malkiel, Burton, and Yexiao Xu. 2002. "Idiosyncratic risk and security returns." Working paper, University of Texas at Dallas.

Malenko, N., and Y. Shen. 2016. "The role of proxy advisory firms: Evidence from a regression-discontinuity design." *Review of Financial Studies* 29: 3394–3427.

Mandelbrot, B. 1963. "The variation of certain speculative prices." *Journal of Business* 36: 394–419.

Markowitz, Harry. 1952. "Portfolio selection." *Journal of Finance* 7: 77–91.

Marsh, T. A., and E. R. Rosenfeld. 1983. "Stochastic processes for interest rates and equilibrium bond prices." *Journal of Finance* 38: 635–646.

Mayers, D., and E. M. Rice. 1979. "Measuring portfolio performance and the empirical content of asset pricing models." *Journal of Financial Economics* 7: 3–28.

McLean, David, and Jeffrey Pontiff. 2016. "Does academic research destroy stock return predictability?" *Journal of Finance* 71: 5–32.

McLean, R. D. 2010. "Idiosyncratic risk, long-term reversal, and momentum." *Journal of Financial and Quantitative Analysis* 45: 883–906.

Mech, T. S. 1993. "Portfolio return autocorrelation." *Journal of Financial Economics* 34: 307–344.

Mehra, R., and E. Prescott. 1985. "The equity premium: A puzzle." *Journal of Monetary Economics* 15: 145–162.

Mehra, Rajnish, Sunil Wahal, and Daruo Xie. 2016. "The demand for diversification in incomplete markets." Working paper, Arizona State University, Tempe.

Mei, J. 1993. "Explaining the cross-section of returns via a multi-factor APT model." *Journal of Financial and Quantitative Analysis* 28: 331–345.

Menzly, Lior, and Oguzhan Ozbas. 2010. "Market segmentation and cross-predictability of returns." *Journal of Finance* 65(4): 1555–1580.

Merton, Robert C. 1973. "An intertemporal capital asset pricing model." *Econometrica* 41: 867–887.

Merton, Robert C. 1980. "On estimating the expected return on the market: An exploratory investigation." *Journal of Financial Economics* 8: 323–361.

Merton, Robert C. 1987. "A simple model of capital market equilibrium with incomplete information." *Journal of Finance* 42: 483–510.

Merton, Robert C., and Roy D. Henriksson. 1981. "On market timing and investment performance II: Statistical procedures for evaluating forecasting skills." *Journal of Business* 54: 513–534.

Michaud, R. O. 1989. "The Markowitz optimization enigma: Is 'optimized' optimal?" *Financial Analysts Journal* 45(1): 31–42.

Milhoj, A. 1987. "A multiplicative parameterization of ARCH models." Working paper, Department of Statistics, University of Copenhagen.

Miller, Edward M. 1977. "Risk, uncertainty and divergence of opinion." *Journal of Finance* 32: 1151–1168.

Mitton, T., and K. Vorkink. 2007. "Equilibrium underdiversification and the preference for skewness." *Review of Financial Studies* 20: 1255–1288.

Modigliani, F., and M. H. Miller. 1958. "The cost of capital, corporation finance and the theory of investment." *American Economic Review* 48: 261–297.

Moneta, F. 2015. "Measuring bond mutual fund performance with portfolio characteristics." *Journal of Empirical Finance* 33: 223–242.

Moskowitz, Tobias J. 2003. "An analysis of covariance risk and pricing anomalies." *Review of Financial Studies* 16: 417–457.

Moskowitz, Tobias J., Christopher J. Malloy, and Annette Vissing-Jorgensen. 2009. "Long-run stockholder consumption risk and asset returns." *Journal of Finance* 64: 2427–2479.

Mossin, J. 1968. "Optimal multiperiod portfolio policies." *Journal of Business* 41: 215–229.

Moulton, B. R. 1987. "Diagnostics for group effects in regression analysis." *Journal of Business & Economic Statistics* 5: 275–282.

Mullins, W. 2014. "The governance impact of index funds: Evidence from regression discontinuity." Working paper, Sloan School of Management, Massachusetts Institute of Technology, Cambridge, MA.

Muthaswamy, J. 1988. "Asynchronous closing prices and spurious correlation in portfolio returns." Working paper, University of Chicago.

Nagel, S., and K. J. Singleton. 2010. "Internet appendix for estimation and evaluation of conditional asset pricing models." Working paper, Stanford University, Stanford, CA.

Nelson, D. B. 1991. "Conditional heteroskedasticity in asset returns: A new approach." *Econometrica* 59(2): 347–370.

Nelson, Daniel. 1992. "Filtering and forecasting with misspecified ARCH models I: Getting the right variance with the wrong model." *Journal of Econometrics* 52: 61–90.

Nelson, Daniel B., and Dean Foster. 1990. "Filtering and forecasting with misspecified ARCH Models II: Making the right forecast with the wrong model." *Journal of Econometrics* 67: 157–190.

Newey, W., and K. D. West. 1987a. "A simple, positive semidefinite heteroskedasticity- and autocorrelation-consistent covariance matrix." *Econometrica* 55: 703.

Newey, W., and K. D. West. 1987b. "Hypothesis testing with efficient method of moments." *International Economic Review* 28: 777–787.

Newey, W., and K. D. West. 1994. "Automatic lag selection in covariance matrix estimation." *Review of Economic Studies* 61: 631–653.

Ni, S. 1995. "An empirical analysis on the substitutability between private consumption and government purchases." *Journal of Monetary Economics* 36: 593–605.

Ni, Shawn. 1997. "Scaling factors in estimation of time-nonseparable utility functions." *Review of Economics and Statistics* 79: 234–240.

Nickell, Stephen. 1981. "Biases in dynamic panel models with fixed effects." *Econometrica* 49: 1417–1426.

Noreen, E. W. 1989. *Computer-Intensive Methods for Testing Hypotheses*. New York: Wiley.

Novy-Marx, R. 2013. "The other side of value: The gross profitability premium." *Journal of Financial Economics* 108: 1–28.

Novy-Marx, R. 2014. "Predicting anomaly performance with politics, the weather, global warming, sunspots, and the stars." *Journal of Financial Economics* 112: 137–146.

Officer, Robert R. 1973. "The variability of the market factor of the New York Stock Exchange." *Journal of Business* 46: 434–453.

Pagan, A. R., and G. W. Schwert. 1990. "Alternative models for conditional stock volatility." *Journal of Econometrics* 45: 267–290.

Parker, J. A., and C. Julliard. 2005. "Consumption risk and the cross section of expected returns." *Journal of Political Economy* 113: 185–222.

Parkinson, M. 1980. "The extreme value method for estimating the variance of the rate of return." *Journal of Business* 53(1): 61–65.

Pástor, Lubos, and R. F. Stambaugh. 2000. "Comparing asset pricing models: An investment perspective." *Journal of Financial Economics* 56: 335–381.

Pástor, Lubos, and R. F. Stambaugh. 2002. "Investing in equity mutual funds." *Journal of Financial Economics* 63: 351–380.

Pástor, Lubos, and R. F. Stambaugh. 2009. "Predictive systems: Living with imperfect predictors." *Journal of Finance* 64: 1583–1628.

Pástor, Lubos, R. F. Stambaugh, and L. A. Taylor. 2015. "Scale and skill in active management." *Journal of Financial Economics* 116: 23–45.

Penaranda, Francisco, and Enrique Sentana. 2016. "Duality in mean-variance frontiers with conditioning information." *Journal of Empirical Finance* 38: 762–785.

Penaranda, Fransisco, and Liuren Wu. 2017. "Predictability and performance." Working paper, City University of New York.

Pesaran, M. H., and Alan Timmerman. 1995. "Predictability of stock returns: Robustness and economic significance." *Journal of Finance* 50: 1201–1228.

Petersen, M. A. 2009. "Estimating standard errors in finance panel data sets: Comparing approaches." *Review of Financial Studies* 22: 435–480.

Peterson, David R., and Adam R. Smedema. 2010. "The return impact of realized and expected idiosyncratic volatility." Working paper, Florida State University, Tallahassee.

Petkova, R. 2006. "Do the Fama-French factors proxy for innovations in predictive variables?" *Journal of Finance* 61: 581–612.

Piazzesi, M., M. Schneider, and S. Tuzel. 2007. "Housing, consumption and asset pricing." *Journal of Financial Economics* 83: 531–569.

Polk, C., S. Thompson, and T. Vuolteenaho. 2004. "New forecasts of the equity premium." NBER Working Paper w10406, National Bureau of Economic Research, Cambridge, MA.

Pontiff, J., and A. Woodgate. 2008. "Share issuance and cross-sectional returns." *Journal of Finance* 63: 921–945.

Pontiff, Jeffrey, and L. Schall. 1998. "Book-to-market as a predictor of market returns." *Journal of Financial Economics* 49: 141–160.

Pontiff, Jeffrey A. 2006. "Costly arbitrage and the myth of idiosyncratic risk." *Journal of Accounting and Economics* 42: 35–52.

Poterba, J. M., and L. H. Summers. 1988. "Mean reversion in stock prices: Evidence and implications." *Journal of Financial Economics* 22: 27–59.

Powell, J. G., J. Shi, T. Smith, and R. Whaley. 2006. "Political regimes, business cycles, seasonalities and returns." Working paper, Vanderbilt University, Nashville.

Pyun, Sungjune. 2018. "Variance risk in aggregate stock returns and time-varying return predictability." Working paper, University of Southern California, Los Angeles.

Rachwalski, Mark, and Quan Wen. 2016. "Idiosyncratic risk innovations and the idiosyncratic risk-return relation." *Review of Asset Pricing Studies* 6: 303–328.

Radner, Roy. 1972. "Existence of equilibrium plans, prices, and price expectations in a sequence of markets." *Econometrica* 40: 289–303.

Ramponi, Valentina, Cesare Robotti, and Paolo Zaffaroni. 2016. "Testing beta pricing models using large cross-sections." Working paper, Imperial College London.

Rapach, D. E., J. K. Strauss, and G. Zhou. 2010. "Out-of-sample equity premium prediction: Combination forecasts and links to the real economy." *Review of Financial Studies* 23: 821–862.

Rauh, J. D. 2006. "Investment and financing constraints: Evidence from the funding of corporate pension plans." *Journal of Finance* 61: 33–71.

Richardson, M. 1993. "Temporary components of stock prices: A skeptic's view." *Journal of Business & Economic Statistics* 11: 199–207.

Roberts, M. R., and A. Sufi. 2009. "Control rights and capital structure: An empirical investigation." *Journal of Finance* 64: 1657–1695.

Rogers, W. 1993. "Quantile regression standard errors." *Stata Technical Bulletin* 2(9): 16–19.

Roll, R. 1977. "A critique of the asset pricing theory's tests—part 1: On past and potential testability of the theory." *Journal of Financial Economics* 4: 129–176.

Roll, R. 1978. "Ambiguity when performance is measured by the security market line." *Journal of Finance* 33: 1051–1069.

Roll, R. 1980. "Performance evaluation and benchmark errors I/II." *Journal of Portfolio Management* 6: 5–12.

Roll, R. 1992. "A mean/variance analysis of tracking error." *Journal of Portfolio Management* 8(4): 13–22.

Roll, R., and S. A. Ross. 1994. "On the cross-sectional relation between expected returns and betas." *Journal of Finance* 49: 101–121.

Roll, Richard, and S. A. Ross. 1980. "An empirical investigation of the arbitrage pricing theory." *Journal of Finance* 35: 1073–1103.

Roll, Richard R. 1984. "A simple implicit measure of the effective bid-ask spread in an efficient market." *Journal of Finance* 39: 1127–1140.

Rosenberg, B., and J. Guy. 1976. "Prediction of beta from investment fundamentals: Part one." *Financial Analysts Journal* 32(3): 60–72.

Rosenberg, B., and V. Marathe. 1975. "The prediction of investment risk: Systematic and residual risk." In working paper, "Proceedings of the Center for Research in Security Prices Seminar," University of Chicago.

Ross, Stephen A. 1976. "The arbitrage theory of capital asset pricing." *Journal of Economic Theory* 13: 341–360.

Ross, Stephen A. 1977. "Risk, return and arbitrage." In *Studies in Risk and Return*, ed. I. Friend and J. Bicksler, 189–218. Cambridge, MA: Ballinger.

Ross, Stephen A. 2005. "Mutual fund separation in financial theory—The separating distributions." In *Theory of Valuation*, 309–356. London: World Scientific Press.

Ross, Stephen A. 2015. "The recovery theorem." *Journal of Finance* 70: 615–648.

Roussanov, Nikolai. 2014. "Composition of wealth, conditioning information, and the cross-section of stock returns." *Journal of Financial Economics* 111: 352–380.

Rozeff, M. S. 1984. "Dividend yields are equity risk premiums." *Journal of Portfolio Management* 10: 68–75.

Rubinstein, M. 1974. "An aggregation theorem for securities markets." *Journal of Financial Economics* 1: 225–244.

Rubinstein, Mark. 1976. "The valuation of uncertain income streams and the pricing of options." *Bell Journal of Economics and Management Science* 7: 407–425.

Samuelson, P. A. 1965. "Proof that properly anticipated prices fluctuate randomly." *IMR; Industrial Management Review* (pre-1986) 6(2): 41.

Samuelson, P. A., and S. Swamy. 1974. "Invariant economic index numbers and canonical duality: Survey and synthesis." *American Economic Review* 64: 566–593.

Santa-Clara, P., and R. Valkanov. 2003. "The presidential puzzle: Political cycles and the stock market." *Journal of Finance* 58: 1841–1872.

Santos, T., and P. Veronesi, 2005. "Labor income and predictable stock returns." *Review of Financial Studies* 19: 1–44.

Sapp, T., and A. Tiwari. 2004. "Does stock return momentum explain the 'smart money' effect?" *Journal of Finance* 59: 2605–2622.

Sargan, J. D. 1959. "The estimation of relationships with autocorrelated residuals by the use of instrumental variables." *Journal of the Royal Statistical Society, Series B* 21(1): 91–105.

Sarkissian, S. 2003. "Incomplete consumption risk sharing and currency risk premiums." *Review of Financial Studies* 16: 983–1005.

Schipper, K., and R. Thompson. 1981. "Common stocks as hedges against shifts in the consumption or investment opportunity set." *Journal of Business* 1: 305–328.

Schipper, Katherine, and Rex Thompson. 1983. "The impact of merger-related regulations on the shareholders of acquiring firms." *Journal of Accounting Research* 21: 184–221.

Scholes, M., and J. Williams. 1977. "Estimating beta from nonsynchronous data." *Journal of Financial Economics* 5: 309–327.

Schultz, Ellen E. 1996. "Vanguard bucks trend by cutting fund fees." *Wall Street Journal*, January 30, c1–c25.

Schwert, G. W. 1990. "Stock volatility and the crash of '87." *Review of Financial Studies* 3: 77–102.

Schwert, G. W. 2003. "Anomalies and markct efficiency." In *Handbook of the Economics of Finance*, ed. George M. Constantinides, Milton Harris, and Rene M. Stulz, 575–603. Amsterdam: Elsevier Science Publishers and North Holland.

Schwert, G. W., and Paul Seguin. 1990. "Heteroskedasticity in stock returns." *Journal of Finance* 45: 1129–1155.

Scott, R. C., and P. A. Horvath. 1980. "On the direction of preference for moments of higher order than the variance." *Journal of Finance* 35: 915–919.

Scruggs, J. 1998. "Resolving the puzzling intertemporal relation between the market risk premium and conditional market variance: A 2-factor approach." *Journal of Finance* 53: 575–603.

Sentana, E. 1995. "Quadratic ARCH models." *Review of Economic Studies* 62: 639–661.

Shanken, J. 1982. "The arbitrage pricing theory: Is it testable?" *Journal of Finance* 37: 1129–1140.

Shanken, J. 1985a. "Multivariate tests of the zero-beta CAPM." *Journal of Financial Economics* 14: 327–348.

Shanken, J. 1985b. "Multi-beta CAPM or equilibrium-APT?: A Reply." *Journal of Finance* 40: 1189–1196.

Shanken, J. 1986. "Testing portfolio efficiency when the zero beta rate is unknown." *Journal of Finance* 41: 269–276.

Shanken, J. 1987. "Multivariate proxies and asset pricing relations: Living with the Roll critique." *Journal of Financial Economics* 18: 91–110.

Shanken, J. 1990. "Intertemporal asset pricing: An empirical investigation." *Journal of Econometrics* 45: 99–120.

Shanken, J. 1992. "On the estimation of beta-pricing models." *Review of Financial Studies* 5: 1–33.

Shanken, J., and Guofu Zhou. 2007. "Estimating and testing beta pricing models: Alternative methods and their performance in simulations." *Journal of Financial Economics* 84: 40–86.

Sharathchandra, G. 1993. "Asset pricing and production: Theory and empirical tests." *Research in Finance* 11: 37–63.

Sharpe, W. F. 1964. "Capital asset prices: A theory of market equilibrium under conditions of risk." *Journal of Finance* 19: 425–442.

Sharpe, W. F. 1966. "Mutual fund performance." *Journal of Business* 39: 119–138.

Sharpe, W. F. 1977. "The capital asset pricing model: A 'multi-beta' interpretation." In *Financial Decision Making under Uncertainty*, ed. H. Levy and M. Sarnat, 127–135. New York: Academic Press.

Sharpe, W. F. 1992. "Asset allocation: Management style and performance measurement." *Journal of Portfolio Management* 18: 7–19.

Shefrin, H. 2008. "Risk and return in behavioral SDF-based asset pricing models." *Journal of Investment Management* 6: 1–18.

Shefrin, H. 2010. "Behavioralizing finance." *Foundations and Trends in Finance* 4: 1–184.

Shiller, Robert. 1981. "Do stock prices move too much to be justified by subsequent changes in dividends?" *American Economic Review* 71: 421–436.

Shleifer, Andrei, and Robert W. Vishny. 1997. "The limits of arbitrage." *Journal of Finance* 52: 35–55.

Shukla, R., and C. Trzcinka. 1994. "Persistent performance in the mutual fund market: Tests with funds and investment advisers." *Review of Quantitative Finance and Accounting* 4: 115–135.

Simin, T. 2008. "The poor predictive performance of asset pricing models." *Journal of Financial and Quantitative Analysis* 43: 355–380.

Sims, C. A. 1980. "Comparison of interwar and postwar business cycles: Monetarism reconsidered." Working Paper 430, National Bureau of Economic Research, New York.

Sirri, E. R., and P. Tufano. 1998. "Costly search and mutual fund flows." *Journal of Finance* 53: 1589–1622.

Smith, Simon, and Allan Timmerman. 2017. "Predictive panel regressions with regime shifts." Working paper, University of California, San Diego.

Snow, Karl N. 1991. "Diagnosing asset pricing models using the distribution of asset returns." *Journal of Finance* 46: 955–983.

Solnik, Bruno. 1993. "The unconditional performance of international asset allocation strategies using conditioning information." *Journal of Empirical Finance* 1: 33–55.

Solomon, D. H., E. Soltes, and D. Sosyura. 2014. "Winners in the spotlight: Media coverage of fund holdings as a driver of flows." *Journal of Financial Economics* 113: 53–72.

Spiegel, M., and X. Wang. 2006. "Cross-sectional variation in stock returns: Liquidity and idiosyncratic risk." Working paper, Yale University, New Haven, CT.

Stambaugh, Robert F. 1982. "On the exclusion of assets from tests of the two-parameter model." *Journal of Financial Economics* 10: 235–268.

Stambaugh, Robert F. 1983a. "Testing the CAPM with broader market indexes: A problem of mean deficiency." *Journal of Banking and Finance* 7: 5–16.

Stambaugh, Robert F. 1983b. "Arbitrage pricing with information." *Journal of Financial Economics* 12: 357–369.

Stambaugh, Robert F. 1988. "Information in forward rates: Implications for models of the term structure." *Journal of Financial Economics* 21: 41–70.

Stambaugh, Robert F. 1999. "Predictive regressions." *Journal of Financial Economics* 54: 375–421.

Stambaugh, Robert F., Jianfeng Yu, and Yu Yuan. 2015. "Arbitrage asymmetry and the idiosyncratic volatility puzzle." *Journal of Finance* 70: 1903–1948.

Stambaugh, Robert F., and Yu Yuan. 2016. "Mispricing factors." *Review of Financial Studies* 30: 1270–1315.

Stanton, R. 1997. "A nonparametric model of term structure dynamics and the market price of interest rate risk." *Journal of Finance* 52: 1973–2002.

Starks, L., and S. Sun. 2016. "Economic policy uncertainty, learning and incentives: Theory and evidence on mutual funds." Working paper, University of Texas, Austin.

Stathopoulos, A. 2016. "Asset prices and risk sharing in open economies." *Review of Financial Studies* 30: 363–415.

Stein, C. 1973. "Estimation of the mean of a multivariate normal distribution." In *Proceedings of the Second Prague Symposium on Asymptotic Statistics*, 345–381. Amsterdam: North Holland.

Stein, E. M., and J. C. Stein. 1991. "Stock price distributions with stochastic volatility: An analytic approach." *Review of Financial Studies* 4: 727–752.

Stock, J. H. 1987. "Asymptotic properties of least squares estimators of cointegrating vectors." *Econometrica* 55(5): 1035–1056.

Stock, J. H., and Mark W. Watson. 2002. "Forecasting using principal components from a large number of predictors." *Journal of the American Statistical Association* 97: 1167–1179.

Stock, J. H., and Mark W. Watson. 2010. "Dynamic factor models." In *Oxford Handbook of Economic Forecasting*, ed. M. P. Clements and D. F. Hendry. Oxford: Oxford University Press.

Stock, J. H., J. H. Wright, and M. Yogo. 2002. "A survey of weak instruments and weak identification in generalized method of moments." *Journal of Business & Economic Statistics* 20: 518–529.

Storey, J. D. 2002. "A direct approach to false discovery rates." *Journal of the Royal Statistical Society* B 64: 479–498.

Subrahmanyam, A. 2008. "Behavioural finance: A review and synthesis." *European Financial Management* 14: 12–29.

Tauchen, G. 1986. "Statistical properties of generalized method-of-moments estimators of structural parameters obtained from financial market data." *Journal of Business & Economic Statistics* 4: 397–416.

Telmer, C. I. 1993. "Asset-pricing puzzles and incomplete markets." *Journal of Finance* 48: 1803–1832.

Theil, H. 1971. *Principles of Econometrics*. New York: Wiley.

Thistlethwaite, D. L., and D. T. Campbell. 1960. "Regression-discontinuity analysis: An alternative to the ex post facto experiment." *Journal of Educational Psychology* 51: 309.

Tierens, Ingrid C. 1993. "Pitfalls in naive implementations of the Hansen-Jagannathan Diagnostic." Working paper, University of Chicago.

Timmerman, A. 2006. "Forecast combination." In *Handbook of Economic Forecasting*, ed. Graham Elliott, Clive W. J. Granger, and Allan Timmerman, 136–196. Amsterdam: Elsevier.

Treynor, J. L. 1965. "How to rate mutual fund performance." *Harvard Business Review* 43: 63–75.

Treynor, J. L., and Fisher Black. 1973. "How to use security analysis to improve portfolio selection." *Journal of Business* 46: 66–86.

Treynor, J. L., and K. Mazuy. 1966. "Can mutual funds outguess the market?" *Harvard Business Review* 44: 131–136.

Van Binsbergen, J., M. Brandt, and R. Koijen. 2012. "On the timing and pricing of dividends." *American Economic Review* 102: 1596–1618.

Van Binsbergen, J., W. Hueskes, R. Koijen, and E. Vrugt. 2013. "Equity yields." *Journal of Financial Economics* 110(3): 503–519.

Van Binsbergen, Jules. 2016. "Good-specific habit formation and the cross-section of expected returns." *Journal of Finance* 71(4): 1699–1732.

Van Nieuwerburgh, S., and L. Veldkamp. 2010. "Information acquisition and under-diversification." *Review of Economic Studies* 77: 779–805.

Vasicek, Oldrich. 1977. "An equilibrium characterization of the term structure." *Journal of Financial Economics* 5: 177–188.

Velu, R., and G. Zhou. 1999. "Testing multi-beta asset pricing models." *Journal of Empirical Finance* 6: 219–241.

Verrecchia, R. E. 1980. "Consensus beliefs, information acquisition, and market information efficiency." *American Economic Review* 70: 874–884.

Vuolteenaho, T. 2002. "What drives firm-level stock returns?" *Journal of Finance* 57: 233–264.

Wachter, Jessica. 2013. "Can time-varying risk of rare disasters explain aggregate stock market volatility?" *Journal of Finance* 68: 987–1635.

Welch, I., and A. Goyal. 2007. "A comprehensive look at the empirical performance of equity premium prediction." *Review of Financial Studies* 21: 1455–1508.

West, K. D. 1997. "Another heteroskedasticity- and autocorrelation-consistent covariance matrix estimator." *Journal of Econometrics* 76: 171–191.

Wheatley, S. M. 1989. "A critique of latent variable tests of asset pricing models." *Journal of Financial Economics* 23: 325–338.

White, H. 1980. "A heteroskedasticity-consistent covariance matrix estimator and a direct test for heteroskedasticity." *Econometrica* 48(4): 817–838.

White, H. 2000. "A reality check for data snooping." *Econometrica* 68(5): 1097–1126.

Wilson, R. 1968. "The theory of syndicates." *Econometrica* 36(1): 119–132.

Wooldridge, Jeffrey. 1990. "An encompassing approach to conditional mean tests with applications to testing nonnested hypotheses." *Journal of Econometrics* 45: 331–350.

Working, H. 1960. "Note on the correlation of first differences of averages in a random chain." *Econometrica* 28(4): 916–918.

Working, Hobrook. 1949. "The investigation of economic expectations." *American Economic Review* 39: 150–166.

Wright, J. H. 2003. "Detecting lack of identification in GMM." *Econometric Theory* 19: 322–330.

Yang, Louis. 2018. "Predicting stock market returns using accounting variables." Working paper, University of Southern California, Los Angeles.

Yao, T. 2008. "Dynamic factors and the source of momentum profits." *Journal of Business & Economic Statistics* 26(2): 211–226.

Yogo, M. 2006. "A consumption-based explanation of expected stock returns." *Journal of Finance* 61: 539–580.

Yule, G. 1926. "Why do we sometimes get nonsense correlations between time series? A study in sampling and the nature of time-series." *Journal of the Royal Statistical Society* 89: 1–64.

Zakoian, J. M. 1994. "Threshold heteroskedastic models." *Journal of Economic Dynamics and Control* 18: 931–955.

Zarowin, P. 1990. "Size, seasonality, and stock market overreaction." *Journal of Financial and Quantitative Analysis* 25: 113–125.

Zawadowski, Adam. 2010. "The consumption of active investors and asset prices." Working paper, Boston University.

Zeldes, Stephen P. 1989. "Consumption and liquidity constraints: An empirical investigation." *Journal of Political Economy* 97(2): 305–346.

Zhang, Chu. 2009a. "On the explanatory power of firm-specific variables in cross-sections of expected returns." *Journal of Empirical Finance* 16: 306–317.

Zhang, Chu. 2009b. "Testing the APT with the maximum Sharpe ratio of extracted factors." *Management Science* 55: 1255–1266.

Zhang, Lu. 2017. "The investment CAPM." *European Financial Management* 23: 545–603.

Zheng, L. 1999. "Is money smart? A study of mutual fund investors' fund selection ability." *Journal of Finance* 54: 901–933.

Zhou, G. 1991. "Small sample tests of portfolio efficiency." *Journal of Financial Economics* 30: 165–191.

Zhou, G. 2008. "On the fundamental law of active portfolio management: What happens if our estimates are wrong?" *Journal of Portfolio Management* 34: 26.

Zhou, G. 2010. "How much stock return predictability can we expect from an asset pricing model?" *Economics Letters* 108: 184–186.

Zhou, G., and Y. Zhu. 2014. "Macroeconomic volatilities and long-run risks of asset prices." *Management Science* 61: 413–430.

Index